Smart Innovation, Systems and Technologies

Volume 22

Series Editors

R. J. Howlett, Shoreham-by-Sea, UK
L. C. Jain, Adelaide, Australia

For further volumes:
http://www.springer.com/series/8767

Anne Håkansson · Mattias Höjer
Robert J. Howlett · Lakhmi C. Jain
Editors

Sustainability in Energy and Buildings

Proceedings of the 4th International Conference on Sustainability in Energy and Buildings (SEB'12)

Volume 2

Editors
Professor Anne Håkansson
The Royal Institute of Technology - KTH
Kista
Sweden

Professor Mattias Höjer
Centre for Sustainable Communications
KTH Royal Institute of Technology
Stockholm
Sweden

Professor Robert J. Howlett
KES International & Bournemouth University
United Kingdom

Professor Lakhmi C. Jain
School of Electrical and Information
  Engineering
University of South Australia
South Australia
Australia

ISSN 2190-3018
ISBN 978-3-642-36644-4
Printed in 2 Volumes
DOI 10.1007/978-3-642-36645-1
Springer Heidelberg New York Dordrecht London

ISSN 2190-3026   (electronic)
ISBN 978-3-642-36645-1   (eBook)

Library of Congress Control Number: 2013932298

© Springer-Verlag Berlin Heidelberg 2013

This work is subject to copyright. All rights are reserved by the Publisher, whether the whole or part of the material is concerned, specifically the rights of translation, reprinting, reuse of illustrations, recitation, broadcasting, reproduction on microfilms or in any other physical way, and transmission or information storage and retrieval, electronic adaptation, computer software, or by similar or dissimilar methodology now known or hereafter developed. Exempted from this legal reservation are brief excerpts in connection with reviews or scholarly analysis or material supplied specifically for the purpose of being entered and executed on a computer system, for exclusive use by the purchaser of the work. Duplication of this publication or parts thereof is permitted only under the provisions of the Copyright Law of the Publisher's location, in its current version, and permission for use must always be obtained from Springer. Permissions for use may be obtained through RightsLink at the Copyright Clearance Center. Violations are liable to prosecution under the respective Copyright Law.
The use of general descriptive names, registered names, trademarks, service marks, etc. in this publication does not imply, even in the absence of a specific statement, that such names are exempt from the relevant protective laws and regulations and therefore free for general use.
While the advice and information in this book are believed to be true and accurate at the date of publication, neither the authors nor the editors nor the publisher can accept any legal responsibility for any errors or omissions that may be made. The publisher makes no warranty, express or implied, with respect to the material contained herein.

Printed on acid-free paper

Springer is part of Springer Science+Business Media (www.springer.com)

Preface

The volume of Smart Innovations, Systems and Technologies book series contains the proceedings of the Fourth International Conference on Sustainability in Energy and Buildings, SEB12, held in Stockholm, Sweden, and is organised by KTH Royal Institute of Technology, Stockholm, Sweden in partnership with KES International.

The International Conference on Sustainability in Energy and Buildings is a respected conference focusing on a broad range of topics relating to sustainability in buildings but also encompassing energy sustainability more widely. Following the success of earlier events in the series, the 2012 conference includes the themes Sustainability, Energy, and Buildings and Information and Communication Technology, ICT.

SEB'12 has invited participation and paper submissions across a broad range of renewable energy and sustainability-related topics relevant to the main theme of Sustainability in Energy and Buildings. Applicable areas include technology for renewable energy and sustainability in the built environment, optimisation and modeling techniques, information and communication technology usage, behaviour and practice, including applications.

This, the fourth conference in the SEB series, attracted a large number of submissions all around the world, which were subjected to a two-stage review process. With the objective of producing a high quality conference, papers have been selected for presentation at the conference and publication in the proceedings. The papers for presentation are grouped into themes. The papers are included in this proceedings.

Four prominent research professors gave interesting and informative keynote talks. Professor Göran Finnveden, Professor in Environmental Strategic Analysis and Vice-President for sustainable development at KTH Royal Institute of Technology, Stockholm, Sweden gave a talk entitled "Sustainability Challenges for the Building Sector". Professor Per Heiselberg Professor at the Department of Civil Engineering at Aalborg University, Denmark, gave a talk about "Buildings – both part of the problem and the solution!". Professor Guðni A. Jóhannesson, Professor in Building Technology, Director General of the Icelandic National Energy Authority, Iceland, and chair of IPGT the International Partnership for Geothermal Technologies spoke on the topic of "Meeting the challenges of climatic change - the hard way or the clever way". Professor Lynne A. Slivovsky, Associate Professor of Electrical and Computer Engineering at California

VI   Preface

Polytechnic State University, San Luis Obispo, California, USA spoke on the topic of "The Questions That Keep Me Up At Night".

Thanks are due to the very many people who have given their time and goodwill freely to make the SEB'12 a success. We would like to thank KTH Royal Institute of Technology for their valued support for the conference. We would also like to thank the members of the International Programme Committee who were essential in providing their reviews of the conference papers, ensuring appropriate quality. Moreover, we would like to thank invited session chairs for their hard work and providing reviewers of the conference papers and upholding appropriate quality. We thank the high profile keynote speakers and panellists for agreeing to come and provide very interesting theme and talks, as well as, inform delegates and provoke discussions. Important contributors to SEB'12 are the authors, presenters and delegates without whom the conference could not have taken place, so we offer them our thanks for choosing SEB'12 conference. We also like to thank the panellists for coming and discussing " Panel: Sustainability: Current and Future with focus on Energy, Buildings and ICT"

The KES International, KES Secretariat staff and the local organising committee worked very hard to bring the conference to a high level of organisation, and we appreciate their tremendous help and we are thankful to them. Finally, we thank City Hall, City of Stockholm Sweden for hosting the reception for the SEB'12 conference and Quality Hotel Nacka for accommodating the conference 3 – 5 of September 2012 International Conference on Sustainability in Energy and Buildings.

We hope that readers will find SEB'12 proceedings interesting, informative and useful rescource for your research.

Assoc Professor Anne Håkansson
SEB'12 General Chair
KTH Royal Institute of Technology,
Sweden

Professor Mattias Höjer
SEB'12 General Co-chair
KTH Royal Institute of Technology,
Sweden

Professor Robert J. Howlett
Executive chair
KES International & Bournemouth University,
United Kingdom

Organization

Honorary Chairs

Göran Finnveden	Strategic Environmental Analysis
	KTH Royal Institute of Technology, Sweden
Björn Birgisson	Division of Highway Engineering
	KTH Royal Institute of Technology, Sweden

and

Professor Lakhmi C. Jain University of South Australia, Australia

General Chairs

Assoc Prof. Anne Håkansson Communication Systems
KTH Royal Institute of Technology, Sweden

and

Professor Mattias Höjer Centre for Sustainable Communications
KTH Royal Institute of Technology, Sweden

Executive Chair

Professor Robert J. Howlett Executive Chair, KES International
Bournemouth University, United Kingdom

Program Chairs

Ronald Hartung Computer Sciences and Mathematics
Franklin University, Columbus, Ohio, USA

VIII   Organization

and

Mark Smith   Communication Systems
                               KTH Royal Institute of Technology, Stockholm,
                               Sweden

Publicity and Support Chairs

Bernhard Huber   Centre for Sustainable Communications (CESC)
                               KTH Royal Institute of Technology, Stockholm, and
                               Sweden

Dan Wu   Communication Systems
                               KTH Royal Institute of Technology, Stockholm,
                               Sweden

Nils Brown   Centre for Sustainable Communications (CESC)
                               KTH Royal Institute of Technology, Stockholm,
                               Sweden

and

Esmiralda Moradian   Communication Systems
                               KTH Royal Institute of Technology, Stockholm,
                               Sweden

Faye Alexander   Conference Operations, KES International, UK
Shaun Lee   Information Systems Support, KES International,
                               UK

and

Russ Hepworth   Business Development Manager, KES Interna-
                               tional, UK

Invited Sessions Chairs and Workshop

Smart Buildings, Smart Grids
Chair: Dr Catalina Spataru

Assessment and Monitoring The Environmental Performance of Buildings
Chair:Dr John Littlewood

Methodology for Renewable Energy Assessment
Chair: Dr Rainer Zah

Improving Office Building Energy Performance
Chair:Dr Emeka Osaji

Multi-Energy Sources
Chair:Prof Aziz Naamane

Technologies and Applications of Solar Energy
Chair: Prof Mahieddine Emziane

Energy Planning in Buildings and Policy Implications
Chair: Dr Eva Maleviti

Sustainable Energy Systems and Building Services
Chair:Prof Ivo Martinac

Sustainable and healthy buildings
Chair: Prof Jeong Tai Kim

Co-chair: Prof. Geun Young Yun

Special selection
Chair: Prof Robert Howlett

REQUEST Workshop

Invited Keynote Speakers

Professor Göran Finnveden	KTH Royal Institute of Technology, Stockholm, Sweden
Professor Per Heiselberg	Aalborg University, Denmark,
Professor Guðni A. Jóhannesson	Icelandic National Energy Authority, Iceland,
Professor Lynne A. Slivovsky	California Polytechnic State University, San Luis Obispo, California, USA

International Programme Committee

Meniai Abdeslam-Hasse	Université Mentouri de Constantine, Algeria
Nora Cherifa Abid	Aix-Marseille University, France
Vivek Agarwal	Indian Institute of Technology Bombay, India
Abdel Aitouche	Lagis Hei, France
Nader Anani	Manchester Metropolitan University, UK
Naamane Aziz	Marseilles Aix Marseille Universite (AMU), France
Messaouda Azzouzi	Ziane Achour University of Djelfa, Algeria
Brahim Benhamou	Cadi Ayyad University of Marrakech, Morroco
Frede Blaabjerg	Aalborg University Inst. Of Energy Technology, Denmark
Saadi Bougoul	Université de Batna, Algerie
Mohamed Chadli	University of Picardie Jules Verne, France
Christopher Chao	Hong Kong University of Science and Technology, China
Zhen Chen	Heriot-Watt University, Scotland
Derek Clements-Croome	Reading University, UK
Gouri Datta	Deshbandhu College, Kalkaji, University of Delhi, India and Stromstad Academy, Stromstad.
Mohamed Djemai	Université de Valenciennes et du Hainaut Cambrésis, France
Philip Eames	Loughborough University, UK
Mahieddine Emziane	Masdar Institute of Science and Technology, Abu Dhabi
Luis Fajardo-Ruano	Uumsnh, Morelia, Mexico
Antonio Gagliano	University of Catania, Italy
Oleg Golubchikov	University of Birmingham, UK
Ahmed Hajjaji	University of Picardie Jules Verne, France
Abdelaziz Hamzaoui	University of Reims Champagne Ardenne, France
Sture Holmberg	Royal Institute of Technology (KTH) Stockholm, Sweden
Robert J. Howlett	Bournemouth University, UK
Bin-Juine Huang	National Taiwan University, Taipei, Taiwan
Kenneth Ip	University of Brighton, UK
Hong Jin	Harbin Institute of Technology, China
Roger Kemp	Lancaster University, UK
Sumathy Krishnan	North Dakota State University, USA
Angui Li	Xi'an University of Architecture & Technology, China
Soren Linderoth	Technical University of Denmark
John Littlewood	Cardiff Metropolitan University, UK
Nacer Kouider M'Sirdi	Laboratoire des Sciences del'Information et des Systèmes, France
Noureddine Manamani	University of Reims, France

Ahmed Mezrhab	University Mohammed 1, Oujda Morocco
Behdad Moghtaderi	University of Newcastle, Australia
Roger Morgan	Liverpool John Moores University, UK
Mostafa Mrabti	Universite Sidi Mohamed Ben Abdellah, Fes, Morocco
Rui Neves-Silva	Universidade Nova de Lisboa FCT/UNL, Portugal
Emeka Efe Osaji	University of Wolverhampton, UK
Frederici Pittaluga	University of Genova, Italy
Giuliano C Premier	University of Glamorgan, UK
Abdelhamid Rabhi	MIS Amiens, France
Ahmed Rachid	University of Picardie Jules Verne, France
Enzo Siviero	University IUAV of Venice, Italy
Shyam Lal Soni	Malaviya National Institute of Technology, Jaipur, India
Catalina Spataru	UCL Energy Institute, Uk
Alessandro Stocco	University of Nova Gorica and partner Progeest S.r.l. of Padua, Italy
Lounes Tadrist	Polytech.univ-mrs, France
Dario Trabucco	IUAV University of Venice, Italy
Mummadi Veerachary	Indian Institute of Technology, Delhi, India
Wim Zeiler	TU Eindhoven, Faculty of the Built Environment, Netherlands
Mohcine Zouak	USMBA FST, Morocco
Rainer Zah	Life Cycle Assessment & Modelling Group, Empa, Switzerland

Keynote Speakers

We are very pleased to have acquired the services of an excellent selection of keynote speakers for SEB'12. These speakers gave a view about technological and scientific activities, relating to sustainability in energy and buildings, taking place in various areas of the world.

Professor Göran Finnveden

KTH Royal Institute of Technology, Sweden

Sustainability Challenges for the Building Sector

Abstract:

The building and real estate management sector is responsible for a significant part of the environmental impacts of our society. The sector's contribution to the threat of climate change for production of heat and electricity for the buildings are of special importance. It is important to consider the full life-cycle of buildings and also consider production and transportation of building materials, construction and waste management. In Sweden, emissions of gases contributing to climate change from heating of buildings have decreased during the last decades as results of strong policy instruments. One the other hand emissions from other parts of the life-cycle of buildings have increased, illustrating the need to have a wide systems perspective in order to avoid suboptimizations. It is also important to consider other environmental threats such as the use of hazardous chemicals, air quality, generation of waste and impacts on ecosystems from production of building materials as well as on building sites.

The building sector has a large potential to reduce its environmental footprint. Many of the most cost-efficient possibilities for mitigation of climate change are related to the building sector. Governmental policies are important for changes to be made. Voluntary instruments such as building rating tools may have an additional role. The ICT-sector may have one of its largest potentials in contributing to a more sustainable society in the building sector. Because of the long life-time of buildings, we are now constructing the future environmental impacts. When looking for cost-efficient solutions, we must therefore also consider the future cost-efficiency. In the presentation also social aspects of sustainability will be discussed including possibilities for the building sector to contribute to a better health and reduced health inequalities.

Biography:

Göran Finnveden is Professor in Environmental Strategic Analysis and Vice-President for sustainable development at KTH Royal Institute of Technology, Stockholm, Sweden. He is a M.Sc. in Chemical Engineering 1989, PhD in Natural Resources Management, Associate Professor in Industrial Ecology 2003 and full Professor 2007. His research has focused on environmental systems analysis tools such as Life Cycle Assessment, Strategic Environmental Assessment and Input-Output Analysis. It has included both methodological development and case studies. Application areas include buildings, energy systems, information and communication technologies, infrastructure and waste management. He has also worked with environmental policy in areas such as environmental policy integration, integrated product policy and waste policy. He is a currently a member of the Scientific Advisory Council to the Swedish Minister of the Environment, an expert in the governmental commission on waste management and a member of the board of directors of the Swedish Waste Nuclear Fund. According to Scopus he has published more than 60 scientific papers and is cited nearly 2000 times.

Professor Per Heiselberg

Aalborg University, Denmark

Buildings – both part of the problem and the solution!

Abstract:

Energy use for room heating, cooling and ventilation accounts for more than one-third of the total, primary energy demand in the industrialized countries, and is in this way a major polluter of the environment. At the same time the building sector is identified as providing the largest potential for CO_2 reduction in the future and many countries across the world have set very ambitious targets for energy efficiency improvements in new and existing buildings. For example at European level the short term goal has recently been expressed in the recast of the EU Building Performance Directive as "near zero energy buildings" by 2020.

To successfully achieve such a target it is necessary to identify and develop innovative integrated building and energy technologies, which facilitates considerable energy savings and the implementation and integration of renewable energy devices within the built environment. The rapid development in materials science, information and sensor technology offers at the same time considerable opportunities for development of new intelligent building components and systems with multiple functions.

Such a development will impose major challenges on the building industry as building design will completely change from design of individual components and systems to integrated design of systems and concepts involving design teams of both architects, engineers and other experts. Future system and concepts solutions will require that building components must be able fulfill multiple performance criteria and often contradictory requirements from aesthetics, durability, energy use, health and comfort. A key example of this is building facades that instead of the existing static performance characteristics must develop into dynamic solutions with the ability to dynamically adjust physical properties and energetic performance in response to fluctuations in the outdoor environment and changing needs of the occupants in order to fulfill the future targets for energy use and comfort. Buildings will also be both consumers and producers of energy, which creates a number of new challenges for building design like identification of the optimum balance between energy savings and renewable energy production. The interaction between the energy "prosuming" building and the energy supply grid will also be an important issue to solve.

The lecture will address and illustrate these future challenges for the building sector and give directions for solutions.

Biography:

Per Heiselberg is Professor at the Department of Civil Engineering at Aalborg University, Denmark. He holds a M.Sc. and a Ph.D. in Indoor Environmental Engineering. His research and teaching subjects are within architectural engineering and are focused on the following topics:

- Energy-efficient building design (Net zero energy buildings, design of low energy buildings - integration of architectural and technical issues, modelling of double skin facades, night cooling of buildings and utilization of thermal mass, multifunctional facades, daylight in buildings, passive energy technologies for buildings, modeling of building energy use and indoor environment)

XIV   Organization

- Ventilation and air flow in buildings (Modelling and measurements of air and contaminants flows (both gas- and particles) in buildings, ventilation effectiveness, efficient ventilation of large enclosures, numerical simulation (computational fluid dynamics) of air and contaminant flows as well as modeling of natural and hybrid ventilation)

Per Heiselberg has published about 300 articles and papers on these subjects.
Currently, Per Heiselberg is leading the national strategic research centre on Net Zero Energy Buildings in Denmark (www.zeb.aau.dk). The centre has a multidisciplinary research approach and a close cooperation with leading Danish companies. He has been involved in many EU and IEA research projects in the past 20 years. He was the operating agent of IEA-ECBCS Annex 35 (1997-2002) and IEA-ECBCS Annex 44 (2005-2009), (www.ecbcs.org). Presently he is involved in ECBCS Annexes 52, 53 and 59.

Professor Guðni A. Jóhannesson

Icelandic National Energy Authority, Iceland

Meeting the challenges of climatic change - the hard way or the clever way

Abstract:

We may not agree on how the possible CO2 driven scenarios of climate change in the future may look like but we all can agree that the anthropogenic increase in CO2 levels in the world atmosphere exposes humanity to higher risks of changes in the environment than we want to face in our, our children's or their children's lifetime.
It is evident that we are now facing a global challenge that we are more often dealing with by local solutions. Our guiding rule is that by saving energy we are also mitigating greenhouse gas emission. Also if we are using renewable energy and substituting fossil fuels we are also moving in the right direction. There are however important system aspects that we should be considering.
The first one if we are using the right quality of energy for the right purpose. A common example is when high quality energy such as electricity or gas is used directly to provide domestic hot water or heat houses instead of using heat pumps or cogeneration processes to get the highest possible ratio between the used energy and the primary energy input.
The second one is if we are obstructing necessary structural changes that could lead to a more effective energy system globally. We have big reserves of cost effective renewable energy sources, hydropower and geothermal energy around the world that are far from the markets and would therefore need relocation structural changes in our industrial production system to be utilized.
The third aspect is if we are using our investments in energy conversion and energy savings in the best way to meet our climatic goals or if we are directed by other hidden agendas to such a degree that a large part of our economical input is wasted.
It is evident that the national and local strategies for energy savings are closely linked to other strategic areas such as industrial development, household economy, mobility.

Also a necessary precondition for investment is that the nations maintain their economic strength and their ability to develop their renewable resources and to invest in new more efficient processes.

The key to success in mitigating the climatic change is therefor to create a holistic strategy that beside the development of technical solutions for energy efficiency and utilization of renewable energy also considers the local and global system aspects. With present technologies for energy efficient solutions, proper energy quality management r and with utilization of cost effective renewable energy sources we have all possibilities to reduce energy related the global CO_2 emissions to acceptable levels.

Biography:

Professor Guðni A. Jóhannesson is born in Reykjavik 1951. He finished his MSc in Engineering physics in 1976, his PhD thesis on thermal models for buildings in 1981 and was appointed as an associate professor at Lund University in 1982. He was awarded the title of doctor honoris causae from the University of Debrecen in 2008 and the Swedish Concrete Award in 2011. From 1975 he worked as a research assistant at Lund University, from 1982 as a consultant in research and building physics in Reykjavik and from 1990 as a professor in Building Technology at KTH in Stockholm and from 2008 an affiliated professor at KTH. His research has mainly concerned the thermodynamical studies of buildings, innovative building systems and energy conservation in the built environment. Since the beginning of 2008 he is the Director General of the Icelandic National Energy Authority which is responsible for public administration of energy research, energy utilization and regulation. He was a member of the The Hydropower Sustainability Assessment Forum processing the Hydropower Sustainability Assessment Protocol adopted by IHA in November 2010 and presently the chair of IPGT the International Partnership for Geothermal Technologies.

Professor Lynne A. Slivovsky

California Polytechnic State University, USA

The Questions That Keep Me Up At Night

Abstract:

This keynote will provide an opportunity for reflection on the work we do. We're here talking about energy and sustainability but we're also talking about a different way of living. We, as a technical field, a society, a world, are on a path of profound technological development. What does it mean to educate someone to contribute to this world? To have a technical education? What does it mean to live in this world? And is it possible that we as designers, innovators, engineers, and scientists can consider these questions in our day-to-day work?

Biography:

Lynne A. Slivovsky (Ph.D., Purdue University, 2001) is Associate Professor of Electrical and Computer Engineering at California Polytechnic State University, San Luis Obispo, California, USA. In 2003 she received the Frontiers In Education New Faculty Fellow Award. Her work in service-learning led to her selection in 2007 as a California Campus Compact-Carnegie Foundation for the Advancement of Teaching Faculty Fellow for Service-Learning for Political Engagement. In 2010 she received the Cal Poly President's Community Service Award for Significant Faculty Contribution. She currently oversees two multidisciplinary service-learning programs: the Access by Design Project that has capstone students designing recreational devices for people with disabilities and the Organic Twittering Project that merges social media with sustainability. Her work examines design learning in the context of engagement and the interdependence between technology and society.
Panel: Sustainability: Current and Future with focus on Energy, Buildings and ICT

Panel:

Sofia Ahlroth, Working Party on Integrating Environmental and Economic Policies (WPIEEP), Swedish EPA
Magnus Enell, Senior Advisor Sustainability at Vattenfall AB, Sweden
Göran Finnveden, Professor, KTH Royal Institute of Technology, Sweden
Danielle Freilich, Environmental expert at The Swedish Construction Federation (BI), Sweden
Catherine.Karagianni, Manager for Environmental and Sustainable Development at Teliasonera, Sweden
Örjan Lönngren, Climate and energy expert, Environment and Health Department, City of Stockholm, Sweden
Per Sahlin, Simulation entrepreneur, Owner of EQUA Simulation AB, Sweden
Mark Smith, Professor, KTH Royal Institute of Technology, Sweden
Örjan Svane, Professor, KTH Royal Institute of Technology, Sweden
Olle Zetterberg, CEO, Stockholm Business Region, Stockholm, Sweden

Contents

Volume 1

Session: Sustainability in Energy and Buildings

1 Transformational Role of Lochiel Park Green Village 1
Stephen Robert Berry

2 Evaluation and Validation of an Electrical Model of Photovoltaic Module Based on Manufacturer Measurement 15
Giuseppe M. Tina, Cristina Ventura

3 Evolution of Environmental Sustainability for Timber and Steel Construction 25
*Dimitrios N. Kaziolas, Iordanis Zygomalas,
Georgios E. Stavroulakis, Dimitrios Emmanouloudis,
Charalambos C. Baniotopoulos*

4 Using the Energy Signature Method to Estimate the Effective U-Value of Buildings 35
Gustav Nordström, Helena Johnsson, Sofia Lidelöw

5 Two Case Studies in Energy Efficient Renovation of Multi-family Housing; Explaining Robustness as a Characteristic to Assess Long-Term Sustainability 45
Vahid Sabouri, Paula Femenías

6 Exploring the Courtyard Microclimate through an Example of Anatolian Seljuk Architecture: The Thirteenth-Century Sahabiye Madrassa in Kayseri 59
Hakan Hisarligil

XVIII   Contents

7   Analysis of Structural Changes of the Load Profiles of the German Residential Sector due to Decentralized Electricity Generation and e-mobility   71
Rainer Elsland, Tobias Boßmann, Rupert Hartel, Till Gnann, Massimo Genoese, Martin Wietschel

8   The Impact of Hedonism on Domestic Hot Water Energy Demand for Showering – The Case of the Schanzenfest, Hamburg   85
Stephen Lorimer, Marianne Jang, Korinna Thielen

9   The Process of Delivery – A Case Study Evaluation of Residential Handover Procedures in Sustainable Housing   95
David Bailey, Mark Gillott, Robin Wilson

10   Sustainable Renovation and Operation of Family Houses for Improved Climate Efficiency   107
Ricardo Ramírez Villegas, Björn Frostell

11   Solar Collector Based on Heat Pipes for Building Façades   119
Rassamakin Boris, Khairnasov Sergii, Musiy Rostyslav, Alforova Olga, Rassamakin Andrii

12   ICT Applications to Lower Energy Usage in the Already Built Environment   127
Anna Kramers

13   Using Dynamic Programming Optimization to Maintain Comfort in Building during Summer Periods   137
Bérenger Favre, Bruno Peuportier

14   Assisting Inhabitants of Residential Homes with Management of their Energy Consumption   147
Michael Kugler, Elisabeth André, Masood Masoodian, Florian Reinhart, Bill Rogers, Kevin Schlieper

15   Raising High Energy Performance Glass Block from Waste Glasses with Cavity and Interlayer   157
Floriberta Binarti, Agustinus D. Istiadji, Prasasto Satwiko, Priyo T. Iswanto

16   A New Model for Appropriate Selection of Window   167
Abdolsalam Ebrahimpour, Yousef Karimi Vahed

17   Improved Real Time Amorphous PV Model for Fault Diagnostic Usage   179
Mehrdad Davarifar, Abdelhamid Rabhi, Ahmed EL Hajjaji, Jerome Bosche, Xavier Pierre

18 An Investigation of Energy Efficient and Sustainable Heating Systems for Buildings: Combining Photovoltaics with Heat Pump 189
Arefeh Hesaraki, Sture Holmberg

19 Assessment of Solar Radiation Potential for Different Cities in Iran Using a Temperature-Based Method 199
Farivar Fazelpour, Majid Vafaeipour, Omid Rahbari, Mohammad H. Valizadeh

20 A Decision Support Framework for Evaluation of Environmentally and Economically Optimal Retrofit of Non-domestic Buildings 209
Taofeeq Ibn-Mohammed, Rick Greenough, Simon Taylor, Leticia Ozawa-Meida, Adolf Acquaye

21 Modeling, From the Energy Viewpoint, a Free-Form, High Energy Performance, Transparent Envelope 229
Luis Alonso, C. Bedoya, Benito Lauret, Fernando Alonso

22 A Mathematical Model to Pre-evaluate Thermal Efficiencies in Elongated Building Designs 239
Alberto Jose Fernández de Trocóniz y Revuelta, Miguel Ángel Gálvez Huerta, Alberto Xabier Fernández de Trocóniz y Rueda

23 Effect of Reaction Conditions on the Catalytic Performance of Ruthenium Supported Alumina Catalyst for Fischer-Tropsch Synthesis 251
Piyapong Hunpinyo, Phavanee Narataruksa, Karn Pana-Suppamassadu, Sabaithip Tungkamani, Nuwong Chollacoop, Hussanai Sukkathanyawat

24 Integration of Wind Power and Hydrogen Hybrid Electric Vehicles into Electric Grids 261
Stephen J.W. Carr, Kary K.T. Thanapalan, Fan Zhang, Alan J. Guwy, J. Maddy, Lars-O. Gusig, Giuliano C. Premier

25 Analysis of Thermal Comfort and Space Heating Strategy: Case Study in an Irish Public Building 271
Oliver Kinnane, M. Dyer, C. Treacy

26 Protection of Ring Distribution Networks with Distributed Generation Based on Petri Nets 281
Haidar Samet, Mohsen Khorasany

27 Real-Time Optimization of Shared Resource Renewable Energy Networks 289
Stephen Treado, Kevin Carbonnier

Contents

28 Evaluation of the LCA Approaches for the Assessment of Embodied Energy of Building Products 299
Ayşen Ciravoğlu and Gökçe Tuna Taygun

29 Exergetic Life Cycle Assessment: An Improved Option to Analyze Resource Use Efficiency of the Construction Sector 313
Mohammad Rashedul Hoque, Xavier Gabarrell Durany, Gara Villalba Méndez, Cristina Sendra Sala

30 Methodology for the Preparation of the Standard Model for Schools Investigator for the Sustainability of Energy Systems and Building Services 323
Hisham Elshimy

31 Latin-American Buildings Energy Efficiency Policy: The Case of Chile 337
Massimo Palme, Leônidas Albano, Helena Coch, Antoni Isalgué, José Guerra

32 Thermal Performance of Brazilian Modern Houses: A Vision through the Time 347
Leônidas Albano, Marta Romero, Alberto Hernandez Neto

Short Papers

33 Energetic and Exergetic Performance Evaluation of an AC and a Solar Powered DC Compressor 357
Orhan Ekren, Serdar Çelik

34 Effectiveness of Sustainable Assessment Methods in Achieving High Indoor Air Quality in the UK 367
Gráinne McGill, Menghao Qin, Lukumon Oyedele

35 A Comprehensive Monitoring System to Assess the Performance of a Prototype House 373
Oliver Kinnane, Tom Grey, Mark Dyer

Session: Smart Buildings, Smart Grids

Invited Sessions

36 *Smart* Consumers, *Smart* Controls, *Smart* Grid 381
Catalina Spataru, Mark Barrett

37 A Qualitative Comparison of Unobtrusive Domestic Occupancy Measurement Technologies 391
Eldar Nagijew, Mark Gillott, Robin Wilson

38 Review of Methods to Map People's Daily Activity – Application for Smart Homes 401
Stephanie Gauthier, David Shipworth

39 Optimizing Building Energy Systems and Controls for Energy and Environment Policy 413
Mark Barrett, Catalina Spataru

40 Towards a Self-managing Tool for Optimizing Energy Usage in Buildings 425
Naveed Arshad, Fahad Javed, Muhammad Dawood Liaqat

41 A Library of Energy Efficiency Functions for Home Appliances 435
Hamid Abdi, Michael Fielding, James Mullins, Saeid Nahavandi

42 Smart Energy Façade for Building Comfort to Optimize Interaction with the Smart Grid 445
Wim Zeiler, Rinus van Houten, Gert Boxem, Joep van der Velden

43 Building for Future Climate Resilience: A Comparative Study of the Thermal Performance of Eight Constructive Methods 453
Lucelia Rodrigues, Mark Gillott

Session: Assessment and Monitoring the Environmental Performance of Buildings

44 Exploring Indoor Climate and Comfort Effects in Refurbished Multi-family Dwellings with Improved Energy Performance 463
Linn Liu, Josefin Thoresson

45 Occupancy-Driven Supervisory Control Strategies to Minimise Energy Consumption of Airport Terminal Building 479
D. Abdulhameed Mambo, Mahroo Efthekhari, Thomas Steffen

46 An Investigation into the Practical Application of Residential Energy Certificates 491
Alan Abela, Mike Hoxley, Paddy McGrath, Steve Goodhew

47 Post-Occupancy Evaluation of a Mixed-Use Academic Office Building 501
Katharine Wall, Andy Shea

48 The Human as Key Element in the Assessment and Monitoring of the Environmental Performance of Buildings 511
Wim Zeiler, Rik Maaijen, Gert Boxem

49 The Effects of Weather Conditions on Domestic Ground-Source Heat Pump Performance in the UK 521
Anne Stafford

XXII   Contents

**50 Asset and Operational Energy Performance Rating of a Modern
    Apartment in Malta** 531
Charles Yousif, Raquel Mucientes Diez, Francisco Javier Rey Martínez

**51 Low Carbon Housing: Understanding Occupant Guidance and
    Training** 545
Isabel Carmona-Andreu, Fionn Stevenson, Mary Hancock

Volume 2

**52 Embodied Energy as an Indicator for Environmental Impacts –
    A Case Study for Fire Sprinkler Systems** 555
*Tom Penny, Michael Collins, Simon Aumônier, Kay Ramchurn,
Terry Thiele*

**53 Understanding the Gap between as Designed and as Built
    Performance of a New Low Carbon Housing Development in UK** 567
Rajat Gupta, Dimitra Dantsiou

**54 Preliminary Evaluation of Design and Construction Details to
    Maximize Health and Well-Being in a New Built Public School in
    Wroclaw** 581
Magdalena Baborska-Narozny, Anna Bac

**55 Comparison of Design Intentions and Construction Solutions
    Delivered to Enhance Environmental Performance and Minimize
    Carbon Emissions of a New Public School in Wroclaw** 591
Magdalena Baborska-Narozny, Anna Bac

**56 An Exploration of Design Alternatives Using Dynamic Thermal
    Modelling Software of an Exemplar, Affordable, Low Carbon
    Residential Development Constructed by a Registered Social
    Landlord in a Rural Area of Wales** 601
*Simon Hatherley, Wesley Cole, John Counsell, Andrew Geens,
John Littlewood, Nigel Sinnett*

**57 Basic Energy and Global Warming Potential Calculations at an
    Early Stage in the Development of Residential Properties** 613
Nils Brown

**58 Passive Cooling Strategies for Multi-storey Residential Buildings in
    Tehran, Iran and Swansea, UK** 623
Masoudeh Nooraei, John Littlewood, Nick Evans

59 An (un)attainable Map of Sustainable Urban Regeneration 637
Linda Toledo, John R. Littlewood

Session: Methodology for Renewable Energy Assessment

60 LCA in The Netherlands: A Case Study 649
Wim Zeiler, Ruben Pelzer, Wim Maassen

61 Comparative Life-Cycle Assessment of Residential Heating Systems, Focused on Solid Oxide Fuel Cells 659
Alba Cánovas, Rainer Zah, Santiago Gassó

Session: Improving Office Building Energy Performance

62 A Sustainable Energy Saving Method for Hotels by Green Hotel Deals 669
Hamid Abdi, Doug Creighton, Saeid Nahavandi

63 The Role of Building Energy and Environmental Assessment in Facilitating Office Building Energy-Efficiency 679
Emeka E. Osaji, Subashini Suresh, Ezekiel Chinyio

64 Improved Personalized Comfort: A Necessity for a Smart Building 705
Wim Zeiler, Gert Boxem, Derek Vissers

65 Reducing Ventilation Energy Demand by Using Air-to-Earth Heat Exchangers: Part 1 - Parametric Study 717
Hans Havtun, Caroline Törnqvist

66 Reducing Ventilation Energy Demand by Using Air-to-Earth Heat Exchangers: Part 2 System Design Considerations 731
Hans Havtun, Caroline Törnqvist

67 The Green Room: A Giant Leap in Development of Energy-Efficient Cooling Solutions for Datacenters 743
Hans Havtun, Roozbeh Izadi, Charles El Azzi

68 The Impacts of Contributory Factors in the Gap between Predicted and Actual Office Building Energy Use 757
Emeka E. Osaji, Subashini Suresh, Ezekiel Chinyio

Session: Multi-Energy Sources

69 Improving Multiple Source Power Management Using State Flow Approach 779
Aziz Naamane, Nacer Msirdi

70 Technical-Economic Analysis of Solar Water Heating Systems at Batna in Algeria. 787
Aksas Mounir, Zouagri Rima, Naamane Aziz

XXIV Contents

71 Design and Control of a Diode Clamped Multilevel Wind Energy System Using a Stand-Alone AC-DC-AC Converter 797
Mona F. Moussa, Yasser G. Dessouky

Session: Technologies and Applications of Solar Energy

72 Analyzing the Optical Performance of Intelligent Thin Films Applied to Architectural Glazing and Solar Collectors 813
Masoud Kamalisarvestani, Saad Mekhilef, Rahman Saidur

73 A Sunspot Model for Energy Efficiency in Buildings 827
Yosr Boukhris, Leila Gharbi, Nadia Ghrab-Morcos

74 Towards 24/7 Solar Energy Utilization: The Masdar Institute Campus as a Case Study 837
Mona Aal Ali, Mahieddine Emziane

75 Effect of Selective Emitter Temperature on the Performance of Thermophotovoltaic Devices 847
Mahieddine Emziane, Yao-Tsung Hsieh

76 New Tandem Device Designs for Various Photovoltaic Applications 859
Mahieddine Emziane

77 GIS-Based Decision Support for Solar Photovoltaic Planning in Urban Environment 865
Antonio Gagliano, Francesco Patania, Francesco Nocera, Alfonso Capizzi, Aldo Galesi

78 Infrared Thermography Study of the Temperature Effect on the Performance of Photovoltaic Cells and Panels 875
Zaoui Fares, Mohamed Becherif, Mahieddine Emziane, Abdennacer Aboubou, Soufiane Mebarek Azzem

79 Integrating Solar Heating and PV Cooling into the Building Envelope 887
Sleiman Farah, Wasim Saman, Martin Belusko

Session: Energy Planning in Buildings and Policy Implications

80 Risk and Uncertainty in Sustainable Building Performance 903
Seyed Masoud Sajjadian, John Lewis, Stephen Sharples

81 Potential Savings in Buildings Using Stand-Alone PV Systems 913
Eva Malevit, Christos Tsitsiriggos

Session: Sustainable Energy Systems and Building Services

82 The Application of LCCA toward Industrialized Building Retrofitting – Case Studies of Swedish Residential Building Stock 931
Qian Wang, Ivo Martinac

83 A Proposal of Urban District Carbon Budgets for Sustainable Urban Development Projects 947
Aumnad Phdungsilp, Ivo Martinac

84 A Study of the Design Criteria Affecting Energy Demand in New Building Clusters Using Fuzzy AHP 955
Hai Lu, Aumnad Phdungsilp, Ivo Martinac

85 Cooling Coil Design Improvement for HVAC Energy Savings and Comfort Enhancement 965
Vahid Vakiloroaya, Jafar Madadnia

86 Sustainable Integration of Renewable Energy Systems in a Mediterranean Island: A Case Study 975
Domenico Costantino, Mariano Giuseppe Ippolito,
Raffaella Riva_Sanseverino, Eleonora Riva_Sanseverino,
Valentina Vaccaro

Session: Sustainable and Healthy Buildings

87 Mobile Motion Sensor-Based Human Activity Recognition and Energy Expenditure Estimation in Building Environments 987
Tae-Seong Kim, Jin-Ho Cho, Jeong Tai Kim

88 Cost and CO_2 Analysis of Composite Precast Concrete Columns 995
Keun Ho Kim, Chaeyeon Lim, Youngju Na, Jeong Tai Kim, Sunkuk Kim

89 A Field Survey of Thermal Comfort in Office Building with a Unitary Heat-Pump and Energy Recovery Ventilator 1003
Seon Ho Jo, Jeong Tai Kim, Geun Young Yun

90 An Analysis of Standby Power Consumption of Single-Member Huseholds in Korea 1011
JiSun Lee, Hyunsoo Lee, JiYea Jung, SungHee Lee, SungJun Park,
YeunSook Lee, JeongTai Kim

91 A Classification of Real Sky Conditions for Yongin, Korea 1025
Hyo Joo Kong, Jeong Tai Kim

92 Influence of Application of Sorptive Building Materials on Decrease in Indoor Toluene Concentration 1033
Seonghyun Park, Jeong Tai Kim, Janghoo Seo

XXVI Contents

93 Perceived Experiences on Comfort and Health in Two Apartment Complexes with Different Service Life 1043
Mi Jeong Kim, Myung Eun Cho, Jeong Tai Kim

94 Impact of Different Placements of Shading Device on Building Thermal Performance 1055
Hong Soo Lim, Jeong Tai Kim, Gon Kim

95 Daylighting and Thermal Performance of Venetian Blinds in an Apartment Living Room 1061
Ju Young Shin, Yoon Jeong Kim, Jeong Tai Kim

96 Environmentally-Friendly Apartment Buildings Using a Sustainable Hybrid Precast Composite System 1071
Ji-Hun Kim, Won-Kee Hong, Seon-Chee Park, Hyo-Jin Ko, Jeong Tai Kim

97 A System for Energy Saving in Commercial and Organizational Buildings 1083
Hamid Abdi, Michael Fielding, James Mullins, Saeid Nahavandi

Session: Special Selection

98 A Comparative Analysis of Embodied and Operational CO_2 Emissions from the External Wall of a Reconstructed Bosphorus Mansion in Istanbul 1093
Fatih Yazicioglu, Hülya Kus

Author Index 1107

Chapter 52
Embodied Energy as an Indicator for Environmental Impacts – A Case Study for Fire Sprinkler Systems

Tom Penny[1,*], Michael Collins[1], Simon Aumônier[1],
Kay Ramchurn[1], and Terry Thiele[2]

[1] Environmental Resources Management Limited, Exchequer Court,
33 St Mary Axe, London, EC3A 8AA
`tom.penny@erm.com`
[2] The Lubrizol Corporation, 29400 Lakeland Blvd, Wickliffe, OH 44092-2298

Abstract. This paper appraises the utility of embodied energy as an indicator of environmental impact through the use of life cycle assessment (LCA). This utility is considered in terms of its use for the preferential selection of materials and for hotspot analysis for the purposes of identifying reduction opportunities. An appraisal of the peer-reviewed LCA of BlazeMaster® CPVC fire sprinkler system and subsequent LCA work commissioned by the Lubrizol Corporation is conducted to investigate the utility of embodied energy as an environmental indicator. Embodied energy is assessed using the Cumulative Energy Demand (CED) method and environmental impacts are appraised using the ReCiPe method. Embodied energy is found to reflect the impact results in terms of preference for CPVC when compared to steel sprinkler systems. However, the inability of CED to identify hotspots consistently, or to provide a reliable measure of relative performance for individual environmental impacts, indicates limited utility.

1 Introduction

Building regulations across the globe require the installation of fire sprinkler systems for fire protection, in buildings above a certain volume or floor area. According to the National Fire Sprinkler Association, in the United States, mercantile buildings over 1,115 square metres or over three stories; high-rise buildings over 22.9 metres high; and residential apartments, except those with individual street exits, have to carry sprinkler systems (NFPA, 2010). Building Regulations in England and Wales are accompanied by documents making specific reference to the use of sprinklers (HM Government, 2010). New residential blocks over 30 metres high; and un-compartmentalized shop and storage areas in buildings over 2,000 square metres must be fitted with sprinklers in order to comply. Corresponding regulations apply to industrial buildings, with the largest permitted compartment without sprinklers being 20,000 square metres. Recent years have seen a growing trend for sprinkler systems to

---

* Corresponding author.

A. Håkansson et al. (Eds.): *Sustainability in Energy and Buildings*, SIST 22, pp. 555–565.
DOI: 10.1007/978-3-642-36645-1_52   © Springer-Verlag Berlin Heidelberg 2013

be installed in smaller buildings, including residential homes, as evidenced by the legal requirement in Wales for all new homes to be fitted with fire sprinkler systems (NAW, 2011).

Sprinkler systems are made up of three components: a water supply system delivering sufficient pressure and flow rate, a water distribution pipe network, and fire sprinklers attached to the pipes. While designing new buildings, or refurbishing existing buildings, architects make a choice regarding sprinkler systems and associated materials. The choice of material has a bearing on environmental impact, installation time and maintenance requirements, lifetime and cost. The primary materials used for fire sprinkler systems include steel, copper and fire-resistant plastics, such as polybutylene and chlorinated polyvinyl chloride (CPVC) with steel being the more traditionally used material.

Chlorinated polyvinyl chloride (CPVC), a thermoplastic, was invented in 1959 by the Lubrizol Corporation, formerly BFGoodrich Performance Materials.

CPVC exhibits corrosion and heat resistance as well as mechanical strength and ductility (Table 1) and is commonly used in pipes and fittings, electrical components, and sheet applications.

In 2010, Lubrizol commissioned Environmental Resources Management to conduct a peer-reviewed life cycle assessment (LCA) in order to establish the environmental impacts of the of the BlazeMaster® Fire Sprinkler System (Aumônier *et al.*, 2010a).

Table 1. Physical and Thermal Properties of CPVC (Tyco 2008)

Property	Value
Specific Gravity	1.53
IZOD Impact Strength (ft.lbs./inch, notched)	3.0
Modulus of Elasticity, at73°F, psi	4.23×10^5
Ultimate Tensile Strength, psi	8,000
Compressive Strength, psi	9,600
Poisson's Ratio	0.35 – 0.38
Working Stress at 73°F, psi	2,000
Hazen Williams "C" Factor	150
Coefficient of Linear Expansion in/(in °F)	3.4×10^{-5}
Thermal Conductivity BTU/hr/ft2/°F/in	0.95
Flash Ignition Temperature	900
Limiting Oxygen Index	60%
Electrical Conductivity	Non Conductor

The study appraised the 'cradle to grave' environmental performance of BlazeMaster® CPVC piping used in light hazard fire protection applications, in accordance with United States National Fire Protection Agency (NFPA). The following environmental impacts were considered in the study:

- Metal depletion
- Fossil depletion

- Terrestrial acidification
- Freshwater eutrophication
- Climate change
- Ozone depletion
- Human toxicity
- Freshwater eco-toxicity
- Terrestrial eco-toxicity
- Photochemical oxidant formation

A subsequent LCA study was conducted to investigate the performance of BlazeMaster® CPVC piping in comparison with other sprinkler pipe materials (Aumônier et al, 2010b). This study included embodied energy (cumulative energy demand) as an environmental indicator of interest.

These two studies enabled the value of cumulative energy demand as an indicator of wider environmental impact to be evaluated in the context of fire sprinkler system material choices.

2 Methods and Data

Life Cycle Assessment (LCA) is a standardized technique for measuring and comparing the environmental consequences of providing, using, and disposing of, a product or a service. The method employed is defined by the International Standards for Life Cycle Assessment (ISO, 2006ab).

LCA attempts to trace back to the environment all of the resources consumed at all stages in the manufacture, use, and disposal, of products. This includes all of the emissions to air, water and land at each of these stages. These data represent an inventory of exchanges of substances between the product and the environment, from the 'cradle' to the 'grave'.

Cumulative energy demand (CED) is an assessment of the energy consumed by a product's life cycle through its consumption of resources. This energy consumption is sometimes referred to as the embodied energy of a product.

Embodied energy is frequently reported in life cycle studies of building products and numerous databases exist which report embodied energy figures (e.g. the University of Bath Inventory of Carbon and Energy Database (University of Bath, 2011). In these cases, embodied energy is frequently provided on a cradle to gate basis and relates to the life cycle up to the point of use.

At the impact assessment stage of an LCA a calculation is made of the potential contribution made by each of these environmental exchanges to important environmental effects such as global warming, acidification, photochemical smog, human- and eco-toxicity, eutrophication and the depletion of non-renewable fossil fuel resources.

Lubrizol commissioned the two LCA studies to explore the cradle to grave environmental impacts associated with BlazeMaster® CPVC fire sprinkler pipes and to benchmark with traditionally used materials, such as steel. LCA was chosen as the most useful tool for assessing the fire sprinkler pipes due to the ability to provide a holistic view of environmental impacts.

The LCAs were undertaken to provide insight into the environmental hotspots, to identify reduction opportunities, and to benchmark the BlazeMaster® product with traditionally used materials.

The relationship between embodied energy and other environmental impacts is discussed in this context.

The functional unit used for the two studies was: *1,000 feet (304.8 metres) of piping installed and used in a high rise multi-residential dwelling in the United States (US) for a 50 year time period*. Figure 1 details the life cycle stages included in the cradle to grave assessment. Hangers, screws and solvents were included; however the sprinkler nozzles and water used by the sprinkler system were excluded from the assessment.

Fig. 1. Life Cycle Overview of BlazeMaster® Pipe

The products assessed include both a CPVC and steel pipe product as described below.

Table 2. Products Considered in the Assessment

Product	Description
BlazeMaster®	SDR 13.5 Iron Pipe Size (IPS) CPVC pipe
Steel pipe	Schedule 10 IPS carbon steel pipe

The assessment included cradle to grave life cycle stages including extraction of raw materials, manufacturing, distribution, installation, use, and disposal. The burdens for removal of the sprinkler systems were excluded from the study.

To assess the environmental impacts of the BlazeMaster® pipe, primary data were collected from Lubrizol production operations and key suppliers. Published life cycle data were used for the remainder of the life cycle stages (Aumônier et al, 2010).

The assessment of steel pipes was based on published life cycle inventories, combined with sprinkler system installation requirements provided by Lubrizol Corporation.

The mass of materials included for the two systems to describe the functional unit is detailed in Table 2.

The system excludes the fire sprinkler nozzles and any materials and energy consumed during use of the sprinkler systems.

SimaPro[1] Life Cycle Assessment software was used to model the product life cycles. Cumulative Energy Demand (CED) and ReCiPe impact assessment methods were used to assess the products. CED accounts for the energy consumed by the

---

[1] PRé Consultants bv, Amersfoort, The Netherlands.

Table 3. Mass of Piping Materials Considered (per 304.6m ~ 1,000 ft)

Component	BlazeMaster® Pipe Mass of Materials (kg)	Steel Pipe Mass of Materials (kg)
Pipes	118.87	763.6
Fittings	19.24	52.3
Solvent cement	1.85	-
Hangers and screws	11.34	54.5
Total	151.30	870.4

system throughout all life cycle stages considered from raw material extraction to disposal. The CED method reports energy consumption as a total and broken down by energy source i.e. non-renewable fossil, non-renewable nuclear, non-renewable biomass, renewable biomass, renewable wind, solar, geothermal and renewable water. Results for CED are summed into non-renewable and renewable energy categories and presented in the following sections (Hischier *et al* 2010).

ReCiPe was developed as a life cycle impact assessment method which comprises harmonized category indicators at midpoint and endpoint level. The method includes three cultural perspectives based on the approach previously employed in Eco-indicator 99. The hierarchic view was considered and the impact categories assessed included metal depletion, fossil depletion, terrestrial acidification, freshwater eutrophication, climate change, ozone depletion, human toxicity, freshwater eco-toxicity, terrestrial eco-toxicity and photochemical oxidant formation (Goedkoop, 2009).

No weighting or end point assessments were undertaken as part of the assessment.

3 Results

Table 4 and Table 5 present the impact assessment and embodied energy results for the BlazeMaster® and steel pipe fire sprinkler systems.

Table 4. ReCiPe Results for the BlazeMaster® Pipe and Steel Pipe (per 304.6m ~ 1,000 ft)

Impact category	Unit	BlazeMaster® Pipe System	Steel Pipe System
Metal depletion	kg Fe eq	1,780	4,090
Fossil depletion	kg oil eq	314	459
Terrestrial acidification	kg SO_2eq	4.61	8.96
Freshwater eutrophication	kg P eq	0.0109	0.214
Climate change	kg CO_2eq	874	1,790
Ozone depletion	kg CFC-11 eq	1.74E-04	1.26E-04
Human toxicity	kg 1,4-DB eq	109	624
Freshwater eco-toxicity	kg 1,4-DB eq	4.42	34.1
Terrestrial eco-toxicity	kg 1,4-DB eq	0.0844	0.405
Photochemical oxidant formation	kg NMVOC	2.88	7.82

Table 5. Cumulative Energy Demand Results for the BlazeMaster® Pipe and Steel Pipe (per 304.6m ~ 1,000 ft)

Impact category	Unit	BlazeMaster® Pipe System	Steel Pipe System
Non-renewable energy	MJ	15,600	22,600
Renewable energy	MJ	710	1,060
Total	MJ	16,310	23,660

Results by life cycle stage for the BlazeMaster® piping system are presented below. The ReCiPe method is presented in Figure 2 and Table 6. The CED method is presented in Figure 3 and Table 7. Results are shown by life cycle stage to identify the high level 'hotspots' of the system.

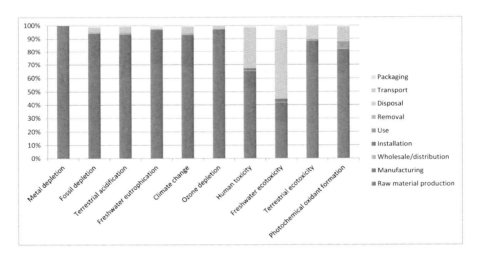

Fig. 2. ReCiPe Normalised Results for the BlazeMaster® Pipe (per 304.6m ~ 1,000 ft)

52 Embodied Energy as an Indicator for Environmental Impacts

Table 6. ReCiPe Results for the BlazeMaster® Pipe (per 304.6m ~ 1,000 ft)

Impact category	Unit	Raw material production	Manufacturing	Whole-sale/distribution	Installation	Use	Disposal	Transport and packaging
Metal depletion	kg Fe eq	1,780	1.58	1.3E-02	2.8E-05	-	4.3E-04	8.5E-02
Fossil depletion	kg oil eq	236	58.9	3.03	1.1E-02	-	2.2E-01	15.7
Terrestrial acidification	kg SO$_2$eq	2.97	1.33	6.8E-02	2.4E-04	-	5.3E-03	2.3E-01
Freshwater eutrophication	kg P eq	7.7E-03	2.8E-03	7.6E-05	1.7E-07	-	8.9E-07	2.8E-04
Climate change	kg CO$_2$eq	601	211	10.9	0.35	-	8.16	42.8
Ozone depletion	kg CFC-11 eq	1.6E-04	5.9E-06	3.0E-07	3.9E-09	-	8.2E-08	5.2E-06
Human toxicity	kg 1,4-DB eq	57.3	13.9	0.693	1.43	2.8E-02	33.6	1.98
Freshwater ecotoxicity	kg 1,4-DB eq	1.21	6.3E-01	2.9E-02	9.7E-02	3.4E-06	2.29	1.7E-01
Terrestrial ecotoxicity	kg 1,4-DB eq	6.0E-02	1.4E-02	7.2E-04	2.6E-04	5.6E-06	6.2E-03	2.8E-03
Photochemical oxidant formation	kg NMVOC	1.83	5.2E-01	2.6E-02	4.9E-04	1.5E-01	1.1E-02	3.5E-01

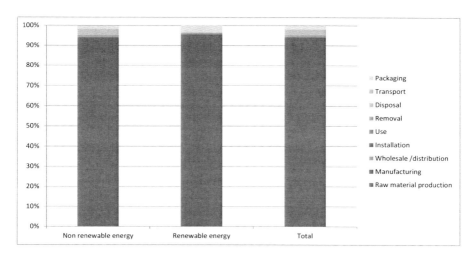

Fig. 3. Cumulative Energy Demand Normalized Results for the BlazeMaster® Pipe (per 304.6m ~ 1,000 ft)

Table 7. Cumulative Energy Demand Results for the BlazeMaster® Pipe (per 304.6m ~ 1,000 ft)

Impact category	Non renewable energy	Renewable energy	Total
Unit	MJ	MJ	MJ
Raw material production	11,300	560	11,860
Manufacturing	3,370	118	3,488
Wholesale /distribution	174	6.05	180.05
Installation	0.542	0.0221	0.5641
Use	-	-	-
Removal	-	-	-
Disposal	11	0.352	11.352
Transport	461	0.663	461.663
Packaging	236	27.8	263.8
Total	15,600	710	16,310

4 Discussion

One aim of the analysis was to investigate CED as an indicator of scale for environmental impact. The relative difference between CED and other environmental life cycle impacts for the BlazeMaster® and steel pipe systems was investigated to understand the ability of CED to represent environmental impact.

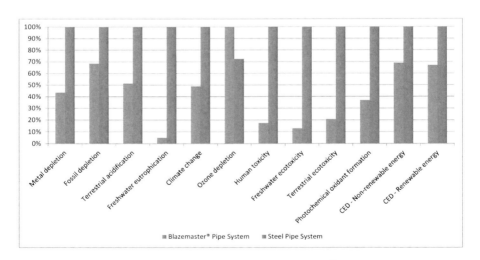

Fig. 4. Normalised ReCiPe and CED Impacts for BlazeMaster® and Steel Pipe Systems

For the majority of environmental impacts assessed, the preference for BlazeMaster® correlates with the results for CED, with the exception of ozone depletion. When determining high level conclusions to identify an environmentally preferable product this suggests it is reasonable to use CED as an indicator.

When comparing the relative difference between the CED of the products, BlazeMaster® is 30% lower, and does not correlate with the relative difference for the individual impact categories. From this comparison it is clear that, although CED may provide a useful initial indicator of preference, it is neither foolproof nor accurate in its ability to reflect relative performance.

The relative difference between products for fossil depletion impacts does appear to correlate with CED. This can be explained by the use of fossil resources as a fuel for energy generation. However, should fossil resources not be the main energy feedstock for the systems, then these relationships would likely be significantly different.

The second aim of the analysis was to investigate the use of CED for hotspot identification. ReCiPe and CED methods have both been utilized to analyze the BlazeMaster® sprinkler system to identify the hotspots i.e. the main contributing life cycle stages and processes. Hotspot identification is often the first step in identifying reduction opportunities. The percentage contribution for each life cycle stage to CED and the impact categories considered are shown in Table 8 and Table 9.

Table 8. BlazeMaster® System Percentage Contribution by Life Cycle Stage per ReCiPe Impact Category

Impact category	Raw material production	Manufacturing	Wholesale / distribution	Installation	Use	Removal	Disposal	Transport	Packaging
Metal depletion	100.0%	0.1%	0.0%	0.0%	0.0%	0.0%	0.0%	0.0%	0.0%
Fossil depletion	75.2%	18.8%	1.0%	0.0%	0.0%	0.0%	0.1%	3.5%	1.5%
Terrestrial acidification	64.4%	28.9%	1.5%	0.0%	0.0%	0.0%	0.1%	4.3%	0.8%
Freshwater eutrophication	70.7%	26.1%	0.7%	0.0%	0.0%	0.0%	0.0%	0.0%	2.5%
Climate change	68.8%	24.1%	1.2%	0.0%	0.0%	0.0%	0.9%	3.6%	1.3%
Ozone depletion	93.7%	3.4%	0.2%	0.0%	0.0%	0.0%	0.0%	2.8%	0.2%
Human toxicity	52.6%	12.8%	0.6%	1.3%	0.0%	0.0%	30.8%	0.4%	1.4%
Freshwater eco-toxicity	27.4%	14.2%	0.7%	2.2%	0.0%	0.0%	51.8%	0.3%	3.5%
Terrestrial eco-toxicity	71.4%	16.7%	0.8%	0.3%	0.0%	0.0%	7.3%	2.6%	0.7%
Photochemical oxidant formation	63.5%	17.9%	0.9%	0.0%	5.1%	0.0%	0.4%	11.1%	1.1%

564   T. Penny et al.

Table 9. BlazeMaster® System Percentage Contribution by Life Cycle Stage per CED Impact Category

Impact category	Raw material production	Manufacturing	Wholesale / distribution	Installation	Use	Removal	Disposal	Transport	Packaging
Non renewable energy	72.4%	21.6%	1.1%	0.0%	0.0%	0.0%	0.1%	3.0%	1.5%
Renewable energy	78.9%	16.6%	0.9%	0.0%	0.0%	0.0%	0.0%	0.1%	3.9%
Total	72.7%	21.4%	1.1%	0.0%	0.0%	0.0%	0.1%	2.8%	1.6%

An investigation into the BlazeMaster® system identifies that raw material production is the most significant contributor to all impact categories, with the exception of freshwater eco-toxicity. Manufacturing is the second most significant contributor, accounting for up to 29% of the total impact. Disposal of CPVC in landfill is a significant contributor to the human and aquatic toxicity impacts. The impact of wholesale, transport, and packaging is very low relative to the other life cycle stages.

The results from the LCA broadly align with the embodied energy results presented in the CED method in that they identify raw material production (73%) and manufacturing (21%) as significant contributors. However, there is significant variation between impacts in the contribution of each life cycle stage. CED is a poor indicator of hotspots for human toxicity, freshwater toxicity and photochemical oxidant formation. There are few similarities as to the magnitude of impact contribution from each life cycle stage across the different impact categories.

5   Conclusions

For a high level assessment to understand whether products are environmentally preferable overall, the use of embodied energy appears to enable similar conclusions to those of LCA to be drawn. However, this would appear to be a consequence of the systems assessed, and their particular hotspots, rather than a universal conclusion. The danger of using CED as an indicator of more detailed results for individual environmental impacts, such as climate change, is highlighted by the inability of CED to identify hotspots consistently or to provide a reliable measure of relative performance. It is likely that CED will become even less reliable an indicator of environmental impact as the use of fossil fuel declines and energy consumption is decoupled from some environmental impacts. Climate change is a specific example of this as, with the advent of carbon capture and storage (CCS), the CED of products is likely to increase due to loss of energy efficiency but greenhouse gas emissions are likely to reduce due to their long term storage.

References

Aumônier, S., Collins, M., Hartlin, B., Penny, T.: Life Cycle Assessment (LCA) of BlazeMaster® Fire Sprinkler System (2010a),
http://www.pvc4pipes.com/en/pdfs/BlazeMaster-LCA.pdf
(accessed April 29, 2012)

Aumônier, S., Collins, M., Hartlin, B., Penny, T.: Streamlined Life Cycle Assessment (LCA) of BlazeMaster® Fire Sprinkler System in Comparison to PPR and Steel Systems (2010b),
http://www.ribaproductselector.com/Docs/8/23138/external/COL436184.pdf (accessed April 29, 2012)

Goedkoop, M.: A life cycle impact assessment method which comprises harmonized category indicators at the midpoint and endpoint level, Report 1: Characterisation (January 2009),
http://www.lcia-recipe.net/ (accessed April 29, 2012)

Hischier, R., et al.: Implementation of Life Cycle Impacts Assessment Methods, ecoinvent report No. 3 (July 2010)

HM Government. The Building Regulations, Approved Document B (Fire Safety), vol. 2 – Buildings other than Dwelling houses (2010)

International Standard Organisation (ISO). Environmental management Life cycle assessment Principles and framework (2nd edn, 2006-07-01). Geneva, Switzerland (2006a)

International Standard Organisation (ISO).Environmental management Life cycle assessment Requirements and guidelines (1st edn, 2006-07-01). Geneva, Switzerland (2006b)

National Assembly for Wales (NAW), Domestic Fire Safety (Wales) Measure 2011 (nawm 3). Stationery Office Limited (2011)

National Fire Protection Association, NFPA: Standard for the Installation of Sprinkler Systems, 2010 edn. (2010)

Tyco. BlazeMaster® Installation Handbook (2008),
http://www.tyco-rapidresponse.com/docs/literature/blazemasterguideih_1900.pdf?sfvrsn=2 (accessed April 29, 2012)

University of Bath. Embodied Energy and Carbon in Construction Materials (2011),
http://opus.bath.ac.uk/12382/ (accessed April 29, 2012)

Chapter 53
Understanding the Gap between 'as Designed' and 'as Built' Performance of a New Low Carbon Housing Development in UK

Rajat Gupta and Dimitra Dantsiou

Low Carbon Building Group, School of Architecture, Oxford Brookes University, Oxford
`rgupta@brookes.ac.uk`

Abstract. This paper investigates forensically the discrepancy between 'as designed' and 'as built' performance of exemplar low carbon housing in UK, since this performance gap has the potential to undermine zero carbon housing policy. Driven by the UK Government Technology Strategy Board's *Building Performance Evaluation (BPE)* competition, a BPE study is undertaken by the authors during post construction and initial occupancy stage of a Code level 5 housing scheme (Swindon, UK) focusing on two house typologies. The performance of the building envelope and service systems are evaluated through a detailed review of design and construction specifications and processes, fabric performance evaluation, observation of handover processes and mapping of occupant satisfaction. This reveals unintended fabric losses, installation and commission issues associated with low carbon technologies, lack of co-ordination and proper sequencing of the building works, and complexity of control interfaces. Lessons from different elements of the BPE study are pointed out and recommendations are drawn for councils, developers, house builders, designers and equipment suppliers to reduce the gap between 'as designed' and 'as built' performance of future low carbon housing developments.

Keywords: Building Performance Evaluation, Energy efficiency, Low carbon housing.

1 Introduction

Energy used in domestic housing in the UK produces over a quarter of UK's total carbon dioxide emissions which drive climate change (DECC, 2010). Under the scope of an 80% CO_2 reduction target that the UK government aims to achieve by 2050, a lot of effort has been put in towards the improvement of the efficiency of new-build and existing housing. This is why the Government has set ambitious targets for incremental changes to building regulatory standards, which are intended to achieve zero carbon new housing from 2016 onwards (DECC, 2011). With the application of improved fabric measures (such as better insulation), improved efficiencies in building services (including more efficient heating and hot water systems, lights and appliances and better controls) and the addition of low and zero carbon renewable

A. Håkansson et al. (Eds.): *Sustainability in Energy and Buildings*, SIST 22, pp. 567–580.
DOI: 10.1007/978-3-642-36645-1_53 © Springer-Verlag Berlin Heidelberg 2013

energy generation, this is theoretically possible. However, many of these solutions are at present untested, and there is growing concern within the housing industry that, in practice, even current energy efficiency and carbon emission standards are not being achieved i.e. that there is a gap between as-built performance and design intent. Furthermore there is concern that this performance gap has the potential to undermine zero carbon housing policy (Zero Carbon Hub, 2011).

Despite the growing interest and the abundance of design solutions for the construction of zero and low carbon housing in the UK, there is limited knowledge on the way homes perform post construction, except on air-tightness (Bordass and Leaman, 2005, Gupta and Chandiwala, 2010). Initial studies, such as Low Carbon Housing: Lessons from Elm Tree Mews suggests a deficiency can exist between design and actual performance of the building fabric and services (Bell et al, 2010). Further research by Wingfield et al (2008) has shown many of the new homes in the sample they tested were not achieving their design energy and ventilation performance standards. There is also growing concern regarding the lack of a systematic feedback approach to monitor, assess and benchmark the modern construction methods and technology.

To address the growing concern that performance of buildings in use is highly variable and does not match the predictions, UK Government's Technology Strategy Board (TSB) launched its Building Performance Evaluation (BPE) competition in 2010 aiming to understand the cause and effects of this discrepancy and enable improvements in the performance of new built low carbon dwellings (TSB, 2010b). A total of £8m was allocated to fund the costs of new built domestic and non-domestic buildings during their construction, initial occupation and in-use phase and create a knowledge base which will enable setting up performance driven standards for the design of sustainable homes.

This paper investigates the impact of the design and build processes on the discrepancy between the as designed and as built performance of low carbon dwellings, using a 'post completion and initial occupation' BPE study of a Code Level 5 housing scheme called Malmesbury Gardens located in Swindon (UK). The aim of the study was to evaluate and compare the performance of the building envelope and systems (before the residents move into their new homes) with the design intent, and correlate it with the resident experience during their first weeks of occupancy.

2 Background to the Low Carbon Housing Development

Malmesbury Gardens is a social housing scheme that was intended to provide an innovative approach to affordable mixed tenure housing design, procurement and finance, with large scale delivery potentials. It consisted of 13 council houses built to Code for Sustainable Homes level 5 criteria aiming to a low energy performance along with space flexibility and design excellence.

The general design approach targeted a Code level 5 rating by a combination of an optimized and airtight building envelope supported by an innovative space and water heating solution and the use of renewables. The construction was based on the application of hempcrete cast into a timber frame in order to achieve high thermal mass levels in combination with optimized U-Values and a maximum airtightness of 2 $m^3/(h.m^2)$

under pressure of 50 Pa. Heating, hot water and mechanical ventilation was provided by an Exhaust Air Heat Pump (EAHP) while the systems were supported by a solar thermal and pre-heat hot water cylinder and Photovoltaic panels on the roof (Table 1).

Table 1. Design and occupancy characteristics of the case study dwellings

Case study design characteristics	
Location	Swindon, UK
Dwelling type(s)	1 End terrace, 5 bedroom, 9 person
	3 End terrace, 3 bedroom, 6 person
	3 Mid terrace, 3 bedroom, 6 person
	6 Mid terrace, 2 bedroom, 4 person
Main construction elements	Walls: Prefabricated timber frame wall panels with solid cast in situ hempcrete around the frame (Design U-value 0.18 W/m^{2}K)
	Roof: Timber panels with hemp fibre insulation bats within depth of beams (Design U-value: 0.15 W/m^{2}K)
	Windows: Timber frame, double glazed (Design U-value: 1.4 W/m^{2}K)
Space heating system	Heat is supplied from an exhaust air heat pump to the underfloor heating coils via a flow and return heating pipe circuit.
Hot water system	A solar thermal system installed in each property will pre-heat the domestic hot water supplied to the EAHP. Two roof mounted vacuum tube heat pipe solar collectors of 4m^2 are placed in each house.
Air tightness	2.0 m^3/(h.m^2) when pressure tested at 50 Pascal
Ventilation strategy	Mechanical ventilation through exhaust air heat pump. NIBE fighter 410 P product with 260% efficiency.
Renewables	Average of 4 kWpk of Photovoltaics in each property.
Sustainability rating	Code for Sustainable Homes level 5
Age/ Date of completion	March 2011
Other information	The 3 and 5-Bed houses use the roof space as a living area adding an extra floor to the design.
Occupancy	All the 13 houses are fully occupied by families or single parents with children.
Photos	

3 Building Performance Evaluation Study

To identify any deviation from the design intent and map the initial occupants' reactions, the BPE study was organised in two key stages covering the *design and construction stage* as well as the *post construction and early occupancy* phase of the new homes. The starting point was a review of the design and specification records that coincided the construction while right after completion the fabric and systems' performance and the occupants' first reactions were established. The parties involved in the process consisted of the research team, the owner (Swindon Borough council) the design team and the contractors (Swindon Commercial Services) owned by SBC.

Table 2. The evaluation framework and methods used in the different stages of the TSB Building Performance Evaluation post construction and initial occupancy study on the case study dwellings (TSB, 2010a)

Stage of investigation	Project activities	Methods
Design	Design and construction audit	SAP calculation review
		Drawings and specifications review
		Semi-structured interviews with design team
		Walkthroughs with client and developer
		Observation and review of control interfaces
		Initial walkthrough with the developer (client)
Post construction	Fabric testing	Co-heating test
		Air permeability test
		Thermography
		In situ U-value measurement
	Review of commissioning process	Review of systems performance and commissioning
Early Occupancy	Observation of the handover process	Review of Home User Guide
		Observation of handover process
	Occupancy evaluation	BUS questionnaire survey
		Walkthrough and interviews with occupants

3.1 Design and Construction Stage

An important insight on the project's design and procurement process as well as its intended performance aspiration was provided by the design and construction audit that took place in the beginning of the BPE study. The initial audit revealed key differences between the final performance and design intent which were investigated with those responsible for delivering the dwellings so as to establish the reasons for any deviations.

Through a SAP calculation review, different ratings came out between the existing and recalculated ratings for the same dwellings. The drawings review along with a walkthrough in the dwellings, identified issues with usability of the air source heat pumps and their controls. The interview with the design team and walkthrough with the developer also revealed key lessons concerning the use of Hempcrete and the provision of drying period, as well as the impact of successful liaison between the different parties involved in design and construction. Finally a critical review of control interfaces addressed the issue of the usability of heating system controls.

Review of design specification and procedures
A review of the design documentation was undertaken which included a variety of drawings, reports and specification documents that were provided by the design team. The study was complemented with a walkthrough that revealed several issues that were not obvious just from the drawings. Unplanned extrusions intersecting with door and window frames trying to cover ventilation system ducting was one apparent notice along with inability to access the handle of the roof windows without the use of a poll or a small ladder. The latter was a result of the PV panels fitted on the roof on the space right underneath the window.

SAP review
The Standard Assessment Procedure (SAP) is the official UK Government's methodology for assessing and comparing the energy and environmental performance of residential dwellings (Department of Energy and Climate Change, 2011). A review of the SAP calculations aimed to ensure that the existing estimates accurately reflected the design of the dwelling and identify any design aspects that were not captured accurately by the calculations. The SAP calculations were repeated for four different dwelling types and correlated with the existing specification notes, drawings and reports. By the time of the study, the initial calculations were not updated to changes in the design resulting in a discrepancy between the existing and recalculated ratings (Table 3) that in the case of one dwelling type it lead to its degradation to level 'B' of the rating scale (Table 4).

Table 3. Discrepancies between SAP calculations and specification notes

SAP	Specification notes
PVC framed windows (Design U-value 1.3 W/m^{2}K)	Timber framed windows
Hot water cylinder volume 170ltr in all houses	Hot water cylinder volume 210 ltr (2-bed) or 300 ltr (3-bed and 5-bed)
In the 3Bed-End terraced house the east side wall and windows were not calculated	
Internal partitions are stated as masonry partitions with plasterboard on dabs	Internal partitions are plasterboards on timber stud

Table 4. Discrepancy between the design and recalculated SAP rating

Dwelling type	Provided SAP rating	STROMA SAP rating Recalculated by an OBU researcher as a part of the BPE
3-bed mid-terrace	99 (A)	88 (B)
3-bed end-terrace	98 (A)	92 (A)

Design team interviews

In order to gain background knowledge on the scheme and gather information that could not be gleaned through direct involvement of the research team, the key members of the design and development team were interviewed. The qualitative interview approach comprised of open-ended semi structured questions providing a significant insight on the basic aspirations and targets of the design team, the effectiveness of certain strategies and experiences gained through the project development and the examination of emerging issues that came through the design documentation review. The unplanned extension of the construction phase was highlighted by the interviewees as a result of the tight funding program, difficulties with the assembly of the timber frame and a considerable delay in hempcrete drying. The lack of a consistent communication base between the contractors, specialists and design team led to changes of the initially chosen suppliers and specialists ending up to a restart of coordination processes and delay in construction.

Review of control interfaces

Control interfaces are the part where the users meet the technology of the building. The usability of local controls for lights, heating, cooling and ventilation largely dictate the performance of a house in various ways such as energy efficiency, occupant satisfaction and thermal comfort (BSRIA, 2007). A review of the control interfaces took place some days before the official handover to investigate the relationship between the design and usability of controls and the potential effect that they could have during the dwelling's occupancy. The use of two different design approaches for the house masterstat and the room thermostat (Table 5) along with an oversimplified interface lacking clear labelling and indication of system's response has set questions on their future usability and degree of control.

Table 5. Usability grading of a room thermostat

Room thermostat		Image
Usability criteria	**Ranking** Poor Excellent	
Clarity of purpose		
Intuitive switching		
Labeling and annotation		
Ease of use		
Indication of system response		
Degree of fine control		

3.2 Post-construction Stage and Early Occupancy Stage

An important objective of the post construction phase was to understand the physical performance of the building fabric and services since it is a fundamental factor for the final energy consumption. Two selected case study houses in the development, went through co-heating and airtightness test to provide a clear image of the post-construction fabric performance. An assessment of the installation and commissioning processes was undertaken to understand where variances from design stage to installation stage had occurred and to determine whether manufacturers' guidelines had been followed. The next step was to assess the induction process through the observation of the handover sessions and the Home User Guide evaluation. Qualitative information were gathered through a BUS questionnaire survey, semi-structured interviews and walkthroughs that were carried out with selected households to capture the residents' experience and understand the behavioural impact to the energy performance of their houses. The findings gathered from the previous activities provided the evidence base for the second study phase, a two year in-use study which aims to explore the energy and environmental performance of the dwellings as well as the behavioural and social patterns developed on the long term.

Fabric testing (Co-heating test, Air permeability test, Party wall bypass test, thermographic survey)

The fabric performance of Malmesbury Garden houses was determined by two whole house co-heating tests on a three-bed end of terrace (Plot 7) and a three-bed mid terrace (Plot 6) dwelling (Fig. 1) complemented by air-permeability tests and additional measurements of the physical properties of the study dwellings according to ATTMA standards (ATTMA, 2010). The equipment used is summarised in Table 6.

Fig. 1. Site plan highlighting the co-heated plots 6 and 7 as well as the just heated plot 5

The main objective of a co-heating test is to measure the total house heat losses in W/h (fabric loss and background ventilation loss) in attribute to an unoccupied building. The main finding of the whole house heat-loss test was the discrepancy of approximately 20%

between the simplified SAP prediction and the one calculated after the co-heating test for both plots (Table 7). In order to quantify the air-leakage rate through the building envelope two standard air leakage tests were carried out before and after the co-heating tests. In both cases the values were well above the design target of 2 $m^3/(h.m^2)$ suggesting noticeable heat losses due to air leakage paths (Table 8).

Table 6. Fabric testing equipment information

Test	Description	Manufacturer
Coheating test set(s)	Tripod mounted test set(s) comprising of an air temp/rh transmitter, electrical power meter and pulse transmitter, CO_2 sensor and transmitter, circulation fan(s) and a fan heater (s)	Vaisala / Radio-Tech / BSRIA
	External air temperature / humidity transmitter (located at same position as weather station)	Radio-Tech
	Wi5 data hub	Radio-Tech
	Weather station and data logger recording wind speed / direction, rainfall, barometric pressure temperature / humidity & solar radiation	Davis / Kipp & Zonen / Grant
Heat flux monitoring	Data loggers with Hukseflux and temperature sensors	Eltek and Hukseflux
Thermographic survey	P640 model, 640x480 pixel, 0.04K thermal resolution infrared camera set on Rainbow colour palette	FLIR
	Type PT100 digital thermometer	Testo
	Weather station	Davis

The calculation of the hempcrete wall U-values also fell short of expectations with heat flux levels rising well above those intended for the hempcrete walls at the design stage (Table 9). A series of thermographs confirmed the high heat loss values providing a quick and visual indication of heat losses and air leakage paths through the roof apex, poor fitting of openings and skirting boards and not properly sealed system pipes detailing further investigation of those elements (Figure 2). A limitation related to this method was the difficulty of keeping the dwellings unoccupied during the three week test period due to cost implications and construction related activities. This resulted in distorting the monitoring values which needed to be 'cleaned' afterwards by keeping an uninterrupted sample of days.

Table 7. Summary of calculated heat loss coefficients

Plot No.	Measured heat loss coefficient (Co-heating) (W/K)	Designed heat loss coefficient predicted SAP (W/K)	Designed heat loss coefficient corrected SAP (W/K)
6	115.79	133.65	94
7	146.69	151	123.7

53 Understanding the Gap between 'as Designed' and 'as Built' Performance

Table 8. Summary of the air permeability measurements at ±50 Pa

Air-permeability values	Plot 6	Plot 7
Design air permeability m^3/(h.m^2)	2.00	2.00
Pre-coheating test final average measured air permeability m^3/(h.m^2)	3.30	4.69
Post-coheating test final average measured air permeability m^3/(h.m^2)	3.70	4.72
Energy Saving Trust Recommendation for CSH Level 5 (EST, 2008)	3.00	3.00
UK Building Regulation Best practice	5.00	5.00
UK Building Regulation Good practice	10.00	10.00

Table 9. Summary of the designed U-values and the measured U-values

Location	Wall design U-value [W/(m^{2}K)]	Wall final measured U-value Average of two sensors [W/(m^{2}K)]	Energy Saving Trust 100% and zero carbon solutions	UK Building Regulation Typical scenario
Plot 7 external north facing wall	0.180	0.471	0.15	0.25
Plot 6 to plot 5 party wall	Not stated on drawings	0.292	-	-

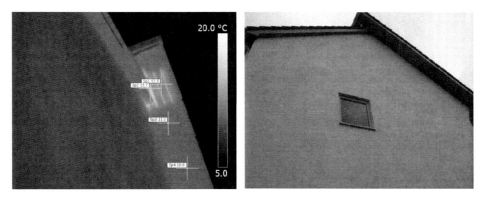

Fig. 2. Left: Plot 7 side upper elevation thermogram indicating heat losses through the roof apex. Right: Plot 7 side upper elevation digital photo.

Observation of commissioning process
A commissioning review undertaken in one of the properties aimed to ensure that the services and equipment's commissioning had been complete and the design and operational strategy was capable of creating the desired conditions. The onsite inspection revealed that most of the systems remained to be fully commissioned while for those that had undergone pre-commissioning there was no documented evidence on site. In addition, all the installation, commissioning and H&S checklists were missing by the time of the inspection and could not be provided by the site manager. The PV's

inverter was still to be installed while the underfloor heating system was out of operation on the first floor and the related control circuit disconnected. A number of installation issues were highlighted, mostly associated with co-ordination and accommodation of the services within the structure.

Evaluation of handover process and Home user guide
The evaluation of the homes handover process involved the observation of the two handover sessions and the evaluation of the Home User Guide documentation. A walk through demonstration of the building features a few weeks before the move in day intended to familiarise the occupants with their homes and systems before use to ensure their proper operation. During the walkthrough, the tenants seemed to be sceptical of the thermostat interface and found the linear scale rather confusing. A Home User Guide was provided in two phases, a week prior to the move in and during the second training day. It was generally clearly organised but lacking of informative illustrations, maps and diagrams that would help an easier content take up. The Occupant's training day helped most of the households to adjust their heating system and address the first in-use issues revealing a series of commissioning faults and a lack of understanding of the heating and hot water system.

Arup Bus Questionnaire survey
Occupants' satisfaction was measured using the established Building Use Studies (BUS) methodology analysis of which was covered centrally by the Technology Strategy Board (TSB, 2010a). The BUS analysis method is a quick and thorough way of obtaining feedback data on building performance through a self-completion occupant questionnaire (Bordass et Leaman, 2005) the results of which are compared against an annually updated benchmark database based on results of other building. The survey took place approximately six weeks after the official move-in mapping the first impression of the tenants rather than long term occupation findings. Within this short tenancy period, the houses proved to be very successful with the tenants being appreciative of the location, space, layout and overall appearance. However, a high 'forgiveness factor' stemming from the dwellings' design values appeared to positively 'trade off' against more functional aspects such as the lack of storage space and the limited control over their mechanical ventilation system.

Initial walkthrough and interviews with the occupants
Approximately six weeks after they moved in their new homes five different households went through a semi-structured interview followed by a walkthrough to map the occupants' first impression of their houses. The interviews and walkthrough observations revealed a common approach pointing clearly the positive aspects of the houses as well as some needing further improvement. The houses have been less successful in delivering adequate storage space and an effective troubleshooting mechanism. Most of the tenants were complaining about finishing issues, system fine-tuning problems and construction mistakes which were yet to be resolved.

4 Discussion of Findings

The post completion and early occupation study of 'Malmesbury Gardens' provides insights into the issues that emerge and lessons to be learned in relation to the *design, construction, commissioning, handover* and *occupation* of a low carbon new built housing.

The difference between design aspiration and as-built performance was highlighted by most of the study elements and the findings were found to be correlated to each other and passed through the different construction phases. Table 10 summarises the key findings from the BPE activities undertaken and the key messages that they convey. The main findings were mostly related to the difficulty of the contractors to engage with new materials and technologies, complications in the communication between the different project parties, lack of proper system commissioning and inadequacy of the induction process.

Table 10. Key findings of the Building Performance Evaluation study and main reasons for discrepancy between as designed and as built

BPE Study Elements	Findings	Key messages
SAP calculation review	- Discrepancy between SAP calculations and specification notes	
Drawings and specifications review	- Unplanned plasterboard soffits to cover ductwork and cables	
Semi-structured interviews with design team	- Lack of effective communication base between design team, contractors and suppliers - Difficulty of contractors to engage with new materials and technologies	
Walkthroughs with client and developer	- Tight access to services leading to unplanned design solutions - Discrepancy between hempcrete laboratory and as-constructed performance enclosures	Lack of effective liaison between different parties
Observation and review of control interfaces	- Lack of clear labelling and annotation of room thermostat and house masterstat - EAHP and DHW interface not accessible by tenants	
Co-heating test	- 20% discrepancy between the as designed and as built heat loss coefficient values	
Air permeability test	- Air-permeability rate for both dwellings fail to meet the design criteria	Difficulty in engaging with new materials and technologies
Thermography	- Air leakage paths behind various construction elements (skirting boards, panels and penetrations, fittings, system cupboards)	
In situ U-value measurement	- Measured wall U-value is double from the material specifications	

Table 10. (*continued*)

BPE Study Elements	Findings	Key messages
Review of systems performance and commissioning	- Incomplete system commissioning - Missing installation, commissioning and H&S checklists - Tight access to plant, pipework and pipes for the EAHP and DHW system - Incorrect installation of room thermostat - Negligence to insulate primary pipework - Operation of ventilation system during sanding activities	Lack of proper system commissioning
Review of Home User Guide	- Inadequate graphic presentation (lack of informative illustrations, maps, diagrams)	
Observation of handover process	- Tenants were uncertain and confused from the demonstration of the control interface - Lack of Home User guide or relevant documentation	
BUS questionnaire survey	- Lack of storage space - Limited control over MVHR	
Walkthrough and interviews with occupants	- Unavailability of troubleshooting mechanism - Complains on finishing issues, system fine-tuning and construction mistakes	Inadequate induction process

The difficulty of the contractors to engage with new materials and technologies had obvious implications from the beginning of the project with a delay in the timber frame construction and the drying of Hempcrete walls. The lack of proper training of the workforce in combination with a poor liaison with the design team and system specialists resulted in significant construction faults, unplanned design solutions and wrong system commissioning highlighted in the fabric testing and commissioning process review. Finally, the difficulty to fine-tune the newly installed systems and the construction faults were pointed out through the walkthroughs and interviews with the occupants. The lack of understanding of systems and control interfaces could also be attributed to the inability of the induction process to provide a clear and comprehensive guidance to the new houses and systems.

5   Conclusions and Key Lessons for Future

As there is a growing evidence for the energy underperformance of low carbon houses in the UK the need to uptake the lessons learned from the relevant studies is imperative if a change is to be made. Through the findings of the Malmesbury

Gardens the following recommendations are drawn for councils, developers, house builders and equipment suppliers, affecting the design and construction stage of a new housing development.

- **Design:** a robust design framework where design specifications and dwelling layouts consider the accommodation of services within the structure, reflect the needs of the occupants and provide intuitive and easy to use control interfaces would be the first step for the improvement of the design of low carbon housing. The design drawings and energy calculations should be carefully updated to reflect any changes in all design stages.
- **Construction:** The case study experience has revealed that proper selection, coordination and liaison of the project's design team, contractors and system specialists is implied from the initial design stages to avoid faults and omissions during construction. The proper training of the construction workforce is implied for the proper application of new innovative materials while an early liaison with the system's specialists is important to procure for adequate space to accommodate systems within the properties (cupboards, shafts, etc.). Greater co-ordination from first-fix onward between trades should be encouraged and need for better management, control and constant feedback loop between the different parties related to the project from its initial conception till the period after the handover.
- **Handover:** the establishment of a comprehensive easy-to-use guidance and troubleshooting framework could prove valuable in addressing the induction needs of residents once they move into their new houses.

It is vital that these evidence-based lessons on the as built performance of new low carbon housing are taken on board on an iterative basis and embedded into knowledge management systems of councils, developers, house builders, designers and equipment suppliers.

References

ATTMA. Technical Standart 1-Measuring air permeability of building envelopes, ATTMA (2010), `http://www.bindt.org/downloads/ATTMA%20TSL1%20Issue%201.pdf` (accessed January 5, 2011)

Bell, M., Wingfield, J., Miles-Shenton, D., Seavers, J.: Low Carbon Housing, Lessons from Elm Tree Mews, Joseph Rowntree Foundation (2010), `http://www.jrf.org.uk/publications/low-carbon-housing-elm-tree-mews` (accessed January 15, 2011)

Bordass, B., Leaman, A.: Making Feedback and Post Occupancy Evaluation routine 1: a portfolio of Feedback techniques. Building Research & Information 33(4), 347–352 (2005)

Bordass, B., Leaman, A., Bunn, R.: Controls for End Users, a guide for good design and implementation (2007)

Bordass, B., Leaman, A., Willis, S.: Control Strategies for Building Services: the role of the user. In: CIB/BRE, ed. Buildings and the Environment, May16-20 (1994)

Bryman, A.: Social Research Methods, 2nd edn. Oxford University Press, Oxford

BSRIA, Controls for End Users, a guide for good design and implementation, BCIA (2007)

CIBSE, Testing buildings for air leakage, CIBSE Technical Memorandum 23. CIBSE, London (2000)

Dam, S., Bakker, C., Hal, J.D.: Home energy monitors: impact over the medium term. Building Research & Information 5(38), 458–469 (2010)

Department for Business Innovation and Skills, HM Treasury The Plan for Growth (March 2011)

Department for Energy and Climate Change (DECC), The Carbon Plan: Delivery over low carbon future (2011a)

Department for Energy and Climate Change (DECC), The Government's Standard Assessment Procedure for Energy Rating of Dwellings (2011b)

Department of Communities and Local Government, DCLG (2010) Code for Sustainable Homes, Technical Guide. RIBA Publishing, London (November 2010)

E.S.T. Energy efficiency and the Code for Sustainable Homes, Levels 4&5, London (2008)

Gupta, R.: Moving towards low-carbon buildings and cities: experiences from Oxford, UK. International Journal of Low-Carbon Technologies 4, 159–168 (2009)

Gupta, R., Chandiwala, S.: Using a learning by doing post-occupancy evaluation approach to provide evidence-based feedback on the sustainability performance of buildings. In: Deemers, C., Potvin, A. (eds.), pp. 222–227. Les Presses de l'Universite´ Laval, Quebec City (2009)

Gupta, R., Chandiwala, S.: Understanding occupants: feedback techniques for large-scale low-carbon domestic refurbishments. Building Research & Information 38(5), 530–548 (2010)

Spataru, C., Gillot, M., Hall, R.M.: 'Domestic energy and occupancy: a novel post occupancy evaluation study. International Journal of Low-Carbon Technologies (5), 148–157 (2010)

Stevenson, F., Bahadur Rijal, H.: Post Occupancy evaluation of the environmental and building performance of the Sigma Home. Oxford Brookes University, OISD:Architecture (2009)

Stevenson, F., Rijal, H.: Developing occupancy feedback from a prototype to improve housing production. Building Research & Information 5(38), 549–563 (2010)

Technology Strategy Board (TSB). Building Performance Evaluation, Domestic Buildings, Technology Strategy Board (2010a),
`https://ktn.innovateuk.org/c/document_library/`
`get_file?p_l_id=787754&folderId=1479937&name=DLFE-15333.pdf`

Technology Strategy Board (TSB). Building Performance Evaluation Competition for Funding May 2010-2012, The Technology Strategy Board (2010b),
`http://www.innovateuk.org`

The Royal Academy of Engineering, Heat: degrees of comfort? Options for heating homes in a low carbon economy, London (2012)

Treasury, H.: The Plan for Growth (2011)

Usable Buildings Trust (UBT). The Softlandings Framework, for better briefing, design, handover and building performance in-use, BSRIA (2009),
`http://www.bsria.co.uk/download/soft-landings-framework.pdf`
(accessed February 9, 2011)

Way, M., Bordass, B.: Making feedback and post-occupancy evaluation routine 2: Soft landings - involving design and building teams in improving performance. Building Research & Information 4(33), 353–360 (2005)

Zero Carbon Hub, Carbon compliance for tomorrow's new homes, Topic4, Closing the gap between designed and built performance, London (2010)

Chapter 54
Preliminary Evaluation of Design and Construction Details to Maximize Health and Well-Being in a New Built Public School in Wroclaw

Magdalena Baborska-Narozny and Anna Bac

Faculty of Architecture, Wroclaw University of Technology, Poland, B. Prusa Str. 53/55
{magdalena.baborska-narozny,anna.bac}@pwr.wroc.pl

Abstract. The paper presents preliminary evaluation of a design approach and solutions applied in the first sustainable school complex in Wroclaw, Poland. The scope of this paper is focused on health and wellbeing of the occupants. The building completed in the year 2009 is planned for an in-depth POE to start this year – the first such broad evaluation project to be carried out in Poland. Measurements taken already and the feedback from the occupants received so far indicate whether certain design intentions have been met. Selected usability problems that have already occurred are discussed as well as the way the occupants cope with them. Selected details that proved to be successful are also presented. An overview of the process of the building delivery, handover and maintenance is also presented as in the authors opinion it has a major impact on the building's overall performance. The paper concludes that most usability problems are lessons to be learned indicating improvements that can be made in a building's life early stages.

1 Introduction

Sustainability discourse introduced analysis of a building in its whole life cycle: briefing, design, construction, handover, occupancy and possible demolition (Preisner and Schramm 2012). Well-being and user satisfaction with a building are most influenced by delivery stages in which the actual occupants do not participate. Unless participative design is introduced, everything is decided for the users by the specialists, basing on their knowledge and experience. The evidence for participative design being an efficient method for finding successful solutions is growing (Day et al. 2011). On the other hand the benchmarking system and growing amount of viable data from research on building's performance evaluation put designers in a good position to treat many solutions as double checked and "satisfaction guarantee" type. Regardless of the method selected for the design process the user's perspective should be fundamental for a building's performance evaluation, as many claim today (Baird 2010).

This paper analyses the design solutions introduced to enhance well-being and health in a public school complex recently built in Wroclaw, Poland. The solutions introduced are grouped according to a design goal they apply to. The user's perspective

A. Håkansson et al. (Eds.): *Sustainability in Energy and Buildings*, SIST 22, pp. 581–590.
DOI: 10.1007/978-3-642-36645-1_54 © Springer-Verlag Berlin Heidelberg 2013

is present through the results of a questionnaire survey for pupils, analysis of all the records documenting the defects claimed by the users since the building's opening in September 2009 and semi structured interviews with crucial occupants (head of school, head of preschool, staff representatives, including the building technician responsible for the building maintenance). Measurements of noise and lux levels in classrooms were also taken.

Employment of green features into the analyzed building made the whole delivery process challenging and not fully "as usual", as it was the first school in the region to rely partly on renewable energy sources. Launching a competition for the school's design was the first sign of a 'special treatment', as it is not a daily procedure in Poland. Most school designs are commissioned through tender with the lowest price as the main choice criterion. The construction and maintenance cost for Suwalska school was to be kept at an average Polish level.

2   The Case Study Building

The 6000-square-meters building accommodates 450 pupils and 100 preschool children. There are 80 employees in the building including: teachers, administration and technical staff. The schools population is divided between two classroom wings with shared entrance lobby, library, sport and dining facilities (see Fig. 1). The third wing with a separate entrance is dedicated to the preschool. The dining area serves as a connection point between the school and the preschool. The school is located on a greenfield site at Suwalska Street in a currently developing suburban housing district of Wroclaw.

The process of delivering the building was complex. The major groups involved were the Department of Municipal Development (DMD), Department of Education (DE), the design team, contractors, building authorities and commissioners. The main groups using the building are the school staff, pupils, their parents and local community. Keeping everyone focused on the wellbeing of the building's users as an ultimate objective of the whole process was not easy. Introducing the Softlandings procedures might have improved the process. The brief was to be developed solely between the DMD and DE. Head of the school and local community council representative were invited for presentations on the current state of design. Though the architects insisted that the future occupants comment on the development of the brief their influence was little. Anna Bac and Krzysztof Cebrat from Grupa Synergia architectural office were commissioned the design in result of presenting the winning competition entry. It included clear signs that the users' well-being and integration of the new development with the local community were among the most important design goals. However, as they admit today, the design was not fully integrated as specialist teams worked parallel with the architect coordinating the whole process. The contractors were selected in a tender process according to the lowest price offered. The commissioning process during pre-occupation stage was under strong time pressure, as the building had to open with the beginning of a school year and no delays were possible. The handover stage did not leave the occupants with full understanding of the functioning of the systems that were to assure both thermal comfort and savings in CO_2 emissions. A place for improvements at each stage of the building's delivery results in certain usability problems further described.

Fig. 1. Visualization of the school complex. © Grupa Synergia

3 The Context and the Stage of Data Collection

The building has been occupied since September 2009. In the year 2009 a questionnaire survey for the children of 9-12 years of age was personally distributed and collected by the architects to learn about the reception of selected issues of the school's design. In 2012 it was repeated with the 12-year-olds. A total of 104 questionnaires were collected for analysis in 2009 and 26 questionnaires in 2012. The questionnaire was developed by the architects themselves based on Rittelmeyer's research on children's perception of the school environment (Rittelmeyr 1994). It comprised of one page with 13 questions: 4 open ended questions and 9 closed-ended questions with a possibility to add explanation of the choice made. Table 1 summarizes key findings form the pupils' questionnaire (see Table 1). The questionnaires were accompanied by semi-structured interviews with 7 key representatives of the technical, administrative and teaching staff. The records of all faults claimed by the occupants covering time from the handover to February 2012 were shared with the authors by the Head of School. It became an important source of information on usability issues. The selection of design targets and details introduced is based on an in-depth knowledge of the building's design (Bac 2005) and procurement phases. Further research is planned however and a Polish translation of Probe questionnaire of users satisfaction developed by Usable Buildings Trust, UK is to be licensed and used to compare the building's perceived performance against collected benchmark examples from other countries (Baborska-Narozny 2011). As a part of planned POE research IAQS measurements of CO_2 levels, humidity and temperature will be taken. Building's energy consumption is analyzed in a chapter "Comparison of design intentions and construction solutions delivered to enhance environmental performance and minimize carbon emissions of a new public school in Wroclaw".

Table 1. Pupil satisfaction and comfort

Parameters	Criteria	Nos. of pupils rating this at top of scale (1-7) 2009	2012
Design overall	satisfactory	92%	90%
Meeting my needs	very well	85%	84%
Lasting impression	very good	89%	91%
Safety inside	very high	79%	82%
Safety outside	very high	69%	70%
Space use	effectively	80%	83%
Functional program	adequate	54%	59%
Comfort overall	satisfactory	62%	72%
Light comfort overall	satisfactory	43%	69%
Noise comfort overall	satisfactory	79%	80%

Source: Adapted from Stevenson and Humphris (2007: 42-44)

4 Health and Well-Being Design Targets and Solutions Applied

To provide an efficient learning environment Anna Bac referred to her precedent research. In her PhD thesis (Bac 2003) she made an insight into various education systems and their influence on school architecture, psychology of child development and perception, psychology of architecture and the presence of schools in urban structure (Posch and Rauscher 1996). The foreseen preferences of future space users were an important design motive. To enhance user's health and well-being through architectural form, functional distribution and materials the following design targets areas were identified:

- *target:* Safety and easy orientation

design: distinct division between public and restricted access zones and enabling insight into most spaces, extensive glazing, keeping visual links between inside and outside (see Fig. 2). Clear functional disposition with distinct color patterns for different sections of the building to facilitate orientation (see Fig. 3).

Fig. 2. The main school hall – keeping visual links between inside and outside

54 Preliminary Evaluation of Design and Construction Details 585

Fig. 3. Clear functional disposition with distinct color patterns for different sections of the building to facilitate orientation. The following colors are used in corridors and classes: space for pupil from classes 4-6 in yellow, classes 0-3 in orange, kindergarten in green.

- *target:* Individual character of architecture to induce a feeling of identity of place and community and prevent vandalism

  design: vivid colors, various surface textures and gloss and not all perpendicular shapes across the outside and inside of the building to prevent bore and enliven the space, custom detailing of selected load bearing structure elements, heater covers.

- *target:* Sharing selected school facilities with the local community. An integration of local community with the school was proposed at the competition stage through inclusion of several public functions into the school: a public library, community club, local community council and a city guard office. The sport facilities were also to be let to local community after school working hours.

Fig. 4. Distinct volumes with separate entrances for different functions shared with local community a) View towards entrance area, b) School and community shared library

design: Separate entrance to sport facilities, library and a multi purpose room. The main lobby may also function as a space for public events after the school is closed (see Fig. 4).

- *target:* Design adjusted to the child-scale wherever possible

  design: the complex is no higher than two floors (one floor in the preschool part), windows with external views are designed at eye level in all areas including preschool, appropriate size of furniture was selected.

- *target:* Exploiting the educational potential of the substance of the building

design: wherever possible the construction, electrical and ventilation systems are left visible to show how the building functions, porthole windows in most internal doors to allow the children an insight into most spaces including ancillary technical rooms, staff rooms and classrooms (see Fig. 5).

usability issues: the insight into some of the spaces i.e. head of the school's office, technical rooms is sealed.

Fig. 5. Across the building the construction, electrical and ventilation systems are left visible to show how the building functions. a) Dining area b) Entrance hall.

- *target*: All parts of the building wheelchair accessible, sanitary facilities for handicapped users

design: no stairs at the entrance areas, internal lift.

usability issues: a telephone connection was not included into the brief for the school and thus the school was not equipped with a telephone. In result the lift was not to be commissioned as long as an emergency phone was not installed inside. The problem is not yet solved.

All the above solutions are generally very successful and appreciated by all users. The questionnaire surveys among the children show their satisfaction with the individual character, appearance and functional disposition of the building (see Table 1). The school did not suffer from any vandalism acts so far. There are many local community events organized indoors and outdoors. The only claims concern the limited number of outside benches.

- *target:* Visual comfort in day-lit interiors

The use of daylight affects both energy demand as it restricts the need for artificial lighting and also enhances the occupants well being.

design: wherever possible all internal spaces are day-lit, including changing rooms adjacent to sport facilities and technical rooms. All classrooms have south facing extensive glazing.

In the preschool area there is a shed roof allowing direct eastern daylight into the rooms and also into half of toilets (see Fig. 6). The rest of the toilets receive dispersed light through translucent upper part of partition walls. The partition walls are glazed from 2m up. Thus the need to use artificial light in the toilets is limited.

Fig. 6. In the preschool area there is a shed roof allowing direct eastern daylight into the rooms and also into half of toilets. a) the shed roof from outside, b) the natural light penetrating the classrooms, c) the restroom illuminated by natural light

how it works: the artificial light in the day-lit toilets is sometimes on even when it is sunny – the lighting scheme needs fine tuning. The modeling of direct light penetrating the interiors, particularly classrooms and sport halls, performed at design stage were insufficient; they proved the exclusion of direct sunlight by fixed horizontal sun louvers on the 20th of June at midday only. Lower angles of solar light were not taken into account. In result in all the classrooms and sport halls glare and over heating became an instant problem. Internal blinds were installed and are in constant use. In the sports hall were the blinds have not been mounted so far the windows are temporarily sealed with paper. Another thing is that shading by trees was an important aspect of solar protection scheme. As the trees were only planted when the school opened the plan doesn't work yet.

Since the handover the school suffered from faulty functioning of external roller blinds that stay open regardless the weather conditions – they were to react to sun to protect the interiors from overheating and the BMS control was to open them in case of strong wind. A probable cause of malfunctioning is wrong location of the wind sensor. The contractor for the blinds is to solve the problem.

- *target:* Acoustic comfort

design: acoustic ceilings Herakustik covering ca. 82 % of ceilings area across the building.

usability issues: The section on acoustic comfort requires some explanation. Even though research data shows that the noise in schools reaches a level of 90 dB is it an issue neglected in all school designs to date in Poland (EIAS 2011). Polish building standards do not include requirements concerning the maximum noise level in school circulation areas. In Polish schools the only acoustically protected areas are sport halls. For Suwalska school the architects insisted, opposing the DE, to perform a study of acoustic quality of the design and in result acoustic ceiling were installed across the school. The measurements of noise levels taken in the classrooms with a group of 25 pupils actively participating in the class was 65 dB and when they worked individually it was 46 dB. At the corridor playing at recess children produced noise at the level of 76 dB. The only acoustic problem claimed was the noise from mechanical ventilation.

- *target:* Assuring thermal comfort and good indoor air quality

design: Two ground source heat pumps serve as a heating source to the building, one providing the preschool with space heating and domestic hot water and the other pump serving the school premises. The total heating capacity of the source is 187 kW. Ground heat exchanger comprises of the 26 vertical piles providing 140 kW maximum. Additionally, solar collectors are used to supplement hot water generation. Space heating distribution divided into 5 zones, all equipped with variable speed pumps and radiators with thermostatic valves. Each zone has a heat meter connected to BMS. The building is divided into 5 ventilation zones. Supply ventilation provides minimum $25m^3/h$ per person and is controlled in function of programmed occupancy schedule and door locks. All parameters of the building controls can be defined in the BMS. The return air path is arranged via transfer grilles into the corridors and to the air handling units (AHU). All AHUs are equipped with the plate heat recovery units and air source heat pumps, which control supply temperature and recover the heat from rejected air. Excellent acoustic performance was to be achieved by lowering the air velocity in the ducts below 5m/s and by applying noise attenuation in the AHUs.

usability issues claimed: A problem claimed by all interviewees is low personal control of mechanical ventilation. The manual control for each classroom is possible at the hight of over 2.5m and requires a screwdriver and assistance of technical staff. In result the teachers experiencing a problem tend to cut the MV off and open the windows rather then try to adjust the air flow rate. In the summer achieving comfort temperature in the classrooms requires a maximum air flow rate that causes disturbing noise. The temperature in the dining area and sport halls falls down to 13°C when the outside temperature drops below -10°C. The problem is yet to be solved. MV in the kitchen was not working and there was no user guide to control it. It took a few months to solve the problem. In the server room the designed air exchange was too little and the equipment overheated. Local air conditioning was introduced to solve the problem. Due to frequent short power outages the mechanical ventilation units are faulty. Precise measurements of air temperature, moisture and CO_2 level are yet to be taken as a part of a planned POE project.

5   Other Usability Issues

Most of the usability problems reveal some weaknesses of the building's delivery process. Initial lack of coordination of public transport stops location and pedestrian crossings with the new school's site is one of them. The improvements made were a result of the official complaints of the school community.

Some faults, particularly the rising damp and penetrating damp, are the result of basic faults at construction stage that could have been easily avoided. They are now very difficult to repair and if persisting present danger to the occupants health. Minor faults with exterior plaster and fitting of sinks were gradually repaired. Roof leaks are almost all repaired, only one problem spot remaining. The other group of faults that influence the usability of the building are most likely the result of installing cheaper equivalents of various elements instead of the ones designated in the design or the cheapest elements available if there was no precise specification. Lack of warm tap water is one of the issues claimed by all users interviewed as the most inconvenient and still not answered; using different than specified heaters are the most suspected reason behind this problem and checking that out is the next step. Solar collectors for water heating were installed in the preschool. Initial lack of their fine tuning was responsible for heating tap water in the preschool up to 80°C. This problem was efficiently solved. Unlike repeating problems with doors and door-locks.

Other issues seem to be a result of lack of efficient communication between DE and the school staff. On request of the DE shower taps for sport facilities were installed as fixed to the wall with no individual adjustments possible. The water was spilling in all directions. The time limit set for three minutes was perceived as too short. The DE refused to agree to change the wall mounted tabs. In result the showers are locked and out of use.

6   Conclusions

The school's design and landscaping proves to be very much appreciated by its users. The building is not free from faults however and that causes certain problems with its usability. An in depth post occupancy evaluation is planned to be performed this year. The planned POE is to become the first element to build a benchmarking system that would enable comparison and evaluation of what has been built. Seeking the reasons behind certain usability problems shows area for improvements in a public building's delivery process and maintenance.

Acknowledgments. The authors would like to thank the head of the school at Suwalska Ewa Glinska for her co-operation and support to research activities concerning the building. The authors would also like to thank all the other staff and pupils who kindly gave their time for interviews or completed the questionnaires. The authors would like to express their deepest gratitude to prof. Fionn Stevenson from the University of Sheffield for sharing her knowledge on POE and BPE research.

References

Bac, A.: Wytyczne projektowania wspolczesnych szkol. In: Wiadomosci - Izba Projektowania Budowlanego 1, 1232-1541:15-18 (2005)

Posch, P., Rauscher, E.: Schul-Raume als gebaute Paedagogik. In: Chramosta, W. (ed.) Das neue Schulhaus. Schueleruniversum und Stadtpartikel. A.F. Koska, Wien-Berlin, pp. 16–23 (1996)

Baird, G.: Sustainable buildings in practice. Routledge, Oxon (2010)

Ruminska, A. (ed.): 101 najciekawszych polskich budynkow dekady. Agora, Warszawa: 130 (2011)

Rittelmeyr, C.: Schulbauten positiv gestalten. Wie Schueler Farben und Formen erleben. Bauverlag GmbH, Wiesbaden, Berlin (1994)

Baborska-Narozny, M.: Oceny POE i BPE – postulowany standard w brytyjskiej praktyce projektowej w okresie transformacji do architektury zero emisyjnej. In: Kasperski, J. (ed.) Dolnoslaski Dom Energooszczedny. Oficyna Wydawnicza Politechniki Wrocławskiej, pp. 24–29 (2011)

Preisner, W., Schramm, U.: A process Model for Building Performance Evaluation. Mallory_Hill S (ed) Enhancing Building Performance, pp. 19–30. Wiley_Blackwell (2012)

Day, L., Sutton, L., Jenkins, S.: Children and Young People's Participation in Planning and Regeneration (2011), http://www.crsp.acuk (accessed April 12, 2012)

EIAS. Ecophon International Acousticians' Seminar (2011), http://www.acousticbulletin.Com/EN/2011/07/ecophon_international_acoustic_1.html (accessed March 27, 2012)

Humphris, M., Stevenson, F.: A post occupancy evaluation of the Dundee Maggie Centre (2007), http://sust.org/pdf/new_maggiecentre.pdf (accessed September 25, 2012)

Miejskich, Z.I.: Konkurs na Zespol Szkolno-Przedszkolny na Maslicach we Wrocławiu (2006), http://www.zim.wroc.pl/przetarg/Przetargi-dokumenty/20060929013834rozstrzygniecie.pdf (accessed September 6, 2009)

Bac, A.: Wybrane zagadnienia projektowania szkol na przykladzie szkol wiedenskich lat 90. XXw. Dissertation, Wroclaw University of Technology (2003)

Chapter 55
Comparison of Design Intentions and Construction Solutions Delivered to Enhance Environmental Performance and Minimize Carbon Emissions of a New Public School in Wroclaw

Magdalena Baborska-Narozny and Anna Bac

Faculty of Architecture, Wroclaw University of Technology, Poland, B. Prusa Str. 53/55
{magdalena.baborska-narozny,anna.bac}@pwr.wroc.pl

Abstract. The paper presents an analysis of the first public school complex in Wroclaw, Poland to use renewable energy sources with an introductory summary of low emissions constructions in Poland. Described is the process of the building's delivery and a preliminary evaluation of selected design solutions. The building, completed in the year 2009, is planned for an in-depth POE to start this year – the first such broad evaluation project to be carried out in Poland. An in depth knowledge of the design and construction process, measurements taken already, feedback from the occupants, and tracks of all faults reported so far, allow a preliminary evaluation of the building's environmental performance. A comparison of total energy consumption and CO_2 emissions between the analyzed building and two other selected schools from Wroclaw is included. It is based on energy bills for all fuel sources used. It indicates energy efficiency of the building and relatively high CO_2 emissions due to its sole dependence on electricity.

1 Introduction

Architecture following the triple bottom line guidelines is only now emerging in Poland. High energy efficiency and low carbon emissions are not yet major concerns for architects and clients. They are neither required by building regulations nor economically viable so far. The year 2010 saw first commercial developments receive BREEAM and LEED certificates[1]. The first detached house certified by Passivhause Institut (Bac 2006) was completed in 2006 in Smolec, near Wroclaw and has no successors so far. The first public building within passive standard - a sports hall adjacent to a secondary school building in Slomniki, Little Poland, was completed in February 2011. The first passive school in the country is currently under construction in a village of Budzow in Lower Silesia. Only single public investments use renewable energy sources (RES) so far. The building analyzed in this paper is one of them. There are

---

[1] BorgWarner Turbo Systems Poland – LEED Silver (PGS Software 2012) and Trinity Park III BREEAM very good (Grontmij 2012).

A. Håkansson et al. (Eds.): *Sustainability in Energy and Buildings*, SIST 22, pp. 591–600.
DOI: 10.1007/978-3-642-36645-1_55   © Springer-Verlag Berlin Heidelberg 2013

signs of interest in both low and high tech ecological solutions among individual clients but still at a limited scale. Polish construction industry, the clients and occupants are all at an early learning stage in terms of low emissions building, though Poland as a member of the EU is obliged to implement the EU energy efficiency objectives i.e. the 20-20-20 targets.

The early learning stage means that there are numerous green-bling solutions available, strongly promoted by their manufacturers, promising zero-energy construction at hand, but little practice in their actual delivery and performance, maintenance requirements, usability for the occupants combined with scarce financial or legal incentives for building green and no support for microgeneration. No in-depth POE or BPE research has been conducted on the few green buildings constructed so far.

2 The Background for the Case Study Building Design

Having all that in mind, it is easier to understand the environmental targets set by the Wroclaw City Council's vote in the year 2006 for the analyzed public school building, though limited and general: the school was to make use of RES, must nevertheless be regarded as a 'green avant-garde'. They were the first and so far the last environmental targets set for a public investment in Wroclaw to go beyond the legal requirements in that respect. An architectural competition followed that vote but no further details in terms of environmental performance were articulated: the school was to make use of renewable energy sources but the type, extent or energy targets were not specified (Zarzad Inwestycji Miejskich 2006). An integration of local community with the school was proposed at the competition stage through inclusion of several public functions into the school: a public library, community club, local community council and a city guard office. The sport facilities were also to be let to local community after school working hours. A greenfield site at Suwalska street, in a currently developing suburban housing district of Wroclaw was designated. The 'sustainable' direction for the development was taken when the jury selected the winner of the 1st prize – a design by Grupa Synergia architectural office, led by Anna Bac and Krzysztof Cebrat, experts in sustainable construction and landscaping at Wroclaw Faculty of Architecture.

3 The Choice of Case Study Example

The choice of the case study example reflects the fact that it is the first, and so far the last, public investment completed in Wroclaw that was to make use of renewable energy sources. Dissemination of evidence based evaluation of its performance may have a major influence on both public opinion and the shape of local policies towards sustainable architecture. The building is presented in a book on recent Polish architecture (Ruminska 2011) and received three local architectural prizes[2].

---

[2] 2010 – II prize in the category of a public building in a competition "Beautiful Wroclaw", 2011 – PLGBC award I prize in the category of "Green building of the Year" and honorable mention in the category "Green Interior of the Year", 2011 - honorable mention in a competition organized by a local branch of Polish Society of Architects SARP in the category "user friendly space".

The competition entry's environmental scheme was developed in cooperation with environmental engineer Wojciech Stec from First Q Amsterdam office.

Grupa Synergia was commissioned all design stages until the handover of the building. The architects invited Eko Energia System for environmental engineering. Eko Energia System developed a feasibility study for the competition entry scheme, which proved many of the proposed 'green' features not be economically viable and in result were dropped from the final design (see Table 1 and Table 2). Department of Municipal Development (DMD) (the client), though strongly represented in the competition jury, had no previous experience in the procurement and management of energy efficient buildings, and in the course of the design proved not to be a partisan of a 'green' design approach. DMD was responsible for the organization of the tender process and as usual the lowest price was the main criterion for the choice of the contractor.

Fig. 1. a) View towards school entrance zone with distinct volumes housing different functions. b) View towards a courtyard between school and preschool wing. A copper volume with dining room is shared by the two functions.

The 6000-square-meters building accommodates 450 pupils and 100 preschool children. There are 80 employees in the building including: teachers, administration and technical staff. The school's population is divided between two classroom wings with shared entrance lobby, library, sport and dining facilities. The third wing with a separate entrance is dedicated to the preschool. The dining area serves as a connection point between the school and the preschool (see Fig. 1).

4 The Context and the Stage of Data Collection

The building has been occupied since September 2009. Some observations, measurements and troubleshooting were performed up to date. In the year 2010 a questionnaire survey for the children of classes 3-6 was personally distributed and collected by the architects to learn about the reception of selected issues in the school's design. After the first winter the leading environmental engineer Andrzej Bugaj was commissioned a year of continuous supervision and fine tuning of the building's environmental systems. The results of his supervision is not clear to the building's users, including the head of the school Ewa Glinska, who was not given any written reports on its progress and outcome. Some problems that had been occurring persisted. An in-depth

POE research is thus planned to start this year using PROBE investigation techniques developed by the UK's Usable Buildings Trust (Baborska-Narozny 2011). It is intended to deliver evidence-based evaluation and indicate ways for improvements. A three year contractor's warranty ends in June 2012, thus it is vital for the head of the school to trace and repair construction defects and systems malfunctions before that date. In result a thermo-graphic survey has already started. The presented evaluation is based on an in-depth knowledge of the building's design and procurement phases, semi structured interviews with crucial occupants (head of school, head of preschool, staff representatives, including the building technician responsible for the building maintenance), questionnaire survey mentioned above and detailed analysis of all records documenting the defects claimed by the users to DMD, who manages construction warranty.

5   Design Targets and Solutions as Built

Comparison of early design stages and as built design solutions to lower the building's energy demand and environmental footprint is presented in Table 1. It is followed by selected usability issues. Table 2 includes preliminary and as built solutions concerning sustainable landscaping and the selection of construction materials.

Table 1. Design targets and solutions as built

Design solution	Competition entry	As built
Improved energy performance of the building's envelope		
Building's envelope thermal insulation exceeding building regulations requirements. (see point A following Table 1)	V	V
Building's air tightness	V	X
Outdoor staircases and glazed roofs covering entrance areas are separate from the construction of the building to avoid thermal bridging. (see point B following Table 1)	V	V
Additional insulation framing windows	V	V
Functional disposition to incorporate solar gains and losses		
Double glass facades on south-facing classrooms	V	X
Reduced heat loss through north-facing walls with smaller openings in the corridors	V	V
The use of exposed thermal mass	V	V
Excessive solar gain and glare protection		
Fixed horizontal brise solei above south-facing openings. (see point C following Table 1)	V	V
Skylights and east and west facing glazing of public areas and administrations facilities with external textile roller blinds	V	V

55 Comparison of Design Intentions and Construction Solutions Delivered 595

Table 1. (*continued*)

Design solution	Competition entry	As built
Use of renewable energy sources to cover energy demand for heating/cooling and hot water		
Heat recovery from a major municipal sewage collector that crosses the site underground	V	X
Heat pump with geothermal loop vertical	V	V
Earth tubes	V	X
Solar collectors for water heating (see point D following Table 1)	V	V
Floor heating in the preschool, dining and sport hall	V	X
Ventilation system		
Natural ventilation with the stack chimneys and double glass facades	V	X
Automatic controls to enhance natural ventilation	V	X
Mechanical ventilation with heat recovery to support natural ventilation when required, particularly in the summer and winter periods	V	X
Mechanical vent. with heat recovery (see point E following Table 1)	X	V
Energy management system		
Partly automated individual comfort management integrated with natural ventilation	V	X
BMS system with the possibility to make individual adjustments in most interiors (see point F following Table 1)	X	V
Use of daylight		
All interiors day-lit (excepts for school toilets)	V	V
Daylight entering classrooms from two sides: direct with views outside and indirect through clerestory widows opening to the corridors	V	V
In the preschool extensive glazing to the south and clerestory windows to the east, allowing daylight also to the toilets	V	V
In the two story communication zone daylight reaches the lower floor through openings in the floor slab under the roof lights	V	V
Controlled use of artificial lights		
Occupancy sensor control in the restrooms	V	V
Zoning of artificial lighting scheme according to the distance from the windows	V	X

The users' comments and faults which were reported concerning selected design solutions listed in Table 1 are as following:

A) A severe problem i.e. frost on the inside of wall and window occurred at one point during the first winter. A thermographic survey revealed a major gap in the thermal insulation layer that was soon repaired. A second thermographic survey carried out this winter revealed only one minor defect in the consistency of thermal insulation.

B) Thermography proved those details to be efficient (see Fig. 2).

Fig. 2. a) Thermographic picture taken at outside temperature -5 °C, inside +19° C, wind 1 m/s, at 11 p.m. on 27 Feb.2012, low moisture, dry surfaces, camera Varioscan 3021ST. b) Daytime view.

C) Glare was an immediate problem in the classrooms from the early occupancy. Internal roller blinds were installed. The modeling of daylight performed at design stage proved to be insufficient.

D) The solar collectors are supplemented by electric water heaters. Even so there are persistent problems with warm water across the school. The problem is still to be solved.

E) Mechanical ventilation produces noise that is perceived by the teachers as disturbing and annoying. Limiting the air flows in turn lowered the air quality.

F) Precise measurements of air flows, internal temperatures and indoor air quality are yet to be taken. The overall evaluation indicates poor usability of the system. It was meant to be controlled by the BMS, but lack of facility manager and lack of competence of the users causes frustration. Separate controls of the air flow rates are provided for each classroom, however their location at the corridor at ceiling height and the need to use a screwdriver makes it impossible for the teachers to make adjustments in air flow rates. The need to call for technical assistance often results in asking for turning the ventilation off and opening the windows (see Fig. 3).

Fig. 3. a) Classroom with MV ducts visible, b) Corridor with MV ducts visible, c) Location of the control gear for MV at ceiling high

55 Comparison of Design Intentions and Construction Solutions Delivered   597

Table 2. Sustainable materials and landscape design targets and solutions as built

Design solution	Competition entry	As built
Sustainable water management		
Water retention ponds as a landscape feature	V	X
Rain water for watering the plants on site	V	V
Gray water for flushing toilets	V	X
Using water permeable surfaces	V	V
Time-limited taps	V	V
Sustainable landscaping		
Local plants	V	V
Shading the building through proper planting	V	V
Experimental garden for the pupils	V	V
Extensive green roof	V	X
Light color of roof membranes and landscape surfaces for heat island effect reduction	V	V
Local materials with low embedded energy where possible		
Use of wood for facades	V	X
Use of wood and wood products for structure and interior, mineral wool for thermal insulation	V	V
Gabions filled with local stones as the base for outside benches	V	V

The school was commissioned and designed as a highly innovative building, making use of renewable energy sources, thus lowering its environmental footprint. It is equipped with sophisticated installations that were to deliver healthy indoor environment while being energy efficient. The change form preliminary design stages towards the built result brought a shift form a co-existence of passive and active methods of IAQ control towards focusing on active ones. What seems to be missing from the user's perspective is a lack of handover stage that would leave them with awareness of the systems installed and the technical skills for their control. There is no professional facility manager employed. Lack of proper fine-tuning of systems installed during the early occupancy and lack of awareness of the expected results in terms of overall comfort result in staff frustration. Detailed measurements of building energy performance are yet to be taken however at this stage of data collection and problem solving it is probable that the building is not yet making full use of its potential in terms of energy efficiency and user's comfort.

6   The Building's Total Energy Consumption and CO_2 Emissions as Compared to Two Other Schools in Wroclaw

A robust check whether the aim to reach energetically and environmentally efficient building is met can be performed by comparing the bills for media of school at

Suwalska Street (see Fig. 4a), with two other schools in Wroclaw. The bills for water are not included as there are too many factors to take into account to explain the results. The focus is on energy consumption and CO_2 emissions (see Table 3). The two other schools selected represent two building types. The school at Rumiankowa Street is a recently retrofit building from the seventies of the XX century to meet "highest European standards" (Hussak 2012) (see Fig. 4b). It is heated with local gas heating. The school at Aleja Pracy is a historical building built in 1934 (Harasimowicz 1998) that is not yet retrofitted (see Fig. 4c). It is heated with co-generation. The first type indicates what can be achieved through retrofitting of the existing building stock and the second represents the many historical buildings before any major modernization. Data on type and amount of energy used by each school were collected and shared by the City Council Office.

Reference carbon emissions for electricity production in Poland is 0,812 Mg CO_2/MWh = 225.44 kg CO_2/GJ (as in June 2011 – applicable for calculations in the year 2012) (KOBiZE 1 2011). It is among the highest in Europe. CO_2 emissions factor for co-generation heat in Poland is 93.97 kg CO_2/GJ, and for gas it is 55.82 kg CO_2/GJ (KOBiZE 2 2011).

Table 3. Comparison of total energy consumption and CO_2 emissions for three selected schools in Wroclaw

	2009	2010	2011	Average
1. Suwalska	1626 kWh	182,817 kWh	252 130 kWh	0.12 GJ/m^2/ pa
5800 m^2	(Sep-Dec)	(658 GJ/pa)	(908 GJ/pa)	
450 pupils	(5.8 GJ)(*)			
Annual CO_2 emissions	1320 kg CO_2/pa	148 371 kg CO_2/pa	204 625 kg CO_2/pa	26 kg CO_2/m^2/pa
2. Rumiankowa	1842 GJ	1813 GJ	1286 GJ	0.33 GJ/m^2/ pa
5000 m^2				
407 pupils				
Annual CO_2 emissions	143 555 kg CO_2/pa	139 027 kg CO_2/pa	110 797 kg CO_2/pa	17.6 kg CO_2/m^2/pa
3. Aleja Pracy	2 744 GJ	3 478 GJ	2 721 GJ	0.84 GJ/sq.m/pa
3567 m^2				
165 pupils				
Annual CO_2 emissions	298 478 kg CO_2/pa	348 kg CO_2/pa	365 278 042 kg CO_2/pa	88 kg CO_2/m^2/pa

(*) The electricity meter for the school was faulty in 2009 and underestimated the energy demand. The difference between the real and the calculated energy consumed was added to the bills for 2011 hence the substantial difference in energy demand shown in the bills for 2010 and 2011. Source: Energy consumption based on unpublished data shared with the authors by the Municipal Department of Education, Wroclaw (Municipal Department of Education, Wroclaw 2011)

Fig. 4. a) School complex at Suwalska, b) School complex at Rumiankowa, c) High school at Aleja Pracy

7 Conclusions

The analyzed school in Wroclaw is built according to energy standards well exceeding even current building regulations. It is equipped to make use of renewable energy sources. Its environmental impact was a major decision factor for the architects. The novelty of many design solutions proposed together with the feasibility study performed were the main reasons for changes introduced to the initial scheme for the building's energy management. A comparison of the resulting building's energy demand and CO_2 emissions with data for two other selected schools in Wroclaw proves it's excellent quality in terms of energy efficiency and at the same time relatively high CO_2 emissions due to coal-fired electricity covering 100% of the building's energy demand. Retrofit school heated by gas, though consuming almost three times more energy, produces 32% less CO_2 emissions than the analyzed building heated by ground heat pump. A more detailed evaluation of the whole building's performance must wait for the results of the planned POE research.

Acknowledgments. The authors would like to thank the head of the school at Suwalska Ewa Glinska for her co-operation and support for research activities concerning the building. The authors would also like to thank all the other staff and pupils who kindly gave their time for interviews or completed the questionnaires. The comparison of energy consumption would not be possible without the data kindly shared with the authors by Municipal Department of Education.The authors would like to express their deepest gratitude to prof. Fionn Stevenson from the University of Sheffield for sharing her knowledge on POE and BPE research.

References

Bac, A.: Budynki pasywne: wymagania techniczne i projektowanie. Wiadomości Projektanta Budownictwa 6, 1232–1541, 30–32 (2006)

Harasimowicz, J. (ed.): Atlas architektury Wrocławia. Wydawnictwo Dolnośląskie: 139 (1998)

Ruminska, A. (ed.): 101 najciekawszych polskich budynkow dekady. Agora, Warszawa, 130–131 (2011)

Baborska-Narozny, M.: Oceny POE i BPE – postulowany standard w brytyjskiej praktyce projektowej w okresie transformacji do architektury zero emisyjnej. In: Kasperski, J. (ed.) Dolnoslaski Dom Energooszczedny. Oficyna Wydawnicza Politechniki Wrocławskiej, pp. 24–29 (2011)

Grontmij (2012),http://www.grontmij.com/mediacenter/Pages/First-Polish-building-certified-under-BREEAM-scheme!.aspx (accessed March 13, 2012)

Hussak, U.: Opis obiektu Zespołu Szkolno-Przedszkolnego(2012), http://innowacyjnyekolog.pl/szkoly/14/szkola-podstawowa-nr-27-w-zespole-szkolnoprzedszkolnym-nr-10-we-wroclawiu Accessed 22 March 2012

KOBiZE 1 Referencyjny wskaźnik jednostkowej emisyjności dwutlenku węgla przy produkcji energii elektrycznej do wyznaczania poziomu bazowego dla projektów JI realizowanych w Polsce (2011), http://www.kobize.pl (accessed April 16, 2012)

KOBiZE 2 Wartości opałowe (WO) i wskaźniki emisji CO_2 (WE) w roku 2009 do raportowania w ramach Wspólnotowego Systemu Handlu Uprawnieniami do Emisji za rok 2012 (2011), http://www.kobize.pl/index.php?page=materialy-do-pobrania (accessed April 16, 2012)

PGS software, http://www.pgs-soft.com/first-building-in-poland-received-leed-certification.html (accessed March 13, 2012)

Miejskich, Z.I.: Konkurs na Zespol Szkolno-Przedszkolny na Maslicach we Wrocławiu (2006), http://www.zim.wroc.pl/przetarg/Przetargi-dokumenty/20060929013834rozstrzygniecie.pdf (accessed September 6, 2009)

Municipal Department of Education, Wroclaw, Data on energy consumption in schools and preschools (2011) (unpublished)

Chapter 56
An Exploration of Design Alternatives Using Dynamic Thermal Modelling Software of an Exemplar, Affordable, Low Carbon Residential Development Constructed by a Registered Social Landlord in a Rural Area of Wales

Simon Hatherley, Wesley Cole, John Counsell, Andrew Geens,
John Littlewood, and Nigel Sinnett

Cardiff Metropolitan University

Abstract. Pembrokeshire Housing Association (PHA) a registered social landlord, based in Haverfordwest, Wales, UK, have developed six low carbon houses to meet Code for Sustainable Homes (CfSH) level four, as part of an exemplar scheme for the Welsh Government's CfSH pilot project. A tried and tested methodology was adopted in developing the PHA's pilot project houses that meant alternative low and zero carbon design methods were not fully explored. This paper employs comparative analysis to evaluate the final PHA scheme against other design options in order to assess alternative low energy approaches that might have been considered during the design of the project. Dynamic thermal modelling is used to assess and compare the design options in which the following are considered: building form; use of the thermal mass within the building fabric; design of the external envelope; and passive solar design strategies. The discussion considers the implications of the results with regard to approaches to low carbon design, as part of a doctoral research project, by the lead author on to develop innovative, affordable, low carbon housing in rural areas of Wales, UK.

1   Background

In 2010 Pembrokeshire Housing Association (PHA) completed a development of six two bedroom, semi-detached houses on Britannia Drive, in Pembrokeshire, west Wales, UK, built to the Code for Sustainable Homes level 4. The project was developed as part of the Welsh government's Code for Sustainable Homes (CfSH) pilot project to promote the construction of low and zero carbon housing through the Registered Social landlord (RSL) framework (Welsh Government 2011). The Britannia Drive design team took the decision to use a tried and tested methodology for the design and construction of the houses and utilised photovoltaic panels to meet the requirements for CfSH code level four.

The use of a tried and tested methodology meant that alternative design options for the low carbon design of the houses were not fully examined and this provides research opportunities to explore alternative approaches. In addition, dynamic thermal

A. Håkansson et al. (Eds.): *Sustainability in Energy and Buildings*, SIST 22, pp. 601–611.
DOI: 10.1007/978-3-642-36645-1_56   © Springer-Verlag Berlin Heidelberg 2013

modelling was not used to examine the energy performance of the design. The fact that design options were not explored and dynamic thermal modelling was not used on this scheme provides opportunities to assess the thermal performance of different passive design approaches through dynamic thermal modelling.

2   Methodology

Dynamic thermal modelling predicts the energy performance of buildings using mathematical models to determine the interplay of heat exchange (Jankovic 2012). Dynamic thermal modelling software is frequently used to provide a prediction of the final energy usage of a scheme based on range of inputs including local climate, patterns of occupancy, building geometry, and building fabric (Morbitzer et al. 2001). However, the aim of this study was not to predict or compare the energy efficiency of the built project against a model, but rather to create variations of a control model, based on the as-built project; to assess design solutions with the potential to minimise energy consumption for space heating.

A comparative analysis approach was used whereby different design options were benchmarked against the original scheme rather than against an absolute standard, such as a representative dwelling built to a zero carbon standard (Bryman 2008; Creswell and Clark 2011). The disadvantage of the comparative analysis approach was that it was difficult to set the building design options within the broader context of aspirations by the British and Welsh Governments to develop zero carbon dwellings (Welsh Assembly Government 2009). However, the use of comparative analysis meant that design solutions developed for the study would be relevant to PHA's current approach and would not significantly deviate from the affordability and build-ability of the original scheme.

Two studies were undertaken; one exploring the design of individual dwellings, and a second investigating alternative approaches to the development of all six residential units. The first study investigated the impact of upgrades and adjustments to building fabric, such as introducing thermal mass, providing additional glazing on the south façade and improving the u-value of the building fabric. For the first study, the physical shape of the building, such as plan form and overall massing was not adjusted. The second study investigated the impact of developing the dwellings as a terrace rather than semi-detached houses. For this second set of studies the buildings' form was adjusted; however, the floor area of the original scheme was maintained as a constant, since any significant increase in floor area would undermine the affordability of any alternatives and any reduction would reduce the ability to compare results.

The designs from these two studies with the lowest sensible heating load were then amalgamated into two final schemes. These two final schemes represented optimum solutions, within the scope of the study, with regard to reducing sensible heating load and maintaining affordability.

2.1 Climate, Heating and Cooling and Occupancy Profiles

The site for Britannia Drive is located in Pembroke Dock, in Pembrokeshire; therefore Aberporth Example Weather Year was used as the weather file because it provided a data set relevant to the location in west Wales. Weather files are used by dynamic thermal modelling software to provide a context for thermal calculations and an Example Weather Year matches the characteristics of a year to the average monthly values for a number of years of data (University of Exeter 2010).

A continuous occupancy profile, whereby the property would be considered occupied twenty-four hours, a day was adopted for these studies for two reasons. Firstly it was decided to minimise the multiple independent variables where possible. Secondly from discussions with PHA staff it was apparent that near continuous occupancy is not uncommon in social housing. However, future models will be calibrated with data from thermal monitoring of PHA's properties to allow occupancy profiles to be based on actual tenancy occupancy patterns.

With regard to annual heating profiles the heating period was set from 1st September to 30th April. A heating set point was set at 19 degrees centigrade (°C); thus, when internal temperatures fell below this threshold during the heating period the boiler was activated. The cooling set point was set at 22°C; thus, when internal temperatures rose above this threshold from 1st May to August 31st cooling was provided by additional natural ventilation through opening windows.

2.2 Building Geometry

The design of the Britannia Drive development had the principle elevations on the north and south facades. A south façade with three windows faces the street, and a north façade with four windows faces the back garden and the east and west facades, while significant in area, contain no glazing (see figures 1 and 2 below). This north-south orientation was maintained for all of the models.

All of the projects were subjected to a Suncast analysis in the Integrated Environmental Solutions (IES) dynamic thermal modelling software package. This component of the software package takes account of building orientation and solar shading to calculate solar gain. PHA's design for the dwellings at Britannia Drive had windows located immediately under the eaves (see figure 1 below) and this design feature was accounted for in the models, to allow Suncast to analyse the effect of shading provided to windows by this feature (see figure 2 below).

Fig. 1. Britannia Drive south elevation as-built, before installation of photovoltaic panels

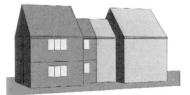

Fig. 2. IES model north elevation

2.3 Building Fabric Specification

The make-up of a building's fabric can significantly affect a building's thermal performance, but not just with regard to thermal conductivity as in u-values, but also through its thermal mass (Tuohy et al. 2009). Four alternative fabric specifications were explored for this study with variations in u-values and thermal mass.

The first specification was based on PHA's standard wall construction, which consisted of 140mm deep timber studs in-filled with mineral wool insulation for the inner skin and a rendered 100mm blockwork outer skin which provided a u-value of 0.21 W/m^{2}K (see table 1 below). The second specification investigated the option of reversing the standard PHA building fabric, so that the blockwork element was on the inside of the insulation; thereby following passive solar design principles (Lowndes, 2008). Moreover, this does not increase the quantity of thermal mass within the building fabric; however it does affect the admittance value and decrement values. The admittance value is the ability of an element to exchange heat with the environment when subjected to cyclic variations in temperature and the decrement factor is the ratio between the cyclical temperature on the inside surface compared to the outside surface (The Concrete Centre 2009). Reversing the building fabric reduced the decrement factor, and increased the admittance value of the wall build-up by a factor of three (see table 1 below).

Table 1. Britannia Drive as-built before, the installation of photovoltaic panels

Building Fabric	U value W/m²K	Internal Heat Capacity KJ/m²K	Admittance Value	Decrement factor
Standard External Wall	0.21	19.95	1.45	0.41
Reversed Standard Wall	0.21	134.07	4.37	0.34
PIR Insulation External Wall	0.16	19.95	1.49	0.39
Reversed Wall w/ PIR insulation	0.16	134.07	4.37	0.32
Standard Internal Wall	0.37	11.97	0.90	0.99
Blockwork Internal Wall	1.62	79.00	3.28	0.59
Standard First Floor Ceiling	0.11	11.97	1.10	0.81
PIR Insulation First Floor Ceiling	0.08	11.97	1.03	0.61

Two further specifications considered the impact of replacing the mineral wool insulation between the interior wall studs with higher performance polyisocyanurate (PIR) insulation which is commonly used in the construction industry as a substitute for mineral wool insulation. This substitution of materials would have cost implications, as solid slab insulation can be as much as three times more expensive than an equivalent amount of mineral wool insulation (Davis Langdon 2009). Nevertheless, it represented a means of upgrading PHA's building fabric while maintaining their current approach of using a 140mm deep timber stud wall. As Table 1 (above) shows this substitution of materials raised u-values from 0.21 W/m²K to 0.16 W/m²K, but did not significantly impact the admittance and decrement values. The fourth fabric specification combined a reversed building fabric with PIR insulation thus the exterior wall benefited from increased admittance and u-values.

3 Results

Two principle outputs were examined to gain an understanding of space heating and internal thermal comfort. Sensible heating load was used as the principle measure of operational energy and was considered both as a total measure and as a percentage reduction over the control building. With regard to thermal comfort, peak dry resultant temperature and number of hours over 25°C were taken as the most significant measures. Heating season temperatures were generally governed by the heating set point, which ensured that internal temperature were maintained within the comfort range; thus, it was only in summer (1st May to 31st August) that internal temperatures exceeded 25°C. In addition, because summer cooling relied on natural ventilation provided by opening windows the houses are reliant on the building fabric to keep temperatures within the comfort range of 19°C to 25°C.

3.1 Modelling of the Individual Buildings

An initial set of models (models A to G in table 2) considered the impact of individual improvements such as increasing solar gain, increasing thermal mass to interior spaces or improving the insulation. The results from these initial models demonstrates that reversing the building fabric (as in model D) can provide modest (2.0%) reductions in sensible heating load, reduce the peak internal temperature by 1.45°C and reduce the number of hours that the house was warmer than 25°C by 34 hours (see Table 2 and graph 1, below). Increasing the thermal mass with the addition of internal concrete blockwork walls in model E doubled this percentage reduction in heating load and eliminated internal temperatures above 25°C (see Table 2 and graph 1, below).

Increasing glazing on the south facade by 1.68m² (model B) produced a reduction in sensible heating load of just 0.6%. Increasing the glazing also increased the number of hours where temperatures were above 25°C by 44hours (see table 2, and graph 1, below). Additional glazing (a further 1.68m² on the south façade, as in model C) aggravated the problems and produced a peak internal dry resultant temperature of 29.74°C, 2.19°C higher than the original model, (see Table 2, and graph 1, below). These results indicate that passive solar gain can be advantageous in providing 'free' space heating and reducing heating loads; however, in lightweight construction temperatures above comfort can occur when measures are not taken to address overheating.

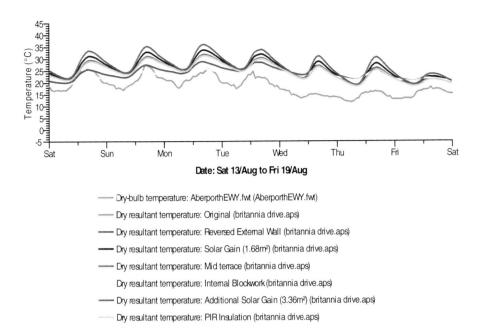

Graph 1. Britannia Drive individual improvements peak dry resultant temperatures for main bedroom (south facing) during peak (external) dry bulb temperature week

Table 2. Britannia Drive single house, individual improvements table of results

Scheme	Annual Sensible Heating Load (MWh)	% reduction in sensible heating load	Peak internal Temperature (°C)	Time/Date of peak internal temperature	No. of hours above 25°C
Original (Model A)	4.51	0.0	27.55	17:30 15th Aug	53
Solar Gain (1.68m²) (Model B)	4.48	0.6	28.48	17:30 15th Aug	97
Additional Solar Gain (3.36m²) (Model C)	4.48	0.7	29.74	17:30 15th Aug	209
Reversed External Wall (Model D)	4.42	2.0	26.10	16:30 15th Aug	19
Internal Blockwork (Model E)	4.32	4.0	24.84	13:30 16th Aug	0
Improved Insulation (Model F)	4.06	10.0	27.82	17:30 15th Aug	69
Mid Terrace (Model G)	4.03	11.0	27.86	17:30 15th Aug	78

56 An Exploration of Design Alternatives Using Dynamic Thermal Modelling Software

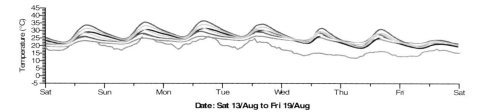

— Dry-bulb temperature: AberporthEWY.fwt (AberporthEWY.fwt)
— Dry resultant temperature: PIR Insulation and Thermal Mass (britannia drive.aps)
— Dry resultant temperature: PIR Insulation, Thermal Mass and Solar Gain (1.68m²) (britannia drive.aps)
— Dry resultant temperature: Original (britannia drive.aps)
— Dry resultant temperature: Thermal Mass and Solar Gain (3.36m²) (britannia drive.aps)
   Dry resultant temperature: Thermal Mass and Solar Gain (1.68m²) (britannia drive.aps)
— Dry resultant temperature: PIR Insulation and Solar Gain (3.36m²) (britannia drive.aps)
--- Dry resultant temperature: PIR Insulation, Thermal Mass and Solar Gain (3.36m²) (britannia drive.aps)
--- Dry resultant temperature: PIR Insulation and Solar Gain (1.68m²) (britannia drive.aps)

Graph 2. Britannia Drive composite improvements peak dry resultant temperatures for main bedroom (south facing) during peak (external) dry bulb temperature week

Table 3. Britannia Drive single house, composite improvements table of results

Scheme	Annual Sensible Heating Load (MWh)	% reduction in sensible heating load	Peak internal Temperature (°C)	Time/Date of peak internal temperature	No. of hours above 25°C
Original (Model A)	4.51	0.0	27.55	17:30 15th Aug	53
PIR Insulation and Thermal Mass (Model H)	3.90	13.6	26.08	13:30 16th Aug	21
Thermal Mass and Solar Gain (1.68m²) (Model I)	4.12	8.6	25.46	13:30 16th Aug	10
Thermal Mass and Additional Solar Gain (3.36m²) (Model J)	4.06	10.1	25.94	13:30 16th Aug	20
PIR Insulation and Solar Gain (1.68m²) (Model K)	4.04	10.5	28.5	16:30 15th Aug	105
PIR Insulation and Solar Gain (3.36m²) (Model L)	4.04	10.5	29.14	16:30 15th Aug	171
PIR Insulation, Thermal Mass and Solar Gain (1.68m²) (Model M)	3.76	17.9	25.94	13:30 16th Aug	11
PIR Insulation, Thermal Mass and Solar Gain (3.36m²) (Model N)	3.71	18.0	25.97	13:30 16th Aug	22

A second set of models (models H to L) combined approaches such as additional south facing glazing and thermal mass to internal walls (models I and J); or thermal mass with higher performance insulation (model H); or additional glazing on the south façade with higher performance insulation (Models K and L) (see Table 3). The most significant energy savings from this set of models was achieved through the combination of thermal mass with PIR insulation which not only reduced the sensible heating load of the original scheme by 13.6% and reduced the number of hours over 25°C and peak internal temperature to 21hours (see table 3).

The results of table 3 indicate that the successful application of passive solar gain to reducing heating load is related to the provision of thermal mass both for addressing thermal discomfort but also for storing and releasing heat when required. This was demonstrated by the fact that the difference in heating reduction between the lightweight models of the initial study (models B and C) with 1.68m² and 3.36m² of extra glazing was 0.6% and 0.7%; however, for the higher thermal mass composite models the respective reduction in sensible heating load was 8.6% (model I) and 10.1% (model J) (see tables 2 and 3).

The use of higher performance insulation and passive solar gain without thermal mass achieved a 10.5% reduction in sensible heating load (see table 3). However, this approach produced a significant increase in internal temperatures over the original model with an additional 52 hours of temperatures in excess of 25°C over the control model (model A) (see table 3). Maintaining the lightweight building fabric, but increasing the glazing from 1.68m² to 3.36m² with higher performance insulation (see models K and L) produced almost no decrease in sensible heating load and substantially increased the number of hours which internal temperatures were above 25°C to 171 hours (model L) (see table 3). These results suggest that there is an upper limit at which passive solar gain can be employed as a strategy for reducing space heating without the aid of thermal mass to store heat and moderate overheating.

Combining all of the passive strategies (models M and N); thermal mass, additional glazing and insulation proved to be a useful strategy achieving a 17.9% and 18% reduction in sensible heating load compared to the control model (model A) (see table 3). The evidence suggests that the use of thermal mass successfully addressed any problems that might have occurred with regard to overheating as the internal temperatures and the number of hours where temperatures were above 25°C was lower than the original model at 11 hours in model M (see table 3).

3.2   Modelling of the Site

The second study took advantage of the fact that the gable ends on the east and west elevations did not have doors or windows and thus, in principle, could be butted against each other to create a terrace (model 2). In practice it would be problematic to abut the houses against each other due to the requirement for the provision of car parking spaces between the units; therefore, a terraced option was explored that utilised a shallow, but wide floor plate (5.04m x 10m) for each dwelling, which located the houses in a 5.05m zone between a front parking zone and rear gardens (model 3, figure 3). A further terraced model (model 4, figure 4) was developed based on a cube form, with a square floor plate (7.1m x 7.1m), to examine the impact of further minimising the ratio of external surface area to internal volume.

Fig. 3. IES visualisation of site model - narrow terrace south façade (model 3)

Fig. 4. IES visualisation of site model - compact terrace south façade (model 4)

Table 4. Britannia Drive site model table of results

Scheme	Annual Sensible Heating Load (MWh)	% Energy Improvement over Base Model	Peak internal Temperature (°C)	Time/Date of peak internal temperature	No. of hours above 25°C per house
Original (Model 1)	27.05	0.0	28.56	17:30 15th Aug	61
Terrace (Model 2)	25.07	7.3	28.75	17:30 15th Aug	73
Narrow Terrace (Model 3)	21.4	20.9	30.17	17:30 14th Sep	303
Compact Terrace (Model 4)	21.5	20.8	29.10	17:30 15th Aug	90

The results from the first terrace model (model 1) indicated the benefit of minimising the external surface and produced a 7.3% saving in overall sensible heating load for all six houses (see table 4 above). Whilst, a compact form, minimising the external fabric surface area was one of the most significant factors in reducing heating it was the narrow terrace (model 3) that achieved the lowest annual sensible heating load of 21.5MWh, slightly lower than the compact terrace (Model 4) with 21.4MWh. This result was due to the fact because of the shallow floor plan it was possible to put almost all of the windows on the south facade. This result confirms the findings from the composite models studies (models H to N) which identified the benefits of combining passive design strategies; however, the fact that the majority of the windows were on the south façade meant that this design had significantly more hours (303 hours) where internal temperatures were above 25°C than the original design (see table 5).

3.3 Amalgamated Models

A final set of options took the site model (model 3) and the individual building model (model M) with the lowest sensible heating load to create an optimum solution within the limited context of the two studies. These amalgamated models (see 3Ma and 3Mb in table 5) combined a narrow compact terrace form with PIR insulation, thermal mass and 3.36m² of additional glazing. Amalgamated model 3Mb achieved a 40% decrease in sensible heating load for a mid-terrace option and model 3Mb achieved a 36% decrease in sensible heating load for an end terrace over the original model

(model A). However, this model did have siginifcant problems with regard to overheating with 348 hours where internal temperatures would be higher than 25°C and peak temperatures of 32.99°C. Further development of the design would be required to address overheating issues to take advantage of energy and carbon savings without reducing thermal comfort.

Table 5. Britannia Drive Amalgamated model results

Scheme	Sensible Heating Load (MWh)	% Energy Improve-ment over Base Model	Peak internal Tempera-ture (°C)	Time/Date of peak internal tempera-ture	No. of hours above 25°C
Amalgamated End Terrace (Model 3Ma)	2.89	36	31.82	15:30 14th Sep	384
Amalgamated Mid Terrace (Model 3Mb)	2.71	40	32.99	15:30 14th Sep	454
Affordable End Terrace (Model 3Ea)	3.40	25	27.62	15:30 14th Sep	65
Affordable Mid Terrace (Model 3Eb)	3.11	30	28.50	15:30 14th Sep	99

An affordable solution was also developed that omitted PIR insulation, maintained the glazing at the same quantity as the original scheme but increased thermal mass (as model E). This design produced a reduction in sensible heating of 25% for an end terrace and 30% for a mid-terrace (see table 5). In addition, the peak internal temperature of 27.62°C for the end terrace was only 0.17°C higher than the original scheme, which indicates that this solution would not require significant design development to address overheating problems.

4   Conclusion

The study indicates that significant energy savings (25% - 40%) in sensible heating load can be achieved even on an exemplar low carbon project with little or no additional capital cost. Measures such as specifying thermal mass to internal walls, considering the ratio of external area to internal volume and the massing of the buildings on the site can provide considerable savings in heating load with apparently little or no addition to the capital cost of a project. The use of designs features to allow internal rooms to capture passive solar gain can also be used to reduce heating load; however it can be problematic with regard to maintaining thermal comfort unless measures to mitigate overheating (Orme et al. 2010) and store heat within the fabric are considered.

The study also indicates that overheating will increasingly become a problem as buildings become more become more effective at reducing sensible heating load through building form and fabric based approaches. Thus, considering measures to address overheating, such as the specification of thermal mass, as in this study, and perhaps ventilating strategies like night cooling or stack ventilation, will become more important in low carbon house design.

The study indicates the benefits of combining approaches rather than being overly reliant on one strategy; however the interplay of these approaches requires careful consideration to maintain reasonable levels of internal comfort and dynamic thermal modelling should be considered to investigate any potential issues. The results indicate that the application of thermal mass, passive solar gain, high levels of insulation and a terrace form should be considered to produce significant savings in energy in dwellings that comply with social housing design standards in Wales, UK.

Acknowledgements. The doctoral research project being undertaken by the first author is funded by Cardiff Metropolitan University, PHA and the Knowledge Economy Skills Scholarship which is supported by the European Social Fund through the European Union's Convergence programme and administered by the Welsh Government.

References

Bryman, A.: Social research methods. University Press, Oxford (2008)

Creswell, J.W., Clark, V.L.P.: Designing and conducting mixed methods research, 2nd edn. Sage, London (2011)

Langdon, D.: Spons: Architects and Builders price book 2010, 1st edn., Oxon, UK (2009)

Jankovic, J.: Designing Zero Carbon Buildings Using Dynamic Simulation Methods. Routledge, Oxon (2012)

Lowndes, S.: Design for passive heating. In: Hall, K. (ed.) The Green Building Bible, vol. ch. 1.12, 1. Green Building Press, Llandysul (2008)

Morbitzer, C., Strachen, P., Webster, J., Spires, B., Cafferty, D.: Intergration of Building Simulation into the design process of an architecture Practice. In: Seventh International IBPSA Conference, Rio de Janeiro, Brazil, August 13-15 (2001) (accessed September 25, 2011)

Orme, M., Palmer, J., Irving, S.: Control of Overheating in Well-insulated Housing. Faber Maunsell Ltd. (2010), `http://www.cibse.org/pdfs/7borme.pdf` (accessed July 18, 2011)

The Concrete Centre, Thermal Mass Explained: Thermal Mass: what it is and how it's used (2009), `http://www.concretecentre.com/online_services/publication_library/publication_details.aspx?PublicationId=681` (accessed July 18, 2011)

Tuohy, P., McElroy, L., Johnstone, C.: Thermal mass, Insulation and Ventilation in Sustainable Housing – an Investigation across climate and occupancy. Strathclyde: University of Strathclyde (2005), `http://www.esru.strath.ac.uk/Documents/05/bs05-paper-pt.pdf` (accessed July 18, 2011)

University of Exeter, Weather Files for Current and Future Climate. University of Exeter, Exeter (2010), `http://emps.exeter.ac.uk/media/universityofexeter/schoolofengineeringmathematicsandphysicalsciences/research/cee/1chsmodule1notes/Weather_Files.pdf` (accessed December 12, 2011)

Welsh Government, 2011. Code for Sustainable Homes Pilot Project Interim Report. Welsh Government, Cardiff (February 2011)

Welsh Assembly Government (WAG), Ministerial Planning Policy Statement - Planning for Sustainable Buildings. Welsh Assembly Government, Cardiff (2009)

Chapter 57
Basic Energy and Global Warming Potential Calculations at an Early Stage in the Development of Residential Properties

Nils Brown

Department of Urban Planning and Environment,
The Royal Insitute of Technology (KTH), Drottning Kristinas Väg 30,
100 44 Stockholm, Sweden
nils.brown@abe.kth.se

Abstract. In this paper three different structural alternatives (wooden frame, solid wood and concrete) for multifamily buildings are compared in terms of global warming potential (GWP) due to material production and bought energy-in-use from a life-cycle perspective using the ENSLIC tool [1]. The work has been performed in the pre-programming phase of a real construction project, aiming at achieving passive house standard and certification with the Swedish environmental rating tool Miljöbyggnad (MB).

The results suggest that the wooden structural alternatives are better in terms of GWP (1.8 to 1.9 kg CO_2-e/m^2, year) compared to the concrete alternative (4.9 kg CO_2-e/m^2, year). Having said that, there is considerable uncertainty in key input parameters in the calculation. Firstly, construction contractors in question could not supply standardized data for GWP and lifetime for their structural elements, and a combination of generic data and assumptions were used. Secondly, GWP for different energy sources was not available in such a way that it could be analyzed for reliability.

1 Introduction

Key barriers to the application of life-cycle assessment (LCA) in the process of developing and constructing new buildings have been identified in [1]. Practitioners cite the complexity and arbitrariness of the procedure (in spite of the well-known international standardization of the LCA process) as well as the lack of LCA tools that are integrated with standardized software applications that are used by architects. It is further pointed out in [2] that in the building sector LCA methodology is perceived as data- and knowledge-intensive and too time consuming.

A recommendation made in [1] is that the identified barriers may be overcome with a simplified LCA-tool that nevertheless highlights significant environmental impacts over important life-cycle stages for the building in question.

This paper presents and discusses a case study the of the aim of which was to improve knowledge for decision-support in the early stages of a residential new build development in Greater Stockholm, Sweden with respect to life-cycle greenhouse gas emissions from building material production and total demand for bought energy-in-use.

The objective of the work is to apply the ENSLIC tool (ENergy Saving through the Promotion of LIfe Cycle Assessment in Buildings, developed in the project of the same name, [1]) for comparing lifetime GWP for three structural alternatives that are being considered for the project. These alternatives are solid wood, wooden frame and concrete.

The work is of particular interest because it is performed in a real development project with high environmental goals. These goals are expressed by the fact that the developer has established that the buildings achieve a high rating according to the Swedish environmental assessment tool Miljöbyggnad (MB) [3] and be classifiable as a Passive House according to the Nordic definition of the term (bought energy-in-use of 45 kWh/m^2 (HFA),year not including user electricity [4]).

The work presented here has been supported by the developer in question as well as by the EU FP7 project LoRe-LCA - Low Resource consumption buildings and constructions by use of LCA in design and decision making. The author is grateful for the collaboration in this work to the developer, architect and local authority that are working on this project.

2   Method

2.1   The ENSLIC Tool

The ENSLIC tool was established to calculate life-cycle global warming potential (GWP) due to material production and energy demand during the use phase for a building [1].

Required input data for the tool are basic dimensional data for the building (perimeter of outer walls, number of storeys, percentage glazed surface area), materials used and dimensions for important building elements (outer walls, inner walls, slabs, roof, windows) as well as lifetimes for said materials.

The ENSLIC tool contains default data for GWP for material production (cradle to gate) for the most common building materials. A large proportion of these values have been established by collaboration with Swedish product manufacturers during the EcoEffect project [5]. For materials where this data is not available, data has been collected from a Danish database established by the Danish Building Research Institute (SBi) or from the EcoInvent database [6]. The tool also allows users to input values specific for a given manufacturer or sub-class of material.

Also included in the tool are default data for GWP for energy-in-use (in kg CO_2-e /kWh, such as Swedish or Nordic electricity mix, electricity from thermal coal power, Swedish average district heating). Custom values may also be inputted by the user here.

Output from the tool is the building's energy demand in kWh/m^2 (HFA), year (HFA – heated floor area) and GWP due to the material production and energy-in-use in kg CO_2-e/year.

2.2 Overall Procedure for Case Study

Simple but accurate architectural sketches were drawn for two types of building – one with a rectangular plan cross section (here called "long", see Figure 1) and the other with a square plan cross section (here called "square"). Dimensions for both building types are shown in Table 2.

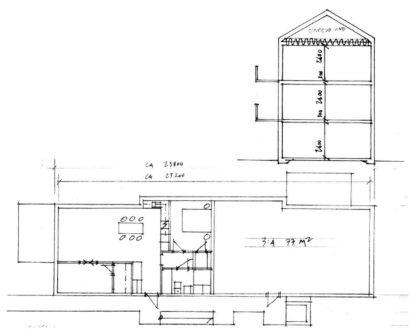

Fig. 1. Architectural sketch for plan cross-section "long" including elevation sketch

In parallel, during November 2011 one specific construction contractor for each of the three structural alternatives in question was contacted and requested to provide information about materials and dimensions for the following building elements: floor/basement slab, load bearing internal wall, non-load bearing internal wall, external walls, floor structure, attic/roof structure. External doors and windows were assumed to be the same in all structural alternatives. Each of the three contractors were also asked to provide data describing the lifetime of particular elements evaluated according to procedures given in ISO 15686-2 Buildings and constructed assets - Service life planning - Part 2: Service life prediction procedures [7] or data for cradle-to-gate GWP from material production according to ISO 21930:2007 Sustainability in building construction -- Environmental declaration of building products [8] (environmental product declarations – EPDs). If such lifetime and GWP data was not available, building lifetimes of 50 and 100 years, and GWP data for materials from the ENSLIC database were used by default. Each of the three contractors has a well-established national presence.

Dimensions for building elements based on the architectural sketches and on information from construction contractors were inputted into the ENSLIC tool. Thermal properties for the elements calculated by ENSLIC based on these dimensions are shown in Table 1.

Table 1. U-values for external building elements calculated by ENSLIC based on dimensional data (thicknesses and material types) from contractors

	U-value, W/K, m^2		
	Solid wood	Wooden frame	Concrete
Foundation	0.09	0.08	0.09
Attic/roof	0.05	0.05	0.05
External wall	0.10	0.12	0.14
Windows	1.00	1.00	1.00
External Doors	1.20	1.20	1.20

In the standard version, ENSLIC uses a degree-day method in calculating active space heating demand. Since such a standard method is not relevant for a passive house, a new module for the ENSLIC tool was created and linked appropriately in the ENSLIC spreadsheet. This module is described further in section 3.3 below.

Table 2. Dimensions for each building type

	Structural alternative (see section 3.2)	
	Square	Long
Heated Floor Area, m^2	383	452
Total facade area m^2	428	548
Total window area, m^2	68	90
Area, external doors, m^2	4	5
Area, attic floor, m^2	128	151
Area, basement slab, m^2	128	151
Area internal walls, m^2	358	343
Number of storeys	3	3
Attic	Unconditioned attic space	Unconditioned attic space
Floor plan per storey	1 x 5 bedroom flat	2 x 2 bedroom flats
Total window area as proportion of total facade, %	15.9	16.5
Glazed area as proportion of total HFA, % (based on requirements in MB)	15	15

The ENSLIC tool calculated kWh/m^2 (HFA), year and kg CO_2-e/year (from material production and energy-in-use). In total six alternatives were compared, based on the three construction alternatives (solid wood, wooden frame and concrete) that were used for two specific building designs ("long" and "square").

2.3 Energy Demand

The following balance was used to calculate the active space heating demand:

(Active space heating) = (transmission losses through climate envelope) + (Ventilation losses) + (Infiltration losses) − (User electricity) − (Solar gains) − (Heat gains from occupants)

$$\text{Eq. 1}$$

Data from peer-reviewed sources relevant for Swedish conditions were used as input data in the balance, as shown in Table 3 ([9] and [10]).

Transmission losses through the climate envelope are calculated according to the building surface area, average U-value for the surface area and monthly average outdoor temperatures for Stockholm [11]. Ventilation losses are calculated based on an airflow of 0.5 ACH in kg/h, a maximum heat recovery efficiency in the ventilation system of 85 % and monthly average outdoor temperatures for Stockholm [11]. Heat loss from infiltration is based on the standards established for Nordic passive houses [12] and monthly climate data for Stockholm [11].

Other parameters used to calculate the demand for active space heating as shown in Eq. 1 are given in Table 3. The indoor temperature in all cases was assumed to be 21 °C.

Table 3. Energy demand and internal gains used in the methodology

User electricity	Assumed to amount to 28 kWh/m^2 (HFA). see for example [10]
Property electricity	Assumed 7 kWh/m^2 (HFA) [9]
Domestic hot water (DHW), from district heating	Assumed 17 kWh/m^2 (HFA) [9]
Solar radiation	Solar radiation into the buildings is calculated from data for solar radiation on vertical surfaces in different directions (North, South, West and East) [11] and an average solar heat gain factor for the windows of 0.38.
Heat gains from occupants	According to [12] the sum of user electricity and heat gains from occupants is a constant 4 W/m^2, (HFA). Based on this requirement, with user electricity at 28 kWh/m^2 (HFA) as shown in the row above, heat gain from occupants has a value of 7 kWh/ m^2 (HFA), year.

2.4 GWP Due to bought Energy Demand

In consultation with project developers, it was assumed that all electricity is provided by Swedish hydropower (this is the current supply mix available from the local electricity distribution monopoly). Likewise, all bought heat (active space heating and DHW) is provided by the local district heating network. Specific GWPs for these energy sources are shown in Table 4.

618   N. Brown

A sensitivity analysis was performed where average Stockholm district heating (107 g CO_2-e/kWh) and average Nordic electricity mix (85 g CO_2-e/kWh) were substituted for the values shown in Table 4, based on [13].

Table 4. GWP for energy sources used in the study

Energy type	GWP	Source
Swedish Hydroelectricity	5.2 g CO_2-e/kWh	[14]
Heat from Local district heating network	5.2 g CO_2-e/kWh	Data from energy supply company

3   Results

None of the 3 construction contractors contacted could supply element lifetime data according to the ISO 15686-2 [7] or GWP due to material production for their specific building elements based on an EPD [8]. As such ENSLIC calculations were performed using default data as described in sections 3.1 and 3.2.

Figure 2 shows the yearly bought energy demand for each of the 6 cases evaluated using the ENSLIC tool. As pointed out in section 3.3, the only part of the above energy demand that is calculated (rather than assumed, see Table 3) is for active space heating. The tool shows that the concrete alternatives have a higher demand for active space heating than either of the wooden alternatives. This is simply due to the fact that the average U-value for wooden alternatives is lower due to more insulation (see also Table 1). Figure 2 also shows that the space heating demand is lower in the "square" than the "long" alternatives due to fact that the square design the climate envelope is smaller in ratio to the heated floor area. Figure 2 shows that each of the alternatives is below or slightly above the energy requirement for passive houses in the Nordic region (see section 2 and [12]) , validating that such requirements can be met by any and all of the structural alternatives with at most slight changes to certain building elements.

Figure 3 shows that base case alternatives with wooden construction have dramatically lower GWP than the concrete alternative. Figure 3 further shows that this is mainly due to the fact that impacts from production of building materials are significantly greater for the concrete alternative. That impacts from bought energy demand are so much lower than for production of building materials is due to the fact that in these cases, bought energy demand from the use phase is renewable (hydropower and biofuel-based district heating, see Table 4), in contrast to material production where it may be assumed that a greater proportion of fossil-based energy is used. This also explains the contrast with studies such as [15] and [16] where a high proportion of fossil based energy is assumed for energy-in-use. The result on the other hand is coherent with [17] where renewable energy-in-use is also assumed. Impacts for a lifetime of 100 years are shown in Figure 4. Not surprisingly the GWP in kg CO_2-e/m^2 (HFA), year reduces significantly compared to Figure 3. Such a difference highlights the importance that the building lifetime may play in terms of assessment of lifetime GWP.

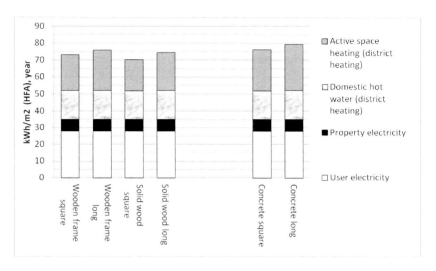

Fig. 2. Yearly bought energy demand for calculated alternatives (assumed constant over building lifetime, no start year defined)

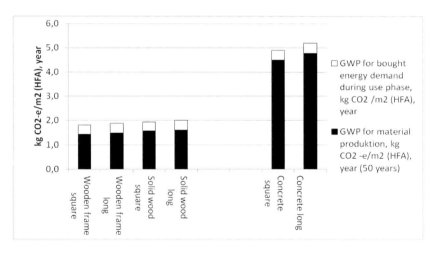

Fig. 3. GWP for production of building material and bought energy during the use phase for base case alternatives (50 year lifetime, assumed constant over building lifetime, no start year defined)

The sensitivity analysis for GWP from energy supply alternatives (see section 3.4) obviously showed considerably higher GWP from each of the structural alternatives, and furthermore (for an assumed lifetime of 100 years) the concrete alternative was demonstrated to have only 30 % higher total GWP (9.6 kg CO_2-e/m^2(HFA), year for concrete versus about 7.5 kg CO_2-e/m^2(HFA), year for the wooden alternatives).

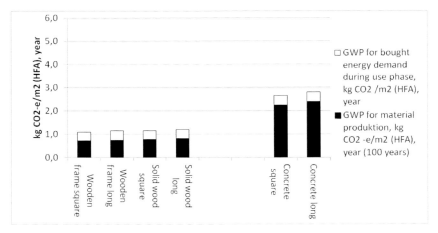

Fig. 4. GWP for production of building material and bought energy during the use phase for base case alternatives (100 year lifetime, assumed constant over entire lifetime, no start year defined)

3.1 Discussion

The results suggest that that the use of EPDs [7] and standardized lifetime measurements for building elements [8] to communicate about products is not commonplace amongst well-established contractors such as those contacted.

This case study shows the power of the ENSLIC tool in demonstrating the differences between considered alternatives at an early stage in the construction process. However, the specific GWPs for energy sources were not established according to a uniform standard and are *very* low, even by the standards of Swedish renewable energy. Since the value used for GWP from district heating was supplied by the district heating company itself, without supporting information it is difficult to analyze its accuracy and precision or how it might compare to GWP for other heat sources. Such information is important for decision making in the ongoing development project in question.

Taken at face value, results such as those presented in figures 2, 3 and 4 suggest that lower GWP may be achieved by reducing impacts due to material production (e.g. by decreasing the quantity of insulation) at the expense of increased impacts due to energy in the use phase. Having said that, in this case this is not possible due to the requirement that the buildings reach passive house standard. This highlights the importance of establishing goals in the development process for both energy-in-use and total GWP.

In light of the demonstrated uncertainties that exist in this calculation process, it is important to discuss what stakeholders in the development process could do to improve the decision support that is provided by such calculation procedures. It seems firstly important to establish comparable standards by which the GWP of a unit of bought energy is calculated in the process. This was not possible here, though an attempt has been made by the Swedish Association of Municipal Housing Companies to apply a common methodology for electricity production and district heating networks in Sweden [13] that was used in the sensitivity analyses.

Other key input data that construction contractors could provide comprises lifetime data for key building elements based on e.g. ISO 15686-2 [7] as well as the more widespread use of EPDs [8]. It may be attractive to certain construction contractors to apply these standards to establish a market advantage.

4   Conclusions

The work presented here has demonstrated how a relatively simple calculation tool (ENSLIC) can provide valuable information about the life-cycle GWP for different structural alternatives in the early stages of a construction project in Greater Stockholm, Sweden. The results suggest that wooden alternatives are favorable to concrete. However there are many uncertainties in input data for specific GWP for a unit of bought energy-in-use and the GWP due to production of specific materials from specific contractors.

Experience from the work suggested that construction contractors are not used to providing standardized data (based on [7] and [8] in this case). Such data would assist greatly in life-cycle calculations of this sort and further strengthen the usefulness of the results obtained.

An interesting comparison to the results presented here could be with those from an otherwise identical study where building components are specially selected on the basis that they have low GWP according to a standardized EPD [7], and where lifetime has been assessed according to ISO 15686-2 [8]. This could be added to by selecting specific GWPs for energy sources based on well-established standardized projections for future energy supply over the buildings' lifetimes rather than simply using current energy mix.

References

1. Malmqvist, T., Glaumann, M., Scarpellini, S., Zabalza, I., Aranda, A., Llera, E., Diaz, S.: Life cycle assessment in buildings: The ENSLIC simplified method and guidelines. Energy 36(4), 1900–1907 (2011), doi:10.1016/j.energy.2010.03.026
2. Blengini, G.A., Di Carlo, T.: The changing role of life cycle phases, subsystems and materials in the LCA of low energy buildings. Energ Buildings 42(6), 869–880 (2010), doi:10.1016/j.enbuild.2009.12.009
3. Malmqvist, T., Glaumann, M., Svenfelt, A., Carlson, P.O., Erlandsson, M., Andersson, J., Wintzell, H., Finnveden, G., Lindholm, T., Malmstrom, T.G.: A Swedish environmental rating tool for buildings. Energy 36(4), 1893–1899 (2011), doi:10.1016/j.energy.2010.08.040
4. Forum för Energieffektiva Byggnader, Kravspecifikation för passivhus i Sverige - Energieffektiva bostäder. Forum för Energieffektiva Byggnader, Gothenburg (2008)
5. Assefa, G., Glaumann, M., Malmqvist, T., Kindembe, B., Hult, M., Myhr, U., Eriksson, O.: Environmental assessment of building properties - Where natural and social sciences meet: The case of EcoEffect. Build Environ. 42(3), 1458–1464 (2007), doi:10.1016/j.buildenv.2005.12.011

6. Frischknecht, R., Jungbluth, N.: EcoInvent, Overview and Methodology. EcoInvent Centre for Life-cycle Inventories. (2007)
7. ISO. Buildings and constructed assets – Service life planning – Part 2: Service life prediction procedures. vol ISO 15686-2. International Organization for Standardization, Geneva (2001)
8. ISO. Sustainability in building construction – Environmental declaration of building products. vol ISO 21930:2007. International Organization for Standardization, Geneva (2007)
9. Janson, U.: Passive houses in Sweden. From design to evaluation of four demonstration projects. Lund Institute of Technology, Lund, Sweden (2010)
10. Johansson, D., Bagge, H.: Simulating energy use in multi-family dwellings with measured, non-constant heat gains from household electricity. Paper Presented at the Building Simulation 2009. 11th International IBPSA Conference, Glasgow, Scotland, July 27-30 (2009)
11. Taesler, R.: Klimatdata för Sverige. Swedish Building Research Institute, Stockholm (1972)
12. FEBY, Kravspecifikation för passivhus i Sverige - Energieffektiva bostäder. Forum för Energieffektiva Byggnader (2008)
13. SABO, Miljövärdering av energianvändningen i ett fastighetsbestånd (Environmental rating of energy demand in a buidling stock). Sveriges Allmännyttiga Bostadsföretag Stockholm (2010)
14. Vattenfall, Livscykelanalys av Vattenfalls el i Sverige. Vattenfall AB (2005)
15. Scheuer, C., Keoleian, G.A., Reppe, P.: Life cycle energy and environmental performance of a new university building: modeling challenges and design implications. Energ Buildings 35(10), 1049–1064 (2003), doi:10.1016/S0378-7788(03)00066-5
16. Suzuki, M., Oka, T.: Estimation of life cycle energy consumption and CO2 emission of office buildings in Japan. Energ Buildings 28(1), 33–41 (1998)
17. Wallhagen, M., Glaumann, M., Malmqvist, T.: Basic building life cycle calculations to decrease contribution to climate change - Case study on an office building in Sweden. Build Environ. 46(10), 1863–1871 (2011), doi:10.1016/j.buildenv.2011.02.003

Chapter 58
Passive Cooling Strategies for Multi-storey Residential Buildings in Tehran, Iran and Swansea, UK

Masoudeh Nooraei, John Littlewood, and Nick Evans

The Ecological Built Environment Research & Enterprise Group,
Cardiff Metropolitan University, Cardiff School of Art & Design,
Cardiff, CF5 2YB, UK

Abstract. Like the UK, the residential sector in Iran has a significant share of the national energy consumption. Therefore, efforts are needed to reduce energy use and greenhouse gas (GHG) emissions from dwellings. This paper discusses two case studies for residential apartment buildings and explores the cooling strategies which could be adopted to reduce energy usage and the associated GHG emissions. For case study one (in Tehran) results from dynamic thermal modelling and simulation tests are presented that assess the effectiveness of a number of design cooling strategies. These include appropriate orientation, solar shading and thermal mass with night time ventilation. These strategies are seen as effective methods to control heat gain and to dissipate excess heat from the residential apartment building during the summer. For case study two (in Swansea) pilot results from spot tests undertaken during interviews with apartment occupants are presented. These spot results illustrate that dynamic thermal modelling should have been undertaken by the design team to inform the design decisions for this building, which was completed and occupied in November 2011, since internal conditions exceed recommended comfort conditions for level four (of the code for sustainable homes) dwellings. Furthermore, measures such as solar shading may need to be retrofitted and combined with a change to the ventilation strategy to reduce overheating during the year. The basis for the paper is to compare the results of two residential apartment buildings that both experience similar problems of overheating, even though they are located in two different countries and adopt different methodologies for recording the data. Lessons adopted as part of case study one to reduce overheating are being considered as potential solutions for case study two.

1 Introduction

It is widely acknowledged that buildings, including dwellings, are major contributors to climate change caused by GHGs and in particular carbon dioxide. Irrespective of climate and location, the majority of dwellings depend on non-renewable energy sources to provide comfortable indoor conditions; whilst traditionally, human ingenuity was able to create architecture in response to the climatic conditions of various regions on the planet (Krishan 2001). Residential buildings in Iran and the UK and the activities undertaken within them, are responsible for approximately 33% and 25% of

A. Håkansson et al. (Eds.): *Sustainability in Energy and Buildings*, SIST 22, pp. 623–636.
DOI: 10.1007/978-3-642-36645-1_58   © Springer-Verlag Berlin Heidelberg 2013

total energy consumption and therefore a significant amount of CO_2 emissions (Building and Housing Research Centre of Iran 2010 and National Statistics 2010). This paper discusses two case studies (one in Iran and one in the UK) for residential apartment buildings and explores the cooling strategies, which could be adopted to reduce energy usage and the associated GHG emissions. Case study one in Tehran, Iran explores effective cooling strategies in a climate with hot summers and cold winters; whilst case study two in Swansea, UK documents an emerging overheating problem in a new (completed in November 2011) airtight, highly insulated, multi-storey residential building. Problems of overheating in new low carbon residential buildings in the UK needs to be addressed, particularly when considering climate change predictions. The Tehran case study is presented as a reference to cooling strategies for the Swansea case study.

2 Context to Case Studies

2.1 Case Study One – Tehran, Iran

Tehran the capital of Iran, is the largest city in the Middle East, is the 16th most populated city globally with over 8 million residents (Statistics Centre of Iran 2009). Urban development has a fast pace in Tehran because of the growing number of immigrants seeking better education, job opportunities and life conditions; thus, there has been a continuous increase in the demand for residence (Tehran Municipality 2012). From the first author's experience of working in Tehran as an Architect she has observed that buildings new and existing consume substantial energy for heating and power. This is in part due to inefficient building fabric and inadequate implementation and enforcement of building regulations, especially in residential buildings. The climate makes the situation worse; in addition to the need for 1.5 million dwellings by 2015 in Iran and the majority of which are in Tehran (Young Cities Project website 2010). Current residential buildings have poor environmental performance and therefore improvements will have a significant impact on reducing energy consumption and associated carbon emissions. In this paper, a simulated box with two exposed walls representing a typical apartment in Tehran was modelled as a case study to be used for parametric tests.

2.2 Analysis of Tehran's Climate

Tehran is situated in the north of the central plateau of Iran, latitude 35.7° and longitude 51.2°. According to the Koppen map, the Tehran climate is semi-arid and continental (Szokolay 2008). Weather data from Mehrabad weather station in Tehran for 1961-1990 period (Meteonorm v. 6) shows that the annual average air temperature in Tehran is 18.4°C with monthly average swings of 17.2K. The coldest month of the year is January and the hottest month is July, with the average air temperatures of 4.9°C and 31°C and average daily swing of 17.1K and 17.3K respectively. In winter temperatures can reach as low as -4°C, but may peak as high as 40°C in summer. Average humidity fluctuates during a year from almost 60% during cold months to as low as 30% during June, July and August. Direct solar radiation is high during each climatic year and diffuse solar radiation is a noticeable amount (Fig. 1). Three climatic periods can be defined in this city. Firstly, a heating season during which the outdoor temperature is below comfort temperatures (18-32°C was considered as the comfort zone according a study of thermal comfort of people in Tehran by Heidari 2008) and solar and internal gains cannot bridge the difference between indoor and comfort temperatures unless specific measures are undertaken, such as space heating. Secondly, a cooling season during which the outdoor temperature is above comfort temperatures, or the gains from sun or internal heat sources make the internal temperature uncomfortably warm, unless specific measures are taken, such as mechanical cooling systems are used. Thirdly, a free-running season during which the indoor temperature is comfortable and there is no need for additional heating or cooling (Yannas 2000). Thus, the aim of a low to zero carbon dwelling in Tehran should be the extension of the free-running season, while the possible conflict between passive heating and cooling strategies should never be neglected.

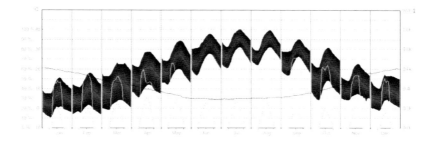

Fig. 1. Monthly diurnal averages for Tehran (Source: Meteonorm v. 6, Weather tool)

High solar radiation in Tehran can cause overheating in hot months if not avoided. Moreover, internal heat gains from occupants and equipments may result in the extension of the cooling season. Since the main source for cooling energy in Tehran is electricity produced from fossil fuels, design for summer comfort needs special attention to reduce cooling energy demand.

2.3 Case Study Two – Swansea, UK

Swansea is the second largest city in Wales, UK with a population of 232,500 (City and County of Swansea 2012). There has been a great deal of urban regeneration in Swansea since 2000, particularly in the dockland area known as the waterfront at SA1. There are plans for business, retail, leisure and residential developments in the Swansea Bay region with an estimated value of over £2 billion until 2022 and thus Swansea has been named the fastest growing city in the UK for business (Swansea Bay 2012[a]). Swansea Bay accounts for 15% of Wales' total land area and with 547,000 residents in 2008, over 18% of its total population. Furthermore, 1.5 million people live within one hour of Swansea. With the estimated £2 billion of investment, it has led to a considerable number of new apartment buildings (ibid). The majority of residential buildings in

Swansea were mainly built before 1919 (Hopper et al 2012) and therefore have a poor environmental performance. Works to improve their environmental performance are discussed in Hopper et al (2012). Since 2010, much of the development of new dwellings in Swansea has been undertaken by social housing developers, including Coastal Housing Group (Coastal Housing Group 2012). Indeed, Coastal have built a number of apartment buildings to meet a range of enhanced environmental assessment standards. Case study two is of a timber frame construction and was designed and built to level four of the code for sustainable homes; includes 66 two bedroom and 2 one bedroom apartments across five floors and occupation commenced in November 2011 (ibid). Case study two is one of a number of case studies being monitored as part of Cardiff Metropolitan University's contribution to work package six (monitoring the performance of low carbon buildings and products) of the Low Carbon Built Environment project (Anon, 2010, Littlewood et al 2011 and Littlewood, 2013). Since occupation internal conditions have been recorded by researchers from Cardiff Metropolitan University, which exceed recommended comfort conditions; where occupants observe that the circulation corridors and their apartments are uncomfortably warm. Therefore, case study two provides a good example to compare with case study one, in terms of the need for cooling strategies, albeit post completion.

2.4 Analysis of Swansea's Climate

Swansea is situated in south-west Wales at latitude of 51.6° and longitude of -3.9° (Travelmath 2012). Fig 2 illustrates the average monthly temperatures and average monthly sunshine for Swansea Bay, UK (Swansea Bay 2012[b] and 2012[c]). On average the warmest month is August; the highest recorded temperature was 41 °C in 1980, the average coolest month is January; the lowest recorded temperature was -18 °C in 1985; and the most precipitation occurs on average in November (ibid and The Weather Channel 2012). Whilst, the temperatures are on average much cooler than Tehran in Iran the climate in the UK is becoming more unstable, with unpredictably warm weather at unusual times of the year. For example, in 2012, March was the third warmest March since records began in 1910, where the highest UK temperature was 23.6 °C and in Swansea was 19.8 °C (Met Office 2012 and Weather Online 2012).

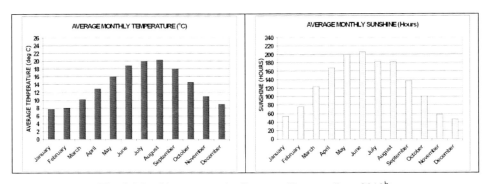

Fig. 2. Monthly averages for Swansea (Swansea Bay, 2012[b])

3   Test Methods

3.1   Case Study One - Modelling

Thermal sensitivity studies were undertaken using dynamic thermal simulation software (EDSL Tas v.9.1.3a). A 10*10*3 metre cube was created as the base model to represent an apartment, with two opposite exposed walls and the construction specifications according to the building culture and regulations in Tehran as follows:

- Model Area: 100 m²; Orientation: South; Façade Area: 60 m²; External Wall Construction: (inside)110mm brick+20mm cavity+110mm brick;
    Internal Wall Construction: (inside) 110mm brick+20mm cavity+110mm brick (no heat transfer between adjacent flats); Ceiling Construction: (inside) 150mm concrete, 30mm concrete, 10mm tile; Window-to-floor ratio: 15%; Window: Double glazed uncoated air-filled; Infiltration: 0.35 ach; Ventilation: 0 ach; Glazing U-value: 2.7 W/m²K; External Wall U-value: 0.85 W/m²K; Internal Gains (from intermittent occupancy of four occupants and appliances): 3700 KWh/yr; Cooling set point temperature: Internal T>32°C; Shading: None; Windows opening schedule: None.

Based on the climate analysis, passive cooling strategies were tested on the study model through parametric tests to investigate suitable strategies for hot summer conditions.

3.2   Case Study Two - Interviews and Spot Measurements

Ethics approval was granted in April 2012, for researchers from Cardiff Metropolitan University to interview all the occupants in the 68 apartments at case study two, every three months between May 2012 and April 2013. The aim of the interviews is to collect information on climatic data, internal comfort conditions in each apartment and circulation spaces, occupant behaviour and occupant attitudes towards the apartment building and individual apartment design and construction; comfort; control of heating, appliances and lighting; hot water and energy use; noise and control of noise; security; natural daylight and control of daylight; household bills; health and wellbeing; and home management. The U values for the key construction components at case study two are 1.8 $W/m^2 \, °C$ (interior door and windows), 0.2 $W/m^2 \, °C$ (exterior wall), 0.26 $W/m^2 \, °C$ (separating floor and wall) and 0.2 $W/m^2 \, °C$ (roof) (Littlewood 2013). Interviews commenced in May 2012 and each occupant was provided with an information sheet and consent form explaining the study and was asked a series of questions. In addition, researchers recorded the air temperature, CO_2 emissions, solar radiation, air movement and daylight levels in each apartment, on the exterior balcony, in the circulation corridor immediately outside the apartment, the stairwells and exterior to the building at street level. Data was recorded using the instruments illustrated in Table 1 below.

Table 1. Instrumentation used to take spot measurements at case study two

Variable	Instrument	Accuracy
Air temperature	Vent check plus	±0.1 °C
CO_2 emissions	Vent check plus	±30ppm, ±5%
Solar radiation	TM-206 Solar power Meter	±10W/m2 or ±5% (highest value is correct)
Air movement	PCE-TA-30	±3%, ±0.2 m/s
Daylight	Testo 540	±3%
Relative Humidity	Vent check plus	±3% (10~90%), ±5% (others)

4 Results

4.1 Case Study One Results - Heat Gain Control through Orientation, Shading and Thermal Mass

Orientation, shading and thermal mass are passive strategies tested to establish how effective they are in reducing cooling energy demand, for case study one in Tehran. Simulations were undertaken on different models – to north, south, east and west orientations, without and with shading- to test the effect of orientation and shading upon the internal comfort conditions on an apartment building in Tehran. Fig. 3 below, illustrate that east and west oriented spaces have much higher temperatures compared to north and south orientations, with approximately 17.0 °C and 13.0 °C difference between indoor maximum and outdoor temperature respectively.

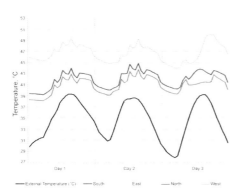 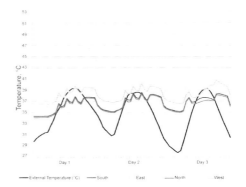

Fig. 3. (Left) Indoor temperatures of models with unshaded and closed windows to north, south, east and west orientations during three hot days-21st to 23rd July. (Window-to-floor ratio: 25%).

Fig. 4. (Right) Indoor temperatures of the same models after applying an external venetian blind with 29% of solar transmittance

58 Passive Cooling Strategies for Multi-storey Residential Buildings   629

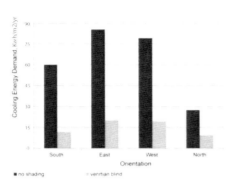
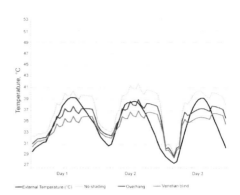

Fig. 5. (Left) Difference between cooling energy demand of spaces in the two conditions tested

Fig. 6. (Right) Difference between indoor temperatures as an effect of various shading conditions (windows are closed)

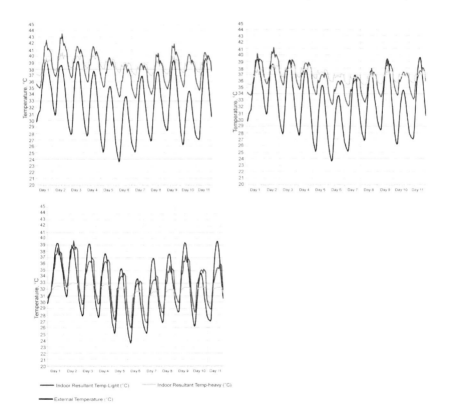

Fig. 7. Indoor thermal conditions in low and high mass models during an 11-day hot period with changing outdoor conditions- windows closed and unshaded (Above left), windows closed and shaded with an overhang (Above right), shaded windows are open from 10pm till 7am (Below left) (After Givoni, 1994).

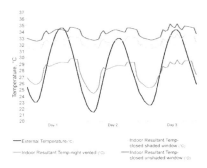

Fig. 8. (Below right) Indoor thermal conditions in a high mass model during three typical summer days

The indoor temperatures in the two models with north and south windows have roughly similar trend with much less elevation above the outdoor temperature. The north oriented apartment has the lowest temperature. Fig. 4 shows that shading using exterior Venetian blinds on the windows is able to minimize overheating and the temperature differences between various orientations. However, east and west oriented apartments still have higher temperatures. Providing shading to apartments orientated towards the south leads to similar performance to apartments orientated north and 50% less cooling energy demand than east and west oriented apartments (Fig. 5). An external Venetian blind can reduce indoor temperatures by 1.0 °C more than an exterior window overhang during most of the three days tested (almost from 7:00 to 24:00). This is because blinds cut more direct, indirect and diffused solar radiation and also improve the thermal resistance of the glazing, compared with overhangs (Fig. 6).

A third parametric test was undertaken to study the performance difference between a low thermal mass model and a high thermal mass (after a test by Givoni (1994) on an experimental box). In the former, 110 mm brick layer in the walls was replaced by 15 mm plaster boards and the concrete of the ceiling is also covered by 15mm plaster boards; thus, there is no available mass in floor or ceiling while the latter is made of brick walls and concrete ceiling –common building materials in Tehran. Fig 7 illustrates the performance of the two above models in three different conditions, closed windows without shading, closed windows with shading and also shaded windows with night ventilation (between 21st till 31st of July, with daily temperature differences of 9 K to 12 K. This period was chosen because of combined hot days). 12% window-to-floor ratio is due south and 3% due north to provide cross ventilation. Results are similar to those reported by Givoni (1994 & 1998) and Shaviv et al (2001) where low thermal mass buildings respond quickly to external temperature fluctuations. This is because they have a low thermal capacity; compared with high thermal mass buildings. It is clear that low thermal mass buildings are more likely to overheat during hot summer days and therefore would need more cooling, prompting the occupants to use cooling devices powered by electricity, generated from fossil fuels. Thereby, increasing CO_2 emissions and contributing to climate change. However, high thermal mass buildings are less affected by the temperature swings, providing more constant and stable internal temperatures. These buildings are heated slowly, since some of the building elements store heat and they also cool down slowly. This is why the indoor temperature is closer to the average outdoor temperature, when the building is night vented. Heavy weight buildings with night ventilation can provide cooler spaces in hot days, since they absorb the heat gains of the building during day and dissipate it to the outdoors during the night when the external temperature is lower than internal temperature. Applying thermal mass with night time ventilation results in up to 7 K reduction in the indoor maximum temperature in days 9 and 11.

The temperature patterns of the model with closed shaded windows show that thermal mass without night ventilation cannot bring the indoor temperature comfortably below the outdoor maximum temperature, especially during less warm days – days 5 and 6- within the hot period (see fig 7 above right). Another important issue is minimizing the solar gain during summer that can lessen the indoor temperature as much as almost 2K in both low and high-mass models. Fig. 8 shows the thermal performance of the high thermal mass model during three typical summer days in Tehran, 14th to 16th of June. It is understood that the reduction in outdoor maximum temperature is lower compared to the hot days (almost 4K compared to 7K). The temperature swing in both mentioned periods are more or less the same, approximately 12K; which shows that high thermal mass is more helpful in hotter days.

4.2 Case Study Two Results

Spot measurements recorded using the instrumentation illustrated in Table two above, were undertaken during interviews with occupants at apartments A, B and C and illustrated in Table 2 to 4 below. These results are initial findings from the piloting stage of a questionnaire interview process commenced in May 2012 and which will be completed in April 2013. Therefore, the results presented are a snap-shot of the problems encountered thus far, which will be supported by a more detailed set of results (from a 12 month monitoring and interview process) and will be documented in further publications.

Table 2. Spot measurements recorded in the corridors across five floors, at case study two on 29/05/12

Corridor. Date: 29/05/2012						
Variable	Ground floor	1st floor	2nd floor	3rd floor	4th floor	Exterior
Air temperature °C	25.1	30.3	30.5	31.3	31.6	22.4
CO2 level ppm	640	520	1142	621	677	438
Relative humidity %	47.9	33.2	32.9	33.7	33.3	55.7

Table 3. Spot measurements recorded in the central staircase across five floors, at case study two on 29/05/12

Central staircase. Date: 29/05/2012						
Variable	Ground floor	1st floor	2nd floor	3rd floor	4th floor	Exterior
Air temperature °C	23.2	23.0	23.5	24.7	28.7	22.4
CO2 level ppm	540	575	539	530	584	438
Relative humidity %	53.6	54	51.3	47.4	39.8	55.7

Table 4. Spot measurements recorded in apartments A, B and C on the 1st and top floor at case study two, on 29/05/12

Apartments. Date: 29/05/2012				Exterior		
Variable	Ap. A* (at 12.06)	Ap. B** (at 14:13)	Ap. C*** (at 15:15)	at 12.06	at 14:13	at 15:15
Air temperature °C	25.8	22.2	23.2	21	21.1	22.4
CO2 level ppm	479	507	577	441	442	438
Solar radiation W/m²	33	4	7.2	809	821	Above 1999
Air movement m/s	0	0	0	3.67	2.15	3.17
Daylight lux	130	96	176	90430	Above 99999	96420
Relative humidity %	42.7	55.9	54.6	55.6	59.6	55.7

*east orientation & top floor; ** north orientation & 1st floor; *** north-west orientation and 1st floor

5 Discussion

The climatic analysis and thermal simulation of case study one revealed a number of effective passive cooling design strategies in Tehran. However, some strategies suitable for the cooling season may conflict with the energy requirements in the heating season. Table 5 illustrates how each cooling strategy may negatively affect the thermal performance of a residential building in the heating season (Tehran). The difference between the two seasons encourages adjustable shading to allow maximum and minimum solar radiation in winter and summer, respectively.

When the spot measurements were taken at each of the apartments at case study two, the heating systems were not activated and the internal windows had all been open (24 hours a day) since March 2012. This is because the occupants stated that they found the apartments were always warm. This is not surprising when observing the temperatures recorded in the corridors on the 1st floor and top floor are between 30.3 and 31.6 °C and in the central staircase on the 1st floor and top floor between 23.0 to 28.7 °C. Yet, the external temperature was only between 21.0 and 22.4 °C. The reasons for this might be heat gain from piped hot water (communal heating) in the corridors; as well as high air tightness values, high thermal insulation, low thermal mass in the construction elements and lack of adequate ventilation, especially in the corridors. These internal temperatures indicate the need for further investigation, which will include detailed monitoring 24 hours per day (at ten minute intervals) 365 days per year, for three apartments, to commence in July 2012. In addition, the spot measurements will be re-recorded on three further occasions when the interviews are repeated. Furthermore, dynamic thermal simulation will be undertaken to investigate potential cooling strategies, since the results indicate that the ventilation and heating strategy for the building is not currently satisfactory as internal conditions exceed recommended maximum comfort temperatures of 25°C and even 28°C (CBSE, 2006). Indeed, overheating is unexpected in such moderate outdoor temperatures. Potential solutions to be investigated include optimising the passive ventilation strategy of the building, the communal heating systems and retrofitting external solar shading and thermal mass into the building. Indeed, the continued and more detailed monitoring at

Table 5. Summary of the tests results and recommendations considering heating season

Tested parameters	Passive strategies for cooling season	Considerations for heating season	Recommendations
Orientation	North performs better	North receives the least solar radiation	South with external venetian blinds to minimise overheating in summer
Shading	external venetian blinds to cut unfavorable solar radiation	Fixed shading is an obstacle to receiving solar radiation	Adjustable: allows winter solar radiation & reduction of summer solar radiation
Window area	10%* Window-to-floor ratio	Small windows let less solar radiation in. 25% window-to-floor performs the best	25%* window-to-floor with external venetian blinds, thermal mass and night ventilation to minimise overheating in summer
Thermal inertia	High (if night ventilated)	High (if sun-oriented)	High with sun-oriented windows
Night Ventilation	Helpful	NA	Activate during war summer nights

* only 10%, 15%, 20% and 25% window-to-floor ratio were tested in this research

case study two will lead to further publications, to document the extent of the over-heating, the occupant attitudes and behaviour as a result of the overheating, and the methods adopted to reduce the overheating.

6   Conclusion

The case study one has recognised heat gain control and excess heat dissipation as effective passive design strategies for minimizing thermal stress in multi-storey resi-dential buildings in Tehran, Iran. Most of the strategies to reduce the cooling energy demand may extend the heating period; so a balance between heating and cooling demands should always be considered. Secondly, flexibility in design (e.g. adjustable shading) is encouraged, so the strategies for the cooling season do not have negative effects in the heating season. Thirdly, design for summer comfort in Tehran in the multi-storey housing typology can be simple if considered at early stages of design, as passive strategies can make considerable reductions in energy consumption of these buildings while providing the designers with freedom in design. The initial results presented for case study two could indicate that the design and modelling of the build-ing and heating systems were not detailed enough. As a result of applied heating strategies, regardless of the thermal conditions of the cooling season, the internal temperatures of the case study two are closer to what happens in warmer climates, such as in case study one and if the future weather in Swansea is to be warmer, as an effect of the climate change, there will be even more similarities between the two cases. It is notable that both case studies have to deal with severe winter under heating as well as summer overheating that shows the importance of finding optimised strate-gies. Considering all the above, the next step in case study two would be investigat-ing passive cooling strategies through dynamic thermal simulation similar to the case study in Tehran, so as to propose solutions to uncomfortable thermal conditions during warm periods of each year.

Acknowledgements. Case study one is based on findings of the first author's MSc dissertation in Sustainable Environmental Design at the Architectural Association School of Architecture (AA) – Design guidelines for low energy multi storey housing in Tehran, Iran- in 2010 which was funded by British Foreign and Commonwealth Office through the Chevening Scholarships programme. Case study two is part of Cardiff Metropolitan University's contribution to Work Package six of the Low Car-bon Built Environment project, funded by Cardiff Metropolitan University, Coastal Housing Group and the European Research Development Fund's Convergence, Re-gional Competitiveness and Employment programmes; administered by the Low Car-bon Research Institute for the Welsh Government.

References

Anon. Low Carbon Built Environment project. Low Carbon Research Institute, Cardiff, UK (2010),
   `http://www.lcri.org.uk/lcri-convergence/pdfs/LCBE_brochure.pdf`
   (accessed February15, 2012)

Autodesk Ecotect 2010. Building and Housing Research Centre of Iran Website (2012), `http://www.bhrc.ac.ir` (accessed May 2, 2012)

CIBSE, Environmental Design: CIBSE Guide A. CIBSE Publications, Norwich, Norfolk, UK (2006)

City and County of Swansea. Population statistics for Swansea (2012), `http://www.swansea.gov.uk/index.cfm?articleid=28567` (accessed May 2, 2012)

Coastal Housing Group. Coastal's First Eco Friendly Building - Harbour Quay (2012), `http://coastalhousing.blogspot.co.uk/2012/05/coastals-first-eco-friendly-building.html` (accessed May 15, 2012)

EDSL, Thermal Analysis Software, v.9.1.3a. Environmental Design Solutions Limited, Cranfield, UK (2010)

Givoni, B.: Passive and Low Energy Cooling of Buildings. Van Nostrand Reinhold, London (1994)

Givoni, B.: Climate Considerations in Building and Urban Design. Van Nostrand Reinhold, London (1998)

Heidari, S.: Thermal Comfort of People in the City of Tehran. Honarhaye Ziba 38, 5–14 (2008) (Persian)

Hopper, J., Littlewood, J.R., Geens, A.J., Karini, G., Counsell, J.A., Thomas, A.: Assessing retrofitted external wall insulation using infrared thermography. Structural Survey – Special edition on Existing Buildings 30(3), TBC (2012)

Littlewood, J.R.: Chapter six - Testing the thermal performance of dwellings during the construction process. In: Emmitt, S. (ed.) Architectural Technology: Research and Practice, pp.TBC. ISBN: TBC. Wiley Blackwell, Oxford (2013) METEONORM (v 6) Meteotest

Krishan, A., et al. (eds.): Climate Responsive Architecture: a design Handbook for Energy Efficient Buildings. Tata McGraw Hill, New Delhi (2001)

MET Office. (March 2012), `http://www.metoffice.gov.uk/climate/uk/2012/march.html`(accessed May 2 2012)

Shaviv, E., Yezioro, A., Capeluto, I.G.: Thermal Mass and Night Ventilation as Passive Cooling Design Strategy. Renewable Energy 24, 445–452 (2001)

Szokolay, S.: Introduction to Architectural Science. The basis of sustainable design. Architectural Press (2008)

Swansea Bay. Swansea Bay facts and figures (2012a), `http://www.abayoflife.com/en/available-property/`(accessed May 15, 2012)

Swansea Bay. Swansea Bay weather guide (2012b), `http://www.welshholidaycottages.com/weather/swansea-weather-guide.htm` (accessed May 15, 2012)

Swansea Bay. Swansea Bay weather and tide (2012c), `http://visitswanseabay.com/index.cfm?articleid=33068` (accessed May 15, 2012)

Tehran Municipality. Atlas of Tehran Metropolis (2012), `http://atlas.tehran.ir/Default.aspx?tabid=272` (accessed June 8, 2012)

The Weather Channel. Monthly averages for temperature in Swansea, UK (2012), `http://www.weather.com/weather/wxclimatology/monthly/graph/USSC0335` (accessed May 15)

Travelmath. the latitude and longitude of Swansea, United Kingdom (2012), `http://www.travelmath.com/cities/Swansea,+United+Kingdom` (accessed May 15)

Weather Online. Swansea (March 2012),
http://www.weatheronline.co.uk/weather/maps/city?WEEK=02&MM=
03&YY=2012&WMO=03609&LANG=en&SID=0360956776c421b303457c2c81a2
0340c9294a&ART=MAX&CONT=ukuk&R=150&NOREGION=0&LEVEL=150&REGIO
N=0003&LAND=___ (accessed May 3)
Yannas, S. (ed.): Design for Summer Comfort: Building Studies. AA EE, London (2000)

Amendments to Meet the Referees' Comments

Referee One and Two

1. The paper has been updated to better discuss the validity of comparing the methods and results from case study one with those of case study two. This includes updates to the abstract, introduction, case study two results, and the discussions sections.
2. The paper has been updated to better explain why case study two's results are not discussed to the same depth as case study one. This includes amendments to case study two results and the discussion section.
3. The discussion section has been updated to discuss why the authors believe the corridors are overheating in comparison to the external temperature.
4. The discussion section has been further updated to 'unpack' the statement that some of the strategies used to prevent overheating can conflict with the energy requirements in the heating season.
5. The discussion section has been updated to discuss how the results from case study two will lead to further publications.

Chapter 59
An (un)attainable Map of Sustainable Urban Regeneration

Linda Toledo and John R. Littlewood

The Ecological Built Environment Research & Enterprise Group (EBERE),
Cardiff Metropolitan University, Cardiff School of Art & Design,
Cardiff, CF5 2YB, UK

Abstract. Reuben et al. (2010) suggests that 'before we can effectively change a system, we must first improve our understanding of the system'. In this spirit, the paper attempts to evaluate the knowledge obtained from interviews with key stakeholders engaged on an urban regeneration project in Swansea, UK known as 'Urban Village'. Urban regeneration is an activity that is largely characterised by complexity, uncertainty and ambiguity. The tacit knowledge acquired from the interviews with the stakeholders of the Urban Village project have been mapped out with IDEFØ language in order to record data and processes that have characterised the project. The intermediate goal of this effort is to recover the connections missed by a first fragmentary subdivision of the collected interviews. In order to achieve this goal, we have re-mapped out the previously collected responses, by so creating a decision-making process tool aimed to orient professionals. In the process we have managed to create a device for the guidance and assessment of the decision making process in urban regeneration. The ultimate goal will be hence reflected in the provision of a tool that provides more scope for auditing the decision process in urban regeneration rather that leaving it up to expert individuals.

This paper fits into the sub-theme 'validation of design and environmental assessment tools and modelling with performance in use - through physical and/or social assessment', at SEB12 conference.

Keywords: sustainability assessment tools, validation, environmental assessment tools, urban regeneration, IDEFØ.

1   Introduction

The mapping of urban regeneration is a matter as complex as the activity of urban regeneration itself. By enclosing it into a rational systematisation, it is not only fragmented its urban organic reality is often scattered and disconnected and therefore it is also possible to miss the strategy (or series of correlations) behind the project (Lefebvre 1998). On the spirit of this challenge, this paper discusses the development of a decision-making process tool aimed at orientating professionals and stakeholders in their decision making on urban regeneration projects. The tool presented in this paper has been developed from a) the tool presented in Fox et al. (2011) and b) the analysis of interviews with the stakeholders of the Urban Village regeneration project in Swansea (UK) using IDEFØ language. This paper is targeted at the sub-theme 'validation of design and environmental assessment tools and modelling with performance in use - through physical and/or social assessment'; of the invited session 'assessment and monitoring of the environmental performance of buildings', at the SEB12 conference.

1.1 The SURegen Project

The SURegen project is an Engineering Physical Sciences Research Council (EPSRC) funded research project, which has been undertaken between 2008 and 2012 in the UK. In this context, researchers from Cardiff Metropolitan University (CMU) have been working with the stakeholders engaged in the ongoing urban regeneration project in the city centre of Swansea (Fox et al. 2011).

The SURegen project has aimed to provide support for all those who are engaged in urban regeneration, through a web based application (SURegen 2012a). This application, or set of integrated decision support tools in the form of a *regeneration workbench*, is aimed to help new and professionals working in the urban regeneration field to make critical decisions, and help those who are new to the field acquire the skills needed to meet the challenges (SURegen 2012a). This workbench can be defined as an online prototype that provides a framework for a regeneration programme 'based on a regeneration knowledge framework, glossary of terms, ontology and underlying data models' (SURegen 2012b).

The focus of the collaboration between SURegen and the stakeholders engaged in the Urban Village - Swansea regeneration project is on making an explicit explanatory guide for people new to the "visioning" stage of urban regeneration. However, experienced people have knowledge, which is tacit and implicit and not explained, so the major challenge and first stage of this process was to tease out that tacit knowledge; which is documented in Fox et al. (2011) and is illustrated in fig. 1 below. The next stage was to have the validity of the tacit knowledge, which has been made explicit, confirmed by Coastal Housing Group stakeholders.

1.2 The Urban Village Case Study

Since the early 2000s, Coastal Housing Group has been in the process of regenerating an area of the city centre of Swansea (UK). Coastal Housing Group has packaged this regeneration scheme into a series of projects, including "Urban Village", situated in the Lower Super Output Area (LSOA) of Castle Two. Castle Two is ranked 11 out of 1896 for the most deprived areas in Wales, according to the Welsh Index of Multiple Deprivation, 2008 (Welsh Government 2008). For more information on the Urban Village case study see Fox et al. (2011).

The Urban Village project is a multi-purpose inner-city regeneration scheme, including residential, office, retail and leisure facilities and has been designed and is being built following sustainable and one planet living principles (Fox et al. 2011 and Coastal Housing Group 2012).

2   Methods

Between 2010 and 2011, researchers from CMU conducted interviews with a number of key stakeholders engaged in Coastal's Urban Village project in order to establish the visioning behind this project. In addition, the objectives were to determine the *key triggers* and issues for the project to create a decision map for sustainable urban regeneration. The decision map was to be the first stage in developing a tool to allow regeneration practitioners when visioning other urban regeneration projects. This initial coding of the interviews, presented in the first EBERE Decision map is reflected in Fig.1 below, which can be classified as o*pen coding*. Open coding consists of breaking down and categorising data; and it is largely influenced by the reflection of the researcher's perspective (National Institute of Standards and Technology, 1993). The EBERE decision map aimed to make explicit the tacit knowledge in the Urban Village project. Within the EBERE decision map the five key triggers in the Urban Village project are: a) personal motivation, b) housing demand, c) finance, d) competition, f) deprivation. These triggers acted as a starting point for the final vision of the Urban Village project.

Following a presentation of the first EBERE Decision map at a SURegen workshop in February 2011; to validate the map, the feedback provided by some of the stakeholders was that the map was too complex and cluttered and did not accurately reflect decision making on the Urban Village project. Therefore, in 2012 the interviews have been re-analysed and the data has been reassembled, in order to bring coherence to the coded data; leading to re-coding the triggers. This second phase of coding is referred as *axial coding* (National Institute of Standards and Technology, 1993). Hence, firstly, *axial coding*[1] has been performed by the authors of this paper in order to map out the missing interconnections in the first open coding (first EBERE Decision map, fig.1) of the tacit knowledge collected from the interviews, in 2010/11. More specifically, axial coding is conducted to recover the interconnections between different triggers as shown in fig. 2. One of the outcomes of this re-codification of interviews data has led to the refinement of triggers. However, the corresponding links remain of the EBERE decision map (i.e. key questions) have been lost; which demonstrates that the axial coding, in this case, was not sufficient. Therefore, this created the opportunity for the next step in the development of this work. This led to the use of IDEFØ which is an advanced level of sophistication for knowledge encoding. (National Institute of Standards and Technology 1993)

---

[1] *Axial coding* is the type of coding practice that brings together the different categories (separated by the previous open coding) by making connections between those categories (National Institute of Standards and Technology, 1993).

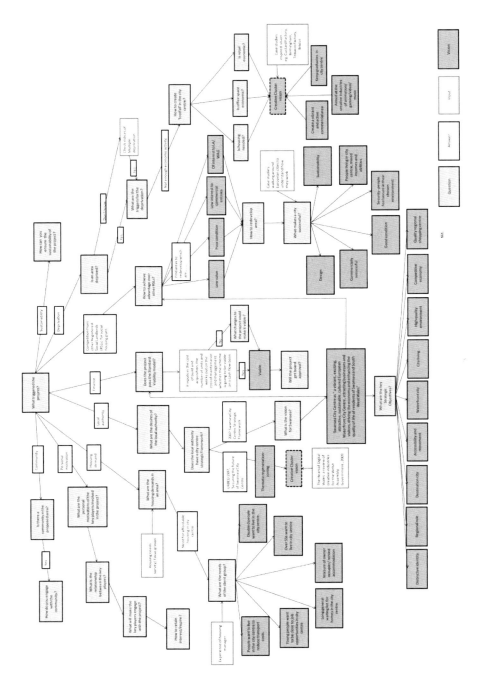

Fig. 1. EBERE Decision map for sustainable urban regeneration (Fox et al. 2011)

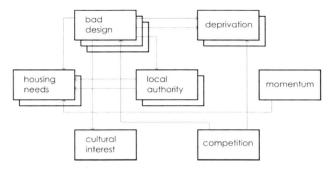

Fig. 2. Axial coding of the triggers in urban regeneration

Secondly, IDEFØ[2] is been used to create a *data flow tool* with the knowledge collected from the interviews. IDEFØ standard offers a language able to model complex functions and thus the use of IDEFØ is justified by the fact that one of its most important features is that it gradually introduces greater levels of definitions (or detail) within a complex process. Furthermore, IDEFØ has been demonstrated to be useful in establishing the scope of an analysis, in this case represented by *urban regeneration decision-making process*. In fact, whether as a communication tool, IDEFØ enhances involvement and consensus decision-making through simplified graphical devices, at the same time as an analysis tool, it assists the modeller in identifying what functions are performed, and what is needed to perform those functions, and (consequentially) it highlights what the current system does, and what the current system does not. (Knowledge Based Systems Inc., 2010).

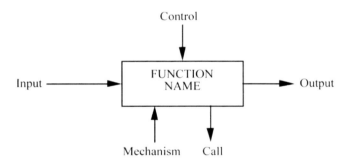

Fig. 3. Elementary information of a basic IDEFØ diagram. (Knowledge Based Systems Inc., 2010)

---

[2] As stated in the IDEF website, *IDEFØ* is a 'method designed to model the decisions, actions, and activities of an organization or system... Effective IDEFØ models help to organize the analysis of a system and to promote good communication between the analyst and the customer'. (Knowledge Based Systems Inc. 2010).

The translation of the interview data into an IDEFØ map has made use of the process map of urban regeneration as documented in the SURegen workbench. (SURegen 2012c). In order to achieve this, firstly the information contained in the SURegen workbench has been sectioned into a *data flow,* with IDEFØ (see fig. 4 to 7), through the identification and sorting of information and whether it relates to an input, or output, or control or mechanism of a function, as per IDEFØ language (see fig. 3). This procedure has enabled the creation of the upper levels of the diagram, in which, on second place, the specific interviews data for Swansea case study is framed, as shown in fig. 8. This is the refined EBERE decision map using IDEFØ language.

3 Results

The initial axial coding performed a review of the interviews to obtain the interrelationships between the triggers such as a) previous *bad design*, that has disrupted the organic patterns of the city; b) *deprivation*; c) *housing need*, including lack of residential activity or suitable accommodation missing for certain age ranges; d) interest from the *local authority* to revitalise the city; e) *cultural interest* on one or more buildings; f) desire of *competition*, within the developers; g) the seize of an opportunity (*momentum*). In fact, most of the respondents have indicated the fact that there is more than one trigger concurring at the same time, such as the concurrence of the housing need and the interest from the local authority to revitalise the city. Therefore, these triggers cannot be considered as separated entities since each trigger influences and at the same time is influenced by another trigger (see fig. 2). This example is represented by the trigger *housing need* which is influenced by the trigger *previous bad design*. Another revealing point of the axial coding was that triggers do not have a pre-established succession; this succession changes according to each respondent's point of view. Also it became clear that some triggers were more influencing than others triggers such as *previous bad design* more influencing that *housing need*, although *housing need* being the most recurring trigger, according to the stakeholder's point of view and profession.

The subsequent creation of the IDEFØ map, which is composed by the figures 4 to 7, led to the creation of a contextual/generalised map (a) and then to the creation of a specific map for the Swansea case study (b), by transferring the knowledge collected from the interviews in the appropriate level of detail of the more generic diagram (a). The generic map (a) is actually a system of different diagrams (fig. 4, 5, 6 and 7) whereas the correspondent of the EBERE Decision map (as per fig. 1) is now represented in fig. 8. In detail:

- Level 0 in fig. 4, shows the activity under analysis and discussion of the SURegen project, which is urban regeneration; (Knowledge Based Systems Inc. 2010; SURegen 2012a).
- Level 1 in fig. 5, shows the level below level 0, which is the SURegen categorisation of urban regeneration; (SURegen 2012c).

- Level 2 in fig. 6, shows the level below level 1, which is the expansion of the strategy formulation stage; (SURegen 2012c).
- Level 3 in fig. 7, shows the level below the level 2, which is the expansion of the key triggers of urban regeneration; (SURegen 2012c).
- Level 3 in fig. 8, is illustrated with the re-codified interview data from the Swansea case study.

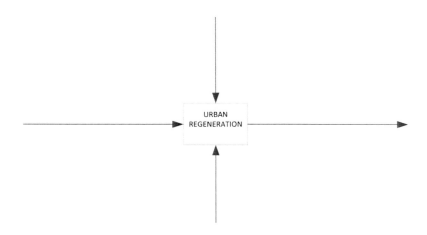

Fig. 4. Level 0: *urban regeneration* as the SURegen scope of analysis, (Knowledge Based Systems Inc., 2010; SURegen 2012a)

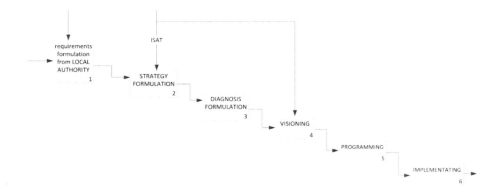

Fig. 5. Level 1: the SURegen categorisation of urban regeneration

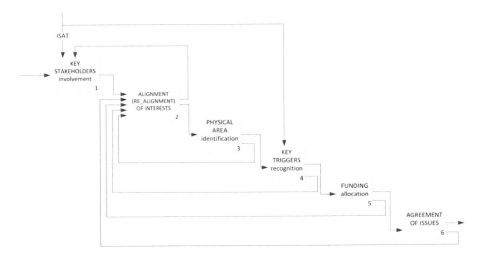

Fig. 6. Level 2: expansion of the *strategy formulation* stage

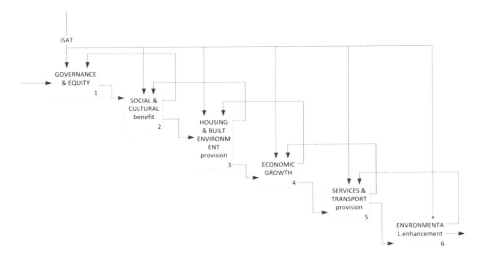

Fig. 7. Level 3: expansion of the *key triggers* of urban regeneration

59 An (un)attainable Map of Sustainable Urban Regeneration 645

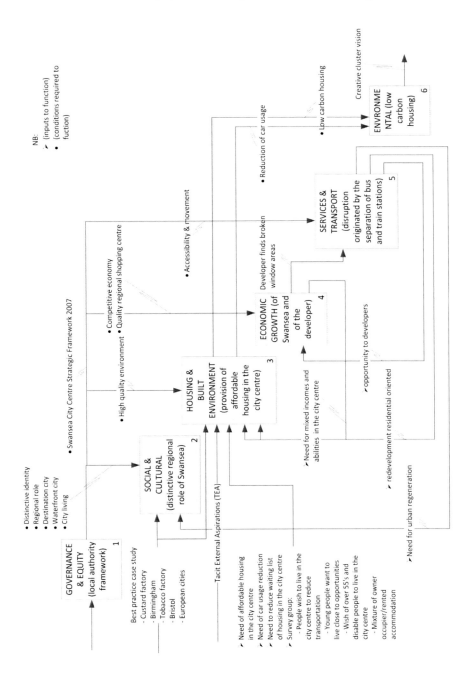

Fig. 8. Level 3: expansion of the *key triggers* of urban regeneration with the re-codified interview data from the Swansea case study

4   Discussion

The triggers of urban regeneration do not have a pre-established succession and they change according to each respondent's point of view, as demonstrated by the axial coding (fig. 2). This underlines the fact that priorities within an urban regeneration project can vary subjectively, according to the people and professional background, but also to their role within the urban regeneration project. Nevertheless, although axial coding clarified the interrelationships between the triggers, it was not sufficient as a method to analyse all of the information correlated to each trigger. In addition, the fact that some triggers appear to be more influencing means that unless urban regeneration is carefully considered, triggers can remain unsolved and escalate; and therefore they will cause further need for urban regeneration. An example of where triggers contradict the needs of each trigger can be found in the case emerging from the interviews discussed above; where the most influencing trigger in Swansea in the late 1990s was the disruption of the organic patterns of Swansea's. This fracture in the city was driven by the physical separation of the train station and bus station and therefore led to depravation, in this part of Swansea, known as Castle two. This deprivation in one of Swansea's key parts of the city (High Street) consequently led Coastal Housing Group to commence Urban Village in the early 2000s.

However, the fragmentation of a complex reality, such as urban regeneration conducted with IDEFØ presents a reality in which the processes are less encapsulated, hence more open to variations (see fig. 7 and 8). Also it shows how an idealised map (see fig. 7 and 8) can be manipulated to better represent a specific reality, without obliterating the first general categorisation shown in fig. 7.

These results are confronted with some specific theoretical literature concerning models and tools for sustainability; and this has enabled to define the limits of validity of the new EBERE decision map, for both generic and Swansea case study cases. In fact, these facts confronted with specific studies on sustainability confirm positions that embrace the *science of the city* as something irreducible to a simple model. Indeed, by studying the complexity of sustainable development, Reuben *et al.* (2010) argue that the complexity of cities cannot solve their 'sustainability' by reaching a balance between a number of identified factors. Instead the position of Reuben *et al.* (2010) remains on the perspective of unpredictable control, 'as an emergent property resulting from the quality and pattern of relationships within the systems of sustainability'. For this reason Reuben *et al.* (2010) recognise a parallel between sustainable development and complex adaptive systems (CAS). This helps to embrace sustainability not as a goal to be reached through the achievement of balance but as a dynamic process of continuous evaluation, action and re-evaluation, something better called as a "continuous cycle of co-evolution" (Reuben *et al.* 2010), as is successfully represented by the transposition of the interview data into an IDEFØ map (fig. 8).

5   Conclusions

The first EBERE decision map confirms the criticism of coding, which is the loss of context following fragmented data analysis (fig.1), and the axial coding subsequently performed (fig.2). In addition, the axial coding initially performed has not sufficiently made sense of the data collected from the responses.

IDEFØ sufficiently enabled the codifying of urban regeneration as an activity and the interview responses, without creating a static decision making tool. Indeed, it has been argued through the review of literature that within complex sciences one must avoid to assume that reality is static (Reuben et al, 2010). This is because disturbance is a natural component of ecosystems that promotes diversity and renewal processes (Scheffer et al. 2011). Therefore, only process oriented approaches will work and the EBERE IDEFØ map of urban regeneration in Swansea (fig.8) successfully embraces that permeability. More generally, one of the major obstacles is that traditional approaches to sustainable development focus on the achievement of sustainability and typically assume that once sustainability is reached, the problem is solved. The process oriented approach to sustainable urban development finds place on the IDEFØ map for urban regeneration, where uncertainty is not analytically reduced but encouraged, and taken as a learning source of reality as it unfolds.

Acknowledgements. This research has been funded by the EPSRC SURegen project in collaboration with Coastal Housing Group. The authors wish to thank the SURegen researchers that have contributed with their time and intellectual input to this research: Prof. Stephen Curwell (University of Salford), Dr. Mohamed El-Haram (Dundee University) and Prof. Ian Cooper (Eclipse Research). Thanks are also extended to all the interviewees.

References

Bryman, A.: Social Research Methods. Oxford University Press, Oxford (2008)
Coastal Housing Group. City living at the heart of the urban village (2012),
    http://www.coastalhousing.co.uk/pdfs/Spring2012.pdf
    (accessed: May 1, 2012)
Fox, C., Littlewood, J.,R., Counsell, J.: An analysis of the visioning process of the Inner City Community Sustainable Swansea project. Paper Presented as an oral paper and published in the proceedings for the SB11 Helsinki World Sustainable Building Conference, Finnish Association of Civil Engineers, Helsinki, Finland, October 18-21 (2011)
Knowledge Based Systems Inc. IDEFØ (2010), http://www.idef.com/IDEF0.htm
    (accessed: May 23, 2012)
Lefebvre, H.: Writings on cities. Blackwell Publishers, Oxford (1905)
National Institute of Standards and Technology Integration definition for function modelling (IDEFØ). Federal Information Processing Standards Publications, 183 (1993),
    http://www.idef.com/pdf/idef0.pdf (accessed May 23, April 5 , 2012)
Reuben, R., McDaniel Jr., Lanham, H.J.: Sustainable development: Complexity and the Problem of Balance. In: Moore, S. (ed.) Pragmatic Sustainability: Theoretical and Practical Tools, pp. 51–66. Routledge, London (2010)
Scheffer, M., Carpenter, S., Fole, J.A., Folke, C., Walker, B.: Catastrophic shifts in ecosystems. Nature 413, 591–596 (2001)
SURegen. SURegen- urban knowledge for regeneration (2012a),
    http://www.suregen.co.uk/index.html (accessed: April 25, 2012)

SURegen. SURegen Glossary V5 (2012b), `http://www.suregen.co.uk/documents/glossary/SURegen_glossary_V5.pdf` (accessed: April 25 , 2012)

SURegen. Regeneration check list (2012c), `http://www.suregen.co.uk/surdsm4t/` (accessed: May 1 , 2012) (unavailable to the general public until October 1, 2012)

Welsh Government 2008. WIMD (2008), `http://dissemination.dataunitwales.gov.uk/webview/maps/wimd08_eng/atlas.html` (accessed: May 3, 2012)

World Commission on Environment and Development, Our Common Future. Oxford University Press, Oxford (1987)

Chapter 60
LCA in The Netherlands: A Case Study

Wim Zeiler[1], Ruben Pelzer[1], and Wim Maassen[2]

[1] TU/e, Technical University Eindhoven, Faculty of Architecture, Building and Planning
[2] Royal Haskoning consulting engineers

Abstract. The Municipality of Venlo is building new municipal offices based on the guiding principles of C2C. A case study is presented on the townhall Venlo preformed in the LCA tool GreenCalc+ V2 (GC+) in a comparison with two GC+ studies on the offices of TNT and townhall Utrechtse Heuvelrug. In concluding, the MIG (Environmental Index Building) score of townhall Venlo, TNT Green Office and townhall Utrechtse Heuvelrug were calculated which showed that making a good comparison is very difficult. This indicates that is important to know how to deal with specific characteristic aspects when calculating the environmental assessment score.

Keywords: Sustaianble assessment tools, LCA, GreenCalc, C2C.

1 Introduction

The Municipality of Venlo is building new municipal offices, however Venlo is not building just any building. Venlo is building a city hall that is characterized by excellent services to the inhabitants and companies of the municipality of Venlo and based on the guiding principles of C2C. So very important are pleasant and healthy workplaces for all the employees of the municipality of Venlo. It should become an icon of the cradle-to-cradle design (C2C), a building that exudes what the municipal organization wants to be: open, transparent, accessible and sustainable. The new city hall is designed by Kraaijvanger Urbis architects of Rotterdam. In 2009 this firm of architects won the European public tender in which Venlo did not ask for a detailed design but for a vision. One of the advantages of this is that the municipality of Venlo will be at the table right at the start of the design process and is not bound by the choice of a sketch which could determine the final picture to a great extent. The assignment for technical system advisor was definitely awarded in April 2010 to the international agency Royal Haskoning. The technical systems adviser gives advice on how the various cradle to cradle aspects can be given shape from a technical perspective in the new design. This includes such things as advice in the area of type of systems, new developments in the area of systems technology, insulation, etc. but also calculating whether the design and the materials and systems do in fact result in a responsible internal climate and whether the building also satisfies the C2C ambition. Royal Haskoning (RH), has embraced the C2C concept as C2C inspired design and is currently designing City hall Venlo, as C2C inspired office building for the municipality of Venlo. In addition, the life cycle analysis (LCA) recently emerged in the

---

A. Håkansson et al. (Eds.): *Sustainability in Energy and Buildings*, SIST 22, pp. 649–658.
DOI: 10.1007/978-3-642-36645-1_60   © Springer-Verlag Berlin Heidelberg 2013

building sector. It is a method which can measure the sustainability of a building by calculating the environmental impact of its life cycle through data analysis. In fact, the program GreenCalc+ (GC+) is a well-known LCA program in the Netherlands, which RH could use to analyze the life cycle of this office. The development of GreenCalc started in 1997. The Greencalc+ assessment method is a questionnaire which allows you to estimate how much land it takes to run and maintain your office. It that can be used to calculate what the developers call the environment index of a building. This is done by calculating the environmental impact of the buildings by Life Cycle Analysis (LCA). The GreenCalc+ software consists of four modules, each representing a different aspect of the building characteristics; mobility, materials, water and energy. The input values for this program are divided in the following four groups: Materials: Energy: Water: Travel to and from work:

Detailed information of the office design on these four aspects needs to be imported in the computer program. The program is able to calculate the amount of emissions, rate of exhaustion, land use, and the degree of nuisance involving these aspects. These values are then expressed in environmental hidden costs using Euros per year and per square meter. Furthermore an environmental index is given, that expresses the relation between the newly imported (environmental friendly) design and an automatically derived reference design. The score is given within a scale from 1 to 2,000. The reference design shows how the building traditionally was designed in 1990 having an index of 100. With the introduction of GreenCalc+ in 2005 program became more user friendly, resulting in a time reduction of importing the data of the building construction into the computer program (13). Despite these developments, RH did not use LCA on City hall Venlo in their consulting and design strategies, because they questioned the methods of GC+. In this article, a LCA study on City hall Venlo is presented performed in GreenCalc+, one of the most popular sustainability assessment tools in the Netherlands. In addition, the outcome of the study was compared with two other offices which were studied in GreenCalc+; TNT Green Office and City hall Utrechtse Heuvelrug.

2 Methodology

The LCA is an approach to make a life cycle analysis of a product by addressing the environmental impact which is created during its life time. Most LCA tools are based on the international standard ISO 14040/14044. In addition, for buildings, special building LCA tools are developed of which GreenCalc+ (GC+) is an eminent example. In general, LCA tools are difficult to compare because of large individual differences. The EPA(1) defines a LCA as a technique to assess the environmental aspects and potential impacts associated with a product, process, or service by:

- Compiling an inventory of relevant energy and material inputs and environmental releases;
- Evaluating the potential environmental impacts associated with identified inputs and releases;
- Interpreting the results to help you make a more informed decision (1).

The ISO guideline describes the framework and terminology of the LCA, by defining four phases; goal and scope, inventory analysis, impact assessment and interpretation phase, viewed in figure 1. In short, the ISO guidelines are also applicable for buildings; GC+ is based on these guidelines (2).

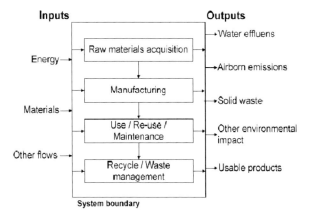

Fig. 1. An example of ISO framework; The goal and scope and inventry analysis (LCI) (3)

The LCA of a building starts with an inventory analysis of the materials and energy needed for the construction, service and demolition of the building. Then, the in- and outputs of the inventory analysis are sorted by characterization models to impact categories. In fact, several characterization models can be used to weight the in- and output towards a specific impact category. Lastly, all impact categories are normalized towards one value, an environmental impact value; several normalization methods can be applied.

In addition, a numerous of different impact categories and environmental impact normalization value types can be found in literature and in international guidelines. For instance, in an extensive normalization study presented by Sleeswijk et.al. (2007), which studied 860 environmental interventions, 15 impact categories were found (4). While the LCA-CML2 model, used in the Dutch harmonized method, only reports on 14 impact categories (5). In addition, the LCA tool GreenCalc+ accounts for 17 impact categories and in the European guideline CEN/TC 350 (ENSLIC model), presented in Malmqvist et al. (6), 12 impact categories were suggested. Also, Malmqvist et al. (6) suggested that the impact categories should also address the recycling potential of a building product and the embodied water, in contrast to most sources who did not address these items. Moreover, in the extended literature study of Finnveden et al. (7) it is found that the impact of land use and water use and weighting methods are still discussed in scientific literature.

In GC+, the inventory analysis is based on a material, water and energy inventory, followed by the weighing of the inventories toward different impact categories and eventual normalization of the impact categories towards environmental impact costs. Since buildings can differ in lifetime and building component can be replacement during service, the environmental impact costs are expressed per year to be able to compare building. The three modules (material, water and energy) are separately inventoried and computed. The materials in the building are inventoried by indentifying the quantities in square meter material or counting the numbers of building elements in the building, based on the GC+ product catalogue. In addition, water use is estimated by using the 'Water Prestatie Normering' model created by the bureau Opmaat en Boom. The energy inventory is done either by using the internal computation module of GC+, based on norm NEN 5128:2004, or the by entering the EPC of the building (based on NPR 5129).

The characterization is based in several characterization models. The eventual normalization of the impact categories is based on the TWIN-model developed by NIBE. In GC+, the impact categories are normalized toward the environmental impact cost of a building. These are the fictional costs that represent the cost for resolving the negative environmental impacts of the construction, service and demolition of the building to an acceptable level. Lastly, GC+ uses the environmental impact cost to calculate a sustainability score for a building, based on a reference which is generated (2).

Based on the works of A. Dobbelsteen, GC+ expresses the sustainability of a building in a score rather than in the environmental impact costs with is initially calculated. According to the PhD thesis of A. Dobbelsteen, a building should be characterized by a reference based score rather than by an absolute value, such as the total environmental impact cost (8). He argues that representing the sustainability of a building through a score has several additional advantages in relation to an absolute value, such as the environmental impact score.

First, he argues that a score can be expressed in a value with a low number of digits, whereas the environmental impact cost are expressed in more digits. Secondly, he states that a score can provide insight into the environmental performance of the building towards a specific target or show an environmental improvement with respect to a reference or situation. Lastly, Dobbelsteen argues that a score would be easier to use in the categorization and labeling of buildings. In the following section, the score computation for office buildings is further explored.

In order to calculate a score, a reference is generated by GC+, based on a reference building model. The reference generation method allows the computation of the MIG (NL: *Milieu-Index-Gebouw*). The MIG score is computed by comparing the imported office with the generated reference office by using an *average* number of occupants. Therefore, the score calculates the degree of sustainability of the imported office independent of the occupants. An average occupant is estimated to use 28 m^2 GFA/FTE (Gross Floor Area per Full-Time Equivalent) (8). The MIG score use the reference office model of GC+, except that the given input of the reference office model varies per score type. Based on the calculated environmental impact cost (Ecost) of both the imported office and the generated reference office, the score is calculated. In fact, the MIG score is calculated by the same equation;

$$Score_i = Ecost_{i;ref}/Ecost_i \cdot 100 \tag{1}$$

In eq.1, the score computation of an random office i is expressed with respect to the corresponding reference (stated by the subscript $_{i;ref}$). The Ecost are expressed in $e€/yr$. In addition, the value obtained as final score is a dimensionless number.

3 Case Study: City Hall Venlo

City hall Venlo is part of Maaswaard, a redevelopment project which include multiple building projects with the city of Venlo as client. Although the building is mainly composed out of offices, the building is divided in a low-rise, high-rise and basement section, were the main function is designated as meeting function, office function and parking function respectively.

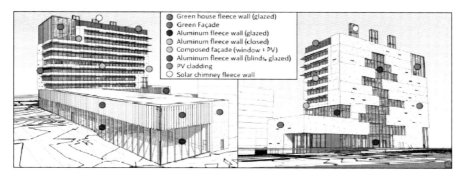

Fig. 2. Two views on the facades; the eastern orientated view (left) and western orientated view (right) (9)

Below the ground floor are located an additional basement layer and three parking layers. Since, GC+ does not take into account the parking space in the reference office, the parking layers are not taken into account. The structural elements of the building are designed as concrete floors and columns. The elevator shaft is conducted as a stiff concrete core to counter wind forces which act on the building. The floors are conducted as bubble deck floors which save on the average 35% mass by using special HDPE (High Density Polyethylene) balls in the floors in order to conserve concrete. The concrete columns in the building create an open spaces in the offices spaces where no additional interior walls are present allowing a flexible layout. In addition to the flexibility and material saving concepts in the structure elements, the cradle to cradle concepts are applied into climate control and materialization system of the building design, the main points are addressed.

Firstly, the office's hybrid ventilation concept consists out of two parts, the summer and winter situation, represented in Fig. 3. In both situations, the air is mechanically pumped into the building and naturally ventilated through the green façade out of the building. In addition, a small amount of the air is extracted through the sanitation facilities and kitchen. In the summer, the solar chimney, located on the roof of the office, is employed to extract additional air from the building due temperature differences. In the winter, the solar chimney is not used, but thermal energy from the air is recovered through the air handlings unit (AHU). The air is blown into the building by swirl diffusers vents.

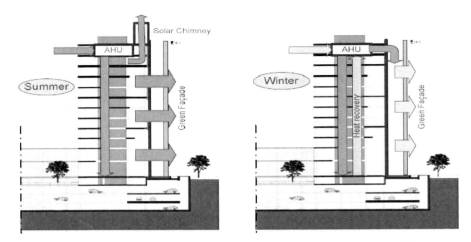

Fig. 3. (Left) The ventilation concept for the summer situation; (Right) The ventilation concept for the winter situation

Approximately 50% of all the working spaces (e.g. offices, meeting rooms, public working rooms, call-centre) are fitted with climate panels which both heat and cool these spaces. In addition, glass covered facades in the high-rise and low-rise section are heated with radiators in order to prevent cold down-drought. The base heat and cooling demand in the low-rise section of the building is being met through concrete core activation (CCA).The thermal energy distribution through the building is based on water. The thermal energy needed for the climate control is extracted through a heat pump from collective geothermal storage. The thermal storage both used to supply the cooling and heating needs for the building. There is an air-water heat pump is present as auxiliary to provide extra capacity during peak hours. By using control values in the climate ceiling, in the offices spaces are functioning as flexible workspace format; the framework allows for customization within a grid of every 1.80 m and controllability every 3.60 m.

City hall Venlo is partly self-sufficient in by generating electric and thermal energy by ca. 25 m^2 solar collectors and 1000 m^2 PV cells. The heat generated by solar collectors is used to heat tap water. The PV cells are used to power the low-voltage grid. The PV cells are applied onto the building as external solar blinds and cladding on the façade facing south and on the roof of the high-rise section. The facades are designed with a R_c of 5 m^2K/W and the windows have an U of 5 W/ m^2K. The water management system in the building for the primary water use is capable of using rainwater and reusing the grey water in multiple cycles resulting in less annual water demand and lower strains on the collective sewer system, represented.

4 Comparison City hall Venlo with TNT Green Office and City hall Utrechtse Heuvelrug

In the case study, the project City hall Venlo is calculated by GC+ and now can be compared with previously executed GC+ case studies on TNT Green Office and City hall Utrechtse Heuvelrug on their MIG scores. For the inventory of Stadkantoor Venlo, the data for the three modules (energy, water and materials) is needed; therefore several data sources are used. Firstly, the inventory of the materials and geometry was adapted from the final architectural drawings of the architect (Kraaijvanger - Ubis). The architectural drawings were scaled to 1:200. In addition, the installation specifications were adapted from the concept and final report for the building services produced by Royal Haskoning (9,10).

The case study on TNT Green Office was conducted by the company DGMR (11) and study on City hall Utrechtse Heuvelrug by the company Peutz (12).

An overview of the used literature resources used for obtaining the inventory for City hall Venlo and the computations of the other offices is presented in table 1.

Table 1. The literature resources used for the inventory and case

Case	Element	Source	Reference
City hall Venlo	Basic information (e.g. occupants, GFA, NFA)	Final report on building service	(9)
	Installation inventory	Concept cost estimation and Final report on building service	(10,11)
	Geometry and materials count	Final architectural sketches	(12)
	Energy consumption	EPC from the final report on building service	(9)
	Water management and consumption	Concept report on building service	(10)
	Replacement of foundation	Equations of the reference office model in GC+v2.20	(2)
TNT Green Office	Output GreenCalc+ study	Conceptual report on sustainable design	(11)
City hall Utrechtse Heuvelrug	Output GreenCalc+ study	Final report on building physics	(12)

During the research it was found that due to limitations of the GC+ catalogue, numerous installation elements could not be addressed during the material inventory. In this inventory only 18 of the 61 elements inventoried could be addressed in the inventory of GC+.

5   Results

The general specifications of all offices and the results are expressed the overview in table 2.

Table 2. The basic information of the three offices compared. (10, 11, 12)

	City hall Venlo	TNT Green Office	GK Utrechtse Heuvelrug
Construction	Estimated 2013	2010	Estimated 2013
GFA [m^2]	13,701	17,956	6,053
NFA [m^2]	11,762	16,136	5,174
Layers [-]	12	6	3
Occupants [p]	800	873	216
Lifespan [yr]	50	35	35
Energy use [MJ/yr]	3,706,069	0	1,319,458
Water use [m^3/yr]	1020	3,618	976
MIG score [-]	250	1,030	235

In conclusion, the two GC+ studies found in literature and Stadkantoor Venlo were compared on the MIG score. The MIG score of City hall Venlo, TNT Green Office and City hall Utrechtse Heuvelrug was calculated at 250, 1030 and 235, respectively. Due to limitations of the GC+ catalogue it was found that numerous installation elements could not be addressed during the material inventory of City hall Venlo.

6 Discussion and Conclusions

During the research it was found that limitations of the GC+ catalogue, numerous installation elements could not to be addressed during the material inventory of City hall Venlo. Also, the materials of the energy generation facilities in both studies from literature were not addressed, due to these limitations. In general, references used in score computations should be constant values or average values found by statistical analysis or, if statistical analysis is not possible, should be estimated by the best possible method. However, when mathematical models are used to generate a reference based on imported data of a building, deviations in the characteristics (variables such as GFA, Ecost, FTE) of a building can occur, since the equations in reference model can change proportions of these characteristics. In fact, the equations in the reference model are hypothetical, so these deviations, could only lead to greater deviation from the actual characteristics of the building which are used to calculate the score. As long as the reference method results in a score computation in which the reference is a constant used to select one of the characteristics of the building to compute to the score, either by division or multiplication, no deviations can occur.

GC + provides the freedom in both the water and energy module to apply corrections on the water and energy use, if the user disagrees with the calculations for the corresponding module. This approach offers a possibility when the inventory method of GC + is too restrictive. Therefore, it may help to import the correct data of energy or water use in the program by using another determination method and apply a correction. However, this also gives the user a large amount of freedom in the interpretation of the inventory of both modules. In addition, the material module is much more restrictive, since it only offers a customization of building products based on the GC+ catalogue; no custom input is possible. Therefore, the PV cells and bioWKK which were used in the studies from the literature and case study could not be addressed in the material module. Consequently it is unknown how the energy generating facilities would affect the total Ecost of an office. However, the results showed that when the influence of local energy generation is disregarded, the effects in the final MIG score can be huge. For instance, the score of TNT Green Office changed to a score which is 20%-25% of the MIG original score. Therefore it is advised that in future studies that the positive contribution of energy generating facilities due to energy savings should only be included in the calculation when the Ecost associated with the materials can also be included in the calculation; either by using the GC+ catalogue or by using a custom input or correction value. In addition, based in the secondary inventory of the installations services in City hall Venlo, the inventory showed that the v2.2 of GC+ can only be used to inventory a small amount of the installation services. Since, only 14 out of 61 elements of the inventory could be entered in the material inventory of GC+, the quality of the output is questioned.

References

1. EPA. LCA assessment | Risk Management Research | US EPA. EPA homepage. [Online] US environmental protection Agency, (2011),
   `http://www.epa.gov/nrmrl/lcaccess/` (cited: May 8, 2011)
2. GreenCalc+. GreenCalc+ help file v2.20. [Compiled HTML Help file] (2009)
3. International Organization for Standardization. ISO 14040:2006. Environmental management – Life cycle assessment – Principles and framework. [International Standard] (2006)
4. Sleeswijk, A.W., et al.: Normalisation in product life cycle assessment: An LCA of the global and European economic systems in the year 2000. Science of the Total Environment 4, 227–240 (2007)
5. SBK. Bepalingsmethode milieuprestatie Gebouwen en GWW-werken. Stichting Bouw Kwaliteit, Rijswijk (2010)
6. Malmqvist, T., et al.: Life cycle assessment in buildings: The ENSLIC simplified method and guidelines. Energy 4(36), 1900–1907 (2010)
7. Finnveden, G., et al.: Recent developments in Life Cycle. Assessment. Journal of Environmental Management 91, 1–21 (2009)
8. Dobbelsteen, A.: The Sustainable Office. [PhD Thesis]. s.l.: Delft Technical University (2004)
9. Kraaijvanger - Ubris. Kraaijvanger - Ubris (2011),
   `http://www.kraaijvanger.urbis.nl/nl/projects/`
   `architectuur/projects/cityhall_te_venlo`
10. Hoes, P., et al.: Definitief Ontwerp Technische Installaties Nieuwebouw City hall Venlo. Nijmegen: Royal Haskoning, Final Design Report (May 27, 2011)

11. Knaap, A., Cremers, E.W.W.: TNT Green Office. s.l.: DGMR Bouw BV, Concept Report GreenCalc+ Calculation (2008)
12. Peutz. Bouwfysische en akoestische beoordeling Defintief Ontwerp City hall Utrechtse Heuvelrug. s.l. : Peutz (July 21, 2010)
13. Entrop, B., Brouwers, J.: The relation between the adoption of sustaianble measures and the composition of an environmental assessment tool for buildinggs. In: Proceedings SASBE 2009, Delft (2009)

Chapter 61
Comparative Life-Cycle Assessment of Residential Heating Systems, Focused on Solid Oxide Fuel Cells

Alba Cánovas[1], Rainer Zah[2], and Santiago Gassó[1,3]

[1] Universitat Politècnica de Catalunya, Barcelona, Spain
alba.canovas@estudiant.upc.edu
[2] Life Cycle Assessment and Modelling, EMPA, Switzerland
[3] Barcelona Supercomputing Center, Centro Nacional de Supercomputación, Spain, and
Laboratory of Environmental Modeling, Universitat Politècnica de Catalunya

Abstract. This study aims to analyze a Solid Oxide Fuel Cell (SOFC) for residential heating applications by applying Life Cycle Assessment (LCA). To do so, three perspectives have been chosen: the *producer*, the *user* as an individual and the *user* intended as the heating demand of a building, applied by default in Switzerland. This SOFC is compared to other systems which fulfill the same function and are already inventoried in international databases (Ecoinvent). That are: a Stirling engine, three types of heat pump, a polyelectrolyte membrane fuel cell (PEMFC) and a gas boiler. The results of these analyses in *SimaPro* software have shown that from the perspective of the producer, impacts reduction should come from reducing the metallic parts included in the inverter unit and parts made of copper, there is a 7.5% reduction potential if recyclable parts are properly managed. From the user perspective, the choice among different heating systems depends strongly on the electricity mix of the country. From the buildings perspective, the SOFC is best suited to a family house type like the SIA-380/1 (Schweizerischer Ingenieur- und Architektenverein) building Swiss standard, which consumes less energy than the current average Swiss family houses.

1 Introduction

The building sector accounts for approximately 40% of primary energy use and 36% of total CO_2 emissions in most developed countries[1]. Then, this sector has to put efforts to design strategies that can help satisfy the same needs with less primary energy consumption. Co-generation systems, such as fuel cells, produce electricity and heat and can be applied locally to satisfy the building needs. This novel technology must be though assessed environmentally through comparison with conventional heating systems in order to see if it can contribute to a reduction of the impacts associated to the building sector.

LCA studies about fuel cells have been continuously adapted over in the last decades, since the fuel cells are an emerging technology that it is nowadays still under development [2]. In this sense, literature has evolved from large-scale fuel cell applications, such as co-generation plants, to micro-scale technologies like mobile

A. Håkansson et al. (Eds.): *Sustainability in Energy and Buildings*, SIST 22, pp. 659–668.
DOI: 10.1007/978-3-642-36645-1_61 © Springer-Verlag Berlin Heidelberg 2013

applications [3], [4]. In the specific case of residential combined heat and power, fuel cells are nowadays also based on new materials in order to reach higher efficiencies [5–7].

Life-cycle assessment results depend strongly on the inventory (LCI), thus data collected to describe the processes involved must be consistent with the time-period on study. Therefore, LCA studies must be done continuously adapting to the new available technologies and updating data inputs, if available. Available international databases just contain few examples of micro-cogeneration systems, which represent partially the average market of products available, thus consistent updating of these datasets must be done.

Most of the studies available are not building-focused or they lack impact indictors other than GHG emissions or primary electricity consumption [8], [9]. When they are building-focused, the approach is often not LCA-based [10–12]. Other studies show only the use phase of the fuel cell and do not include its production nor disposal, so that not all of the life-cycle is represented [10], [11], [13]. In these studies the co-production of electricity and heat has been treated with allocation, and the main focus was electricity.

Therefore, this study aims to compare several micro-scale heating systems for residential applications based on a LCA methodology, in order to promote further comparative studies and to give new inputs to international databases, thus contributing to the global LCA community.

2 Material and Methods

2.1 Life-Cycle Assessment Approach

In this study, an attributional LCA has been chosen, since it is the most broadly applied method and because modeling consequences of decisions is somewhat pointless when there are no decisions to be made [13].

In order to deal with the co-produced electricity, system expansion, also called avoided burden, has been considered instead of allocation, which is a novel approach in the field of LCA of fuel cells.

2.2 Goal and Scope Definition

The goals of this study are: (a) to study a SOFC specific product in order to find out hot spots of its life-cycle; (b) to compare this SOFC with other heating supply systems and in different country scenarios; and (c) to assess the suitability of this SOFC for different building energy demand scenarios.

In all cases the geographical representativeness set by default has been Switzerland. For goal (b), the SOFC is compared to other systems which fulfill the same function and are already inventoried in international databases (Ecoinvent). That are: a Stirling engine, three types of heat pump, a polyelectrolyte membrane fuel cell (PEMFC) and a gas boiler. The functional unit for each equipment is "1 kWh of heat, delivered by the system, at facility" except from the goal (c), which has been set as "heat demand of the building, in $kWh/m^2 \cdot y$".

The scenarios in which this fuel cell has been analyzed include the disposal, the country of its use and the energy demand of different buildings. Regarding the disposal, 4 scenarios have been defined: one (1) with 70% weight recycling and the rest is sent to a landfill, which would be the current recycling potential of separable materials; another (2) with 50%-50% proportion of landfill and incineration; and two scenarios of 100% landfill and 100% incineration, respectively (3 and 4).

Regarding the geographical representation, the fuel cell and the other heating systems have been analyzed as if they were connected to the electricity mix and natural gas mix of 4 countries: Switzerland, Germany, Austria and The Netherlands.

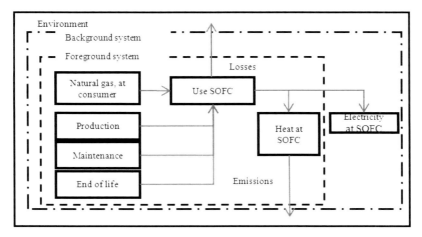

Fig. 1. Scope of the SOFC system assessed in this study

The three types of household energy demand are based in another study [14] and they are a Passive House following the German standards, the Swiss SIA-380/1 construction standard, and a Swiss average building. All of them have the same structure (single-family house) and 188.8 m^2 of usable floor.

System boundaries are different for each goal: to study the SOFC hot spots, thus goal (a), it covers from the production of raw materials which are sent to the SOFC manufacturer, until the fuel cell has been used and disposed to the environment (Fig. 1). However, when comparing systems, so for the rest of goals (b, c), disposal is not taken into account, and the focus is put on the production and use phases of the fuel cell. This is mainly due to lack of information about current disposal scenarios for neither the SOFC nor the rest of the systems. The SOFC also includes a back-up burner but this is not inventoried for the use phase, since it should be assessed following a dynamic model based on the energetic demand.

2.3 Inventory Analysis

Data has been obtained mainly from two sources: firstly, site specific data from the SOFC manufacturer and also from some of their raw materials suppliers; and literature or assumptions made when no information was available. Secondly, data from the *Ecoinvent v2* database has been used for background processes.

2.4 Impact Assessment

SimaPro 7.3.2. Analyst software has been used for calculating the LCA. Impact assessment methods used were *Eco-indicator 99* hierarchist (version 2.08), for midpoint and endpoint impact categories, and also greenhouse gas emissions (GHG) set by IPCC factors and cumulative energy demand (CED), as midpoint categories.

3 Results and Discussion

Normalized results about the production of the SOFC assessment in Fig. 2 show that the most relevant impact are carcinogens (46%), which affect to the human health area of protection. The part of the SOFC impacting the most in all impact categories is the *fuel processor* unit (61%), which includes: the heat exchanger, the back-up burner, the electrical inverter, the desulphurization unit for the natural gas with a catalyst, and some other parts, like fans, pumps, wires and gas piping. From this unit, the pieces affecting the most are the inverter (33%) and the copper (35%).

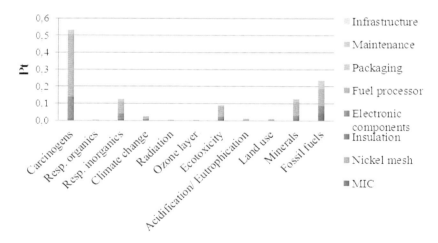

Fig. 2. Production of SOFC analysis with EcoIndicator 99 (H): normalized results

Regarding the whole life-cycle of the SOFC, in the characterization analysis of the disposal scenarios comparison shown in Fig. 3 it is observable that scenario 1 is the best choice for the environment compared to landfill and/or incineration: recycling a 70% of the fuel cell avoids carcinogens emissions (-100% instead of +40% approximately), mineral extractions for new pieces, because these are recycled (-100% instead of +10%), and also ecotoxicity (-60% less than other scenarios). Categories barely affected by the choice of the disposal scenario are the ozone layer and fossil fuels (<1% of difference), because these depend on the use of energy. However, these results represent not a realistic situation, since in scenario 1 no energy has taken into account for recycling the materials.

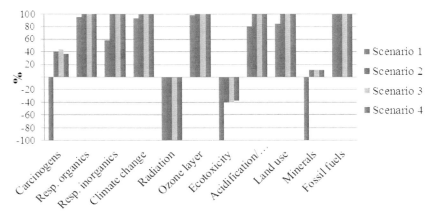

Fig. 3. Life-cycle comparison of disposal scenarios with EcoIndicator 99 (H) method: characterisation results

Results in Fig. 4 show that the most important impact categories observed in the whole life cycle of the SOFC are fossil fuels depletion and carcinogens emissions (due to materials and production process), and that the use phase of the fuel cell accounts for more than 95% of the overall impacts, mainly in the fossil fuel depletion category (due to natural gas consumption). Disposal is responsible for a -7.46% of avoided impacts and the assembly accounts for 12.06% of impacts.

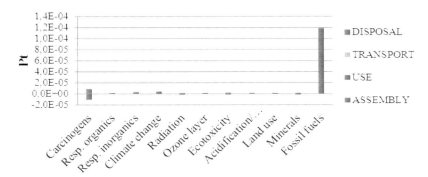

Fig. 4. SOFC life cycle analysis with EcoIndicator 99 (H) method: normalization results

The results of the comparison among different systems are shown in Fig. 5. It might be observed that only co-generation systems have avoided use of energy due to the co-production of electricity, which would substitute the nuclear power in the case of Switzerland (it is about 40% of the electrical mix). For 1 kWh (3.6 MJ) delivered

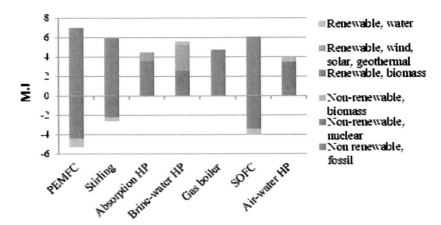

Fig. 5. Comparison among heating systems with Cumulative Energy Demand Method, by source of energy (Switzerland scenario)

by each device, only these co-generation systems have a net energetic benefit, whereas heat pumps and the gas boiler are only net-energy-consuming.

In Fig. 6 *EcoIndicator* Endpoints are shown for the different countries have been analyzed. The results are only shown for Switzerland and The Netherlands that showed the most opposite results. The results for Germany and Austria have not been shown, as they lay in between.

Compared to the Swiss case, in The Netherlands all co-generation devices have less impact, whereas the gas boiler and the absorption heat pump remain more or less the same. In the resources area of protection, the PEMFC impacts reduce from 39 to 18 mPt. The Stirling engine changes from 33 to 21 mPt. For the SOFC the reduction is from 34 to 17, thus a 50% reduction. On the other hand, brine-water and air-water heat pump would have more impact in The Netherlands.

Following the IPCC GHG emission factors and the correspondent method implemented in SimaPro, greenhouse gas impacts are shown in Fig.7, in Switzerland the device emitting most CO_2 equivalent per kWh of heat produced is the heating boiler (100 kg CO_2/kWh), followed by the absorption pump (73 kg CO_2/kWh), the co-generation systems (45 kg CO_2/kWh for Stirling, 44 kg CO_2/kWh for PEMFC and 43 kg CO_2/kWh for SOFC). The air-water pump (30 kg CO_2/kWh) and brine-water heat pump (21.47 kg CO_2/kWh) are the one with less GHG emissions. But as happens with EcoIndicator, in The Netherlands the results differ significantly.

The difference between countries' impacts, for EcoIndicator Endpoints as well as for GHG potential, depends on the electrical mix of the country: Switzerland has a low-carbon profile, with nuclear and hydrothermal power, whereas The Netherlands has a high-carbon electrical mix profile based on natural gas co-generation plants, in which the substitution of the electricity given by the SOFC and fed to the grid is reflected as a positive effect.

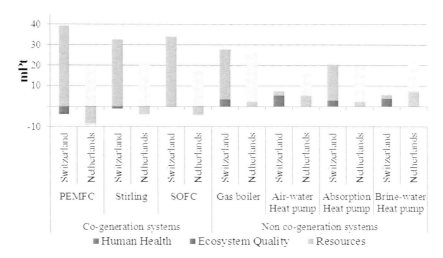

Fig. 6. Comparison among heating devices in different countries scenarios, assessed with EcoIndicator 99 (H) method – Single score (Endpoints)

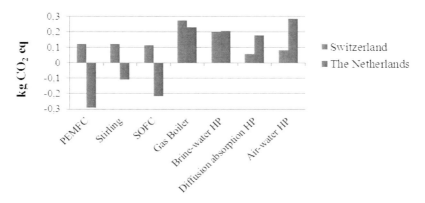

Fig. 7. Comparison among heating devices in different countries scenarios, assessed with Greenhouse Gas potential by IPCC standards

Finally, in the analysis of the SOFC in 3 building demand scenarios, the impacts are represented in discontinuous lines in Fig. 8. Their behavior is linear with respect to the heat demand. In continuous lines the area in square meters that the fuel cell could cover is represented, taking into account the minimum and maximum duration of the SOFC mentioned before. The horizontal green line represents the area considered for all buildings (188.8 m^2).

For the Swiss average house, the SOFC could be able only to cover between 56 and 77 m^2 of usable floor, which is less than the half of the house, so the average Swiss house requires another device with higher nominal power in order to cover its demand. For the other extreme case, which is a house following the German passive standards, the SOFC could cover between 436 and 600 m², because of its low heat demand. This is between 2 and 3 times greater than the usable floor area of the house. This means the SOFC would be only working half of its capacity, so not working efficiently. The best adaptable case study would be the SIA-code building, which represents the Swiss Standard applicable for newly constructed buildings.

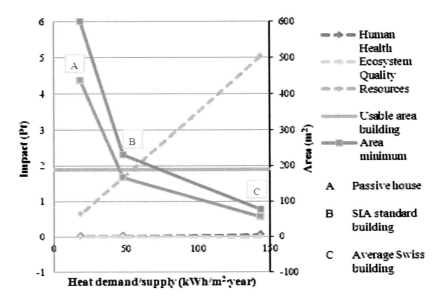

Fig. 8. SOFC impacts in three household energetic demands, with EcoIndicator 99 (H) method – Single score (Endpoints). Representation of heat demand and supply in terms of area of building heated

4 Conclusions

In this study a SOFC specific product from a Swiss company has been assessed environmentally using LCA methodology, looking at its main life-cycle stages (production, use and disposal). Several scenarios have been analyzed within the use of the fuel cell, regarding geographical representation and energy demand of a building, by comparing the SOFC with other conventional heating systems.

It has been seen that from the producer point of view, efforts must be put on reducing impacts on the inverter unit and the copper components. The production step has more impacts on the carcinogens emissions indicator, followed by fossil fuels depletion, minerals depletion, respiratory inorganics and ecotoxicity. The use phase is the most impacting in the whole life cycle, with more than 95% of overall environmental burdens. Given the duration of the fuel cell the production phase takes less importance, but depending on the disposal scenario the overall life cycle impacts can change: a possible recycling of the 70% of weight could lead to -7.5% reduction potential on the overall life-cycle impacts.

From the user perspective, the choice between one heating system or another depends basically on the country electrical mix. It has been proven that co-generation systems have a better energy balance than other systems, but when looking at CO_2-equivalent emissions and overall EcoIndicator impacts, results can vary substantially in the two countries such Switzerland and The Netherlands due to their different electricity mixes.

A stand-alone LCA study is not sufficient for analyzing the suitability of the fuel cell for a specific household, since all impacts follow a linear dependence with the energetic demand. Therefor, future studies in this field should incorporate temporal fluctuations due to user demand and efficiency reduction. From the comparison between demand and supply of different building scenarios, it has been observed that the most suitable single family house would be the according to the Swiss Standard building reference code (SIA 380/1), so it is recommendable that the manufacturer focuses on this type of buildings to sell the product. The present Swiss average demand is not suitable for the SOFC since it would use a lot the additional back-up energy, thus contributing to more emissions and not producing electricity. For a passive building, or a house with very low energetic demand, the analyzed type of SOFC would be over-dimensioned and not work efficiently.

List of Abbreviations

IPCC   International Panel on Climate Change
PEMFC   Polymer Electrolyte Membrane Fuel Cell
LCA   Life-cycle Assessment
LCI   Life-cycle Inventory
SIA   Schweizerischer Ingenieur- und Architektenverein
SOFC   Solid Oxide Fuel Cell

References

[1] Levine, M., Ürge-Vorsatz, D., Blok, K., Geng, L., Harvey, D., Lang, S., Levernore, G., Mongarneli Mehlwana, A., Mirasgedis, S., Novikova, A., Rilling, J., Yoshino, H.: Residential and commercial buildings. In: Climate Change 2007: Mitigation. Contribution of Working Group III to the Fourth Assessment Report of the Intergovernmental Panel on Climate Change, vol. 6, Cambridge University Press, Cambridge (2007)

[2] Pehnt, M., Vielstich, W., Lamm, A., Gasteiger, H.A.: Handbook of fuel cells: fundamentals technology and applications. Fuel cell technology and applications. Pt. 2, vol. 4. Wiley, Chichester (2003)

[3] Hart, D., Bauen, A.: Further Assessment of the Environmental Characteristics of Fuel Cells and Competing Technologies. Department of Trade and Industry, London (1997)

[4] Pehnt, M.: Assessing future energy and transport systems: the case of fuel cells. The International Journal of Life Cycle Assessment 8(6), 365–378 (2003)

[5] Olausson, P.: Life-cycle Assessment of an SOFC/GT Process. Department of Heat and Power Engineering, Lund Institute of Technology, Lund (1999)

[6] Pehnt, M.: Life Cycle Assessment of Fuel Cell Systems. Erscheint in Fuel Cell Handbook (Fuel Cell Technology and Applications) 3 (2002)

[7] Zapp, P.: Ganzheitliche Material- und Energieflußanalyze von SOFC Hochtemperaturbrennstoffzellen (Lifecycle Material and Energy Analysis of SOFC High Temperature Fuel Cells). Universität-Gesamthochschule Essen, Essen (1997)

[8] De Paepe, M., D'Herdt, P., Mertens, D.: Micro-CHP systems for residential applications. Energy Conversion and Management 47(18-19), 3435–3446 (2006)

[9] Bennett, S.J.: Review of Existing Life Cycle Assessment Studies of Microgeneration Technologies. Energy Saving Trust, United Kingdom (2007)

[10] Dorer, V., Weber, A.: Energy and CO2 emissions performance assessment of residential micro-cogeneration systems with dynamic whole-building simulation programs. Energy Conversion and Management 50(3), 648–657 (2009)

[11] Dorer, V., Weber, R., Weber, A.: Performance assessment of fuel cell micro-cogeneration systems for residential buildings. Energy and Buildings 37(11), 1132–1146 (2005)

[12] Alanne, K., Söderholm, N., Sirén, K., Beausoleil-Morrison, I.: Techno-economic assessment and optimization of Stirling engine micro-cogeneration systems in residential buildings. Energy Conversion and Management 51(12), 2635–2646 (2010)

[13] Roselli, C., Sasso, M., Sibilio, S., Tzscheutschler, P.: Experimental analysis of microcogenerators based on different prime movers. Energy and Buildings 43(4), 796–804 (2011)

[14] Dorer, V., Weber, A.: CANMET Energy Technology Centre (Canada), Performance assessment of residential cogeneration systems in Switzerland a report of Subtask C of FC+COGEN-SIM. The Simulation of Building-Integrated Fuel Cell and Other Cogeneration Systems: Annex 42 of the International Energy Agency, Energy Conservation in Buildings and Community Systems Programme. Natural Resources Canada, Ottawa (2008)

Chapter 62
A Sustainable Energy Saving Method for Hotels by Green Hotel Deals

Hamid Abdi, Doug Creighton, and Saeid Nahavandi

Center for Intelligent Systems Research, Deakin University,
Geelong Waurn Ponds Campus, Victoria 3217, Australia
`hamid.abdi@deakin.edu.au`

Abstract. The hospitality industry is the largest business worldwide with an increasing market. The energy used in this industry produces a considerable amount of greenhouse gas emissions. Further improvements are required to address climate change requirements despite various existing technologies for saving energy and water in hotels. Providing more energy saving options can encourage hoteliers to perform further suitability measures. There is a special requirement of methods for energy saving that suits for small and medium size hotels. This paper introduces a sustainable energy saving method for hotels that uses active engagement of the guests in energy saving process. The proposed method provides the guests a direct benefit from the saved energy. The paper presents a novel concept of green hotel deals is provide an initial design for the system. The sustainability analysis of the proposed method shows that the method is able to benefit guests, hoteliers, governments and importantly the environment.

Keywords: Energy saving, Hospitality industry, Energy efficiency, Greenhouse gas emission.

1 Introduction

Climate change is a life-threatening problem specially for next generations. There have been serious impacts of climate change over the past three decades with an increasing environmental pollution as well as extreme weather conditions around the world. The increase of use of fossil fuel for generating electricity and the nature of human life in the 21st century are the main reasons of world pollution and greenhouse gas emissions resulting to the climate change. There are many discussions among researchers, policy makers, media, and individuals to recognize, minimize and/or tackle the impacts of climate change. There has been also a lot of work toward efficiency and saving of energy and water in various applications to reduce greenhouse gas emission as well is pollution to the water and earth. However, much more research and innovative technologies are required to improve the performance of these method towards climate change.

A. Håkansson et al. (Eds.): *Sustainability in Energy and Buildings*, SIST 22, pp. 669–677.
DOI: 10.1007/978-3-642-36645-1_62   © Springer-Verlag Berlin Heidelberg 2013

Similar to other countries, in Australia, large-scale industries such as mining, oil, vehicle and coal power plants are the main greenhouse gas emitters [1]. There are other small and medium enterprises (SMEs) in Australia such as hospitality accommodations with a considerable carbon footprint. These SMEs have not been received enough attention [2] and the business owners are not fully aware of the optimal ways of energy saving as they do not have enough time and financial or scientific supports . The new carbon tax legislation in Australia comes into effect from 2112 will result to further increase of energy price. The legislation provides support for the SMEs to encourage them to move toward eco-sustainability and reduce their energy bill as well as minimize their carbon footprint.

Environmental management systems (EMS) [3] consists of method for controlling the impact of organizations and communities on environment. ISO 14000 is well accepted EMS system consists of a series of international standards and guidelines [4] designed for energy efficiency of the organizations. It is obvious that improving the energy efficiency in different applications and the implementation of energy management systems are two of the fastest ways to reduce greenhouse gas emission with instantaneous and short term results but they are normally costly [5-7] and therefore not sustainable. Most of the existing methods fail to be sustainable with this respect that has a very low acceptance rate into energy saving in small and medium hotels [6]. This paper present a novel method that has sustainability property and therefore it is expected to receive more attention by small and medium hotels.

From the previous studies on energy efficiency and saving for hotels, three categories of studies are observed. The first category includes the statistical or psychological analysis of energy and water use and waste product in this industries [8-14]. The second category discusses different technologies for energy saving for hotels [6-8, 11, 15-18]. The third category address some case studies on energy saving in different hotels around the world [5-9, 13, 17, 19].

None of the aforementioned literature has addressed the economic sustainability of the proposed systems. For example while the second category provides a practical solution for energy saving in hotels, they do not encourage guest for further energy saving. Such problem is observed in the work of Simmons et al [7] where a central power unit system could control the lighting and air conditioning based on the human occupancy. This system is passive as it does not engage the guest in energy saving process and when he is at the room, there is no motivation for the guest to save energy. Tsagarakis et al in [20] addressed tourists attitudes for selecting accommodation and indicated that 86% of the responders would prefer the eco-sustainable hotels. Similar study has been performed by Millar et al. in [21] but the human behavior in energy saving has not been addressed. They indicate that guests would prefer green hotels but the indication of sustainability of the guest energy consumption behavior is not clear. Salon et al in [5] has concluded that 'Even though much work has been carried out in the field of green costumer behavior, the main characteristic of green customer behavior is not clear?'. Our hypothesis is that the existing systems do not provide a systematic control over the guest's behavior. Therefore, in this paper we presents a system and an appropriate control mechanism that has an impact on guest's hotel costs.

This paper is organized in six sections as follows. Section 1 presents an introduction and literature review. In section 2. The hospitality industry of Australia is studied and the energy and water consumption and the produced greenhouse gas emissions of the industry are shown. In section 3, a novel concept for active engagement of hotel guests in energy saving in the hotels is proposed and an system for the implementation of the proposed method is designed in Section 4. Then in section 5, the sustainability of the proposed method is discussed and finally, in section 6 the concluding remarks are presented.

2 Energy Use in Hospitality Industry

There exists previous studies for energy consumption in different hotels in various places around the world including China [19], Hong Kong [11], Greece [17], Germany and Estonia [5], New Zealand [10], North American [12], and Turkey [22]. But there is no specific scientific paper on Australian hotels except the work in [10] that mainly addresses the New Zealand hospitality industry and has a minor overlap with Australia. The increasing price of energy in Australia and its impact on the hospitality industry is the key motivation for this study; however, the proposed method and system is applicable for any hotel anywhere.

2.1 Australia's Tourist Accommodation Industry

Australian Bureau of Statistics (ABS) in the 2011 tourist accommodation report [23] shows that there are 4250 hotels, motels, guesthouses and service apartments with 15 or more rooms/units in Australia. The report shows 2128.8 million dollars taking from hospitality accommodation for the April, May and June with 6.0 percent growth in compare to 2010. This survey included 226,582 rooms with 637,298 beds and shows average occupancy rate of 61.0%-65.5% for April to June of 2011. From the literature, the hotels' energy cost varies between 4% to 10% of a property's revenue we assume an average of 7% in this study. Therefore, the annual energy cost of hotels in Australia is estimated by $596.1 million in 2011. In Australia because of the new carbon tax legislation and the growing tourist accommodation industry, the energy consumption value of this industry will be increased if appropriate energy saving measures are not implemented.

2.2 Energy Saving Barriers in Tourist Accommodation Industry

The ABS 2011 report indicated 109,246 of full time employees in tourist accommodation sector. This indicates an average number of 25.70 employees per each accommodation showing most accommodation can be considered as SMEs. Tourist accommodation SMEs have limited time, financial and information resources and they struggle to survive the business therefore they are not able to do research for improving their energy efficiency. Furthermore, in this industry, the subject of green hotels are new and most of the existing methods for energy saving are borrowed from other industries therefore they are not optimized for this industry.

2.3 Opportunities in Energy Saving for Hotels

Any method to save only 10% energy in hotels has a value of 59.6 million dollars per year only in Australia. For the 226,582 number of hotel rooms, it has an average saving of $263.1 per room per year. Currently, different technologies are used for energy management and saving for the hotels including a very basic system to turn off air conditioning and lights when the guest leaves the room and there are several technologies for energy management in hotels [5, 7, 8, 15, 17, 19]. However, none of these technologies can actively engage the guests in energy saving process and therefore has no influence on the guests' energy consumption behavior.

3 Green Hotels

The green hotels introduced as hotels that are environmentally friendly [12, 21, 24]. Different measures have been proposed or used for the hotels or even for public or governmental buildings [3, 8, 9, 12, 16]. Researchers have also investigated economic sustainability of green hotels, for example Sloan et al. [5] interviewed with hotel executive managers and he has shown an increase in market share and profitability of the hotels, however the payback period of the investment toward energy saving has been long. The result is justified by guests' preferences on green hotels due to the increasing awareness of people from climate change [21].

3.1 Existing Hotel Deals

The importance of human factor in energy saving has been notices in [18], the authors developed a mobile application "EnergyWiz" to provide information channel to the users that include a comparative information between different applicants energy saving. This works focuses on psychology and information channels/feedback. Further engagement of hotel guest in energy saving process needs an innovative method about hotel and customers interactions and behaviors. Currently, the hotels' deals are a fixed value package that include the energy and accommodation. This model of the deal is not able to address the guests energy consumption behavior whether he uses a small amount of energy or bit the model provides a same bill. Such deal is not encouraging guest for energy saving and the guest energy consumption totally depends to his/her energy consumption behavior. Any work on encouraging the guest for energy saving in this model will not result to any direct benefit for the guests and therefore it is not sustainable. Large-scale hoteliers have recognized the complexities of guest behaviors, therefore they have focused on the technologies and training of their employees to save energy.

3.2 Green Hotel Deals

Sustainable energy saving in hotels can be achieved if a new model could address the guest energy consumption behavior in the hotel deals. The new model must have incentive for the guests, hoteliers and environment. Therefore, any energy saving in this structure will be sustainable because all parties will have a direct benefit from the saved energy. In this paper we introduce a novel hotel deal method and we call it "green hotel deal". In this deal, the hotel deal is divided into an accommodation expense and an energy expense where the first part is a fixed value and the second part is calculated form the guest's energy consumption. This will be based on the actual or estimated energy consumption of the guest to address the uncertainties. The main advantage of the green deals is that, if the guests saves energy, then he/she will get lower hotel costing.

3.3 Green Hotel Deals Benefits

This method can reduce the guest's hotel expenses and the energy costs of hotels. This model contributes to more green environment via energy saving and reduction in greenhouse gas emissions in hotels.

4 Implementation

The green deals requires a technology that enables measuring the guests energy consumption and then integrate the information in the hotel billing system. Such technology includes sensors, electronic hardware, a software. The hardware part includes 1) electricity meter, 2) hot water meter, and 3) a communication channel with the hotel billing system. Then the software records the measured information and calculates the energy and water bill. The recorded information includes meters readings and cleaning staffs report. The staffs input includes 1) Reporting the room power and water consumption on guest arrival and vacation and 2) recording of room cleaning works.

Such system includes an electricity meter and a hot water water meter. In our initial design, a power meter such as the one illustrated in Figure 1 can be easily used for this study. This meter has different functionality including clock and alarm function, measuring and displaying room temperature and humidity, measuring line voltage, current and power. The device can also display the total power and greenhouse emissions of the room. Each socket is rated as 10A outlet that is more than the total current use in hotel room appliances. The power meter has been priced $64.90 retail price and can be considerably cheaper for mass use.

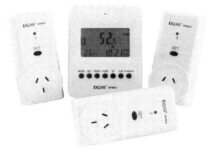

Fig. 1. A wireless power meter with three main sockets from DGSS

In addition to the power meter, a hot water meter is required for this system because the hot water consists of high energy component to heat up the water. A hot water meter with the retail price of $178 is illustrated in Figure 2. It is a rotary water meter with numerical reading and pulse output and 1 pulse per meter. This system can be simply interfaced with a the electricity meter and enables measuring of hot water consumption in hotel room. However, because this work is at the early stage we study we performm manual reading of the meters.

Fig. 2. A hot water rotary water flow measurement with pulse outputs from BMETER

The meters' measurements are recorded on the guest arrival and departure and are reported to the hotel billing system as part of room introduction and vacation procedure. It is possible to have an automated system that can communicate the meter reading with the hotel billing system by a communication channel. However, such channel requires further costing and can only be justified for large-scale hotels and not SMEs. A new software module is added to the hotel billing and management system the uses the guest's arrival and vacation dates, electricity and water meters' readings on arrival and vacation, room cleaning history, and number of clean towels. Then based on the information, it is possible to calculates the guest energy expenses and include it in his final bill.

5   Sustainability Analysis

5.1   Sustainability Chain

For the proposed green deal model in this paper, the hotel consists of four chains (Figure 3) including guest, hotel, government, and environment. The hotel and

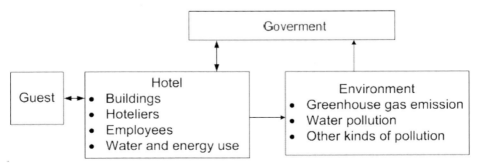

Fig. 3. Four chains for the sustainability analysis of the energy consumption in hotels, the arrows shows the bidirectional or unidirectional interaction between the chains

environment chains consist of number of subsections for example the hotel consists of the building, hotel owners, hotel employees, and energy/water use. The impact of the hotel on environment can be assessed based on produced greenhouse gas emission, water use and other kinds of pollution measures however, in this paper, we only focus on energy consumption.

It is important to show that all the parts of the sustainability chain as well as the elements of each part will have benefit from the implementation of the green hotel deals to ensure that this model will have much greater impact on the proposed energy saving in the industry. The sustainability analysis of this deal is performed here is for the Australian hospitality accommodation industry but is can be easily extended for other countries.

5.2   Sustainability Analysis for Guests

Firstly, in the green hotel deals model, the hotel deal is cheaper than existing deals because it does not include energy cost. Secondly, the guest's expense depends to his/her actual energy consumption and if the guest saves energy, then he/she will get lower hotel costing. Furthermore, the guests prefer green hotels than other hotels. These advantages for the guest provide him the incentive that guarantees the sustainability from the guest perspective.

5.3   Sustainability Analysis for Hotels

The green hotel deals model is beneficial to the hotels owners because guest prefers the green hotels therefore the hotels will have an increased market. Different study in the literature of this paper show that the green hotels means more than just a business from the guest perspective as they consider other benefit of the hotel to the environment and society. The green deal in a green hotel is much more visible than the other green aspects of hotels because it directly influences in the guest's expenses. Furthermore, the hotels will have lower energy consumption that results to lower energy bills. The costing of the metering system and development of a software module is estimated as $250 per room that is reasonable and not expensive for SME hotels in Australia with the room price between $70-$250 per night. Based on this estimated price cost for the system, the payback time will be less than a year if this system only contribute to lowering of 10% energy consumption (note that there was $263 saving per room per year). Furthermore, the existing Australian government offers several support for SMEs to help them to reduce they carbon footprint, such support schemes can help these SMEs in the implementation of the proposed system in this paper.

5.4   Sustainability Analysis for Environment

The green hotel deals model contributes to environmental sustainability by reducing the energy consumption in hotel industry. Such reduction in countrywide scale will result to considerable reduction in greenhouse emissions and water use.

5.5 Sustainability Analysis for Government

Australian government has implemented various schemes to reduce the carbon footprint in Australia. In 2012, a new carbon tax legislation has just started that is expected to have a significant impact on house holders and businesses. Implementation of the green deals model aligns with this government goal and could receive some attention. The government support could be used to minimize the initial investment required for implementation of the hardware and software for the green hotel deal that is the future direction of this research.

6 Conclusions

Small and medium hospitality accommodation has limited time, financial resources and accessibility to information to develop scientific and innovative methods to reduce their footprint on climate change. There is also a lack of specific and sustainable methods for energy saving in hospitality industry. The present paper introduced a novel hotel deal method called "green hotel deals". This model aims to actively engage hotel guests in the energy saving process. Sustainability analysis of the green hotel deals method was performed and it was shown that it has direct benefit for hoteliers, guests, environment, and government. The implementation of the proposed method was discussed and the cost of $250 per room was estimated. In the future, the authors aim to: 1) Improve the initial design of the hardware for measuring; 2) Develop a system to accurately accounts the guest energy consumption including the laundry works, and 3) Perform statistical analysis on the guest energy consumption behaviors before and after the implementation of the green deals.

References

[1] Schaper, M.T., Carlsen, J., Taylor, R.: The Role of Government in Fostering Sectoral SME Development: An Analysis of the Tourism Industry in Australia, Hong Kong and Singapore. Small and Medium Sized Enterprises in East Asia: Sectoral and Regional Dimensions, 1–395 (2008)

[2] Moser, C.: Carbon Mitigation Strategies in the Electricity Distribution Sector in Australia. Lund University

[3] Chan, W.W.: Environmental measures for hotels' environmental management systems: ISO 14001. International Journal of Contemporary Hospitality Management 21, 542–560 (2009)

[4] Tibor, T., Feldman, I.: ISO 14000 a guide to the new environmental management standards. IRWIN professional publishing, BURR RIDGE, IL, USA (1995)

[5] Tooman, H., Sloan, P., Legrand, W., Fendt, J.: Best practices in sustainability: German and Estonian hotels. Advances in Hospitality and Leisure 5, 89–107 (2009)

[6] Bohdanowicz, P., Churie-Kallhauge, A., Martinac, I.: Energy-efficiency and conservation in hotels–towards sustainable tourism. Simpósio Internacional em Arquitetura da Ásia e Pacífico, Havaí, pp. 1–12 (April 2001)

[7] Simmons, M.L., Gibino, D.J.: Energy-saving occupancy-controlled heating ventilating and air-conditioning systems for timing and cycling energy within different rooms of buildings having central power units ed: Google Patents (2002)

[8] Lee, W.S.: Benchmarking the energy efficiency of government buildings with data envelopment analysis. Energy and Buildings 40, 891–895 (2008)

[9] Beccali, M., La Gennusa, M., Lo Coco, L., Rizzo, G.: An empirical approach for ranking environmental and energy saving measures in the hotel sector. Renewable Energy 34, 82–90 (2009)

[10] Becken, S., Frampton, C., Simmons, D.: Energy consumption patterns in the accommodation sector–the New Zealand case. Ecological Economics 39, 371–386 (2001)

[11] Shiming, D., Burnett, J.: Energy use and management in hotels in Hong Kong. International Journal of Hospitality Management 21, 371–380 (2002)

[12] Rahman, I., Reynolds, D., Svaren, S.: How "green" are North American hotels? An exploration of low-cost adoption practices. International Journal of Hospitality Management, 1–8 (2011)

[13] Li, W., Liu, Y.: Present energy consumption and energy saving analysis for star-rated hotels in Xi'an. Architecture, 1–5 (2010)

[14] Deng, S.M., Burnett, J.: Water use in hotels in Hong Kong. International Journal of Hospitality Management 21, 57–66 (2002)

[15] Chun-feng, L.: Analysis of the Approaches of Hotel Energy Saving. Journal of Tourism, College of Zhejiang 1, 1–4 (2009)

[16] Doukas, H., Nychtis, C., Psarras, J.: Assessing energy-saving measures in buildings through an intelligent decision support model. Building and Environment 44, 290–298 (2009)

[17] Zografakis, N., Gillas, K., Pollaki, A., Profylienou, M., Bounialetou, F., Tsagarakis, K.P.: Assessment of practices and technologies of energy saving and renewable energy sources in hotels in Crete. Renewable Energy, 1323–1328 (2010)

[18] Petkov, P., Köbler, F., Foth, M., Medland, R., Krcmar, H.: Engaging energy saving through motivation-specific social comparison, 1945–1950 (2011)

[19] Li, R.: Energy Saving Strategy of China's Green Hotel. Advances in Education and Management, 423–428 (2011)

[20] Tsagarakis, K.P., Bounialetou, F., Gillas, K., Profylienou, M., Pollaki, A., Zografakis, N.: Tourists' attitudes for selecting accommodation with investments in renewable energy and energy saving systems. Renewable and Sustainable Energy Reviews 15, 1335–1342 (2011)

[21] Millar, M., Baloglu, S.: Hotel guests' preferences for green hotel attributes. Hospitality Management, 1–11 (2008)

[22] Erdogan, N., Baris, E.: Environmental protection programs and conservation practices of hotels in Ankara, Turkey. Tourism Management 28, 604–614 (2007)

[23] ABS, Tourist Accommodation, Australia Australian Bureau of Statistics (June 2011)

[24] Chan, W.W., Ho, K.: Hotels' environmental management systems (ISO 14001): creative financing strategy. International Journal of Contemporary Hospitality Management 18, 302–316 (2006)

Chapter 63
The Role of Building Energy and Environmental Assessment in Facilitating Office Building Energy-Efficiency

Emeka E. Osaji, Subashini Suresh, and Ezekiel Chinyio

School of Technology, University of Wolverhampton, WV1 1LY, United Kingdom
e.osaji@wlv.ac.uk, arcosaji@yahoo.com

Abstract. Ten years ago, the primary author developed the Building Energy-Efficient Hive (BEEHive) concept in order to demonstrate in theory that environmental design – which is aimed at addressing environmental parameters – can support the design and operation of energy-efficient office buildings. This was a result of his analysis of the spheroid form's efficiency in nature, and his development of a spheroid-like energy-efficient office built form. The BEEHive incorporates environmental design principles such as: site considerations; built form; ventilation strategy; daylighting strategy; and services strategy. Furthermore, several notable environmental design advocates and practitioners have made significant contributions in order to improve building performance. However, in practice environmental design has had limited success in the attainment of balance and optimisation in all aspects of energy use; hence there is typically a gap between predicted and actual office building energy use. The primary author's previous study established the impacts of contributory factors in the gap between predicted and actual office building energy use. It has contributed to this current study, which is also a part of the primary author's doctor of philosophy (Ph.D) research, and it has established the role of a key contributory factor, that is, the role of building energy and environmental assessment in facilitating office building energy-efficiency. It involved a combination of literature reviews, multiple case study research and comparative studies in order to build theory. It also established the methods and tool to be used in the primary author's Ph.D research for multiple case studies and simulation studies of office building energy-efficiency. Analysis of the literature revealed that the role of building energy and environmental assessment involves assessment of the impacts of environmental design principles, and the impacts of factors that contribute to office building energy use gap decreases, for example: solar gain minimisation orientations; energy-efficient strategies for built forms, ventilation, lighting and services; and decreases in hours of operation and occupancy. Its role also involves assessment of the impacts of factors that contribute to office building energy use gap increases, for example: weather variation and microclimates; and increases in hours of operation and occupancy. There are three key types of building energy and environmental assessment, and these are: building energy use audit method; building energy simulation analysis

A. Håkansson et al. (Eds.): *Sustainability in Energy and Buildings*, SIST 22, pp. 679–704.
DOI: 10.1007/978-3-642-36645-1_63   © Springer-Verlag Berlin Heidelberg 2013

method and tools; and building energy and environmental assessment rating method and tools. Their respective roles include: tracking building energy use over time; predicting future building energy use within multiple environmental design scenarios and parameters; and assessing, rating, and certifying building energy and environmental efficiency. However, limitations of building energy and environmental assessment, and impacts of factors that contribute to office building energy use gap increases need to be addressed in order to achieve: optimum building energy use assessments and predictions; optimum environmental design principles; and building energy use gap decreases for improved energy performance. This study has contributed to ideas for the development of a Building Management System (BMS-Optimum) for Bridging the Gap, which is comprised of optimum conditions and considerations such as: optimum environmental design principles; optimum weather and microclimate considerations; accessibility to reliable office building energy use data; optimum building energy and environmental assessment; optimum hours of operation; and optimum level and nature of occupancy. Future work will include further development of BMS-Optimum, using methods such as: multiple case study research supported by building energy use audits, observations, questionnaire surveys, interviews, benchmarking and comparative studies; building energy simulations within multiple scenarios, parameters and variables, and supported by benchmarking and comparative studies; and peer reviews and focus group sessions. These will also help establish and validate a Framework for Improved Environmental Design and Energy Performance (FEDEP).

Keywords: Energy Assessment, Energy-Efficiency, Environmental Assessment, Environmental Design, Office Building, Simulation.

1 Introduction

Ten years ago, the primary author developed the Building Energy-Efficient Hive (BEEHive) concept in order to demonstrate in theory that environmental design – which is a philosophy aimed at addressing environmental parameters – can support energy-efficient office building design and operation (Osaji, 2002). This was a result of his analysis of the spheroid form's efficiency in nature, and his development of a spheroid-like energy-efficient office built form that encloses and shades the most volume of office space with the least surface area possible (Osaji, 2002). The BEEHive incorporates environmental design principles such as: site considerations (location and weather, microclimate, site layout and orientation); built form (shape, thermal response, insulation and windows/glazing); ventilation strategy; daylighting strategy; and services strategy (plants and controls, fuels and metering) (Osaji, 2002). Furthermore, several notable environmental design advocates and practitioners have made significant contributions in order to improve office building energy performance, and these include: Arup (Cramer, 2011; and Foster and Partners, 2012); Norman Foster (Foster and Partners, 2012); HOK (Beautyman, 2006; and Cramer, 2011); Ken Yeang (Greener Buildings, 2008); and several others. However, many modern buildings apparently do not perform well in terms of their energy use (Leaman et al., 2010). For instance, approximately 18 percent of United Kingdom

(UK) CO_2 emissions are attributable to energy use in the non-domestic building sector, which includes offices (UK-GBC, 2011). Office buildings contribute significantly to energy use and CO_2 emissions, and to the total energy used by buildings globally (Action Energy, 2003; Mortimer et al., 2000; and Wade et al., 2003). For instance, UK office buildings are significant contributors to energy use and CO_2 emissions whereby combined heating and cooling loads contribute about 61 percent to total annual energy use (Mortimer et al., 2000; and Wade et al., 2003). United States of America (US) office buildings use the most energy of all US building types. The 2003 Commercial Buildings Energy Consumption Survey (CBECS) shows that US office building energy use represents 19 percent of all US commercial energy use (EIA, 2011). The expenditure associated with such energy use exceeds 15 billion US dollars (EIA, 2011). This is the highest energy expenditure among all US commercial building types, and constitutes 23 percent of their energy expenditure (EIA, 2011). Unfortunately, there is typically a gap between predicted and actual building energy use (Azar and Menassa, 2010), and such significant gaps have been reported to occur (Bordass et al., 2004). These include gaps between predicted and actual office building energy use as described in: Diamond et al. (2006); EERE (2004c); EERE (2006c); Tuffrey (2005); and Wellcome Trust (2011). According to Demanuele et al. (2010), the significant gap that exists between the predicted and actual energy use of buildings is mainly as a result of a lack of understanding of the factors that affect building energy use.

Therefore, the primary author's previous study sought to address this problem by establishing the impacts of contributory factors in the gap between predicted and actual office building energy use. It revealed two types of gaps, and these are a gap increase and a gap decrease, which are among the impacts attributable to contributory factors in the gap between predicted and actual office building energy use such as: the nature of environmental design measures implemented; weather variation and microclimates; unavailability of reliable building energy use data; limitations of building energy simulation software; level of hours of operation; and level and nature of occupancy. Amongst these, the key contributors to gap increases are increases in hours of operation and occupancy, and weather variation and microclimates. Their respective major impacts are discrepancy between predicted and actual hours of operation and increased energy use, increased heat output and uncertainties, and variable heating and cooling requirements. The key contributors to gap decreases are environmental design measures such as the use of: natural ventilation strategies; daylighting strategies; solar photovoltaic systems; and spheroid-like built forms. Their respective major impacts are: the production of more energy than an office building uses; and energy uses that are below, for instance, Energy Consumption Guide 19 typical and good practice energy use for office type 4, and relevant ASHRAE (American Society of Heating, Refrigerating and Air Conditioning Engineers) standards.

The primary author's previous study has contributed to this current study, which is also a part of the primary author's doctor of philosophy (Ph.D) research, and it has established the role of a key contributory factor, that is, the role of building energy and environmental assessment in facilitating office building energy-efficiency.

2 Aim and Methods of This Study

The aim of this study – which is a part of the primary author's Ph.D research – is to establish the role of building energy and environmental assessment in facilitating office building energy-efficiency. This study has been fulfilled using a combination of literature reviews, multiple case study research and comparative studies in order to build theory. This involved a desk based study in order to establish: the role of building energy and environmental assessment in facilitating office building energy-efficiency; the methods and tool to be used in the primary author's Ph.D research for multiple case studies and simulation studies of office building energy efficiency; and a building management system (BMS-Optimum) for bridging the gap between predicted and actual office building energy use. The following methods were used:

- Data collection from literature of the types and role of various building energy and environmental assessment methods, building energy simulation analysis tools, and building energy and environmental assessment rating tools in facilitating office building energy-efficiency.
- Comparative studies of the number of building energy simulation analysis tools by country in order to establish a case for either proliferation (decentralisation) or non-proliferation (centralisation) of these tools.
- Establishment of Autodesk Ecotect Analysis – out of a choice of twenty key building energy simulation analysis tools – as the preferred choice for the primary author's Ph.D simulation study because of its importance in simulating building energy performance within environmental contexts, including: whole-building energy analysis; thermal performance; solar radiation; daylighting; and shadows and reflections.
- Comparative studies of building energy and environmental assessment rating tools by country in order to establish the most frequently used, that is, the key building energy and environmental assessment rating tools.
- Multiple case study research of BREEAM and LEED certified office buildings in order to establish the role of BREEAM and LEED as key building energy and environmental assessment rating tools.
- Comparative studies of the key building energy and environmental assessment rating tools in order to establish similarities and dissimilarities.
- Drawing conclusions from the deduced role of building energy and environmental assessment methods and tools, and the ways they can be enhanced in order to overcome limitations and facilitate office building energy-efficiency.

3 Findings

This study has established the role of building energy and environmental assessment. Its role includes the assessment of the impacts of environmental design principles, and the impacts of factors that contribute to office building energy use gap decreases, for example: solar gain minimisation orientations; energy-efficient strategies for built forms, ventilation, lighting and services; and decreases in hours of operation and occupancy. Its role also includes the assessment of the impacts of factors that contribute to office building energy use gap increases, for example: weather variation and microclimates; and increases in hours of operation and occupancy. There are three key types of building energy and environmental assessment, and these are: building energy use audit method; building energy simulation analysis method and tools; and building energy and environmental assessment rating method and tools. Their respective roles include: tracking building energy use over time; predicting future building energy use within multiple environmental design scenarios and parameters; and assessing, rating, and certifying building energy and environmental efficiency. However, limitations of building energy and environmental assessment, and impacts of factors that contribute to office building energy use gap increases need to be addressed. This is important in order to achieve: optimum building energy use assessments and predictions; optimum environmental design principles; and building energy use gap decreases for improved energy performance.

3.1 The Role of Building Energy and Environmental Assessment Methods

There are three key types of building energy and environmental assessment methods for: tracking building energy use over time; establishing building energy use baseline data; predicting future building energy use within multiple environmental design scenarios and parameters; undertaking comparative studies and benchmarking of building energy performance results; and assessing, rating, and certifying building energy and environmental efficiency. These three key types of building energy and environmental assessment methods are:

- Building energy use audit (ASHRAE, 2004; Department for Communities and Local Government, 2007; and Opitz, 2011).
- Building energy simulation analysis (Crawley et al., 2008; NREL, 2010; Osaji et al., 2010; Paradis, 2010; and US DOE, 2011).
- Building energy and environmental assessment rating (Association HQE, 2011; BRE Global, 2010; CaGBC, 2011; GBCA, 2011; Roderick et al., 2009; and USGBC, 2011).

3.1.1 The Role of Building Energy Use Audit

Building energy use audit involves using an energy audit system for analysis of historical actual energy use that takes into consideration: operational modifications; climatic extremities; and other factors that affect building energy use and expenditure (ASHRAE, 2004; Department for Communities and Local Government, 2007; and Opitz, 2011). According to Opitz (2011), building energy use audit is important for the following reasons:

- The need to reduce energy use and expenditure based on improved building energy performance.
- The potential for significant sustainability benefits based on reduced building energy use, reduced expenditure, and reduced carbon emissions.

- The need to benchmark building energy performance, and to determine if it is not performing as well as similar building types and similar building energy use categories.
- The need to determine strategies that should be employed in order to improve building energy performance.

In describing appropriate procedures for building energy use audits, ASHRAE (2004) states that these are important for ensuring effective sharing of building energy use and performance data. ASHRAE (2004) describes different levels of building energy use audits that are organised into several categories, and these are:

- Preliminary energy use analysis.
- Level I analysis: walk-through analysis.
- Level II analysis: energy survey and analysis.
- Level III analysis: detailed analysis of capital-intensive modifications.

Opitz (2011) also describes different levels of building energy use audits, and these are:

- Level I (Basic): this is also known as the One-Day/Walk-Through Audit. It involves a quick review of a building and its strategies, and analysis of its energy use and expenditure data in order to determine a preliminary estimate of its energy performance.
- Level II (Intermediate): this involves an in-depth review of a building and its strategies, analysis of its energy use and expenditure data, and system performance testing. A level II (intermediate) building energy use audit provides a description of the nature of building energy use, and its options for potential energy savings. Furthermore, it accounts for occupancy related parameters such as staff occupancy and operational hours, and their impact on energy use and potential savings.
- Level III (Advanced): this is also known as the Investment-Grade Audit. It involves an in-depth review of the potential of any future large capital projects based on the lessons learned from previous level I and level II audits. Furthermore, it involves in-depth data collection from equipment, monitoring of energy use, and energy expenditure performance analysis.

Therefore, the role of the building energy use audit method is to analyse historical actual building energy use. However, the building energy simulation analysis method fulfills a different role.

3.1.2 The Role of Building Energy Simulation Analysis
Building energy simulation analysis involves the modelling of building energy use – also known as calibrated simulation – and the use of complex computer software to predict future building energy use (Crawley et al., 2008; NREL, 2010; Osaji et al., 2010; Paradis, 2010; and US DOE, 2011).

According to Crawley et al. (2008), NREL (2010), Paradis (2010) and US DOE (2011), building energy simulation analysis is important for reasons that support improved building energy performance such as:

- The need to determine appropriate specifications for HVAC equipment and other key equipment.
- The need to predict annual building energy use in order to provide insights into building energy performance and potential energy expenditure savings.
- The need to support reduction in building energy demand.
- The need to support effective decision-making during the whole-life of building energy performance, including from building design to building operations.
- The need to determine the way that building control systems behave and impact on building energy use.
- The need to evaluate environmental design and its potential for achieving building energy-efficiency.

3.1.3 The Role of Building Energy and Environmental Assessment Rating

Building energy and environmental assessment rating facilitates the evaluation of building specification, design, construction and use by using an Environmental Performance Indicator for the award of performance score ratings, and for the certification of building energy and environmental efficiency (Association HQE, 2011; BRE Global, 2010; CaGBC, 2011; GBCA, 2011; Roderick et al., 2009; and USGBC, 2011).

3.2 The Role of Building Energy and Environmental Assessment Tools

There are two key types of building energy and environmental assessment tools, and these are:

- Building energy simulation analysis tools that are used for predicting future building energy use within multiple environmental design scenarios and parameters, as well as undertaking comparative studies and benchmarking of building energy performance results (Crawley et al., 2008; NREL, 2010; Osaji and Price, 2010; Paradis, 2010; and US DOE, 2011).
- Building energy and environmental assessment rating tools that are used for assessing, rating, and certifying building energy and environmental efficiency (Association HQE, 2011; BRE Global, 2010; CaGBC, 2011; GBCA, 2011; Roderick et al., 2009; and USGBC, 2011).

3.2.1 The Role of Building Energy Simulation Analysis Tools

According to Howell and Batcheler (2005), a key factor that facilitates the integration of environmental design principles into building design is technology especially the use of building simulation analysis tools. US DOE, (2011) describes building energy simulation analysis tools as databases, spreadsheets, component and systems analyses, and whole-building energy performance simulation tools that support the evaluation of energy efficiency, renewable energy, and sustainability in buildings.

Appropriate software is important for achieving successful building energy simulation analysis (Paradis, 2010 and US DOE, 2011). According to US DOE (2011), an estimated 392 building energy simulation analysis tools that are used globally are described in the U.S. Department of Energy (DOE) Building Energy Software Tools Directory. These 392 building energy simulation analysis programs are organised according to the following categories: subject; alphabetically; platform; and country (refer to Table 1). Table 1 is based on the US DOE (2011) Building Energy Software Tools Directory, and it shows that there are 392 building energy simulation analysis tools by 28 countries. The US has a 59.43 percent share with 233 building energy simulation analysis tools, which is over six times more than the UK's 9.43 percent share with 37 building energy simulation analysis tools (refer to Table 1). This appears to indicate that the US is more proactive in the development of building energy simulation analysis tools when compared to at least 27 other countries (refer to Table 1).

Table 1. Number of Building Energy Simulation Analysis Tools by Country (Source: US DOE, 2011)

S/N	Country	Number of Building Energy Simulation Analysis Tools
1	Australia	10
2	Austria	1
3	Belarus	1
4	Belgium	3
5	Brazil	1
6	Canada	20
7	Chile	1
8	China	2
9	Czech Republic	1
10	Denmark	4
11	Finland	3
12	France	7
13	Germany	18
14	India	1
15	Ireland	1
16	Israel	2
17	Italy	2
18	Japan	2
19	Netherlands	4
20	New Zealand	2
21	Portugal	1
22	Russia	2
23	South Africa	2
24	Spain	4
25	Sweden	14
26	Switzerland	13
27	United Kingdom (UK)	37
28	United States of America (US)	233
Total	28	392

There appears to be somewhat of a proliferation of building energy simulation analysis tools, and this is based on the occurrence of 392 building energy simulation analysis tools by 28 countries (refer to Table 1). However, this presents an opportunity for the adoption of either of two alternative approaches, and these are:

- A proliferation (decentralisation) approach whereby the development of more building energy simulation analysis tools from every country is encouraged.
- A non-proliferation (centralisation) approach whereby exemplar building energy simulation analysis tools are encouraged to integrate in order to achieve robustness and facilitate greater success.

A starting point for the adoption of a non-proliferation (centralisation) approach is the establishment of the key building energy simulation analysis tools. For instance, 20 of the 392 building energy simulation analysis tools are described by Crawley et al. (2008) as major building energy simulation analysis tools, and these are: BLAST; Bsim; DeST; DOE-2.1E; Autodesk Ecotect Analysis; Ener-Win; Energy Express; Energy-10; EnergyPlus; eQUEST; ESP-r; IDA ICE; IES <VE>; HAP; HEED; PowerDomus; SUNREL; Tas; TRACE; and TRNSYS. One of these 20 major building energy simulation analysis tools, that is, Autodesk Ecotect Analysis is described by Autodesk (2012) as a key tool for simulating building energy performance within environmental contexts. Furthermore, Autodesk Ecotect Analysis is described as a major environmental design tool and building energy simulation analysis tool (Autodesk, 2012; Crawley et al., 2008; and US DOE (2011). Therefore, Autodesk Ecotect Analysis will be used in the primary author's Ph.D research for multiple case studies and simulation studies of office building energy-efficiency. According to Autodesk (2012), Autodesk Ecotect Analysis fulfils the following role:

- Whole-building energy analysis: it supports the calculation of total annual, monthly, daily, and hourly energy use and carbon emissions of purpose-built models using a global weather database.
- Thermal performance: it supports the calculation of heating and cooling loads for purpose-built models, and the analysis of the impacts of occupancy, equipment, internal gains, and infiltration.
- Water usage and cost evaluation: it supports the estimation of building internal and external water use.
- Solar radiation: it supports the visualisation of incident solar radiation on windows/glazing and surfaces.
- Daylighting: it supports the calculation of daylight factors and illuminance levels at any point in purpose-built models.
- Shadows and reflections: it supports the display of the sun's position and path relative to a purpose-built model at any date, time, and location.

As stated earlier, there are two key types of building energy and environmental assessment tools, and these are building energy simulation analysis tools, and building energy and environmental assessment rating tools.

3.2.2 The Role of Building Energy and Environmental Assessment Rating Tools

Building energy and environmental assessment rating tools facilitate the evaluation of building specification, design, construction and use by using an Environmental Performance Indicator for the award of performance score ratings, and for the certification of building energy and environmental efficiency (Association HQE, 2011; BRE Global, 2010; CaGBC, 2011; GBCA, 2011; Roderick et al., 2009; and USGBC, 2011).

This study has reviewed building energy and environmental assessment rating tools by 29 countries (refer to Table 2). Table 2 shows that some of these building energy and environmental assessment rating tools are used by multiple countries. Therefore, this study will focus on the role of five key building energy assessment rating tools based on their importance (Dixon et al., 2007; Madew, 2011; and Reed et al., 2009), and these are:

- BREEAM (Building Research Establishment Environmental Assessment Method) (BRE Global, 2010; Madew, 2011; and Roderick et al., 2009).
- USGBC LEED (U.S. Green Building Council Leadership in Energy and Environmental Design) (Roderick et al., 2009; and USGBC, 2011).
- CaGBC LEED (Canada Green Building Council Leadership in Energy and Environmental Design) (CaGBC, 2011).
- GBCA (Green Building Council Australia) Green Star (GBCA, 2011; and Roderick et al., 2009).
- HQE (High Quality Environmental) Standard (Association HQE, 2011).

3.2.2.1 The Role of BREEAM

BRE Global (2010) and Madew (2011) describe BREEAM as the most widely used energy and environmental assessment rating tool for the measurement of the sustainability of buildings. This is evident in the 200,000 buildings that have certified BREEAM assessment ratings, and the over one million buildings registered for BREEAM assessment (BRE Global, 2010). BREEAM sets the standard for sustainable building design best practice, and is arguably, the de facto measure for building environmental performance description (BRE Global, 2010). BREEAM evaluates the sustainability of building specification, design, construction and use by awarding performance score ratings using an Environmental Performance Indicator (EPI), which represents recognised measures of performance that are set against established benchmarks (BRE Global, 2010).

63 The Role of Building Energy and Environmental Assessment 689

Table 2. Building Energy and Environmental Assessment Rating Tools by Country

S/N	Country	Building Energy Assessment Rating Tool	Data Source and Reliability
1	Australia	NABERS (National Australian Built Environment Rating System); Green Star	GBCA (2011); & NABERS (2010)
2	Brazil	AQUA; LEED	Gomes et al. (2008)
3	Canada	BREEAM; Green Globes System; LEED	CaGBC (2011); Dixon et al. (2007); & Green Globes (2011)
4	China	GBAS (Green Building Assessment System)	Borong et al. (2007)
5	Czech Republic	SBToolCZ (Czech Sustainable Building Tool)	Hajek & Tywoniak (2008)
6	Finland	PromisE	DESA (2010)
7	France	HQE (High Quality Environmental Standard)	Association HQE (2011)
8	Germany	DGNB (Deutsche Gesellschaft für Nachhaltiges Bauen); CEPHEUS (Cost Efficient Passive Houses as EUropean Standards)	GeSBC (2012); & Passive House Institute (2011)
9	Hong Kong	HK BEAM (Building Environmental Assessment Method)	BEAM Society (2011)
10	India	GRIHA (Green Rating for Integrated Habitat Assessment); IGBC Green SEZ Rating System; LEED-INDIA for New Construction	IGBC (2008); & GRIHA (2011)
11	Indonesia	Greenship Existing Building; Greenship New Building	GBC Indonesia (2010)
12	Italy	LEED Italy; Protocollo Itaca	GBC Italy (2011); & ITACA (2011)
13	Japan	CASBEE (Comprehensive Assessment System for Built Environment Efficiency)	JSBC (2006)
14	Jordan	EDAMA (an Arabic word meaning 'sustainability')	EDAMA (2009)
15	Korea	GBCS (Green Building Certificate System)	KGBC (2000)
16	Malaysia	GBI (Green Building Index)	GBI (2011)
17	Mexico	LEED Mexico	Bondareva (2005)
18	Netherlands	BREEAM-NL	DGBC (2011)
19	New Zealand	Green Star NZ; Homestar	NZGBC (2008)
20	Philippines	BERDE (Building for Ecologically Responsive Design Excellence)	PHILGBC (2010); & WGBC (2011)
21	Portugal	LiderA - Sustainable Assessment System	LiderA (2012)
22	Singapore	BCA (Building & Construction Authority) Green Mark Scheme	Building & Construction Authority (2011)
23	South Africa	Green Star SA Rating Tools, for example, Green Star SA – Office v1	GBCSA (2012)
24	Spain	LEED; VERDE Certificate	GBCe (2012a); GBCe (2012b); Hui (2010)
25	Switzerland	DGNB; MINERGIE	GeSBC (2012); & MINERGIE (2012)
26	Taiwan	EEWH (Ecology, Energy Saving, Waste Reduction & Health); Green Building Label	Madew (2011); Southern Taiwan Science Park Administration (2012)
27	UAE	Estidama Pearl Rating System Version 1.0 (Estidama PRS v. 1.0); & Estidama PRS v. 2.0	Abu Dhabi Urban Planning Council (UPC) (2010)
28	UK	BREEAM; SBEM; Code for Sustainable Homes	BRE Global (2010); Madew (2011); & Roderick et al. (2009)
29	US	LEED; Living Building Challenge; Green Globes; International Green Construction Code (IGCC); ENERGY STAR	Madew (2011); Roderick et al. (2009); US DOE & US EPA (2012); & USGBC (2011)

The basis for the BREEAM scheme is an individual building certificate award that is based on points for a set of performance criteria, and provides a categorisation in recognition of the energy and environmental performance of a building (BRE Global, 2010; and Roderick et al., 2009). This certificate award can be used as an environmental statement for the building and its ownership. BREEAM categories and criteria include the following aspects: energy use; water use; health and well-being; pollution; transport; materials; waste; ecology; and management (BRE Global, 2010; and Madew, 2011). As an environmental policy scheme, BREEAM provides best practice guidance on the minimisation of building impacts, and the delivery of comfortable interior environments (BRE Global, 2010; and Roderick et al., 2009). BREEAM provides clients, developers, designers and others with a benchmark, and support for the development of low environmental impact buildings (BRE Global, 2010). BREEAM also involves building assessment by qualified independent assessors that are appointed by the Building Research Establishment (BRE Global, 2010). According to BRE Global (2010), BREEAM 2011 will replace BREEAM 2008 for new building assessment registrations and certifications. The environmental section weightings for BREEAM 2011 are listed in Table 3.

Table 3. BREEAM 2011 Environmental Section Weightings (Source: BRE Global, 2010)

Environmental Section	Weighting (%)
Management	12
Health and Well-being	15
Energy	19
Transport	8
Water	6
Materials	12.5
Waste	7.5
Land Use and Ecology	10
Pollution	10
Innovation (additional)	10

Table 3 shows that Energy has the greatest weighting, which is 19 percent when compared to other environmental section weightings of BREEAM 2011.

3.2.2.2 A BREEAM Excellent Rated Case Study

Greater London Authority (GLA) (2002) identifies BREEAM as an important building energy assessment rating tool for the objective review of the environmental perspective of building design. According to GLA (2002), a BREEAM Design and Procurement Assessment was carried out for City Hall London. For instance, City

Hall London was awarded an EPI of 10 out of 10, and was awarded an excellent score rating of 76 percent under the BREEAM Design and Procurement Assessment (GLA, 2002). However, City Hall London's energy use during the April 2004 to March 2005 period was approximately 50 percent greater than what was envisaged at its design stage (Tuffrey and David, 2005). This occurred despite City Hall London's excellent BREEAM rating, and despite Foster and Partners (2002) using building energy simulation analysis tool(s) to develop City Hall London's low energy use environmental design strategies.

This particular case implies that the award of an excellent BREEAM score rating and the use of building energy simulation analysis tools might not always guarantee the avoidance of a gap increase between predicted and actual office building energy use. Therefore, this indicates that greater success might be achieved by using a building energy and environmental assessment integrated approach that incorporates all of the following:

- Exemplar building energy assessment methods.
    - Exemplar building energy use audit.
    - Exemplar building energy simulation analysis.
    - Exemplar building energy and environmental assessment rating.
- Exemplar building energy simulation analysis tools.
- Exemplar building energy and environmental assessment rating tools.

It is expected that this suggested integrated approach will provide robustness and likely lead to the achievement of greater success in building energy and environmental assessment, and building energy performance.

3.2.2.3 The Role of USGBC LEED

The US Green Building Council (USGBC) (2011a) describes its Leadership in Energy and Environmental Design (LEED) as a globally recognised green building certification system. Furthermore, LEED provides third-party verification that building performance improvement strategies across all key performance related metrics have been incorporated into building design and construction (USGBC, 2011a). These key performance related metrics are: energy savings; carbon emissions reduction; improved indoor environmental quality; water efficiency; and stewardship of resources and sensitivity to their impacts (USGBC, 2011a).

LEED provides a concise framework that supports the identification and implementation of strategies for sustainable building design, construction, operations, and maintenance (USGBC, 2011a). Furthermore, it can be implemented throughout the whole-life of commercial and residential buildings, that is, design and construction, operations and maintenance, tenant fit out, and significant retrofit (USGBC, 2011a).

According to USGBC (2011b), LEED measures the following: Sustainable Sites; Water Efficiency; Energy and Atmosphere; Materials and Resources; Indoor

Environmental Quality; Locations and Linkages; Awareness and Education; Innovation in Design; and Regional Priority. Through the independent Green Building Certification Institute, LEED delivers third-party certification that demonstrates a LEED building has been constructed as intended (USGBC, 2011c).

3.2.2.4 Two LEED Platinum Certified Case Studies

USGBC (2011) describes 7,455 LEED case studies. However, this study will focus on two LEED case studies because they are office buildings and are both among the earliest buildings to achieve LEED Platinum Certification, that is, the highest LEED certification. These two LEED Platinum Certified case studies are:

- The Philip Merrill Environmental Center: its building type is a Corporate Office; it is located in Annapolis, Maryland, US; and it is the first building to achieve LEED Platinum Certification, which is the highest USGBC LEED rating offered (Clark Construction Group, LLC, 2011).
- The Center for Neighborhood Technology (CNT): its building type is a Commercial Office; it is located in Chicago, Illinois, US; and it is the second Chicago project and thirteenth US project to achieve Platinum Certification in the USGBC LEED Rating System for LEED NC 2.1 (USGBC, 2011).

The attributes of these two LEED Platinum Certified case studies (refer to Table 4) include the following:

- The Philip Merrill Environmental Center uses 66 percent less energy than a typical office building with similar volume specifications (Clark Construction Group, 2011). Its energy-efficiency and achievement of LEED Platinum Certification, that is, the highest LEED rating possible involved the implementation of environmental design strategies.
- The Center for Neighborhood Technology (CNT) was renovated in 2003, and its average energy use has been 44 percent less than that of a comparable office building ever since its LEED Platinum Certification in 2005 (USGBC, 2011). However, from 2005 to 2007, its energy use increased probably due to an increase in its number of staff and computers (USGBC, 2011). Its energy-efficiency and achievement of LEED Platinum Certification (highest LEED rating possible) involved the implementation of environmental design strategies.
- Based on evidence from the Philip Merrill Environmental Center, and the Center for Neighborhood Technology (CNT), it is apparent that LEED Platinum Certification facilitates building energy-efficiency and vice versa. This occurs if environmental design principles are successfully implemented. However, if environmental design principles are not successfully implemented at the design stage – and maintained throughout the whole-life of a building – then this might lead to higher building energy use.

Table 4. Attributes of Two LEED Platinum Certified Case Studies (Source: Clark Construction Group, 2011; and USGBC, 2011)

S/N	Office Building Case Study	LEED Certification	Key Environmental Design Strategies Used	Implications
1	The Philip Merrill Environmental Center Built in 2000 Located in Annapolis, Maryland, US	Platinum certification (highest LEED certification)	It uses parallam-strand lumber beams that are made from new growth wood and can be regenerated. It uses rooftop cisterns for rainwater capture that is then used for hand washing and fire suppression. Approximately 30% of its energy load is via renewable energy sources. It uses natural ventilation and does not rely entirely on air conditioning. It uses natural lighting and does not rely entirely on artificial lighting. Its envelope, walls and ceilings use Structural Insulated Panels (SIPS).	Building energy-efficiency, that is, 66% less energy use than a similar office building.
2	The Center for Neighborhood Technology (CNT) Renovated in 2003 Located in Chicago, Illinois, US	Platinum certification (highest LEED certification)	Its envelope is super-insulated. Cooling occurs via stored ice from efficient mechanical equipment such as an ice-chiller system. Its equipment is Energy Star rated. 5% of its electricity is generated via photovoltaic panels. It achieves natural lighting, natural ventilation, and views via operable, energy-efficient windows. It developed a web-based tool, known as Green Intelligence – a Building Performance Analyzer. It calculates and displays energy use and carbon emissions data continuously.	Building energy-efficiency, that is, 44% less energy use than a similar office building.

Findings: It is apparent that LEED Platinum Certification facilitates building energy-efficiency and vice versa, and this occurs if environmental design principles are successfully implemented. However, these principles need to be successfully implemented by design professionals at the design stage and maintained throughout the whole-life of a building. Therefore, integration of exemplar methods and tools that facilitate the successful implementation of environmental design principles throughout the whole-life of a building is important for building energy-efficiency.

3.2.2.5 The Role of CaGBC LEED

Canada Green Building Council (CaGBC) (2011) describes its Leadership in Energy and Environmental Design (LEED) as a Green Building Rating System that facilitates the adoption of environmental design practices globally. CaGBC LEED fulfils these through the development and implementation of internationally accepted tools and performance criteria (CaGBC, 2011). As an internationally accepted benchmark,

CaGBC LEED facilitates the design, construction and operation of energy-efficient buildings, and its key beneficiaries include building owners and operators (CaGBC, 2011). According to CaGBC (2011), CaGBC LEED also facilitates sustainability through a whole-building approach that measures and rates building performance in six key areas, and these are: Sustainable Site Development; Water-Efficiency; Energy-Efficiency; Materials Selection; Indoor Environmental Quality; and Innovation, which is related to sustainable building expertise, performance exemplars, and design/operational strategies. The LEED certification level awarded to a project is based on the overall score rating that such a project achieves after an independent review has been undertaken (CaGBC, 2011). CaGBC LEED has four certification levels, and these are: Platinum; Gold; Silver; and Certified (CaGBC, 2011). USGBC LEED and CaGBC LEED have several similarities (refer to Table 5).

3.2.2.6 *Comparison of USGBC LEED and CaGBC LEED*
USGBC LEED and CaGBC LEED are similar (refer to Table 5) because CaGBC LEED is an adaptation of USGBC LEED. However, USGBC LEED and CaGBC LEED are dissimilar because CaGBC LEED is adapted specifically for the Canadian climate, as well as the Canadian construction industry, practices and regulations.

Table 5. Comparison of USGBC LEED and CaGBC LEED (Source: CaGBC, 2011; USGBC, 2011a; USGBC, 2011b; and USGBC, 2011c)

S/N	Comparison Categories	CaGBC LEED	USGBC LEED
1	Similarities	Certification Levels: CaGBC LEED and USGBC LEED have four certification levels, and in order of highest to lowest LEED rating these are: Platinum; Gold; Silver; and Certified.	
		Key Rating Categories: CaGBC LEED and USGBC LEED facilitate sustainability through a whole-building approach that measures and rates building performance in similar key areas, and these are: Sustainable Sites/Sustainable Site Development; Water Efficiency; Energy and Atmosphere/Energy Efficiency; Materials and Resources/Materials Selection; Indoor Environmental Quality; and Innovation in Design.	
2	Dissimilarities	Key Rating Categories: these exclude Awareness and Education; Locations and Linkages; and Regional Priority.	Key Rating Categories: these include Awareness and Education; Locations and Linkages; and Regional Priority.
		Adaptation: CaGBC LEED is adapted specifically for the Canadian climates, Canadian construction industry, practices and regulations.	Adaptation: USGBC LEED is not adapted specifically for the Canadian climates, and the Canadian construction industry, practices and regulations.

Findings: It is apparent that CaGBC LEED and USGBC LEED facilitate sustainability through a whole-building approach that measures and rates building performance in key areas. These key areas are: Sustainable Sites / Sustainable Site Development; Water Efficiency; Energy and Atmosphere / Energy Efficiency; Materials and Resources / Materials Selection; Indoor Environmental Quality; Locations and Linkages; Awareness and Education; and Innovation in Design. Furthermore, Adaptation is important for building energy assessment methods and tools to function efficiently for specific climates, construction industries, practices and regulations.

3.2.2.7 The Role of GBCA Green Star

Green Building Council of Australia (GBCA) (2011) describes GBCA Green Star as a voluntary environmental rating system, which is comprehensive and national, and evaluates building environmental design and construction. GBCA Green Star is a key building energy assessment rating tool in Australia (Roderick et al., 2009). The Australian property and construction market is a major beneficiary of GBCA Green Star whereby millions of square metres of Australian space have either been Green Star-Certified or Green Star-Registered (GBCA, 2011). According to GBCA (2011), the role of GBCA Green Star includes the following:

- The establishment of a uniform format.
- The establishment of a green building standard of measurement.
- The promotion of an integrated whole-building design approach.
- The recognition of environmental leadership.
- The identification of whole-life building impacts.
- The increased awareness of the benefits of green building.
- Reduction of operating costs.
- Higher return on investment.
- Increased tenant attraction.
- Improved marketability.
- Improved productivity.
- Reduction in liability and risk.
- Healthier spaces and places to live and work.
- Demonstration of Corporate Social Responsibility.
- Future proofed assets.
- Competitive advantage.

According to GBCA (2011), GBCA Green Star has nine categories that are divided into credits, and points are awarded in each credit. Furthermore, GBCA (2011) explains that these nine categories assess a project's environmental impact attributable to its site selection, design, construction and maintenance. These nine categories are: Management; Indoor Environment Quality; Energy; Transport; Water; Materials; Land Use and Ecology; Emissions; and Innovation. GBCA (2011) and Roderick et al. (2009) state that GBCA Green Star has three Certified Rating categories – and in the order of highest to lowest rating – these are:

- 6 Star Green Star Certified Rating: it has a score range of 75 to 100, and it signifies that a project has attained 'World Leadership' in environmentally sustainable design and/or construction.
- 5 Star Green Star Certified Rating: it has a score range of 60 to 74, and it signifies that a project has attained 'Australian Excellence' in environmentally sustainable design and/or construction.
- 4 Star Green Star Certified Rating: it has a score range of 45 to 59, and it signifies that a project has attained 'Best Practice' in environmentally sustainable design and/or construction.

GBCA Green Star has an important feature that the primary author refers to as Adaptation, which is the ability of a building energy assessment method and/or tool to adapt to varying contexts in order to suit specific needs. Furthermore, Adaptation is important for building energy assessment methods and/or tools to function efficiently for specific climates, construction industries, practices and regulations. For instance, GBCA Green Star's environmental weighting factors vary across different Australian states and territories in order to accommodate specific environmental needs.

3.2.2.8 The Role of the HQE Standard

The High Quality Environmental (HQE) Standard is a green building standard in France – for improving environmental quality through certification – that validates environmental schemes in building development (Association HQE, 2011; and Concept BIO, 2008). Association HQE (2011) and Concept BIO (2008) state that the HQE Standard aims to mitigate impacts on the external environment while creating high quality internal environments, and it has the following fourteen targets:

- Mitigating the impacts on the external environment:

  - The harmonious relationship of buildings and their external environment.
  - The choice of integrated construction methods and building materials.
  - The avoidance of pollution by the construction site.
  - The minimisation of energy use.
  - The minimisation of water use.
  - The minimisation of waste in building operations.
  - The minimisation of the need for building maintenance and repair.

- Creating high quality internal environments for comfort and good health:

  - The integration of hygrometric control measures.
  - The integration of acoustic control measures.
  - The attainment of visual appeal.
  - The integration of measures to control smells.
  - The attainment of good sanitary conditions of the indoor spaces.
  - The integration of air quality controls.
  - The integration of water quality controls.

3.2.2.9 Comparison of BREEAM, LEED, Green Star, and the HQE Standard

BREEAM, USGBC LEED, CaGBC LEED, GBCA Green Star, and the HQE Standard have similarities, but they also have dissimilarities (refer to Table 6). BREEAM, USGBC LEED, CaGBC LEED, GBCA Green Star, and the HQE Standard have similarities in their role as key building energy and environmental assessment rating tools (refer to Table 6). Their role involves an ability to describe, rate, and certify building environmental design, construction, energy use and environmental performance. Furthermore, BREEAM, USGBC LEED, CaGBC LEED, GBCA Green Star, and the HQE Standard have similarities in their rating categories whereby Energy Use, Materials, and Water Use are common to all of them (refer to Table 6).

Table 6. Comparison of BREEAM, LEED, Green Star and the HQE Standard (Source: Association HQE, 2011; BRE Global, 2010; CaGBC, 2011; GBCA, 2011; USGBC, 2011a; USGBC, 2011b; and USGBC, 2011c)

S/N	Comparison Categories	BREEAM	USGBC LEED	CAGBC LEED	GBCA Green Star	HQE Standard
1	Similarities					
1.1	Building Energy & Environmental Assessment	BREEAM, USGBC LEED, CaGBC LEED, Green Star, & the HQE Standard are all used for building energy & environmental assessment.				
		It involves the description, rating, & certification of building environmental design, construction, energy use & environmental performance.				
1.2	Rating Categories	Energy Use				
		—	Indoor Environmental Quality			—
		Innovation				—
		Land Use & Ecology	—	—	Land Use & Ecology	—
		Management	—	—	Management	—
		Materials				
		Pollution	—	—	—	Pollution
		—	Sustainable Sites		—	—
		Transport	—	—	Transport	—
		Waste	—	—	—	Waste
		Water Use				
2	Dissimilarities					
2.1	Rating Categories	—	—	—	—	Acoustic
		—	—	—	—	Air Quality
		—	Awareness & Education	—	—	—
		—	—	—	—	Building Maintenance & Repair
		—	—	—	Emissions	—
		Health & Well-being	—	—	—	—
		—	—	—	—	Hygrometric
		—	Locations & Linkages	—	—	—
		—	—	—	—	Olfactive
		—	Regional Priority	—	—	—
		—	—	—	—	Relationship between buildings & their external environment
		—	—	—	—	Sanitary condition of indoor space
		—	—	—	—	Water Quality
		—	—	—	—	Visual

BREEAM, USGBC LEED, CaGBC LEED, GBCA Green Star, and the HQE Standard have dissimilarities in their rating categories (refer to Table 6). For instance, the following rating categories appear to be synonymous with the HQE Standard: Acoustic; Air Quality; Building Maintenance and Repair; Hygrometric; Olfactive; Relationship between buildings and their External Environment; Sanitary Condition of Indoor Space; Water Quality; and Visual.

4 Discussion

The issue of the gap between predicted and actual office building energy use is complex and requires the best strategies in order to successfully bridge it. Therefore, this study has established the role of building energy and environmental assessment in facilitating office building energy-efficiency.

- The role of building energy and environmental assessment methods and tools is to:
  - Assess the impacts of environmental design principles and the impacts of several factors that contribute to gap decreases, for example: solar gain minimisation orientations; energy-efficient strategies for built forms, ventilation, lighting and services; and decreases in hours of operation and occupancy.
  - Assess the impacts of environmental design principles, and the impacts of several factors that contribute to gap increases, for example: weather variation and microclimates; increases in hours of operation; and increases in occupancy.
  - Assess the energy-efficient built form aspect of Biomimicry.

- The role of the building energy use audit method is to:
  - Track office building energy profile over time.
  - Determine office building baseline energy use data.
  - Determine office building energy performance from historical energy use data.
  - Compare and benchmark office building energy use against office building energy use good practice and/or typical practice.

- The role of the building energy simulation analysis method and tools is to:
  - Predict future office building energy use within multiple environmental design related scenarios and key parameters despite the impact of limitations.
  - Assess office building energy expenditure and environmental design benefits.
  - Assess office building energy performance and potential for energy-efficiency.

- The role of the building energy and environmental assessment rating method and tools is to:
  - Assess, rate, and certify office building energy-efficiency.

- The role of building energy and environmental assessment methods and tools can be enhanced by:
  - Overcoming dissimilarities in order to achieve robustness through integration based on similarities.

63 The Role of Building Energy and Environmental Assessment 699

- – Overcoming format and compatibility issues in order to improve interoperability and integration objectives, and the accuracy of office building energy use predictions.
- – Overcoming barriers to accessing energy use data, including the reluctance of several managers to make office building energy use data available for building energy and environmental assessment.
- – Establishing alternative ways to undertake, for instance, a level I (basic) office building energy use audit even when there is limited access to office building energy use data.
- – Overcoming the reluctance of several clients to embrace the use of appropriate building energy and environmental assessment methods and/or tools.
- – Successfully implementing environmental design principles throughout the whole-life of office buildings in order to facilitate energy-efficiency, a BREEAM excellent rating, and a LEED platinum certification.
- – Overcoming limitations in order to fully represent office building parameters and scenarios during energy and environmental assessments.
- – Overcoming discrepancies that might occur during building energy and environmental assessments.
- – Overcoming limitations such as the occurrence of a gap between predicted and actual office building energy use despite using building energy simulation analysis and achieving a BREEAM excellent rating or LEED platinum certification.
- – Overcoming limitations such as: the occurrence of 392 building energy simulation analysis tools by only 28 countries; and a 50 percent gap between the US' 59 percent share of these tools and the UK's 9 percent share of these tools.
- – Ensuring that either the proliferation or non-proliferation of building energy simulation analysis tools translates to implementing environmental design principles for office building energy-efficiency.
- – Achieving Adaptation, which the primary author refers to as the ability of building energy and environmental assessment to adapt to varying contexts in order to function well for specific climates, construction industries, practices and regulations.

- • The impacts of contributory factors in the gap between predicted and actual office building energy use, and limitations in the role of building energy and environmental assessment need to be addressed in order to achieve:

  - – Optimum building energy use assessments and predictions.
  - – Improved environmental design principles.
  - – Gap decreases and improved energy performance whereby office building actual energy use is less than predicted energy use.

5 Conclusion

This study used a combination of literature reviews, multiple case study research and comparative studies in order to build theory, and it established the role of building energy and environmental assessment in facilitating office building energy-efficiency. It involved a desk based study and it also established the methods and tool to be used in the primary author's Ph.D research for multiple case studies and simulation studies of office building energy-efficiency. These are: the building energy use audit method; the building energy simulation analysis method; and Autodesk Ecotect Analysis, which is a key building energy simulation analysis tool.

This study also revealed that the role of building energy and environmental assessment involves assessment of the impacts of environmental design principles, and the impacts of factors that contribute to office building energy use gap decreases, for example: solar gain minimisation orientations; energy-efficient strategies for built forms, ventilation, lighting and services; and decreases in hours of operation and occupancy. Its role also involves assessment of the impacts of factors that contribute to office building energy use gap increases, for example: weather variation and microclimates; and increases in hours of operation and occupancy. There are three key types of building energy and environmental assessment, and these are: building energy use audit method; building energy simulation analysis method and tools; and building energy and environmental assessment rating method and tools. Their respective roles include: tracking building energy use over time; predicting future building energy use within multiple environmental design scenarios and parameters; and assessing, rating, and certifying building energy and environmental efficiency. However, the limitations of building energy and environmental assessment methods and tools, as well as the impacts of factors that contribute to office building energy use gap increases need to be addressed in order to achieve: optimum building energy use assessments and predictions; optimum environmental design principles; and building energy use gap decreases and improved energy performance whereby office building actual energy use is less than predicted energy use.

This study has contributed to ideas for the development of a Building Management System (BMS-Optimum) for Bridging the Gap, which is comprised of optimum conditions and considerations such as: optimum environmental design principles; optimum weather and microclimate considerations; accessibility to reliable office building energy use data; optimum building energy and environmental assessment; optimum hours of operation; and optimum level and nature of occupancy.

6   Future Work

Future work will include further development of the BMS-Optimum, using methods such as: multiple case study research supported by building energy use audits, observations, questionnaire surveys, interviews, benchmarking and comparative studies; building energy simulation within multiple scenarios, parameters and variables, and supported by benchmarking and comparative studies; and peer reviews and focus group sessions. These will also help establish and validate a Framework for Improved Environmental Design and Energy Performance (FEDEP).

References

Abu Dhabi Urban Planning Council (UPC) Pearl Rating System (2010),
    http://www.estidama.org/pearl-rating-system-v10.aspx?lang=
    en-US (retrieved January 10, 2012)
Action Energy, Energy Consumption Guide 19: Energy Use in Offices. Action Energy, Best
    Practice Programme: UK (2003)

ASHRAE, Procedures for Commercial Building Energy Audits. ASHRAE - American Society of Heating, Refrigeration, and Air Conditioning Engineers (2004)

Association HQE. High Quality Environmental (HQE) Standard (2011), `http://assohqe.org/hqe/` (retrieved March 17, 2011)

Autodesk. Autodesk Ecotect Analysis (2012), `http://usa.autodesk.com/adsk/servlet/pc/index?id=12602821&siteID=123112` (retrieved January 10, 2012)

Azar, E., Menassa, C.: A Conceptual Framework to Energy Estimation in Buildings using Agent Based Modeling. In: Proceedings of the 2010 Winter Simulation Conference (2010)

BEAM Society. BEAM - Building Environmental Assessment Method (2011), `http://www.beamsociety.org.hk/general/home.php` (retrieved June 15, 2011)

Beautyman, M.: HOK Earns Sustainable Leadership Award. Interior Design (2006), `http://www.interiordesign.net/article/484923-HOK_Earns_Sustainable_Leadership_Award.php` (retrieved March 25, 2012)

Bondareva, E.S.: Green Building in the Russian Context: An Investigation into the Establishment of a LEED-Based Green Building Rating System in the Russian Federation. Master of Arts Thesis. Cornell University (2005)

Bordass, B., Cohen, R., Field, J.: Energy Performance of Non-Domestic Buildings: Closing The Credibility Gap. In: Proceedings of IEECB 2004 Building Performance Congress, Frankfurt (2004)

Borong, L., Yingxin, Z., Lei, T., Youguo, Q.: Study of Several Factors for Green Building Assessment System in China. In: SB 2007 Seoul: Proceedings of the International Conference on Sustainable Building Asia, Seoul, June 27-29 (2007)

BRE Global. BREEAM: BREEAM 2011. BRE Global (2010), `http://www.breeam.org/page.jsp?id=374` (retrieved June 7, 2011)

BRE Global. BREEAM: What is BREEAM? BRE Global (2010), `http://www.breeam.org/page.jsp?id=66` (retrieved June 7, 2011)

Building and Construction Authority About BCA Green Mark Scheme (2011), `http://www.bca.gov.sg/GreenMark/green_mark_buildings.html` (retrieved January 10, 2012)

Canada Green Building Council (CaGBC) Introduction to LEED (2011), `http://www.cagbc.org/AM/Template.cfm?Section=LEED` (retrieved March 17, 2011)

Clark Construction Group. Philip Merrill Environmental Center (2011), `http://www.clarkconstruction.com/index.php/projects/feature_project/38` (retrieved May 23, 2011)

Concept BIO FAQ - HQE Approach (2008), `http://www.concept-bio.eu/hqe-approach.php` (retrieved June 5, 2011)

Cramer, J.P.: Sustainability Leadership: From Stagnation to Liberation. Design Intelligence (2011), `http://www.di.net/articles/archive/3616/` (retrieved March 25, 2012)

Crawley, D.B., Hand, J.W., Kummert, M., Griffith, B.T.: Contrasting the Capabilities of Building Energy Performance Simulation Programs. Building and Environment 43(4), 661–673 (2008)

Demanuele, C., Tweddell, T., Davies, M.: Bridging the Gap between Predicted and Actual Energy Performance in Schools. In: World Renewable Energy Congress XI, Abu Dhabi, September 25-30 (2010)

Department for Communities and Local Government, A Guide for Businesses: Reducing the Energy Usage and Carbon Emissions from your Air Conditioning Systems (2007)

DESA - Department of Economic and Social Affairs, Trends in Sustainable Development. Towards Sustainable Consumption and Production. United Nations, New York (2010)

DGBC. Dutch Green Building Council (DGBC) - BREEAM-NL (2011),
http://www.dgbc.nl/wat_is_dgbc/dgbc_english (retrieved June 15, 2011)

Diamond, R., Opitz, M., Hicks, T.: Evaluating the Energy Performance of the First Generation of LEED-Certified Commercial Buildings. ACEEE Summer Study on Energy Efficiency in Buildings (2006)

Dixon, T., Gallimore, P., Reed, R., Wilkinson, S., Keeping, M.: A Green Profession? RICS Members and the Sustainability Agenda. Royal Institution of Chartered Surveyors, RICS (2007)

EDAMA. EDAMA - Emerging Objectives (2009),
http://www.edama.jo/static/what_EDAMA.shtm (retrieved June 15, 2011)

EERE. Buildings Database. Caribou Weather Forecast Office (WFO) (2004c),
http://eere.buildinggreen.com/energy.cfm?ProjectID=334 (retrieved July 10, 2011)

EERE. Buildings Database. U.S. EPA Science and Technology Center (2006c),
http://eere.buildinggreen.com/energy.cfm?ProjectID=323 (retrieved July 8, 2011)

EIA. Commercial Buildings Energy Consumption Survey (CBECS) 2003 (2011),
http://www.eia.doe.gov/emeu/cbecs/ (retrieved March 17, 2011)

Foster and Partners, Foster and Partners City Hall. Foster and Partners Limited, London (2002)

Foster and Partners. Norman Foster (2012),
http://www.fosterandpartners.com/Team/SeniorPartners/11/Default.aspx (retrieved March 25, 2012)

German Sustainable Building Council (GeSBC) DGNB - Deutsche Gesellschaft für Nachhaltiges Bauen (2012), http://www.dgnb.de/_en/index.php (retrieved January 10, 2012)

Gomes, V., Gomes, M., Lamberts, R., Takaoka, M.V., Ilha, M.S.: Sustainable Building in Brazil. In: Melbourne: SB 2008 (World Sustainable Building Conference) (2008)

Greater London Authority, Environmental Assessment of City Hall (Report Number: 7). Greater London Authority, London (2002)

Green Building Council España (GBCe) LEED (2012a),
http://www.gbce.es/en/pagina/leed (retrieved January 10, 2012)

Green Building Council España (GBCe) VERDE Certificate (2012b),
http://www.gbce.es/en/pagina/verde-certificate (retrieved January 10, 2012)

Green Building Council (GBC) Indonesia Greenship (2010),
http://www.gbcindonesia.org/greenship.html (retrieved June 15, 2011)

Green Building Council (GBC) Italy Certification (2011),
http://www.gbcitalia.org/certificazione (retrieved June 15, 2011)

Green Building Council of Australia (GBCA) Green Star / Green Star Overview (2011),
http://www.gbca.org.au/green-star/green-star-overview/ (retrieved June 2, 2011)

Green Building Council South Africa (GBCSA) Green Star SA – Office v1 (2012),
http://www.gbcsa.org.za/greenstar/office.php (retrieved January 10, 2012)

Green Building Index (GBI) How GBI Works - The GBI Assessment Process (2011),
http://www.greenbuildingindex.org/how-GBI-works.html (retrieved June 15, 2011)

Greener Buildings. Green Skyscraper: The Basis for Designing Sustainable Intensive Buildings (2008), http://www.greenerbuildings.com (retrieved March 25, 2012)

Green Globes Green Globes - The Practical Building Rating System (2011),
http://www.greenglobes.com/ (retrieved June 15, 2011)

GRIHA Green Rating for Integrated Habitat Assessment. Welcome to GRIHA (2011),
http://www.grihaindia.org/ (retrieved June 15, 2011)

Hajek, P., Tywoniak, J.: Towards Sustainable Building in the Czech Republic. In: Melbourne: SB 2008 (World Sustainable Building Conference) (2008)

Howell, I., Batcheler, B.: Building Information Modeling Two Years Later – Huge Potential, Some Success and Several Limitations. The Laiserin Letter (2005)

Hui, S.C.M. Building Environmental Assessment (2010), `http://www0.hku.hk/bse/sbs/sbs0910-01.pdf` (retrieved January 10, 2012)

Hui, S.C.M. Energy Efficiency in Buildings (II). MEBS6016 Energy Performance of Buildings (2010), `http://www.hku.hk/bse/MEBS6016/` (retrieved March 17, 2011)

IGBC - Indian Green Building Council. IGBC Green SEZ Rating System (2008), `http://www.igbc.in/site/igbc/testigbc.jsp?desc=233674&event=233670` (retrieved June 15, 2011)

IGBC - Indian Green Building Council. LEED-INDIA for New Construction and Major Renovations (2008), `http://www.igbc.in/site/igbc/testigbc.jsp?desc=22905&event=22869` (retrieved June 15, 2011)

ITACA. System of Assessment of Sustainability Energy and Environmental Building - Protocollo Itaca (2011), `http://www.itaca.org/` (retrieved June 15, 2011)

JSBC - Japan Sustainable Building Consortium CASBEE - Comprehensive Assessment System for Built Environment Efficiency. An Overview of CASBEE (2006) , `http://www.ibec.or.jp/CASBEE/english/overviewE.htm` (retrieved June 15, 2011)

KGBC - Korea Green Building Council. Status of GBCS (Green Building Certificate System) (2000), `http://greenbuilding.or.kr/` (retrieved June 15, 2011)

Leaman, A., Stevenson, F., Bordass, B.: Building Evaluation: Practice and Principles. Building Research and Information 38(5), 564–577 (2010)

LiderA. LiderA - Sustainable Assessment System (2012), `http://www.lidera.info/?p=MenuPage&MenuId=29` (retrieved January 10, 2012)

Madew, R.: The WorldGBC and Green Building Rating Tools (2011), `http://www.dgnb.de/fileadmin/downloads/The_WorldGBC_and_greenbuilding_ratingtools.pdf` (retrieved January 10, 2012)

MINERGIE MINERGIE (2012), `http://www.minergie.com/home_en.html` (retrieved January 10, 2012)

Mortimer, N.D., Elsayed, M.A., Grant, J.F.: Patterns of Energy Use in Non-Domestic Buildings. Environment and Planning B: Planning and Design 27(5), 709–720 (2000)

NABERS - National Australian Built Environment Rating System. NABERS - FAQs (2010), `http://www.nabers.com.au/faqs.aspx` (retrieved June 15, 2011)

NREL - National Renewable Energy Laboratory. Energy Analysis and Tools (2010), `http://www.nrel.gov/buildings/energy_analysis.html` (retrieved April 17, 2011)

NZGBC - New Zealand Green Building Council. Green Star New Zealand (2008), `http://www.nzgbc.org.nz/main/greenstar` (retrieved June 15, 2011)

NZGBC - New Zealand Green Building Council. Homestar - New Zealand's Residential Rating Tool (2008), `http://www.nzgbc.org.nz/main/resources/articles/ResidentialRatingTool` (retrieved June 15, 2011)

Opitz, M.: Energy Audits (2011), `http://www.fmlink.com/article.cgi?type=Sustainability&title=Energy%20Audits&pub=USGBC&id=40640&mode=source` (retrieved April 17, 2011)

Osaji, E.E.: The BEEHive (Building Energy-Efficient Hive). Bachelor of Architecture (BArch) Thesis, University of Lagos (2002)

Osaji, E.E., Price, A.D.F., Mourshed, M.: Guest Editorial. The Role of Building Performance Simulation in the Optimization of Healthcare Building Design. Journal of Building Performance Simulation 3(3), 169 (2010)

Osaji, E.E., Price, A.D.F.: Parametric Environmental Design for Low Energy Healthcare Facility Performance. In: 6th International Conference on Innovation in Architecture, Engineering and Construction (AEC), Pennsylvania, June 9-11 (2010)

Paradis, R.: Energy Analysis Tools. National Institute of Building Sciences (2010), `http://www.wbdg.org/resources/energyanalysis.php` (retrieved April 17, 2011)

Passive House Institute. CEPHEUS - Cost Efficient Passive Houses as EUropean Standards (2011), `http://www.cepheus.de/eng/index.html` (retrieved June 15, 2011)

PHILGBC. Philippine Green Building Council (2010), `http://philgbc.org/index.php/home.html` (retrieved June 15, 2011)

Reed, R., Bilos, A., Wilkinson, S., Shulte, K.-W.: International Comparison of Sustainable Rating Tools. Journal of Sustainable Real Estate 1(1), 1–22 (2009)

Roderick, Y., David, M., Craig, W., Carlos, A.: Comparison of Energy Performance Assessment Between LEED, BREEAM and Green Star. In: 11th International IBPSA Conference, Glasgow, July 27-30 (2009), `http://www.ibpsa.org/proceedings/BS2009/BS09_1167_1176.pdf` (retrieved April 17, 2011)

Southern Taiwan Science Park Administration. EEWH - The Green Building Rating System in Taiwan (2012), `http://gsp.stsipa.gov.tw/eng/main03_2.html` (retrieved January 10, 2012)

Tuffrey, M.: Mayor Answers to London: City Hall Energy Consumption. Liberal Democrat Group London Assembly Report (2005)

UK-GBC - UK Green Building Council. Existing Non-Domestic Buildings (2011), `http://www.ukgbc.org/site/info-centre/display-category?id=24` (retrieved April 17, 2011)

US DOE - US Department of Energy. Building Energy Software Tools Directory (2011), `http://apps1.eere.energy.gov/buildings/` (retrieved March 17, 2011)

US DOE and US EPA. About ENERGY STAR (2012), `http://www.energystar.gov/index.cfm?c=about.ab_index` (retrieved January 10, 2011)

USGBC - US Green Building Council. Intro - What LEED Is (2011a), `http://www.usgbc.org/DisplayPage.aspx?CMSPageID=1988` (retrieved March 17, 2011)

USGBC - US Green Building Council. Intro - What LEED Measures (2011b), `http://www.usgbc.org/DisplayPage.aspx?CMSPageID=1989` (retrieved March 17, 2011)

USGBC - US Green Building Council. Intro - What LEED Delivers (2011c), `http://www.usgbc.org/DisplayPage.aspx?CMSPageID=1990` (retrieved March 17, 2011)

USGBC - US Green Building Council. LEED Projects and Case Studies (2011), `http://leedcasestudies.usgbc.org/overview.cfm?ProjectID=468` (retrieved March 17, 2011)

Wade, J., Pett, J., Ramsay, L.: Energy Efficiency in Offices: Assessing the Situation. Association for the Conservation of Energy, London (2003)

Wade, J., Pett, J., Ramsay, L.: Energy Efficiency in Offices: Motivating Action. Association for the Conservation of Energy, London (2003)

Wellcome Trust. Environmental Report 2006 (2011), `http://www.wellcome.ac.uk/About-us/Policy/Policy-and-position-statements/WTX022977.htm` (retrieved June 26, 2011)

WGBC - World Green Building Council. World Green Building Council (2011), `http://www.worldgbc.org/site2/#` (retrieved June 15, 2011)

Chapter 64
Improved Personalized Comfort: A Necessity for a Smart Building

Wim Zeiler[1,2], Gert Boxem[1], and Derek Vissers[1]

[1] TU/e, Technical University Eindhoven, Faculty of Architecture, Building and Planning
[2] Kropman Building Services Contracting

Abstract. The reasons for the inferior performance of many of the current buildings and their related energy systems are diverse and for a major part caused by insufficient attention to the influence of occupant behaviour. In Smart buildings it is necessary to implement new opportunities to integrate human behaviour in the Heating Ventilation and Air-Conditiong process control loop. To realize this strategy we developed an advanced control setup, based on the combination of ubiquitous low cost wireless sensors. The article describes the proof of the principle to take the perceived thermal comfort as leading principles in the comfort/energy process control. The experiments described illustrate the feasibility of the approach.

Keywords: process control, individual comfort control, energy management.

1 Introduction

In the Dutch built environment nearly 40% of the total energy use is used for building systems to provide comfort for building occupants. Traditionally, the process control strategy used in buildings is based on a simplified approach, were a general set point is taken as comfort control parameter for a whole room. This leads often to dissatisfied occupants and additional energy consumption. Users are shown to consistently over-turn actions in response to uncomfortable conditions, causing oscillations that can waste energy and create an uncomfortable environment [2]. The human behaviour can negatively influence the energy consumption by more than 100% [3], so therefore it is necessary to incorporate the human need better in the control strategies. Introducing advanced control algorithms to lighting and shading control can reduce the energy loads with more than 45% [4]. Currently the energy management within buildings is far from optimal and the potential savings of energy due to improved control by better use of ICT technology. Optimised process control is a necessity for the improvement energy performance of buildings [5]. However, in most of the research focusing on improved ICT often overlooks the role of user in reducing the energy conservation.

A. Håkansson et al. (Eds.): *Sustainability in Energy and Buildings*, SIST 22, pp. 705–715.
DOI: 10.1007/978-3-642-36645-1_64   © Springer-Verlag Berlin Heidelberg 2013

With smart energy efficient buildings, the relation between behaviour and energy consumption has become significant, and should be looked into [6].Until now however, the actual building occupant is not included in the control loop of comfort systems. There is a necessity for spatially distributed information about the location and the needs of occupants to provide for them a higher thermal comfort level and to realize energy savings at the same time. Therefore, research was done on possible and measurable critical indicators representing the individual perceived comfort feeling. A possible indicator for the individual thermal comfort is for example the skin temperature difference over time, from a specific part of the human body.

2 Thermal Comfort

The individual user has to become leading in the whole climatisation process, to optimize the necessary use of energy to supply the occupants with their own preferred comfort environment and energy for their activated appliances.Therefore we started from literature to investigate on an individual level, comfort and user behaviour. Thermal comfort for all can only be achieved when occupants have effective control over their own thermal environment [7]. Therefore, Individually Controlled Systems [ICS] with different task-ambient heating/cooling options are required as proposed by Filippini [8] and Watanabe [9]. Arens [10] proposes a distributed sensor network which can predict a person's thermal state from measured skin temperatures sensed through contact or remotely by infrared sensors. At room scale, the control and actuation could take place within the room itself by a kind of remote controller. In the proposed concept user behaviour is only taken into account by an occupancy sensor. Human beings sense the warmth or coolness of an environment through thermoreceptors located in the skin and core. A thermo-receptor adapts to the rate of change of temperature [14]. Traditionally, calculations of human thermal comfort have been based on the Predictive Mean Vote(PMV) /Predicted Percentage of Dissatisfied (PPD) model of Fanger [15]. However the comfort prediction by the PMV/PPD model is only valid for a large population and individual differences are not taken into account [18]. There are differences in the comfort perception of people [16]. The major reasons for those differences are: gender, clothing resistance and body fat.

Recent studies in Heating Ventilation and Air-Conditionings (HVAC) control methods incorporating PMV algorithms, involving the use of smart distributed sensor networks to measure operative temperature, mean air speed and relative humidity which are the localized thermal parameters of a particular occupant [11]. Feldmeier and Paradiso[12] developed a personalized HVAC system consisting of four main components: portable nodes, room nodes, control nodes, and a central network hub. To assess the occupant's comfort level it uses a portable node which senses the local temperature, humidity, light level, and inertial activity level of the user. The actuation of the various heating and cooling systems is achieved via control nodes. This distributed information can be obtained by low-cost wireless sensor networks [10, 11], low-cost infrared sensors [16], and smart badges/portable nodes [12]. Distributed wireless sensors networks have been employed in attempts to assess comfort by measuring PMV values in real-time, Tse and Chan [11] , and Revel and Sabbatini [13] have shown that infrared imaging (i.e. by low cost IR camera's) can be used for real-time estimating heat rate and comfort parameters in a room. Combining these technologies could make it possible to provide real-time comfort-energy management based on the available distributed information.

Thermal environments are often asymmetrical, meaning either spatially non-uniform or transient-changing over time. However, asymmetrical environments may not necessarily be less desirable than thermal neutrality, but might actually produce better comfort than a uniform neutral condition [e.g. the identical cool face sensation may be perceived as more comfortable when the whole body feels warm] [17]. In addition, standards regarding local discomfort might be conservative [18] and fundamental differences between measured and predicted comfort values are observed in field studies [19]. The most recent research on human comfort looks at local sensations and comfort predictions of individual body parts [20]. Recent studies on human thermal comfort have demonstrated that the human head and extremities [hands and feet] dictate overall discomfort [17-21]. These studies have led to the development of individual controlled systems with local HVAC options, creating a thermal microclimate surrounding the human body [8, 9]. Lowering the ambient set point temperature to 18°C [in the 18-30°C ambient dead-band] and applying radiant heating to specific body parts in a micro-climate set-up can result in local discomfort of non-radiated body parts [19]. Wang found that the finger temperature (30°C) and finger-forearm temperature gradient (0°C) are significant thresholds for overall thermal sensation [22]. However Zhang et al. have shown that the 18-30°C ambient death band control zone in combination with task-ambient conditioning is an acceptable range [23]. However, more research is needed on the acceptance of extended ambient temperature ranges at room level, when people are away from their workplace / micro-climate [8].

These results suggest that a personally controlled task-ambient system (TAC) that focuses directly on these body parts may efficiently improve thermal comfort in office environments [23]. Although differences between individual building occupants, the successive step towards measuring and dealing with those differences is not yet realized. The question remains how to couple physiological responses to a thermal comfort assessment on a more physiological basis (i.e. less empirical derived regression formulas), in comparison to current existing modes, e.g.: the 25-node Stolwijk model [24], the Dynamic Thermal Sensation model of Fiala [25] and models of Zhang [26]. The models introduced by Zhang use empirical derived regression formulas, that only account for convective effects (due to applying air-sleeves for local cooling/heating) and not for radiant effects.

The most commonly used indicator of thermal comfort is air temperature, because it is easy to use and most people can relate to it. However, it does not represent the real thermal comfort behaviour of the room [13] and it is certainly not an accurate indicator of thermal comfort or thermal stress as perceived by the user itself [10]. Therefore air temperature should always be considered in relation to the other physical or environmental factors (e.g. radiant temperature, air velocity, and humidity). Additional (low-cost) sensors would make it possible to take into account these environmental factors and make it easier to determine when problems reported by occupants can be resolved automatically. If environments were more completely sensed, it could be possible to provide thermal comfort as efficiently as possible. Infrared imaging (i.e. by low cost IR sensors) can be used for real-time estimating heat rate, and comfort parameters in a room. The interaction between indoor environment and skin temperature, and thus thermal sensation is for normal office conditions largely determined by the mean radiant temperature. Measuring the radiant temperatures by low cost IR sensors makes it possible by image post-processing to estimate energy fluxes and temperature distributions with comfort prediction. The use of thermal images (i.e. provided by infrared sensor/camera) is able to provide distributed information on human body surrounding temperatures which can be used for a more efficient control [13]. Revel and Sabbatini used the information provided by the infrared camera, to derive in real time, the air volume and comfort conditions necessary for the occupants through controlling the actual PMV value. The real-time PMV can be used to control room temperature as proposed by Kang [26]. In the study of Kang, the set-point room temperature was adjusted according to changes in indoor climate by measuring the other environmental variables (radiant temperature, air velocity etc) to maintain the comfort (i.e. defined by PMV) in between certain values. The results have shown that it is possible to maintain thermal comfort and save energy by incorporating a PMV-algorithm.

3   Experimental Set-Up: Individual Thermal Comfort

Well-being, health and productivity of office workers are highly related to the indoor thermal conditions, as shown in literature [32-39]. Due to individual differences, it is not possible to satisfy all office workers with the same indoor thermal conditions [27]. By conditioning the body parts with a relatively small amount of energy, the ambient temperature could be allowed to float in a relatively wide range, generating energy savings up to 40% [28]. However, the benefits of an individual controlled system are only fully achieved when the building occupant understands the system behaviour and can deal with it. User problems with individual temperature control [2, 29] can result in energy wasting behaviour, or can even result in thermal discomfort. For this reason, it is more convenient that the personal climate set-points can automatically adapt to the individual needs. This requires a method to include the human body as a sensor in the control loop of personal climate systems. In this research non-contact infrared sensing is applied to measure the time dependent facial and hands skin temperatures which show high potential for a new personal based energy efficient control strategy, Fig. 1.

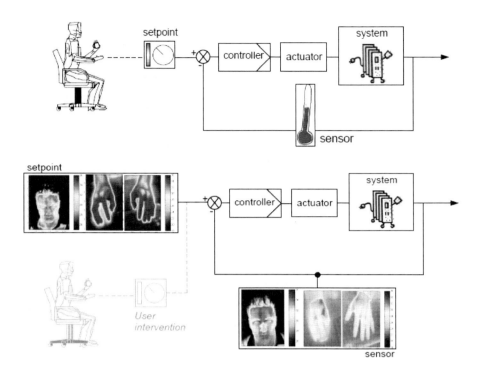

Fig. 1. Traditional process control loop(top) and user in the process control loop approach (below)

To achieve the personal process control the effectiveness of applying infra red sensors was investigated. In the current stage of the research only experiments were done to proof the principle. In the configuration shown in Fig. 2 only one test-person was used for a series of sessions with different temperature scenarios. The total duration of each session was about 3 hours. During the first hour the human subject has no control options; this is the so-called acclimatization phase. Thereafter, the radiant panels are activated to a starting condition ('zero') and the human subject will be asked to adjust his local microclimate by the radiant panels to the most satisfactory thermal environment for a time period of 1.5 hours. Time step printed images: 5min. (measurement time step 20 s).

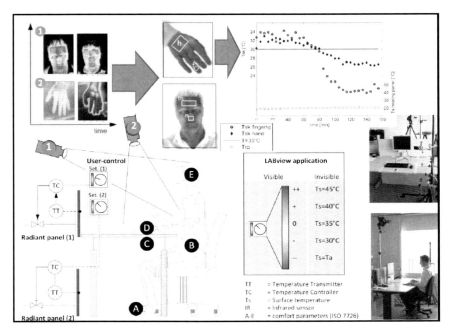

Fig. 2. Experimental workplace setting

4 Results

In the user-controlled experiments the radiant heating system was activated and the subject had the ability to control the panel surface temperature. The room temperature was controlled at 20°C, which is below the thermo-neutral zone of the subject. Figure 3 shows a transition in fingertip skin temperature (t=40min), before the subject performed a control action (t=50min). A time difference of about 10min occurred in between. This indicates that the fingertip skin temperature might be a useful parameter for automatic climate control purposes. The facial skin temperatures show less potential in this, because no clear trend can be recognized in it. Only a few outliers in nose skin temperature were detectable.

Remarkable is that the user still preferred the maximum radiant panel temperature, also when skin temperatures of the upper-extremities, after 100min, were back in the neutral zone.

Analysing all the data is complex, but clearly can be seen that the finger tips are the most sensitive body parts in relation to room temperature and radiation temperature, see Fig. 3. The average response skin temperature is represented by the regression line y=-0.008x+29.2, where as the response of the 4th finger is much faster and represented by y=-0.097x+32.5. The starting moment of the strong change in the finger temperature is the so called transition point. If there is enough time between the moment of occurrence of the transition point and the human control interaction (putting the heater higher), we could take corrective measures even before the occupant would feel discomfort. When a person's whole-body thermal sensation is near neutral (from -0.5–0.5), the finger temperature occurs in a wide range (28–36°C), and the finger-forearm temperature gradient between -2 and 2°C. This is in part due to the 1–2°C fluctuations in finger skin temperature occurring over time periods of a few minutes (2–5) caused by vasomotion. It is also due to people's different thermoregulatory set-points ranging around neutral [30].

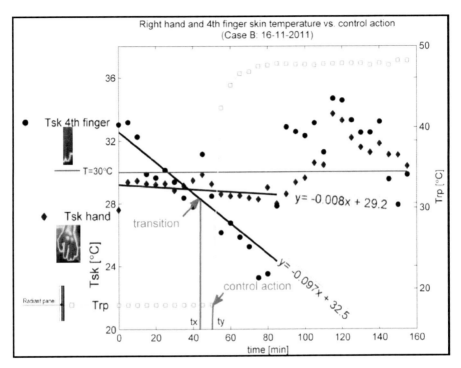

Fig. 3. Effects of cooling on hand and finger skin temperatures and effects of user control action by turning on the heating panels

These results are consistent to the findings of Wang [30] and Humphreys [31]. Facial skin temperature variations show however less correlation with the environmental control actions of the human subject. The goal of the user-controlled experiments (n=5 sessions) was to detect a feed forward transition out of the comfort zone, at time t_x, before the user took any control action at time t_y. First results show that this transition is quite difficult to detect. Standard fluctuations of 2°C in finger skin temperature make it difficult to recognize a clear trend out of the neutral zone. Additionally, in some of the user-controlled experiments a decreasing trend in finger temperature is shown before the user had taken any control action (Fig. 3), while in other sessions the decreasing trend was recognized too late, which means that subject already had taken a control action to compensate for his cool sensations. More experiments are needed to accept or reject the hypotheses.

5   Discussion

The benefits of an individual controllable task-ambient comfort system are only fully achieved when the building occupant understands the system behaviour and can deal with it. Due to the large individual differences between building occupants it is more desirable that the building system can adapt to the individual needs and behaviour of the end-user, realizing highest comfort level and highest energy savings. Therefore the user has to be the leading factor in the control of comfort systems.

Nowadays the focus of comfort control is still on the parameters at room level (i.e. surrounding the human body) and does not include the local comfort perception and actions of the individual building occupant.

Based on the acquired insights a human focussed control strategy, based on the perceived individual comfort, was derived on the level of workplace. Our experiments with the infrared sensors showed that when a person's whole-body thermal sensation is near neutral (from -0.5–0.5), the finger temperature occurs in a wide range (28–36°C), and the finger-forearm temperature gradient between -2 and 2 °C. This means that under neutral conditions, both finger temperature and the finger forearm temperature may not be good indicators of overall sensation. Fortunately environmental control action is usually not needed within the neutral range. The challenge is to detect a transition occurring out of this range. In general, building occupants interact with a building to enhance their personal comfort (e.g. by heating or cooling their local environment to improve their thermal comfort or adjust lighting system or blinds to optimize their visual comfort etc).

Distributed sensors can make it much easier to sense variables of interest directly within the occupied zone. Sensors on a desk, within chairs, or computer keyboard or mouse could measure air temperature and air motion within the occupant's local microclimate. Sensors at various levels on furniture, partitions, and ceiling tiles could detect sources of local discomfort (e.g. vertical stratification). In addition, the increased sensor densities allow measurement errors and sensor faults to be more easily spotted and corrected than is possible at present. However, the focus is still on the comfort parameters at room level (surrounding the human body) and not on the individual building occupant at user level itself. Due to large individual differences in comfort perception and user behaviour, the focus of comfort control has to be on user-adaptive and user-interactive systems instead of only environmental adaptive systems.

6   Conclusion

Due to the large individual differences between building occupants it is more desirable that the building system can adapt to the individual needs and behaviour of the end-user, realizing highest comfort level and highest energy savings. Therefore the user has to be the leading factor in the control of comfort systems. Integration of the distributed information about the user behaviour in the control of building systems results in (a) more inclusion of building occupants in the control loops (human focussed principle), (b) achieving demand-responsive energy management in buildings (energy-saving), and (c) combining all building systems into one efficient system.

Low cost wireless sensors and low cost infra red sensors networks of a smart building provide real-time comfort-energy management on workplace and even personal level. Indicators for thermal comfort on personal level could be the skin- and/or clothing temperatures of the building occupant. This information provided by the distributed sensors can be used to derive real-time set points for task-ambient conditioning (TAC) systems to the minimal required room conditions. Further research is needed on the robustness of the measurable indicators representing the individual comfort feeling of the building occupant.

References

[1] Opstelten, J., Bakker, J.E., Kester, J., Borsboom, W., van Elkhuizen, B.: Bringing an energy neutral built environment in the Netherlands under control, ECN-M-07-062, Petten (2007)

[2] Vastamäki, R., Sinkkonen, I., Leinonen, C.: A behavioural model of temperature controller usage and energy saving. Personal and Ubiquitous Computing 9, 250–259 (2005)

[3] Parys, W., Saelens, D., Hens, H.: Implementing realistic occupant behavior in building energy simulations. In: Proceedings Clima 2010 (2010)

[4] Daum, D., Morel, N.: Assessing the total energy impact of manual and optimized blind control in combination with different lighting schedules in a building simulation environment. Journal of Building Performance Simulation 3, 1–16 (2010)

[5] Yu, Z., Zhou, Y., Deter, A.: Hierarchical Fuzzy Rule-based control of renewable energy building systems. In: Proceedings CISBAT 2007 (2007)

[6] Pauw, J., Roossien, B., Aries, M.B.C., Guerra Santin, O.: Energy Pattern Generator; Understanding the effect of user behaviour on energy systems, ECN-M–09-146 (2009)

[7] van Hoof, J.: Forty years of Fanger's model of thermal comfort: comfort for all? Indoor Air 18, 182–201 (2008)

[8] Filippini, G.J.A.: De mens centraal bij het ontwerp van het binnenklimaat: ontwerp en ontwikkeling van een duurzaam lokaal klimatiseringsysteem, [Dutch], MSc thesis TU Eindhoven (2009)

[9] Watanabe, S., Melikov, A., Knudsen, G.: Design of an individually controlled system for an optimal thermal microenvironment. Building and Environment 45, 549–558 (2010)

[10] Arens, E., Federspiel, C., Wang, D., Huizenga, C.: How ambient intelligence will improve habitability and energy efficiency in buildings. Ambient Intelligence, 63–80 (2005)

[11] Tse, W., Chan, W.: A distributed sensor network for measurement of human thermal comfort feelings. Sensors and Actuators A: Physical 144, 394–402 (2008)

[12] Feldmeier, M., Paradiso, J.: Personalized HVAC Control System, in Proceedings of Internet of Things [IoT], Tokyo Japan (2010)

[13] Revel, G., Sabatini, E.: A new thermography based system for real-time energy balance in the built environment. In: Proceedings Clima 2010 (2010)

[14] Arens, E., Zhang, H.: The skin's role in human thermoregulation and comfort, Center for Built environment, University of California, Berkeley (2006)

[15] Fanger, P.O.: Thermal Comfort: Analysis and Applications in Environmental Engineering, New-York (1970)

[16] Isalgue, A., Palme, M., Coch, H., Serra, R.: The importance of users' actions for the sensation of comfort in buildings. In: Proceedings PLEA 2006, Geneva (2006)

[17] Zhang, H., Arens, E., Huizinga, C., Han, T.: Thermal sensations and comfort models for non-uniform and transient environments. Building and Environment 45, 380–410 (2010)

[18] Arens, E., Zhang, H., Huizenga, C.: Partial- and whole body thermal sensation and comfort – Part I: Uniform environmental conditions. Journal of Thermal Biology 31, 53–59 (2006)

[19] Schellen, L., Loomans, M., Van Marken Lichtenbelt, W., Frijns, A., de Wit, M.: Assessment of thermal comfort in relation to applied low exergy systems. In: Proceedings of Conference: Adapting to Change, New thinking of comfort (2010)

[20] Humphreys, M.A., Nicol, J.F.: The validity of ISO-PMV for predicting comfort votes in every-day thermal environments. In: Proceedings Moving Thermal Comfort Standards into the 21st Century, Windsor, UK (2001)

[21] Arens, E., Zhang, H., Huizenga, C.: Partial- and whole body thermal sensation and comfort – Part II: Non-uniform environmental conditions. Journal of Thermal Biology 31, 60–66 (2006)

[22] Wang, D., Federspiel, C.C., Rubinstein, F.: Modeling occupancy in single person offices. Energy and Buildings 37(3), 121–126 (2005)

[23] Zhang, H., Arens, E., Kim, D., Buchberger, E., Bauman, F., Huizenga, C.: Comfort, perceived air quality, and work performance in low-power task-ambient conditioning system. Building and Environment 45, 29–39 (2010)

[24] Munir, A., Takada, S., Matsushita, T.: Re-evaluation of Stolwijk's 25-node human thermal model under thermal-transient conditions: Predictions of skin temperature in low-activity conditions. Building and Environment 44(9), 1777–1787 (2009)

[25] Fiala, D., Lomas, K., Stohrer, M.: First principles modeling of thermal sensation responses (2003)

[26] Kang, D.H., Mo, P.H., Choi, D.H., Song, S.Y., Yeo, M.S., Kim, K.K.: Effect of MRT variation on the energy consumption in a PMV-controlled office. Building and Environment 45, 1914–1922 (2010)

[27] Page, J., Robinson, D., Morel, N., Scartezzini, J.L.: A generalized stochastic model for the simulation of occupant presence. Energy and Buildings 40, 83–98 (2008)

[28] Zhang, H., Hoyt, T., Lee, K.H., Arens, E., Webster, T.: Energy savings from extended air temperature setpoints and reductions in room air mixing. In: Proceedings International Conference on Environmental Ergonomics (2009)

[29] Karjalainen, S., Koistinen, O.: User problems with individual temperature control in offices. Building and Environment 42, 2880–2887 (2007)

[30] Wang, D., Zhang, H., Arens, E., Huizenga, C.: Observations of upper-extremity skin temperature and corresponding overall-body thermal sensations and comfort. Building and Environment 42, 3933–3943 (2007)

[31] Humphreys, M.A., McCarntney, K.J., Nicol, J.F., Raja, I.A.: An analysis of some observations of the finger temperature and thermal comfort of office workers. In: Proceedings of Indoor Air (1999)

[32] Niemelä, R., Hannula, M., Rautio, S., Reijula, K., Railio, J.: The effect of air temperature on labour productivity in call centres - a case study. Energy and Buildings 34, 759–764 (2002)

[33] Parsons, K.C.: Human thermal environments: The effects of hot, moderate, and cold environments on human health, comfort and performance. Taylor & Francis, London (2003)

[34] Leaman, A., Bordass, B.: Productivity in buildings the killer 'variables'. In: Clements-Croome, D. (ed.) Creating the Productive Workplace. Taylor & Francis, London (2006)

[35] Seppänen, O., Fisk, W.J., Lei, Q.H.: Room temperature and productivity in office work, eScholarship Repository, Lawrence Berkeley National Laboratory, University of California (2006), `http://repositories.cdlib.org/lbnl/LBNL-60952`

[36] Seppänen, O., Fisk, W.J.: Some quantitative relations between indoor environmental quality and work performance or health. International Journal of HVAC&R Research 12(4), 957–973 (2006)

[37] Olesen, B.W.: ThermCo: Thermal comfort in buildings with low-energy cooling – Establishing an annex for EPBD-related CEN-standards for buildings with high energy efficiency and good indoor environment, Deliverable D1.4 (2008)

[38] Kolarik, J., Toftum, J., Olesen, B.W.: Occupant responses and office work performance in environments with moderately drifting operative temperatures (RP-1269). International Journal of HVAC & R Research 15(5), 931–960 (2009)

[39] Lan, L., Wargocki, P., Lian, Z.: Optimal thermal environment improves performance of office work. REHVA Journal (January 2012)

Chapter 65
Reducing Ventilation Energy Demand by Using Air-to-Earth Heat Exchangers
Part 1 - Parametric Study

Hans Havtun and Caroline Törnqvist

KTH Royal Institute of Technology, Dept of Energy Technology,
Div of Applied Thermodynamics and Refrigeration, Stockholm, Sweden
`hans.havtun@energy.kth.se`

Abstract. Air-to-Earth heat exchangers (earth tubes) utilize the fact that the temperature in the ground is relatively constant during the year. By letting the air travel through an air-to-earth heat exchanger before reaching the house's ventilation air intake the air gets preconditioned by acquiring heat from the soil in the winter, and by rejecting heat to the soil in the summer. There are few studies showing how large the energy saving would be by using earth tubes. The existing studies and models are adapted to a warm climate like India and Southern Europe. Few studies are made for a Nordic climate.

To be able to use earth tubes efficiently, different parameters need to be optimized. The parameters that have the largest effect are length, depth, and diameter of the earth tube , as well as the air velocity inside the tube. To analyze this influence, a numerical model has been created in the simulation program Comsol Multiphysics 4.0a. Weather data for Stockholm, Sweden was used for all simulations. The soil type was chosen to be clay and the material of the duct was polyethylene. The parameters were varied one at a time and compared to a base case consisting of a 10 m long duct placed at a depth of 2 m and with a diameter of 20 cm. The air velocity in the duct for the base case is 2 m/s and the corresponding volumetric flow rate is 60 l/s.

Results show that longer heat exchangers with a smaller diameter, lower air velocity and buried at a deeper depth gives a larger energy saving. The increase in efficiency that comes from a deeper placed earth tube levels out at depth over 3.5 m. The decrease in efficiency that comes from an increase of the diameter of the duct levels out at diameters of 60 cm. The total energy saving for one year increased by 70 % for a 20 m long earth tube compared to a 10 m long earth tube. The energy saving for the base case is 525 kWh/year for the heating season and 300 kWh/year for the cooling season. This corresponds to an energy saving of 5 % for heating and 50 % for cooling compared to a case where no earth tube is used.

A. Håkansson et al. (Eds.): *Sustainability in Energy and Buildings*, SIST 22, pp. 717–729.
DOI: 10.1007/978-3-642-36645-1_65   © Springer-Verlag Berlin Heidelberg 2013

1 Nomenclature

Symbol	Description
c_p	Specific heat at constant pressure (J/(kg·K))
d	Duct diameter (m)
Q	Heat flow (W)
ρ	Air density (kg/m³)
t	Temperature (°C)
w	Air velocity (m/s)

2 Introduction

The usage of the earth as a heat source or heat sink is not a new invention. In fact, it has been used for thousands of years in e.g. Persian architecture. At the end of the 1970-ies and the beginning of the 1980-ies, air-to-earth heat exchangers (earth tubes) gained some attention as an alternative to air conditioning. A few systems were installed, but did not receive much attention on the market, as the efficiency was low, and the investment cost was high. Moreover, the air quality yielded from these systems was deemed unsatisfactory.

In recent years, earth tubes have gained renewed attention, mainly due to the increasingly higher requirements for energy conservation. Earth tubes utilize the fact that the temperature in the ground is relatively constant during the year. By letting the air travel through an earth tube before reaching the house's ventilation air intake, the air gets preconditioned by acquiring heat from the soil in the winter, and by rejecting heat to the soil in the summer. There are few studies showing how large the energy saving would be by using earth tubes. The existing studies and models are adapted to a warm climate like India and Southern Europe. Few studies have been made for a Nordic climate.

Several publications have treated earth tubes. However, in many of them simplifying assumptions are made such as a constant duct temperature or that no latent heat exchange takes place in the earth or that they only consider one duct.

Basal and Sodha (1986), Trombe et al. (1991), Thanu et al. (2001), Sharan and Jadhav (2003), Jalaluddin (2011), and Misra (2012) have all used experiments to investigate the performance of earth tubes. None of these investigations treated a Nordic climate.

Tzaferis et al. (1992), Bojic et al. (1997), Ståhl (2002), Lee and Strand (2006), Cucumo et al. (2008), Ascione et al. (2011), and Su et al. (2012) have all published numerical investigations. None of these, except the publication of Ståhl (2002), treated a Nordic climate, and Ståhl did not investigate the effect of latent heat exchange.

Trombe et al. (1994), Mihalakakou et al. (1994a,b, 1996), Kumar et al. (2003), Ghosal et al. (2005), Hollmuller and Lachal (2005), Wu et al. (2007), Bansal et al. (2009),

Trzaski and Zawada (2011), Badescu et al. (2011), and Bansal et al. (2012) all performed studies combining numerical investigations with experimental validation. However, none of these treated a Nordic climate.

Hollmuller (2003), gave examples of earth tube applications. A comparison between the usage of ventilation air heat recovery and earth tubes was made and the earth tubes were found to be inferior to ventilation air heat recovery.

In this paper, based on the Master thesis by Törnqvist (2011), a numerical model has been developed using Comsol Multiphysics 4.0.a in order to perform a parametric study of the influence of duct length, duct depth, duct diameter, and air velocity on the performance of the earth tube. Weather data for Stockholm, Sweden was used for all simulations.

3 Numerical Model

The earth tube has been modeled numerically using Comsol Multiphysics 4a (2010). The modeling process is quite complicated as it involves transient fluid flow and heat transfer analyses. Therefore, the conjugate module in the software was used.

The evaluation and implementation of thermo-physical properties for the earth has been a large part of the work by Törnqvist (2011). Temperature dependent models for the thermal conductivity, specific heat, density, latent heat of melting for the earth (clay) has been derived and implemented in the software. These models are vital for the accuracy of the model as the water in the clay will change phase during the year. The influence of the phase-change on the energy saving of the earth tube will be discussed later in this paper. The choice of using clay as the earth material is due to the fact that it is very common in the Stockholm area. Other thermo-physical properties, e.g. for the air and the duct wall material, have been provided by the software. All materials used have been assumed to be homogeneous. The model is three-dimensional and transient. The time step chosen is 24 hours.

The fluid flow and heat transfer model assumes that the flow is fully turbulent. For turbulence modeling, the built-in k-ε-model of the software has been used. The equations for conservation of mass, momentum, and energy, as well as the equations for the turbulence kinetic energy and its dissipation, are explained in detail in Törnqvist (2011) as well as in the documentation of the software, Comsol (2010) and are therefore omitted here.

3.1 Geometry of the Base Case

The model has the physical dimensions 6 m wide by 6 m deep by 10 m long. The reason for choosing these dimensions is that during a year, the temperature remains almost constant at 6 m below the surface.

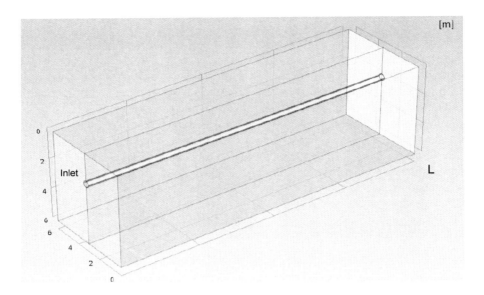

Fig. 1. The geometry of the base case

The earth tube is modeled as a cylindrical duct with a constant diameter through which the air is flowing at a constant mean velocity. The outlet of the duct is modeled as an outflow element which means that heat is only transferred through the surface by convection. The symmetry of the geometry is utilized to save computational time. The symmetry plane is modeled as an adiabatic surface. All remaining outer walls of the geometry except for the top surface, are modeled as adiabatic surfaces. The top surface assumes the ambient temperature at all times and is modeled as a temperature controlled surface which varies with time. In the simulations, the ambient temperature was modeled using a cosine function for the ease of implementation.

3.2 Boundary and Initial Conditions of the Base Case

The variation of the ambient temperature over time is shown in Fig. 2. The velocity at the inlet was modeled as a uniform velocity profile at 2 m/s. The inlet temperature of the air was assumed to be equal to the ambient temperature at all times. The turbulence intensity at the inlet and the turbulence length scale was assumed constant at the inlet at 5 % and 0.07·d respectively.

The initial conditions of the earth were assumed to be equal to the undisturbed temperature distribution of the ground. This temperature distribution was adopted from a model developed by Lee and Strand (2006). The initial temperature of the air in the duct as well as the duct walls was assigned the earth temperature at the depth of 2.0 m for the base case.

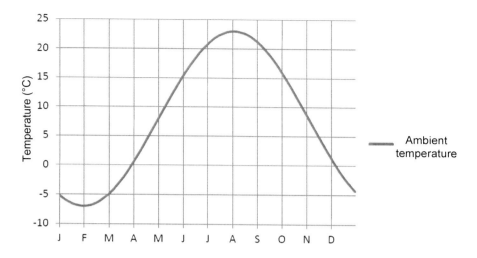

Fig. 2. Ambient temperature variation

3.3 Meshing

A free meshing scheme was implemented by the software. Close to boundaries with fluid flow interaction, wall functions were applied to resolve the viscous sub layer. The y^+ was checked and was never found to exceed the value 11.06. The mesh for the base case is shown in Fig. 3. The mesh was refined successively in order to produce accurate results while not spending unreasonable amounts of computational time. Several simulations were run to determine the final mesh. The difference in temperature between the chosen mesh and the successively finer mesh was 0.01 °C.

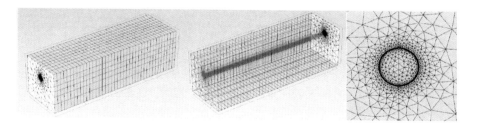

Fig. 3. Mesh used for the base case

3.4 Latent Heat Modeling

The fact that the water in the ground freezes and melts also influence the simulation results. This effect can be considerable. By taking into account the latent heat connected with the freezing and melting of the water in the ground, the energy saving increased by up to 5 % for the heating season and by up to 9 % in the cooling season. The details on how the latent heat was modeled is explained in detail in Törnqvist (2011).The difference between a model with and without latent heat for the base case is shown in Fig. 4.

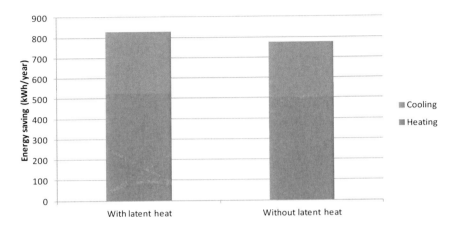

Fig. 4. Difference in energy saving for a model with and without considering latent heat of melting in the ground

4 Data Reduction

The heat that is taken up from or rejected to the earth by using the earth tube can be calculated as:

$$Q_{\text{earth tube}} = \rho \cdot w \cdot \frac{\pi \cdot d^2}{4} \cdot c_{p,\text{air}} \cdot (t_{\text{out, earth tube}} - t_{\text{amb}})$$

This heat flow also represents the ventilation energy saved, as the ventilation system in the building now does not need to change the temperature as much as it would have, if there was no earth tube installed.

The ventilation energy demand of the building considered can be calculated as:

$$Q_{\text{ventilation}} = \rho \cdot w \cdot \frac{\pi \cdot d^2}{4} \cdot c_{p,\text{air}} \cdot (t_{\text{supply}} - t_{\text{amb}}) \quad (1)$$

To put this energy saving into perspective, the relative energy saving is defined as:

$$\frac{Q_{\text{earth tube}}}{Q_{\text{ventilation}}} = \frac{t_{\text{out,earth tube}} - t_{\text{amb}}}{t_{\text{supply}} - t_{\text{amb}}} \quad (2)$$

During the cooling season, the $t_{out, earth\ tube}$ can sometimes be lower than the desired supply temperature of the building, t_{supply}. In these cases, the ventilation energy saving is defined as the energy saved for cooling the air to the temperature, t_{supply}.

It is also important to remember that during certain parts of the year, the ground duct should not be used at all. These cases occur during spring and autumn when the ambient air temperature changes faster than the ground temperature. The effect of the ground duct may hence be contrary to its intention. As an example we can consider a spring case; the ambient air temperature may be 10 °C while the ground temperature is 6 °C. Since the building still needs heating, allowing ventilation air through the ground duct will actually decrease the air temperature resulting in an increased ventilation heating demand. To eliminate this effect, the ventilation air must be allowed to bypass the ground duct system.

In this report, we have eliminated this effect by monitoring the outlet temperature from the ground duct. For the heating season, we set the energy saving equal to zero for each time step when the ambient air temperature is higher than the outlet temperature of the ground duct. For the cooling season, we set the energy saving equal to zero when the ambient air temperature is lower than the outlet temperature of the ground duct.

5 Parametric Study

The parameters varied in this study include the duct depth, duct length, duct diameter, as well as the ventilation air velocity. In table 1, the range of the parameters varied is shown. It is important to remember that not all parameters have been varied for all cases. Instead the base case (one single duct with parameter values marked in bold in table 1) was used as a starting point and one parameter were varied while the others were kept constant and equal to their base case values.

Table 1. Parameters that were varied

Parameter	Values			
Duct depth [m]	1.5	2.0	3.5	4.5
Duct length [m]	10	20	30	
Duct diameter, d, [m]	0.2	0.4	0.6	0.8
Air velocity, w, [m/s]	1.0	1.5	2.0	2.5

6 Results and Discussion

6.1 The Base Case

The base case is used as the reference case. From earlier we recall that the base case is a single duct with an air velocity of 2 m/s, a duct diameter of 20 cm, a duct depth of 2 m, and a length of 10 m. Its thermal performance is shown in table 2.

Table 2. Thermal performance of the base case

Base case	Heating season (6 Oct – 29 Mar)	Cooling season (13 Jun - 18 Sep)
Energy saving [kWh/year]	525	300
Relative energy saving [%]	8.2	52.6
Mean temperature change in earth tube [°C]	1.5	-1.9

The energy saving for the base case is 525 kWh/year for the heating season and 300 kWh/year for the cooling season. This corresponds to an energy saving of 5 % for heating and 50 % for cooling compared to if no earth tube is used.

6.2 Influence of Duct Depth

The duct depth was varied between 1.5 and 4.5 m for the base case (i.e. the duct diameter, length and air velocity was held at 20 cm, 10 m, and 2 m/s respectively). Results showed that the outlet temperatures of the earth duct differed only slightly during the heating season while the difference during the cooling season was larger. This is shown in Fig. 5, where it is obvious that the energy saving during the cooling season is much more dependent on the duct depth as compared to the heating season. The increase in total energy saving that comes from a deeper placed earth tube levels out at depth over 3.5 m.

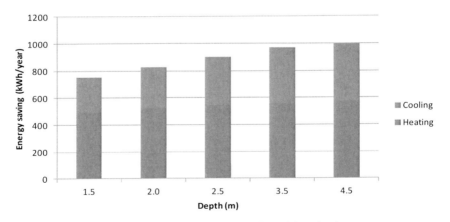

Fig. 5. Energy saving as a function of duct depth

6.3 Influence of Air velocity

The air velocity was varied between 1 and 2.5 m/s for the base case (i.e. the duct diameter, depth, and length was held at 20 cm, 2 m, and 10 m respectively). In Fig. 6, it is clearly shown that an increase in air velocity increases the energy saving.

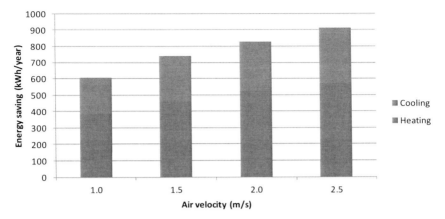

Fig. 6. Energy saving as a function of Air velocity

However, an increased air velocity also implies a higher volumetric flow rate. If the comparison is made for equal volumetric flow rate, the thermal performance of the duct can be shown.

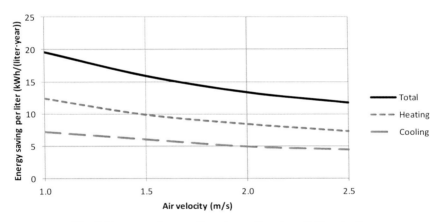

Fig. 7. Energy saving per liter air as a function of Air velocity

As shown in Fig. 7, the energy saving per liter air is decreasing. This means that although the total energy saving increases, it does not increase proportionally with increasing air velocity. Increasing the air velocity by 100 % does hence not increase the energy saving by 100 %.

6.4 Influence of Duct Diameter

The duct diameter was varied between 20 and 80 cm for the base case (i.e. the duct depth, length and air velocity was held at 2 m, 10 m, and 2 m/s respectively). Results showed that the temperature change is larger for the ducts with smaller diameter compared to the larger diameter ducts. However, as the volumetric flow rate is much larger for the larger diameter ducts, the energy saving is larger for larger diameter ducts, see Fig 8.

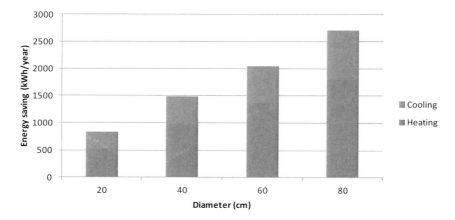

Fig. 8. Energy saving as a function of duct diameter for a constant inlet velocity

As the volumetric flow rate is increasing with increasing duct diameter in Fig. 8, it is interesting to see a comparison of the thermal performance at the same flow rate.

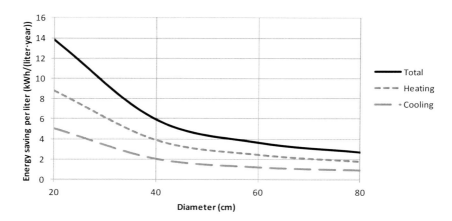

Fig. 9. Energy saving per liter air as a function of duct diameter

In Fig. 9 it is shown that the energy saving per liter air is reduced with increasing duct diameter. The decrease in energy saving per duct that comes from an increase of the diameter of the duct levels out at diameters of 60 cm.

6.5 Influence of Duct Length

The duct length was varied between 10 and 30 m for the base case (i.e. the duct diameter, depth, and air velocity was held at 20 cm, 2 m, and 2 m/s respectively). A longer duct length results in a greater temperature change of the air. This is due to the prolonged time the air spends in the earth tube allowing more heat exchange to take place. The energy saving is also increased with increasing duct length which is shown in Fig. 10.

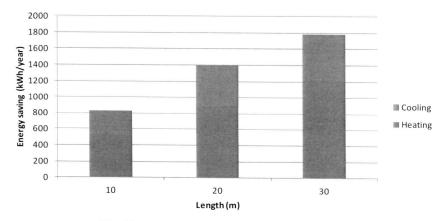

Fig. 10. Energy saving as a function of duct length

As can be seen in Fig. 10, the energy saving is greatly increased with increasing duct length. However, increasing the duct length by 100% does not give an increase in the energy saving by 100%. In fact the total increase is 69 % if the duct length is increased from 10 to 20 m. The energy saving during the cooling season does not increase as much as it does during the heating season. This is explained by the fact that during some time of the cooling season, the outlet temperature from the earth tube is lower than the desired supply temperature of the building. Therefore, there are periods of time where the air from the earth duct is mixed with ambient air to achieve the desired supply temperature.

7 Conclusions

A numerical model has been developed to simulate the thermal performance of an earth tube. The software used was Comsol Multiphysics 4.0.a. The flow was assumed fully turbulent and for turbulence modeling, the k-ε-model implemented in the software was used. A parametric study has been conducted where the influence of duct depth, duct diameter, duct length, and air velocity on the ventilation energy saving resulting from installing an earth tube has been studied.

Results show that longer heat exchangers with a smaller diameter, lower air velocity and buried at a deeper depth gives a higher energy saving. The increase in efficiency that comes from a deeper placed earth tube levels out at depths over 3.5 m. The decrease in efficiency that comes from an increase of the diameter of the duct levels out at diameters of 60 cm. The total energy saving for one year increased by 70 % for a 20 m long earth tube compared to a 10 m long earth tube. The energy saving for the base case is 525 kWh/year for the heating season and 300 kWh/year for the cooling season for a base case with a duct length of 10 m, a duct diameter of 20 cm, a duct depth of 2 m below the surface and an air velocity of 2 m/s. This corresponds to an energy saving of 5 % for heating and 50 % for cooling compared to if no earth tube is used.

Acknowledgements. This paper is based on a MSc thesis written by Caroline Törnqvist under the supervision of Hans Havtun, KTH Energy Technology. The project was initiated by the company Incoord whose support is gratefully acknowledged.

References

Ascione, F., Bellia, L., Minichiello, F.: Earth-to-air heat exchangers for Italian climates. Renewable Energy 36, 2177–2188 (2011)

Badescu, V., Laaser, N., Crutescu, R., Crutescu, M., Dobrovicescu, A., Tsatsaronis, G.: Modeling, validation and time-dependent simulation of the first large passive building in Romania. Renewable Energy 36, 142–157 (2011)

Bansal, N.K., Sodha, M.S.: An Earth-Air Tunnel System for Cooling Buildings. Tunneling and Underground Space Technology 1(2), 177–182 (1986)

Bansal, V., Mishra, R., Agarwal, G.D., Mathur, J.: Performance analysis of integrated earth–air-tunnel-evaporative cooling system in hot and dry climate. Energy and Buildings 47, 525–532 (2012)

Benkert, S., Heidt, F.D., Scholer, D.: Calculation tool for earth heat exchangers GAEA. In: Proceeding of building simulation, vol. 2. Prague: Fifth International IBPSA Conference (1997)

Bojic, M., Trifunovic, N., Papadakis, G., Kyritsis, S.: Numerical Simulation, Technical and Economic Evaluation of Air-To-Earth Heat Exchanger Coupled To a Building. Energy 22(12), 1151–1158 (1997)

Comsol, Heat Transfer Module User Guide, Comsol Multiphysics 4.0.a (2010)

Cucumo, M., Cucumo, S., Montoro, L., Vulcano, A.: A One-Dimensional Transient Analytical Model for Earth-to-Air Heat Exchangers, Taking into Account Condensation Phenomena and Thermal Perturbation from the Upper Free Surface as Well as Around the Buried Pipe. International Journal of Heat and Mass Transfer 51 (2008)

Ghosal, M.K., Tiwari, G.N., Das, D.K., Pandey, K.P.: Modeling and comparative thermal performance of ground air collector and earth air heat exchanger for heating of greenhouse. Energy and Buildings 37, 613–621 (2005)

Hollmuller, P.: Thumb Rules for Design of Earth Channels (2003), http://software.cstb.fr/articles/17.pdf (as accessed on September 15, 2010)

Hollmuller, P., Lachal, B.: Buried pipe systems with sensible and latent heat exchange: Validation of numerical simulation against analytical solution and long-term monitoring (2005), http://www.unige.ch/cuepe/html/biblio/pdf/BuriedPipes_IBPSA_2005.pdf (as accessed on September 19, 2010)

65 Reducing Ventilation Energy Demand by Using Air-to-Earth Heat Exchangers

Jalaluddin, Miyara, A., Tsubaki, K., Inoue, S., Yoshida, K.: Experimental study of several types of ground heat exchanger using a steel pile foundation. Renewable Energy 36, 764–771 (2011)

Kumar, R., Ramesh, S., Kaushik, S.C.: Performance evaluation and Energy Conservation Potential of Earth-Air Tunnel System Coupled with Non-Air Conditioned Building. Building and Environment 38 (2003)

Lee, H.K., Strand, R.K.: Implementation of an earth tube system into EnergyPlus program (2006), http://simulationresearch.lbl.gov/dirpubs/SB06/kwangho.pdf (as accessed September 18, 2010)

Mihalakakou, G., Lewis, J.O., Santamouris, M.: The Influence of Different Ground Covers On the Heating Potential of Earth-To-Air Heat Exchangers. Renewable Energy 7(1), 33–46 (1996)

Mihalakakou, G., Santamouris, M., Asimakopoulos, D.: Modelling the Thermal Performance of Earth-To-Air Heat Exchanger. Solar Energy 53(3), 301–305 (1994a)

Mihalakakou, G., Santamouris, M., Asimakopoulos, D., Papanikolaou, N.: Impact of Ground Cover on the Efficiencies of Earth-to-Air Heat Exchangers. Applied Energy 48 (1994b)

Misra, R., Bansal, V., Agarwal, G.D., Mathur, J., Aseri, T.: Thermal performance investigation of hybrid earth air tunnel heat exchanger. Energy Buildings (2012), doi:10.1016/j.enbuild.2012.02.049

Sharan, G., Jadhav, R.: Performance of Single Pass Earth-Tube Heat Exchanger: An Experimental Study (2003), http://www.builditsolar.com/Projects/Cooling/Earth%20Tubes2003-01-07GirjaSharan.pdf (as accessed on September 6, 2010)

Ståhl, F.: Preheating of Supply Air through an Earth Tube System - Energy Demand and Moisture Consequences (2002), http://www.ivt.ntnu.no/bat/bm/buildphys/proceedings/67_Staahl.pdf (as accessed on September 19, 2010)

Su, H., Liu, X.-B., Ji, L., Mu, J.-Y.: A numerical model of a deeply buried air–earth–tunnel heat exchanger. Energy and Buildings 48, 233–239 (2012)

Thanu, N.M., Sawhney, R.L., Khare, R.N., Buddhi, D.: An Experimental Study of the Thermal Performance of an Earth-Air-Pipe system in Single Pass Mode. Solar Energy 71(6), 353–364 (2001)

Törnqvist, C.: Markkanaler för förvärmning och förkylning av ventilationsluft, MSc thesis in Energy Technology, KTH, Dept. Energy Technology, EGI-2011-027MSC (2011) (in Swedish)

Trombe, A., Serres, L.: Air-earth Exchanger Study in Real Site Experimentation and Similation. Energy and Buildings 21(1994), 155–162 (2011)

Trombe, A., Pettit, M., Bourret, B.: Air Cooling By Earth Tube Heat Exchanger: Experimental Approach. Renewable Energy 1(5/6), 699–707 (1991)

Tzaferis, A., Liparakis, D., Santamouris, M., Argiriou, A.: Analysis of the accuracy and sensitivity of eight models to predict the performance of earth-to air heat exchangers. Energy and Buildings 18, 35–43 (1992)

Wu, H., Wang, S., Zhu, D.: Modelling and Evaluation of Cooling Capacity of Earth-Air-Pipe Systems. Energy Conversion and Management 48, 1462–1471 (2007)

Chapter 66
Reducing Ventilation Energy Demand by Using Air-to-Earth Heat Exchangers
Part 2 – System Design Considerations

Hans Havtun and Caroline Törnqvist

KTH Royal Institute of Technology, Dept of Energy Technology,
Div of Applied Thermodynamics and Refrigeration, Stockholm, Sweden
`hans.havtun@energy.kth.se`

Abstract. Air-to-Earth heat exchangers (earth tubes) utilize the fact that the temperature in the ground is relatively constant during the year. By letting the air travel through an air-to-earth heat exchanger before reaching the house's ventilation air intake the air gets preconditioned by acquiring heat from the soil in the winter, and by rejecting heat to the soil in the summer. There are few studies showing how large the energy saving would be by using earth tubes. The existing studies and models are adapted to a warm climate like India and Southern Europe. Few studies are made for a Nordic climate.

To be able to use earth tubes efficiently, different parameters need to be optimized. A numerical model has been developed using Comsol Multiphysics 4.0.a in order to study earth tubes with multiple ducts. Both the spacing between ducts as well as the number of ducts is simulated. Finally, results have been extrapolated to mimic an installation in a building with a large ventilation demand. Weather data for Stockholm, Sweden was used for all simulations. The soil type was chosen to be clay and the material of the duct was polyethylene.

For the cases where the duct spacing was investigated, results showed that the outlet temperature of the earth ducts changed only marginally for the three cases simulated. The energy saving per duct showed a slight increase as the spacing was increased. For the cases with different number of ducts, the energy saving increases with increasing number of ducts. However, the increase in energy saving is less than the increase in heat transfer area. The case study considering a building with a large ventilation energy demand, several configurations of earth tube installations have been investigated. Results showed that the best configuration is a case with a small velocity, small duct diameters, long ducts installed at as deep in the earth as possible. However, Once the depth goes below 3.5 m, the increase in energy saving is marginal. For the building having a ventilation air flow demand of 1000 liters/s, a configuration of 33 parallel ducts with a duct diameter of 20 cm and a spacing of 1 meter gave the greatest energy saving. For this configuration, a total energy saving of 34.2 % is possible.

A. Håkansson et al. (Eds.): *Sustainability in Energy and Buildings*, SIST 22, pp. 731–742.
DOI: 10.1007/978-3-642-36645-1_66   © Springer-Verlag Berlin Heidelberg 2013

1 Nomenclature

Symbol	Description
c_p	Specific heat at constant pressure (J/(kg·K))
d	Duct diameter (m)
Q	Heat flow (W)
ρ	Air density (kg/m³)
t	Temperature (°C)
w	Air velocity (m/s)

2 Introduction

As discussed in the preceeding paper, Havtun and Törnqvist (2012), the renewed interest in air-to-earth heat exchangers (earth tubes) has increased in recent years due to the increasingly higher requirements for energy conservation. Earth tubes utilize the fact that the temperature in the ground is relatively constant during the year. By letting the air travel through an earth tube before reaching the house's ventilation air intake, the air gets preconditioned by acquiring heat from the soil in the winter, and by rejecting heat to the soil in the summer. There are few studies showing how large the energy saving would be by using earth tubes. The existing studies and models are adapted to a warm climate like India and Southern Europe. Few studies are made for a Nordic climate.

The more recent publications include Jalaluddin (2011), and Misra (2012) who used experiments who investigated the usage of earth tubes experimentally. Ascione et al. (2011), and Su et al. (2012) who performed numerical analysis of earth tubes, as well as Bansal et al. (2009), Trzaski and Zawada (2011), Badescu et al. (2011), and Bansal et al. (2012) who combined experimental and numerical investigations. None of these treated a nordic climate.

Leopold (2006), Brinkley M (2009), Larson L (2009), and REHAU (2010) all describe actual installations of earth tubes. It is well known that one of the major difficulties with installing earth tubes is the air quality of the outlet air during certain periods of the year. The poor air quality is due to a number of reasons, the major one is the fact tha condensation of water from the air is taking place inside the tube during the cooling season. This condensation in turn may cause growth of bacteria or fungi inside the earth tube which is then spread to the indoor air. REHAU (2010) discuss this in detail and offer a solution to the problem.

In this paper, based on the Master thesis by Törnqvist (2011), and the preceding paper by Havtun and Törnqvist (2012) a numerical model has been developed using Comsol Multiphysics 4.0.a in order to study earth tubes with multiple ducts. Both the spacing between ducts as well as the number of ducts is simulated. Finally, results have been extrapolated to mimic an installation in a building with a large ventilation demand. Weather data for Stockholm, Sweden was used for all simulations.

3 Numerical Model

The earth tube has been modeled numerically using Comsol Multiphysics 4.0.a (2010). The modeling process is summarized in Havtun and Törnqvist (2012) and described in detail in Törnqvist (2011). This section is therefore a summary where necessary details for this paper are repeated.

The thermo-physical properties for the earth (clay) in the model are described in Törnqvist (2011). Temperature dependent models for the thermal conductivity, specific heat, density, latent heat of melting for the earth (clay) has been derived and implemented in the software. The choice of using clay as the earth material is due to the fact that it is very common in the Stockholm area. Other thermo-physical properties, e.g. for the air and the duct wall material, have been provided by the software. All materials used have been assumed to be homogeneous.

The fluid flow and heat transfer model assumes that the flow is fully turbulent. For turbulence modeling, the built-in k-ε-model of the software has been used. The model is three-dimensional and transient. The time step chosen is 24 hours.

3.1 Geometry of the Base Case

The model has the physical dimensions 6 m wide by 6 m deep by 10 m long. The reason for choosing these dimensions is that during a year, the temperature remains almost constant at 6 m below the surface.

The earth tube is modeled as a cylindrical duct with a constant diameter through which the air is flowing at a constant mean velocity. The outlet of the duct is modeled

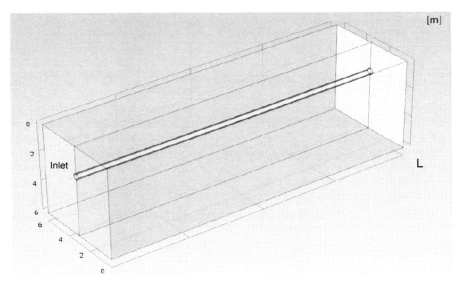

Fig. 1. The geometry of the base case

as an outflow element which means that heat is only transferred through the surface by convection. The symmetry of the geometry is utilized to save computational time. The symmetry plane is modeled as an adiabatic surface. All remaining outer walls of the geometry except for the top surface, are modeled as adiabatic surfaces. The top surface assumes the ambient temperature at all times and is modeled as a temperature controlled surface which varies with time. In the simulations, the ambient temperature was modeled using a cosine function for the ease of implementation.

3.2 Boundary and Initial Conditions of the Base Case

The variation of the ambient temperature over time is shown in Fig. 2. The velocity at the inlet was modeled as a uniform velocity profile at 2 m/s. The inlet temperature of the air was assumed to be equal to the ambient temperature at all times. The turbulence intensity at the inlet and the turbulence length scale was assumed constant at the inlet at 5 % and 0.07·d respectively.

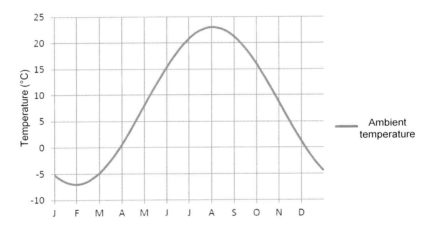

Fig. 2. Ambient temperature variation

The initial conditions of the earth were assumed to be equal to the undisturbed temperature distribution of the ground. This temperature distribution was adopted from a model developed by Lee and Strand (2006). The initial temperature of the air in the duct as well as the duct walls was assigned the earth temperature at the depth of 2.0 m for the base case.

4 Data Reduction

The heat that is taken up from or rejected to the earth by using the earth tube can be calculated as:

$$Q_{earth\ tube} = \rho \cdot w \cdot \frac{\pi \cdot d^2}{4} \cdot c_{p,air} \cdot (t_{out,\ earth\ tube} - t_{amb})$$

This heat flow also represents the ventilation energy saved, as the ventilation system in the building now does not need to change the temperature as much as it would have, if there was no earth tube installed.

The ventilation energy demand of the building considered can be calculated as:

$$Q_{\text{ventilation}} = \rho \cdot w \cdot \frac{\pi \cdot d^2}{4} \cdot c_{\text{p,air}} \cdot \left(t_{\text{supply}} - t_{\text{amb}}\right)$$

To put this energy saving into perspective, the relative energy saving is defined as:

$$\frac{Q_{\text{earth tube}}}{Q_{\text{ventilation}}} = \frac{t_{\text{out,earth tube}} - t_{\text{amb}}}{t_{\text{supply}} - t_{\text{amb}}}$$

During the cooling season, the $t_{\text{out, earth tube}}$ can sometimes be lower than the desired supply temperature of the building, t_{supply}. In these cases, the ventilation energy saving is defined as the energy saved for cooling the air to the temperature, t_{supply}.

It should also be remembered that during some time of the year, the earth tube should not be in operation at all due to the inertia of the earth making the temperature in the ground change more slowly than the change in ambient air temperature. This is explained in Havtun and Törnqvist (2012).

5   Parametric Study

The parameters varied in this study include the distance between ducts and the number of ducts when multiple ducts are used in parallel. In table 1, the range of the parameters varied is shown.

Table 1. Parameters that were varied

Parameter	Values					
Duct spacing [m]	0.5	1.0	1.5			
Number of ducts	1	2	3	4	5	6

6   Results and Discussion

6.1   Influence of Duct Spacing

By placing more than one duct in the ground, a larger energy saving can be achieved. However, placing the ducts too close to each other mean that the ducts are exchanging heat with the same earth, thereby "stealing" heat from each other. The duct spacing between two parallel ducts was varied between 0.5 and 1.5 m for the base case (i.e. the duct diameter, depth, length and air velocity was held at 20 cm, 2 m, 10 m, and 2 m/s respectively). Results showed that the outlet temperature of the earth ducts changed only marginally for the three cases simulated. The energy saving per duct showed a slight increase as the spacing was increased. In Fig. 3, the energy saving per duct is shown. In Fig. 3, the results for a single duct are also shown, the single duct represents a case with infinite spacing between the ducts.

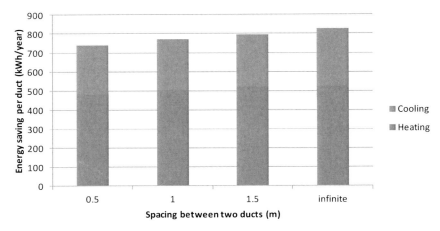

Fig. 3. Energy saving per duct as a function of duct spacing

6.2 Influence of the Number of Ducts

Six ducts was placed at a spacing of 0.5 m while the other parameters were the same as for the base case; 20 cm duct diameter, 2 m duct depth, 10 m duct length, and an air velocity 2 m/s. In Fig. 4, the energy saving is shown. As can be seen, the energy saving increases with increasing number of ducts. However, the increase in energy saving is less than the increase in heat transfer area.

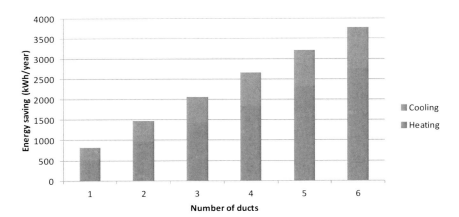

Fig. 4. Energy saving as a function of the number of ducts

66 Reducing Ventilation Energy Demand by Using Air-to-Earth Heat Exchangers

In Fig. 5, the energy saving per duct is shown. As can be seen, the energy saving decreases as the number of ducts increases. Obviously, the ducts interfere with each other as shown in the previous section.

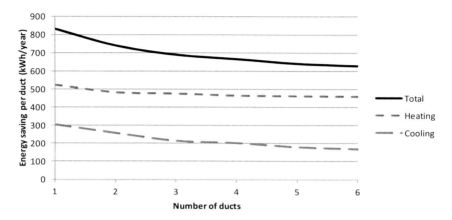

Fig. 5. Energy saving per duct as a function of the number of ducts

In Fig. 6 and Fig. 7, temperature plots are shown. Fig. 6 represents a heating season temperature plot (January) while Fig. 7 represents a cooling season temperature plot (August).

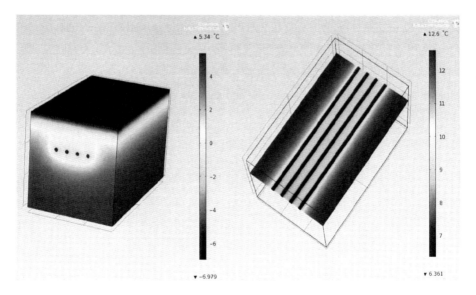

Fig. 6. Temperature plot during the heating season (January)

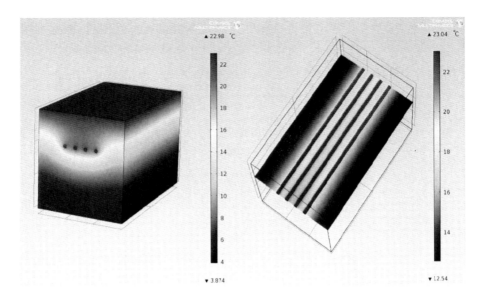

Fig. 7. Temperature plot during the cooling season (August)

6.3 System Design Considerations - Case Study

The results from the preceding parametric study, Havtun and Törnqvist (2012), and the results presented earlier in this section now allow us to consider system design aspects. From earlier results, it has been shown how different parameters influence the energy saving. However, in some cases comparisons have been made between cases with different ventilation demands which consequently results in different energy savings.

In this case study, the focus is set on a building with a larger ventilation energy demand. Different configurations of ground duct systems will be explored to find a configuration with the largest energy savings. In order to find all the necessary combinations of duct diameter, length, depth, spacing, and air velocity, the results from the parametric study above, and from the preceding paper, Havtun and Törnqvist (2012), has been interpolated and extrapolated.

The ventilation flow rate for the building considered is 1000 liter/s and the supply temperature to the building, t_{supply}, is assumed to be 18 °C. Further, it is assumed that the building is located in Stockholm, Sweden with an ambient temperature according to Fig. 2. For these assumptions, the ventilation heating energy demand is 100 MWh/year, and the cooling energy demand is 9.3 MWh/year for all cases shown here. In Table 2 and Table 3, some of the results are shown. The heating, cooling, and total energy saving is presented as a percentage of the energy demand.

Table 2. Heating, cooling, and total energy saving for an earth tube installation with 33 ducts in parallel, a duct diameter of 20 cm, and a velocity of 1 m/s

Energy saving per year

		10	10	20	20	30	30
	Length (m)	10	10	20	20	30	30
	Spacing (m)	0.5	1	0.5	1	0.5	1
	Depth (m)						
Heating (%)	1.5	9.9	10.4	16.9	17.6	22.8	23.9
Cooling (%)		33.5	34.9	55.6	57.9	63.0	65.5
Total (%)		11.9	12.5	20.2	21.1	26.2	27.4
Heating (%)	2	10.9	11.4	18.5	19.3	25.0	26.1
Cooling (%)		37.9	39.4	63.0	65.5	71.3	74.2
Total (%)		13.2	13.7	22.3	23.2	28.9	30.2
Heating (%)	2.5	11.2	11.7	19.0	19.8	25.7	26.8
Cooling (%)		42.9	44.6	71.2	74.0	80.6	83.9
Total (%)		13.9	14.5	23.4	24.4	30.3	31.7
Heating (%)	3.5	11.8	12.3	20.1	21.0	27.2	28.4
Cooling (%)		47.7	49.6	79.2	82.3	89.6	93.2
Total (%)		14.9	15.5	25.1	26.2	32.5	33.9
Heating (%)	4.5	11.9	12.4	20.1	21.1	27.3	28.5
Cooling (%)		48.9	50.8	81.2	84.4	91.9	95.6
Total (%)		15.0	15.7	25.3	26.4	32.8	34.2

As can be seen in Table 2, the heating, cooling, and total energy saving is increasing with the duct length, the duct spacing, and the duct depth. The best case can be found in the lower right hand corner of the table. For an installation with a duct depth of 4.5 m, a duct spacing of 1 m, and a duct length of 30 m, the total energy saving is 34.2 %. However, the difference between duct depths of 3.5 m and 4.5 m is very small, and the extra work needed to excavate an extra meter of earth to install the ducts may not be economically feasible.

By changing the velocity, the number of ducts needs to be changed in order to maintain the set air flow rate of 1000 liter/s. For a velocity of 1.5 m/s, the installation now requires only 22 ducts in parallel. The results for this case are shown in table 3.

Table 3. Heating, cooling, and total energy saving for an earth tube installation with 22 ducts in parallel, a duct diameter of 20 cm, and a velocity of 1.5 m/s

Energy saving per year

	Length (m)	10	10	20	20	30	30
	Spacing (m)	0.5	1	0.5	1	0.5	1
	Depth (m)						
Heating (%)	1.5	8.1	8.5	13.8	14.4	18.7	19.5
Cooling (%)		26.8	27.9	44.5	46.3	50.4	52.4
Total (%)		9.7	10.1	16.4	17.1	21.4	22.3
Heating (%)	2	8.7	9.1	14.8	15.4	20.0	20.9
Cooling (%)		32.5	33.8	53.9	56.0	61.0	63.5
Total (%)		10.7	11.2	18.1	18.9	23.5	24.5
Heating (%)	2.5	9.3	9.7	15.7	16.4	21.3	22.2
Cooling (%)		35.3	36.7	58.5	60.8	66.3	68.9
Total (%)		11.5	12.0	19.4	20.2	25.1	26.2
Heating (%)	3.5	9.6	10.0	16.3	17.0	22.0	23.0
Cooling (%)		41.6	43.3	69.1	71.8	78.2	81.3
Total (%)		12.3	12.8	20.7	21.6	26.8	27.9
Heating (%)	4.5	9.4	9.9	16.0	16.8	21.7	22.7
Cooling (%)		44.0	45.7	73.0	75.9	82.7	86.0
Total (%)		12.4	12.9	20.9	21.8	26.9	28.1

If results from Table 2 and Table 3 are compared, it can be seen that the energy saving presented in Table 2 is larger than in Table 3. This implies that increasing the velocity does not increase the energy saving. Results for the velocities 2 m/s and 2.5 m/s reported in Törnqvist (2011) confirmed this finding. In that report, simulation results for installations using larger diameter ducts were also reported. Results showed that increasing the duct diameter, and thereby decreasing the number of ducts to maintain the air flow rate requirement, made the energy saving smaller than for cases with smaller duct diameters. These findings are in agreement with Hollmuller (2003) and other publications reviewed in the introduction of Havtun and Törnqvist (2012).

7 Conclusions

A numerical model has been developed using Comsol Multiphysics 4.0.a in order to study earth tubes with multiple ducts in parallel. Both the spacing between ducts as well as the number of ducts has been simulated. Finally, results have been extrapolated to mimic an installation in a building with a large ventilation demand. Weather data for Stockholm, Sweden was used for all simulations.

66 Reducing Ventilation Energy Demand by Using Air-to-Earth Heat Exchangers 741

For the cases where the duct spacing was investigated, results showed that the outlet temperature of the earth ducts changed only marginally for the three cases simulated. The energy saving per duct showed a slight increase as the spacing was increased.

For the cases with different number of ducts, the energy saving increases with increasing number of ducts. However, the increase in energy saving is less than the increase in heat transfer area.

The case study considering a building with a large ventilation energy demand, several configurations of earth tube installations have been investigated. Results showed that the best configuration is a case with a small velocity, small duct diameters, long ducts installed at as deep in the earth as possible. However, Once the depth goes below 3.5 m, the increase in energy saving is marginal. For the building having a ventilation air flow demand of 1000 liters/s, a configuration of 33 parallel ducts with a duct diameter of 20 cm and a spacing of 1 meter gave the greatest energy saving. For this configuration, a total energy saving of 34.2 % is possible.

Acknowledgements. This paper is based on a MSc thesis written by Caroline Törnqvist under the supervision of Hans Havtun, KTH Energy Technology. The project was initiated by the company Incoord whose support is gratefully acknowledged.

References

Ascione, F., Bellia, L., Minichiello, F.: Earth-to-air heat exchangers for Italian climates. Renewable Energy 36, 2177–2188 (2011)

Badescu, V., Laaser, N., Crutescu, R., Crutescu, M., Dobrovicescu, A., Tsatsaronis, G.: Modeling, validation and time-dependent simulation of the first large passive building in Romania. Renewable Energy 36, 142–157 (2011)

Bansal, V., Mishra, R., Agarwal, G.D., Mathur, J.: Performance analysis of integrated earth–air-tunnel-evaporative cooling system in hot and dry climate. Energy and Buildings 47, 525–532 (2012)

Bansal, V., Misra, R., Agarwal, G.D., Mathur, J.: Performance Analysis of Earth-Pipe-Air Heat Exchanger for Winter Heating. Energy and Buildings 41, 1151–1154 (2009)

Brinkley, M.: Earth pipe installation. Homebuilding (2009), http://www.homebuilding.co.uk/extra/earth-pipe-installation (as accessed on September 15, 2010)

Comsol, Heat Transfer Module User Guide, Comsol Multiphysics 4.0.a (2010)

Havtun, H., Törnqvist, C.: Reducing Ventilation Energy Demand by Using Air-to-Earth Heat Exchangers – Part 1 – Parametric Study. Paper to the International Conference on Sustainability in Energy and Buildings (2012)

Hollmuller, P.: Thumb Rules for Design of Earth Channels (2003), http://software.cstb.fr/articles/17.pdf (as accessed on September 15, 2010)

Jalaluddin, Miyara, A., Tsubaki, K., Inoue, S., Yoshida, K.: Experimental study of several types of ground heat exchanger using a steel pile foundation. Renewable Energy 36, 764–771 (2011)

Larson, L.: Earth Air Tubes (2009), http://www.earthairtubes.com/index.html (as accessed on September 8, 2010)

Lee, H.K., Strand, R.K.: Implementation of an earth tube system into EnergyPlus program (2006), http://simulationresearch.lbl.gov/dirpubs/SB06/kwangho.pdf (as accessed September 18, 2010)

Leopold, A.: Green Feature: Earth Tubes (2006), http://www.aldoleopold.org/LandEthicCampaign/earth%20tubes.pdf (as accessed on September 8, 2010)

Misra, R., Bansal, V., Agarwal, G.D., Mathur, J., Aseri, T.: Thermal performance investigation of hybrid earth air tunnel heat exchanger. Energy Buildings (2012), doi:10.1016/j.enbuild.2012.02.049

REHAU, AWADUKT Thermo (2010), http://export.rehau.com/files/REHAU_AWADUKT_Thermo_342100_UK.pdf (as accessed on August 30, 2010)

Su, H., Liu, X.-B., Ji, L., Mu, J.-Y.: A numerical model of a deeply buried air–earth–tunnel heat exchanger. Energy and Buildings 48, 233–239 (2012)

Trzaski, A., Zawada, B.: The influence of environmental and geometrical factors on air-ground tube heat exchanger energy efficiency. Building and Environment 46, 1436–1444 (2011)

Törnqvist, C.: Markkanaler för förvärmning och förkylning av ventilationsluft, MSc thesis in Energy Technology, KTH, Dept. Energy Technology, EGI-2011-027MSC (2011) (in Swedish)

Chapter 67
The Green Room: A Giant Leap in Development of Energy-Efficient Cooling Solutions for Datacenters

Hans Havtun, Roozbeh Izadi, and Charles El Azzi

KTH Royal Institute of Technology, Dept of Energy Technology,
Div of Applied Thermodynamics and Refrigeration, Stockholm, Sweden
hans.havtun@energy.kth.se

Abstract. Nowadays, promoting energy-efficient solutions will have a strong return on investment not only economically but also socially and environmentally. As an added value to economic savings, the carbon footprint of the companies will be reduced and contribute to slowing down the environmental degradation and global warming. The IT-sector is no exception in this aspect. Swedish-Finnish Company *TeliaSonera* has taken a giant leap in the development of energy reduction by introducing the *Green Room Concept* which combines not only an energy-efficient cooling production but also an efficient way of distributing the cooling air flow inside the room. Both of these technologies will reduce the energy needed for cooling the equipment on their own, but combining them ensures a very energy-efficient datacenter. Although geothermal cooling or free-cooling would be the preferred choice for cooling production, it is dependent on the geographical location and climate conditions of the site, and investment potential of the company and hence not a possible solution in all cases. For these cases, the Green Room can be installed with a conventional chiller-based cooling production and still reduce the energy consumption.

The main feature of the Green Room concept is that the air coolers are installed along the length of the room parallel to the cabinet rows which will minimize the air flow complications in the cold aisle. This delivers cold air to the cabinet in a straight flow path to the racks and hence avoids the conventional raised floor to deliver cold air. One of the downsides of raised floor air-distribution is that it usually suffers from maldistribution of cooling air meaning that some racks suffer from inadequate cooling. The improved air flow distribution of the Green Room concept consequently leads to a more efficient cooling system. By carefully ensuring that no cold air bypasses the racks, any unwanted mixture of hot and cold air is also eliminated. In conventional datacenters, this problem usually occurs because proper hot/cold aisle separation is often neglected during construction. Furthermore, while many conventional datacenters face numerous problems as a result of messy cablings both inside and outside of the cabinets, the Green Room concept has managed to resolve these issues thanks to an effective cable management.

The research method in this project was to conduct a series of experimental tests to collect as much necessary data as possible. During tests, temperatures inside the cold and hot aile were monitored and recorded. Special emphasis was

---

A. Håkansson et al. (Eds.): *Sustainability in Energy and Buildings*, SIST 22, pp. 743–755.
DOI: 10.1007/978-3-642-36645-1_67   © Springer-Verlag Berlin Heidelberg 2013

put on measuring the temperature distribution in the cold aisle as the temperatures reflect the air distribution. Afterwards, the parameters used to evaluate the efficiency of the system were calculated and in some parts, simulated. Finally, the results were compared with other equivalent data and measurement from other datacenters and conclusions were made based on them. In addition to quantitative results based on a variety of calculations, the qualitative characteristics of this approach are also included to provide the readers with a better outlook on the system while being compared to other solutions currently available for datacenters. To assess the energy efficiency of a datacenter, critical measures such as Power Usage Effectiveness (PUE), defined as the total site power consumption divided by the power consumption of the IT-services, as well as the Coefficient of Performance (COP), defined as the power consumtion of the IT-services divded by the power consumtion needed to operate the cooling system. Results show that the PUE with room operational temperature of 22.5 °C can be as low as 1.05 for for a Green Room with geothermal cooling production, 1.11 for free-cooling, and 1.50 for chiller-based cooling production using old chillers and 1.32 for chiller-based cooling production using modern and more advanced chillers as compared to a PUE of 2.0 for a typical raised floor datacenter.

1 Introduction

To begin with, it seems necessary to describe what it is meant with a *datacenter* and why it is crucial to establish an efficient cooling system for it. A broad definition suggests that *a datacenter is a centralized repository, either physical or virtual, for the storage, management, and dissemination of data and information organized around a particular body of knowledge or pertaining to a particular business.* However in this report, a datacenter is defined as a facility that accommodates computer systems and a wide range of components and devices such as routers, servers, power conditioners and backup systems.

The rising demand for IT services accompanied with the increased density of equipment (fulfilling the prediction of Moore's law) have forced the IT companies to invest in new innovations and/or systems reducing the energy consumption. It has been estimated that data centers worldwide consumed about 1 % of the global electricity use in 2005 and that this electricity use doubled from 2000–2005. At an estimated growth rate of 12 % per year after 2005, the absolute electricity use could be on pace to double again by 2011. (Howard and Holmes 2012). The power consumption of a typical rack of servers in a data center is about 30 kW. Such power will increase nearly to 70 kW within the next decade due to the introduction of ultra-dense computing architectures (Lu et al. 2011). This significant growth in the number, size and power density of datacenters can also be fueled by paradigms such as software as a service (SaaS), cloud computing, a whole gamut of Internet-based businesses, social networking sites and multimedia applications and services (Marvah et al. 2010).

Data center energy consumption in the US has doubled every five years, reaching about 110 billion kWh per year by 2011 and representing an annual electricity cost of 7.4 billion US dollars. The peak load on the power grid from servers of data centers will be 12 GW by 2011, equivalent to the output of 25 baseload power plants. More than half of the electrical power is used by cooling equipment and for power conversion and distribution (Li 2012). On the other hand, the rising concerns about the environmental impact of unsustainable use of energy resources and the consequent carbon footprint have raised the alarm for the IT-sector to move towards green energy management. Regardless of the driving forces above, the aim remains the same; reduction of energy consumption.

Energy costs including cooling and thermal management make up a great portion of the datacenter operational cost. In typical datacenters, nearly half of the electrical power is consumed for cooling purposes and consequently, nearly half of the power consumption cost for datacenters is spent on cooling (Schulz 2009). For TeliaSoneras's sites in Sweden and worldwide, the PUE is approximately 1.7 and 2.0 respectively (Enlund and Lundén 2012).

Today, most of the advanced datacenters use some sort of cold aisle/hot aisle approach to separate the cold and hot exhaust air. Basically, the aim is to cool down the hot exhaust air produced by the equipment in the datacenter and to return it in such a way that the internal temperature of the equipment remain below the maximum safe temperature. Cooling of hot exhaust air into cold air is performed by Computer Room Air Conditioning (CRAC) units or ordinary air conditioners (AC). In conventional datacenters, the cold air usually is delivered to the equipment from the tiled raised access floor between two rows of racks. However, an insufficient supply of cold air leaves the devices close to the top of the cabinets considerably warmer. This is due to several aspects, the most important being mixing of cold and hot airflow which reduce effectiveness of the air distribution and decreases the efficiency of the energy consumption.

Other issues that datacenters often face is that the equipment inside the racks are changing because of renewal plans or due to the equipment layout, sometimes the rack themselves experience a dynamic configuration. Typically, ten percent of the equipment in a data center is replaced each month and the cooling design has to keep pace with this frequent change (Patankar 2010). As Beitelmal and Patel (2007) have touched upon, the CRAC units' total capacity should be adequate for the actual heat load at any given time. Other known problems to cold aisle/hot aisle system are hot-air recirculation and hotspots. Hotspots within the raised floor can increase the inlet air temperatures of the individual server racks. In order to compensate for these hotspots data center managers often choose excessively low temperature set points for the air conditioning, which can have a significant impact on the energy consumption and efficiency of the facility (Lopez and Hamann, 2011). Another problematic condition which usually occurs at conventional cold/hot aisle systems is bypass of cold airflow which occurs when the cold air which is supposed to cool down the equipment, bypasses it and goes directly to the hot aisle. All the problems mentioned can be generally addressed as mal-provisioning and leads to waste of energy (Patel et al. 2002) and (Chao and Kim 2011).

In this paper, which is based on a Master's thesis by El Azzi and Izadi (2012), results from TeliaSonera's Green Room are presented. The Green Room Concept cleverly avoids the downsides with the raised floor cooling system by placing the coolers in a face-to-face position in front of the cabinets. The results are acquired mainly by measurements; however simulations and assumptions were also utilized when necessary. Qualitative characteristics of the Green Room concept are included to provide the readers with a better outlook on the system while being compared to other solutions currently available for datacenters. The energy-efficiency of the Green Room is assessed by presenting measures such as Power Usage Effectiveness (PUE) and the Coefficient of Performance (COP) which are defined below.

2 Measures to Determine the Efficiency of the Cooling System

Basically, there are two important efficiency indicators primarily used to evaluate the progress:

1. The power usage effectiveness (PUE)
PUE = Total facility power / IT equipment power

2. The Coefficient of Performance (COP)
COP = IT equipment power / total cooling power

The IT equipment power includes the load that is associated with all IT equipment (such as servers, storage, and network equipment) and supplemental equipment (such as keyboard, video,and mouse switches; monitors; and workstations or mobile computers) that are used to monitor or otherwise control the data center (Ebbers et al. 2008). Total Facility Power includes the IT equipment as well as their supportive components like Uninterruptible Power Supply (UPS), generators, Power Distribution Units (PDU), batteries, CRACs, pumps, storage nodes and lighting.

A crucial measure to calculate the efficiency of the system is the Power Usage Effectiveness (PUE). The closer the PUE value gets to 1, the more efficient the system will be. In other words, the amount of additional energy needed to accomplish adequate removal of heat from the datacenter to the ambient must be calculated. As mentioned, the total facility power includes the IT equipment power and everything else. To be more specific, the remaining power is used mainly for cooling but also accounts for electricity losses in the UPSs, switchboards, generators, power distribution units, batteries etc. In this paper, only the power consumption of the air purifier will be considered as other power losses in the PUE calculations, and other minor equipment are deliberately omitted.

$$PUE = \frac{P_{IT} + P_{cooling\ production} + P_{Pumps} + P_{Fans} + P_{Losses,\ UPS} + P_{other}}{P_{IT}}$$

The Coefficient of Performance (COP) is also another measure to determine the cooling efficiency of the whole system. It is generally used as a measure of efficiency of heat-pumping systems, but in data centers it is defined as the total IT equipment power to be cooled divided by the total cooling power required.

$$COP = \frac{P_{IT}}{P_{cooling\ production} + P_{Pumps} + P_{Fans}}$$

3 The Green Room Components

TeliaSonera designed and constructed a complete datacenter providing the necessary conditions to have a real-life test environment. The main part of the construction is the main room mostly consisting of the server racks, the coolers and optical fiber distribution frames. In addition, several other rooms have been built to provide supplementary space for installation of the pump rack, the switchboards, DC batteries, UPSs and other indispensable elements to run the test. Besides, to get the instant temperature, pressure and power-load data to control the datacenter environment and also to properly run the test, numerous sensors and measurement devices have been put in place in different parts of the system. These sensors and devices make up a "Datacenter nervous system" providing the crucial information needed for determining the efficiency level of the Green Room.

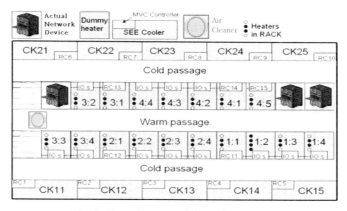

Fig. 1. Green Room and the rack components.

In total, two rows of racks have been symmetrically set up in the main room. Each row consists of 10 cabinets to keep the routers, servers and the dummy loads. The distance between the two rows is 1.5 meters which is the width of the hot aisle. As Figure 1 shows, in the first row of racks, the 1st, 2nd and the 10th cabinets are equipped with real devices. Of the remaining racks, 16 are each equipped with 2 heaters (dummy loads) each producing up to 12 kW of heat. The only remaining rack (number 1:2) is provided with 4 dummy loads which totally generate up to 48 kW of heat. The dummy loads are put in place to simulate the role of the real telecommunication devices in terms of power consumption and heat generation.

Placing the coolers in a face-to-face position in front of the cabinets will minimize the air flow complications in the cold aisle. This delivers cold air to the cabinet in a straight flow path to the racks and hence avoids the conventional raised floor to deliver cold air. The Green Room's design provides a considerably better air distribution in the room leading to a more efficient cooling cycle.

Inside the cabinets, numerous "blind panels" have been used below and above the heaters to minimize the fresh-air loss and unnecessary air flow. Each server rack has a smaller cabinet placed between the main body and the ceiling. Between the top of the cabinets and the ceiling, special rectangular metal plates are designed and tightly put in place to fully isolate the hot and cold aisles from each other. Finally, the doors of the server racks are sufficiently perforated to let the cold air in with optimal flow rate.

4 The Green Room's Cooling System

The cooling system used in the Green Room project comprises two rows of coolers in the main room and a pump rack in a room specifically designed for it. The distance between the rows of coolers and the cabinets (width of the cold aisles) is 1.2 meters. Each row consists of 5 identical SEE HDZ-3 coolers which are considered as one of the most efficient coolers in the world specifically designed for high-density datacenters. The heat from the air absorbed by the SEE coolers is in turn heat exchanged to a water circuit. The water circuit consists of a pump rack system in another room and this pump rack system itself is backed up by an identical redundant one in an adjacent room to support the cooling system in case of malfunction of the first one. Each of these two adjacent rooms mainly contained a pump rack and a Siemens Control system in charge of connecting and processing the thermal data from the whole system. Each pump rack includes one in-operation pump and a backup one. The cooling production for the Green Room during winter is presently a free-cooling system from a nearby lake. The cooling capacity is 60-750 kW. During summer, the cooling production comes from conventional chillers.

One of the main advantages of the Green Room compared to a typical raised floor approach is that the flow rate of each SEE cooler can be controlled and desirably manipulated. This feature is advantageous because a higher air flow rate can be directed towards high-heat-dissipating equipment in the racks. In fact, different equipment with different cooling needs have the chance to receive appropriate levels of air flow rather than being delivered equal amounts of fresh air.

One of the reasons for the high efficiency of the SEE coolers is the absence of a built-in filter to capture tiny particles in the air. However, absence of such a filter inside the coolers needs to be compensated by utilizing a separate air purifier in the room. In this project, an air purifier consuming up to 400 watts was used and included in P_{other} in relevant power calculations for COP and PUE.

The whole cooling system of the Green Room (including the coolers and the pumps) is supplied by a huge cooling production system dedicated to the whole building where the test room is located. To calculate the key values of this test including the COP and PUE, it is necessary to cover all the components contributing to the Green Room's cooling system. To achieve this objective, the electricity consumption of the coolers and the pump racks were precisely measured during the test. Since the Green Room is supposed to be a universal solution, it was necessary to calculate the required electricity consumption for chiller-, free-cooling-, and geothermal-based systems shown in Table 1.

Table 1. Different cooling methods considered in COP and PUE calculations

Cooling Method	Cooling Production Based on:	Internal Cooling System
Entirely Chiller-based	Chillers	Conventional CRAC Units
Test Site Summer	Chillers	SEE Coolers (Green Room)
Test Site Winter	Free Cooling	SEE Coolers (Green Room)
Geothermal Cooling	Geothermal Cooling	SEE Coolers (Green Room)

4.1 Test Description

During the test period, many values including the power load of the dummy heaters, fan speed of SEE coolers and the pump speed were changed in order to get wide variety of data leading to the best conclusions. In general, 25 tests were performed, and some of the results are shown here divided into three groups:

4.1.1 Efficiency Testing: Test 3 – 14 and Test 25

In total, 13 tests were carried out to specifically evaluate the efficiency of the cooling system by applying different values for pump speed, fan speed of the SEE coolers, and heat load. In addition, before beginning Test 13, all the gaps, empty spaces, holes and openings in inside the room were carefully taped up to ensure the minimum air leakage and fluid mixture.

4.1.2 Flow Leakages and Temperature Distribution Testing: Test 12

Test 12 was exclusively conducted to determine which parts of the room were suffering from air flow leakage as a result of sealing imperfections or unwanted hot/cold air mixture. To achieve so, an infra-red camera was employed to provide accurate thermal images of different parts of the room including back and front of the cabinets, cable trays, SEE coolers, hot aisle ceiling, containments and finally the whole cold aisles themselves.

4.1.3 Containment Effect Testing: Test 17 – 22

This series of tests were particularly carried out in order to determine how much the hot/cold air mixture prevention would be effective compared to a half-sealed or entirely unsealed room. To achieve this, certain coverings including the cable racks' cover plates and the cardboard-made vertical and horizontal stripes used on the walls of the room (adjacent to the two coolers' rows) were all removed. The information obtained from these tests will be used to investigate how effective the sealing was in terms of energy efficiency.

In addition to the tests above, other tests were carried out to investigate the thermal response of the Green Room in case of a cooling production failure. These results are presented in El Azzi and Izadi (2012).

5 Results and Discussion

5.1 Undesired Heat Transfer vs. Air Flow

Perhaps one of the largest problems that affect the efficiency of the cooling system in a datacenter is the unwanted heat transfer in the room between the cooling fluid and the environment before reaching the computer cabinets. In order to show how well cold fluid is isolated in the Green Room, the temperature difference between the outlet of the coolers and the inlet of the cabinets was measured for each test. The temperatures were measured by radio loggers which had been calibrated to an error of ± 0.5 °C. The temperature difference was measured on three vertical levels for each row and averaged over the entire aisle. As can be seen in Table 2, the temperature differences are very small and in some cases even smaller than the error of the temperature reading devices. Even if a maximum error would be considered, the temperature difference is very small compared to conventional raised floor cooling systems as will be discussed below.

Table 2. Associated ΔT with position of Radio loggers on the coolers and cabinets

	Row 1 (CKA-CKE)		
Test #	ΔT high	ΔT middle	ΔT low
9	~0	0.78	0.07
10	0.39	0.63	0.3
11	0.27	0.75	0.35
13	0.15	1.09	0.11
14	0.27	0.65	0.16
17	0.14	0.3	0.43

For all the data presented in this section, only a relatively small temperature gain of the cooling air before it reaches the cabinets (a maximum value of about 1 °C) occur. In contrast, in a published paper by TeliaSonera in 2009 (Enlund, 2009), similar measurements were done on five conventional datacenter sites equipped with raised floor cooling systems. Figure 2 illustrates the difference in results obtained from the earlier

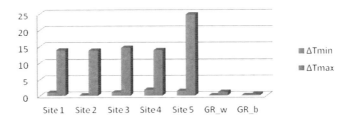

Fig. 2. Temperature differences between the outlet temperature from the cooler and the inlet temperature of the rack of 5 other TeliaSonera datacenters compared with Green Room

TeliaSonera paper compared to the present results. For each site, the smallest and the largest temperature difference between the outlet temperature from the cooler and the inlet temperature of the rack is shown. For the Green Room, two results are shown, the worst case (GR_w) and the best case (GR_b).

As can be seen in Figure 2, the Green Room has very small difference between the largest and the smallest temperature difference regardless if the best or worst case is considered. Looking at the data from conventional datacenters, for all sites there is a large difference between that largest and the smallest temperature difference indicating that the airflow distribution is far from uniform. In contrast, the Green Room provides a much more uniform airflow.

5.2 Calculation of COP and PUE

Based on the COP and PUE results from the tests presented in Table 3.

Table 3. COP and PUE values based on cooling production in Test Site during both summer and winter

Test	Total Load (kW)	IT Load (kW)	Cooling within room	Cooling production (kW)		COP		PUE	
#	(kW)	(kW)		Winter	Summer	Winter	Summer	Winter	Summer
3	160	157.9	1.9	14.5	46.4	9.8	3.31	1.119	1.322
4	159.3	157.6	1.6	14.4	46.2	9.99	3.34	1.114	1.316
5	353.4	346.6	3.2	31.9	102.5	10.06	3.34	1.122	1.326
6	355.6	350.7	3.2	32.1	103.1	10.07	3.34	1.116	1.318
7	351.3	348.5	3.2	31.7	101.9	10.05	3.34	1.109	1.311
8	356.9	350.5	4.2	32.2	103.5	9.81	3.32	1.123	1.327
9	355.9	349.2	5.9	32.1	103.2	9.36	3.26	1.129	1.333
10	355.3	350.7	9.5	32.1	103	8.54	3.16	1.133	1.335
13	324.8	302.5	6	29.3	94.2	9.2	3.24	1.192	1.406

In Table 3, it can be seen that the COP is around 10 during winter and around 3.2 during summer. The corresponding PUE data is around 1.12 and 1.32 for the winter and the summer respectively. One should recall that the cooling production during winter is free-cooling while the cooling production during summer is chiller based. The chiller used in the Green Room is modern and quite advanced which hence give very good performance. For older chillers, the PUE is expected to be around 1.5. If this is compared to a raised floor cooling system for which the PUE usually goes up to and sometimes beyond 2.0, it can be concluded that by changing only the air distribution system to the Green Room Concept, significant energy savings can be achieved.

Based on the results from the tests, the COP and PUE have been estimated for different cooling production technologies and is presented in Table 4.

Table 4. COP and PUE values for cooling production based on geothermal and free cooling

Test	Total Load	IT Load	Cooling within room	Conventional Chillers		Geothermal	
#	(kW)	(kW)	(kW)	COP	PUE	COP	PUE
3	160	157.9	1.9	2.05	1.509	23.94	1.058
4	159.3	157.6	1.6	2.06	1.503	25.12	1.054
5	353.4	346.6	3.2	2.07	1.515	25.55	1.061
6	355.6	350.7	3.2	2.07	1.506	25.59	1.055
7	351.3	348.5	3.2	2.07	1.497	25.5	1.049
8	356.9	350.5	4.2	2.05	1.515	23.97	1.062
9	355.9	349.2	5.9	2.03	1.521	21.48	1.068
10	355.3	350.7	9.5	1.99	1.522	17.6	1.072
13	324.8	302.5	6	2.03	1.605	20.62	1.127

Based on these estimations, the best efficiency was obtained for geothermal cooling production, with a COP reaching up to 25.5 (at a Green Room temperature of 22.5 °C) and a PUE as low as 1.05.

5.3 Heat Leakage and Temperature Distribution in the Cold Aisle

As Figure 3 below show, the higher part of the cold aisle generally receives more warmth compared to the bottom. This is due to heat conduction through the roof of the cold aisle. The hot air return ducts to the SEE coolers are located on top of the cold aisle and heat is conducted through the roof. To illustrate the temperature distribution in the cold aisle, a net was suspended between the SEE coolers (to the right) and the racks (to the left). To be able to see the temperature distribution, some of the net mesh was taped with ordinary paper tape. In this way, the infra-red camera can capture the air temperature. In Figure 3 below, the air temperature can be seen on the net.

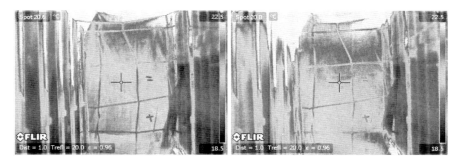

Fig. 3. Thermal images of the 1st cold aisle captured at two different occasions

5.4 Energy Reduction with the Green Room Concept - Economical Implications

As was mentioned before, the Green Room approach has the potential to be utilized in all TeliaSonera's 12,000 Telecom sites operating in Sweden (including datacenters, core sites for mobile operations and exchange stations). According to data provided by TeliaSonera (Enlund and Lundén, 2012) cooling constitutes 17.6 % of the total annual electricity consumption of TeliaSonera.

Based on these facts and assuming an average PUE of 1.7 for the telecom sites before implementation of the Green Room approach, at least 61.5 GWh of electricity can be saved annually if the Green Room is correctly utilized in Sweden. In addition, based on the suggestions made by the company's experts, the average PUE for Telia-Sonera's telecom sites around the world was considered to be 2.0. Based on this assumption, 236.9 GWh of electricity can be saved if all TeliaSonera telecom sites implement the Green Room system. A rough financial study is carried out to estimate the amount of money expected to be saved as the result of implementation of Green Room concept. As a result of these calculations, it is concluded that more than 9 million USD can be saved annually if a PUE of 1.1 is achieved by implementation of the Green Room in all TeliaSonera datacenters in Sweden. In addition, more than 31 million USD can be saved if this technology is applied to all TeliaSonera datacenters worldwide.

6 Conclusions

Different categories of tests with different purposes were carried out but the main aim of most of them was to examine the efficiency of the system through performing COP and PUE calculations. The Green Room was proven to have many advantages which have helped it to outpace many of its market rivals in terms of efficiency. For instance, COP estimated in this paper can rise to 25.5 if the Green Room concept is combined with geothermal cooling production and reaching a PUE as low as 1.05. For free-cooling, the PUE was about 1.12.

Comparing only the air distribution system to a raised floor system, the PUE for the Green Room having a chiller-based cooling production has a PUE of about 1.32 for a modern chiller and 1.5 for an older one. This should be compared to the value of 2.0 which is normally accepted as a standard value for chiller-based raised floor cooling systems.

The advantages of the Green Room approach over many of its rivals in the market are effective aisle sealing, lack of a raised floor, fluid-mix prevention, distinctive layout of CRAC units inside the room, efficient cable management. One of the main advantages of the Green Room over typical raised floor approach is that the flow rate of each SEE cooler can be controlled and desirably manipulated. This feature is advantageous because a higher air flow rate can be directed towards high-heat-dissipating equipment in the racks. In fact, different equipment with different cooling needs have the chance to receive appropriate levels of air flow rather than being delivered equal amounts of fresh air.

The project was focussed on assessing the efficiency of the Green Room concept for high-density datacenters. However, most of the datacenters utilized by TeliaSonera are categorized as mid-density and therefore this study should be expanded to investigate the efficiency of not only mid-density but also low-density datacenters. In addition, it was realized that a great advantage of the Green Room concept is being capable to manipulate the air flow towards the hotspots in the aisle. Since this action will definitely save even more energy by reducing the unnecessary generation of cold air as well as reducing the fan power needed for air distribution.

Acknowledgements. This paper is based on a MSc thesis written by Charles El Azzi and Roozbeh Izadi under the supervision of Hans Havtun, KTH Energy Technology. The project was initiated by the company TeliaSonera whose support is gratefully acknowledged.

References

Beitelmal, A.H., Patel, C.D.: Thermo-Fluids Provisioning of a High Performance High Density Data Center. Distributed and Parallel Databases 21(2-3), 227–238 (2007), doi:10.1007/s10619-005-0413-0

Chao, J., Kim, B.: Evaluation of air management system's thermal performance for superior cooling efficiency in high-density data centers. Energy and Buildings 43(9), 2145–2155 (2011)

Ebbers, M., Galea, A., Schaefer, M., Tu Duy Khiem, M.: The Green Data Center: Steps for the Journey, An IBM Red paper publication (2008), `http://ibm-vbc.centers.ihost.com/site-media/upload/briefingcenter/ebc-2/flyer/Green%20Data%20Center.pdf`

El Azzi, C., Izadi, R.: Green Room: A Giant Leap in Development of Green Datacenters, Master's thesis EGI-2012-012MSC, KTH Royal Institute of Technology, Stockholm, Sweden (2012)

Enlund, S., Lundén, D.: TeliaSonera. Personal communication (2012)

Enlund, S.: Measurements Q 2009 Haninge DC (Project Sauna). TeliaSonera Energy Solutions (2009)

Howard, A.J., Holmes, J.: Addressing data center efficiency: lessons learned from process evaluations of utility energy efficiency programs. Energy Efficiency 5, 137–148 (2012), doi:10.1007/s12053-011-9128-4

Li, K.: Optimal power allocation among multiple heterogeneous servers in a data center. Sustainable Computing: Informatics and Systems 2(1), 13–22 (2012), `http://dx.doi.org/10.1016/j.suscom.2011.11.002`

Lopez, V., Hamann, H.F.: Heat transfer modeling in data centers. International Journal of Heat and Mass Transfer 54(25-26), 5306–5318 (2011), `http://dx.doi.org/10.1016/j.ijheatmasstransfer.2011.08.012`

Lu, T., Lü, X., Remes, M., Viljanen, M.: Investigation of air management and energy performance in a data center in Finland: Case study. Energy and Buildings 43(12), 3360–3372 (2011)

Marwah, M., Maciel, P., Shah, A., Sharma, R., Christian, T., Almeida, V., Araújo, C., Souza, E., Callou, G., Silva, B., Galdino, S., Pires, J.: Quantifying the Sustainability Impact of Data Center Availability. ACM SIGMETRICS Performance Evaluation Review 37(4), 64–68 (2010), `http://dx.doi.org/10.1145/1773394.1773405`

Patankar, S.V.: Airflow and Cooling in a Data Center, 2010, Airflow and Cooling in a Data Center. J. Heat Transfer 132(7), 073001 (2010), http://dx.doi.org/10.1115/1.4000703

Patel, C.D., Sharma, R., Bash, C.A., Beitelmal, A.: Thermal Considerations in Cooling Large Scale High Compute Density Data Centers. In: ITherm 2002. Eighth Intersociety Conference on Thermal and Thermomechanical Phenomena in Electronic Systems, pp. 767–776 (2002) ISBN 0-7803-7152-6

Schulz, G.: The Green and Virtual Data Center, pp. 8–139 (2009) ISBN 978-1420086669

Chapter 68
The Impacts of Contributory Factors in the Gap between Predicted and Actual Office Building Energy Use

Emeka E. Osaji, Subashini Suresh, and Ezekiel Chinyio

School of Technology, University of Wolverhampton, WV1 1LY, United Kingdom
e.osaji@wlv.ac.uk, arcosaji@yahoo.com

Abstract. Ten years ago, the primary author developed the Building Energy-Efficient Hive (BEEHive) concept in order to demonstrate in theory that environmental design – which is aimed at addressing environmental parameters – can support the design and operation of energy-efficient office buildings. This was a result of his analysis of the spheroid form's efficiency in nature, and his development of a spheroid-like energy-efficient office built form that encloses and shades the most volume of office space with the least surface area possible. The BEEHive concept also incorporates several other aspects of the environmental design philosophy, including: site considerations (location and weather, microclimate, site layout and orientation); built form (shape, thermal response, insulation and windows/glazing); ventilation strategy; daylighting strategy; and services strategy (plants and controls, fuels and metering). Furthermore, several notable environmental design advocates and practitioners have made significant contributions in order to improve building performance. However, in practice environmental design has had limited success in the attainment of balance and optimisation in all aspects of energy use; hence there is typically a gap between predicted and actual office building energy use. This study has established the impacts of contributory factors in the gap between predicted and actual office building energy use, and it is a part of the primary author's doctor of philosophy (Ph.D) research. It involved a combination of literature reviews, multiple case study research and comparative studies in order to build theory, and it established the reasons for the gap, as well as the best ways to bridge it for improved office building environmental design and energy performance. Analysis of the literature revealed two types of gaps, and these are a gap increase and a gap decrease, which are among the impacts attributable to contributory factors in the gap between predicted and actual office building energy use such as: the nature of environmental design measures implemented; weather variation and microclimates; unavailability of reliable building energy use data; limitations of building energy simulation software; level of hours of operation; and level and nature of occupancy. Amongst these, the key contributors to gap increases are increases in hours of operation and occupancy, and weather variation and microclimates. Their respective major impacts are discrepancy between predicted and actual hours of operation and increased energy use, increased heat output and uncertainties, and variable heating and cooling requirements. The key contributors to gap decreases are environmental design measures such as the use of: natural ventilation strategies; daylighting

---

A. Håkansson et al. (Eds.): *Sustainability in Energy and Buildings*, SIST 22, pp. 757–778.
DOI: 10.1007/978-3-642-36645-1_68   © Springer-Verlag Berlin Heidelberg 2013

strategies; solar photovoltaic systems; and spheroid-like built forms. Their respective major impacts are: the production of more energy than an office building uses; and energy uses that are below, for instance, Energy Consumption Guide 19 typical and good practice energy use for office type 4, and relevant ASHRAE (American Society of Heating, Refrigerating and Air Conditioning Engineers) standards. This study has contributed to ideas for the development of a Building Management System for Bridging the Gap, otherwise known as 'BMS-Optimum', which is comprised of optimum conditions and considerations such as: optimum environmental design principles; optimum weather and microclimate considerations; accessibility to reliable office building energy use data; optimum building energy and environmental assessment; optimum hours of operation; and optimum level and nature of occupancy. Future work will include further development of BMS-Optimum, using methods such as: multiple case study research supported by building energy use audits, observations, questionnaire surveys, interviews, benchmarking and comparative studies; building energy simulations within multiple scenarios, parameters and variables, and supported by benchmarking and comparative studies; and peer reviews and focus group sessions. These will also help establish and validate a Framework for Improved Environmental Design and Energy Performance (FEDEP).

Keywords: Energy Use, Environmental Design, Heating, Occupancy, Office Building, Ventilation, Weather.

1 Introduction

According to Holm (2006), environmental design involves addressing environmental parameters. Environmental design aims to facilitate the design and operation of energy-efficient buildings while addressing environmental parameters, including: site considerations (location and weather, microclimate, site layout and orientation); built form (shape, thermal response, insulation and windows/glazing); ventilation strategy; daylighting strategy; and services strategy (plants and controls, fuels and metering) (Shu-Yang et al., 2004; Strong, 2004; and Thomas, 2002). Furthermore, environmental design aims to mitigate negative environmental impacts while creating a harmonious relationship between the built, human and natural environments (McLennan, 2004). Such a harmonious relationship is important for building energy-efficiency, and Hui (2002) states that building operations and energy-efficiency are among the green features for environmental design. In defining an energy-efficient building, Majumdar (2002) and Crawley et al. (2008) describe it as one that attains balance and optimisation in all aspects of its energy use that include: building form; building envelope; materials; daylighting and solar; infiltration; HVAC systems; space-conditioning; ventilation; electrical systems and equipment; and renewable energy systems. Thomas (2002) explains the economic importance of designing for energy-efficiency by stating that contrary to sceptical beliefs, the implementation of an environmental design philosophy has both significant short-term and long-term economic benefits for the end user.

Ten years ago, the primary author developed the Building Energy-Efficient Hive (BEEHive) concept in order to demonstrate in theory that environmental design – which is a philosophy aimed at addressing environmental parameters – can support the design and operation of energy-efficient office buildings (Osaji, 2002). This was a result of his analysis of the spheroid form's efficiency in nature, and his development of a spheroid-like energy-efficient office built form that encloses and shades the most volume of office space with the least surface area possible (Osaji, 2002). The BEEHive concept also incorporates several other aspects of the environmental design philosophy (Osaji, 2002). Furthermore, several notable environmental design advocates and practitioners have made significant contributions in order to improve office building energy performance, and these include: Arup (Cramer, 2011; and Foster and Partners, 2012); Norman Foster (Foster and Partners, 2012); HOK (Beautyman, 2006; and Cramer, 2011); Ken Yeang (Greener Buildings, 2008); and several others.

However, many modern buildings apparently do not perform well in terms of their energy use (Leaman et al., 2010). For instance, approximately 18 percent of United Kingdom (UK) CO_2 emissions are attributable to energy use in the non-domestic building sector, which includes offices (UK-GBC, 2011). Office buildings contribute significantly to energy use and CO_2 emissions, and to the total energy used by buildings globally (Action Energy, 2003; Mortimer et al., 2000; and Wade et al., 2003). For instance, UK office buildings are significant contributors to energy use and CO_2 emissions whereby combined heating and cooling loads contribute about 61 percent to total annual energy use (Mortimer et al., 2000; and Wade et al., 2003). United States of America (U.S.) office buildings use the most energy of all U.S. building types. The 2003 Commercial Buildings Energy Consumption Survey (CBECS) shows that U.S. office building energy use represents 19 percent of all U.S. commercial energy use (EIA, 2011). The expenditure associated with such energy use exceeds 15 billion U.S. dollars (EIA, 2011). This is the highest energy expenditure among all U.S. commercial building types, and constitutes 23 percent of their energy expenditure (EIA, 2011).

Unfortunately, there is typically a gap between predicted and actual building energy use (Azar and Menassa, 2010), and such significant gaps have been reported to occur (Bordass et al., 2004). In fact, these include gaps between predicted and actual office building energy use as described in: Diamond et al. (2006); EERE (2004c); EERE (2006c); Tuffrey (2005); and Wellcome Trust (2011).

For instance, City Hall London had a 50 percent gap between its predicted energy use of 250 kWh/m^2 and actual energy use of 376 kWh/m^2 for 2004-2005 (Tuffrey, 2005). This gap occurred despite the use of the Building Research Establishment Environmental Assessment Method (BREEAM), and building energy simulation for City Hall London's energy assessment (Foster and Partners, 2002). This gap also occurred despite City Hall London receiving a BREEAM excellent rating in the Design, Operation and Management categories (Osaji et al., 2007; and Tuffrey, 2005).

30 St Mary Axe London had a presumed 23.56 percent gap increase between its predicted energy use of 174 kWh/m^2 (Foster and Partners, 2006b), and its predicted-actual energy use of 215 kWh/m^2 (Buchanan, 2007; and CTBUH, 2009). Despite the

implementation of environmental design measures (Foster and Partners, 2006b), 30 St Mary Axe London's presumed 23.56 percent gap still occurred. The environmental design measures were meant to ensure 30 St Mary Axe London's use of 50 percent less energy than a similar air conditioned prestige UK office building (Foster and Partners, 2006b).

The Gibbs Building had a 0.65 percent gap between its predicted and actual energy use in 2005. This is based on its predicted energy use of 309 kWh/m^2 and actual energy use of 311 kWh/m^2 in 2005 (Wellcome Trust, 2011). Despite the implementation of commendable environmental design measures (Wellcome Trust, 2011), the Gibbs Building's 0.65 percent gap still occurred. The environmental design measures were aimed at ensuring the Gibbs Building uses significantly less energy than a standard building of similar volume (Wellcome Trust, 2011). The Gibbs Building had a 6.8 percent gap between its predicted and actual energy use in 2006. This is based on its predicted energy use of 309 kWh/m^2 and actual energy use of 330 kWh/m^2 in 2006 (Wellcome Trust, 2011). Despite the implementation of commendable environmental design measures (Wellcome Trust, 2011), the Gibbs Building's 6.8 percent gap still occurred. The environmental design measures were aimed at ensuring the Gibbs Building uses significantly less energy than a standard building of similar volume (Wellcome Trust, 2011).

In the case of the Caribou Weather Forecast Office, it had an 11.85 percent gap increase between its predicted and actual energy use for 2003-2005. This is based on a predicted energy use of 447 kWh/m^2, and an actual energy use of 500 kWh/m^2 from utility bills for 2003-2005 (Diamond et al., 2006; and EERE, 2004c). The Caribou Weather Forecast Office's 11.85 percent gap increase occurred despite the implementation of environmental design measures described in EERE (2004c).

In the case of the U.S. EPA Science and Technology Center, it had a 1.66 percent gap increase between its predicted and actual energy use for 2003-2005. This is based on a predicted energy use of 842 kWh/m^2 from simulation data for 2002, and an actual energy use of 856 kWh/m^2 from utility bills for 2003-2005 (Diamond et al., 2006; and EERE, 2006c). The U.S. EPA Science and Technology Center's 1.66 percent gap increase occurred despite the implementation of environmental design measures described in EERE (2006c).

The cases of City Hall London, 30 St Mary Axe London, the Gibbs Building, the Caribou Weather Forecast Office, and the U.S. EPA Science and Technology Center depict instances where an office building energy use gap increase occurred despite the implementation of environmental design measures.

However, there is the case of the Commerzbank Tower whose actual electricity energy use was about 20 percent less than predicted (Buchanan, 2007). In the case of the Hawaii Gateway Energy Center, it had a 36 percent gap decrease between its predicted and actual energy use for 2006-2007. This is based on a predicted energy use of 136 kWh/m^2 from simulation data for 2003, and an actual energy use of 87 kWh/m^2 from utility bills for 2006-2007 (EERE, 2009c). The Hawaii Gateway Energy Center's 36 percent gap decrease is partly attributable to the implementation of environmental design measures as described in EERE (2009c). Fortunately, the implementation of environmental design measures contributed to the Hawaii Gateway

Energy Center's 36 percent gap decrease between its predicted and actual energy use for 2006-2007. According to EERE (2009c), one of these key environmental design measures is the use of a 20 kW photovoltaic system for generation of 24,455 kWh of electricity annually. These two particular cases depict instances where an office building energy use gap decrease occurs. This implies that actual energy use needs to be less than predicted energy use in order to decrease the gap between predicted and actual office building energy use. Therefore, the best ways to achieve such a gap decrease need to be established. For instance, Menezes (2011) states that initiatives by the Technology Strategy Board (TSB), as well as Royal Institute of British Architects (RIBA) and Chartered Institute of Building Services Engineers (CIBSE) aim to reduce the gap between predicted and actual building energy use. These initiatives are: TSB's Building Performance Evaluation Programme; and RIBA and CIBSE's CarbonBuzz (Menezes, 2011). This is important in order to bridge the gap between predicted and actual UK office building energy use for improved environmental design and energy performance.

According to Demanuele et al. (2010), the significant gap that exists between the predicted and actual energy use of buildings is mainly as a result of a lack of understanding of the factors that affect building energy use. According to Leaman et al. (2010), several modern buildings experience poor performance, and this causes an embarrassment that leads to the non-publication of results and the perpetuation of past mistakes. This then makes it difficult to gain access to reliable data on UK building energy use and CO_2 emissions (UK-GBC, 2011). In fact, attempts by this study to gain access to both predicted and actual UK office building energy use have been challenging. Therefore, Pett et al. (2005) suggests that there is a need for a centralised registry of UK building energy use data or UK building energy-efficiency data.

Furthermore, the causes of the significant gap existing between predicted and actual energy use of office buildings also appear to be factors related to the building itself such as increased occupancy and operational hours (Tuffrey, 2005). Additionally, warm weather (Wellcome Trust, 2011), and the simulation process (Demanuele et al., 2010) are also factors responsible for such gaps. Since such gaps occurred despite the implementation of environmental design measures, it is necessary to establish the impacts of contributory factors in such gaps in order to improve office building environmental design and energy performance.

2   Aim and Methods of This Study

The aim of this study is to establish the impacts of contributory factors in the gap between predicted and actual office building energy use, and it is a part of the primary author's doctor of philosophy (Ph.D) research. This study has been fulfilled using a combination of literature reviews, multiple case study research and comparative studies in order to build theory. This involved a desk based study in order to establish: the impacts of contributory factors in the gap between predicted and actual office building energy use; and a building management system (BMS-Optimum) for bridging this gap. The following methods were used:

- Data collection from literature of predicted and actual office building electricity and/or gas energy use.
- Multiple case study research of predicted and actual office building energy use, including use of energy use audits in order to establish their energy performance, and the several instances whereby gaps have occurred.
- Data collection from literature of the various factors that contribute to the gap between predicted and actual office building energy use.
- Comparative studies of the various factors that contribute to the gap between predicted and actual office building energy use, and the various ways they affect office building energy performance.
- Drawing conclusions from the deduced relationships between the factors that contribute to the gap between predicted and actual office building energy use, and their effect on office building energy performance.

3   Findings

This study has established three possible types of gaps between predicted and actual office building energy use. These are a gap increase, a gap decrease, and a zero gap, which are among the possible impacts attributable to contributory factors such as: the nature of environmental design measures implemented; weather variation and microclimates; unavailability of reliable building energy use data; limitations of building energy simulation software; level of hours of operation; and level and nature of occupancy.

3.1   Nature of Environmental Design Measures Implemented

3.1.1   Type of Orientation

According to Foster and Partners (2002; and 2003), one of the key environmental design measures implemented in City Hall London is its type of orientation. The aim of City Hall London's north-south axis orientation is to save energy (Foster and Partners, 2002; and Foster and Partners, 2003). The glazed facade of City London's assembly chamber faces north in order to minimise solar gain by minimising the amount of direct sunlight falling on it (Foster and Partners, 2002; and Foster and Partners, 2003). City Hall also tilts towards the south where its floor plates are stepped inwards from its apex to base, thereby providing its offices beneath with natural shading against direct solar glare (Foster and Partners, 2002; and Foster and Partners, 2003).

Although City Hall London had a 50 percent gap increase during 2004-2005 (Tuffrey, 2005), its north-south axis orientation – as part of several environmental design measures – contributes to the following impacts:

- City Hall London's energy use is expected to be 25 percent of the energy use of a typical high-specification office building as described in Foster and Partners (2002).

- City Hall London's energy use is 34 percent below the Energy Consumption Guide 19 typical energy use for office type 4 (air conditioned, prestige office) as described in Spring (2005).

3.1.2 Type of Built Form

According to Lane (2004), 30 St Mary Axe London's design is visually and environmentally innovative. One of the key environmental design measures implemented in 30 St Mary Axe London is its type of built form, which is described as unusual by C&A (2005) and a distinctive form by Foster and Partners (2006b). 30 St Mary Axe London's built form is generated by a radial plan with a circular perimeter, and responds to site constraints by widening as it rises and tapering towards its apex (Foster and Partners, 2006b). According to Freiberger (2007), the choice of this built form is aimed at maximising natural ventilation and natural lighting, which is expected to reduce its energy use costs attributable to air conditioning, heating and lighting. This built form – along with its transparency and fully glazed façade – permits the entry of sunlight into its interior, thereby reducing its need for artificial lighting (Foster and Partners, 2006b; and The Economist, 2004). 30 St Mary Axe London's aerodynamic built form also reduces the amount of wind deflected to the ground better than a cuboid tower of similar size (Foster and Partners, 2006b; and Freiberger, 2007). Therefore, this creates external pressure differentials that facilitate a unique system of natural ventilation (Foster and Partners, 2006b). Although 30 St Mary Axe London had a presumed 23.56 percent gap increase between its predicted energy use (Foster and Partners, 2006b) and its predicted-actual energy use (Buchanan, 2007; and CTBUH, 2009), its built form – as part of several environmental design measures – contributes to the following impact:

- Energy use that is expected to be 50 percent below the Energy Consumption Guide 19 typical energy use for office type 4 (air conditioned, prestige office) as described in Foster and Partners (2006b).

3.1.3 Type of HVAC Strategy

There are several cases whereby the type of HVAC strategy used – as part of several environmental design measures – contributes either to an increase or a decrease in the gap between predicted and actual office building energy use.

For instance, in the case of the Commerzbank Tower, its use of natural ventilation – whereby air movement is created and heat even captured without the need for a mechanical system – contributed to the following impact:

- A 20 percent gap decrease in its electricity energy use than originally predicted (Buchanan, 2007).

In the case of the Philip Merrill Environmental Center, its use of natural ventilation – as part of several environmental design measures – contributed to the following impact:

- 66 percent less energy than a typical office building of the same volume (Clark Construction Group, 2011).

As described in Wellcome Trust (2011), the Gibbs Building's use of chilled ceilings and a displacement ventilation system – as part of several environmental design measures – contributed to the following impacts:

- Good comfortable conditions and less energy use than a variable air volume air conditioning system or traditional fan coil system.
- Its total energy use was 10 percent below Energy Consumption Guide 19 good practice energy use for office type 4, and 45 percent below Energy Consumption Guide 19 typical energy use for office type 4 despite a 0.65 percent gap increase between its predicted and actual energy use in 2005.

3.1.4 Type of Lighting Strategy

There are several cases whereby the type of lighting strategy used – as part of several environmental design measures – either contributes to an increase in the gap between predicted and actual office building energy use or to office building energy-efficiency.

For instance, as described in EERE (2004c), the Caribou Weather Forecast Office's lighting strategy includes the use of: an east-west axis floor plan orientation for the efficient use of daylighting; building elements for the redirection of daylight and for the control of glare; north/south roof monitors; and high-efficacy light sources such as high-efficacy T8 fluorescent lamps. The Caribou Weather Forecast Office's lighting strategy – as part of several environmental design measures – contributes to the following impacts:

- Expected annual energy use that is 32 percent less than that of a conventional building designed to ASHRAE 90.1-1999 standards (EERE, 2004c) despite the Caribou Weather Forecast Office's 11.85 percent gap increase between its predicted and actual energy use for 2003-2005.

As described in EERE (2006c), the U.S. EPA Science and Technology Center's lighting strategy includes the use of: building elements for the redirection of daylight and for the control of glare; high-efficacy light sources such as high-efficacy T5 and T8 fluorescent lamps, and high-efficient lighting fixtures; and lighting controls using occupancy sensors. The U.S. EPA Science and Technology Center's lighting strategy – as part of several environmental design measures – contributes to the following impacts:

- It is 25 percent more energy-efficient than other EPA laboratories (EERE, 2006c) despite a 1.66 percent gap increase between its predicted and actual energy use for 2003-2005.

3.1.5 Type of Services Strategy

There are several cases whereby the type of services strategy used – as part of several environmental design measures – either contributes to a decrease or an increase in the gap between predicted and actual office building energy use or contributes to office building energy-efficiency.

For instance, as described in EERE (2009c), the Hawaii Gateway Energy Center's services strategy includes the use of: electric ambient lighting controls using occupancy sensors and photo sensors; a 7.22°C deep seawater cooling system, which is pumped from 3,000 feet below sea level and distributed through cooling coils for passive cooling; and a 20 kW photovoltaic system for generation of more electricity annually than the Hawaii Gateway Energy Center uses, which has led to it becoming a net-exporter of electricity. The Hawaii Gateway Energy Center's services strategy – as part of several environmental design measures – contributes/contributed to the following impacts:

- Its energy use is designed to be approximately 20 percent of that of a comparable building designed in minimal compliance with ASHRAE 90.1-1999 standards (EERE, 2009c).
- A 36 percent gap decrease between its predicted and actual energy use for 2006-2007.

As described in Wellcome Trust (2011), the Gibbs Building's services strategy includes the use of: lighting controls using occupancy sensors; a demand-led BMS (Building Management System), which controls the Gibbs Building's plant in order to facilitate efficient working of its ventilation, heating and cooling so that they only function when required; low NO_x burners on boilers, including the use of boilers that are fitted with high efficiency burners, which is aimed at limiting the production of pollutants; real time energy monitoring of all of the Gibbs Building's incoming utility supplies such as gas, water and electricity; and low flow or low flush toilets, that is, toilets fitted with low flush cisterns that consume only six litres of water per flush instead of the nine litres of water per flush used by standard cisterns. The Gibbs Building's services strategy – as part of several environmental design measures – contributed to the following impacts:

- Its total energy use was 5 percent below Energy Consumption Guide 19 good practice energy use for office type 4, and 41 percent below Energy Consumption Guide 19 typical energy use for office type 4 despite a 6.8 percent gap increase between its predicted and actual energy use in 2006 (Wellcome Trust, 2011).

3.2   Weather Variation and Microclimates

There are several cases whereby weather variation and microclimates contribute to an increase in the gap between predicted and actual office building energy use.

For instance, in Osaji and Price (2010), several simulations of the energy use of purpose-built models were undertaken, and these purpose-built models are representative of a single room as a sample of a building. They had the same: built form; height; volume; time zone; climate zone; hours of operation; design conditions; HVAC system; level and nature of occupancy; and material assignments (Osaji and Price, 2010). However, they were situated in four different cities in England (Osaji and Price, 2010), and these are: Birmingham; London; Manchester; and Newcastle. According to Osaji and Price (2010), the single room experienced weather variation and microclimates because of its location in different cities, which contributed to the following impacts:

- It used the most energy for heating in Newcastle than in Manchester, Birmingham and London. It also used the second most energy for heating in Manchester than in Birmingham and London.
- It used the least energy for heating in London than in Newcastle, Manchester and Birmingham.
- Its maximum heating occurred on four different days between December and February, that is, 13th January, 11th February, 15th February, and 24th December. However, no cooling occurred over a 12 month period between January and December.

In the case of the Gibbs Building, the 2006 summer warm weather that often exceeded 33°C contributed to the following impacts:

- An unusually high cooling load that resulted in the Gibbs Building's electricity consumption rising above the best practice guideline (Wellcome Trust, 2011).
- A 6.8 percent gap increase between the Gibbs Building's predicted and actual energy use in 2006 (Wellcome Trust, 2011).

Therefore, weather variation and microclimates contribute to the following impacts:

- Variable building heating and cooling requirements and variable building energy performance because of exposure to weather variation and microclimates (Osaji and Price, 2010).
- A gap increase between predicted and actual office building energy use because of exposure to weather variation and microclimates (Wellcome Trust, 2011).
- The assumption that a gap decrease between predicted and actual office building energy use will likely occur because of reduced exposure to weather variation and microclimates.

3.3   Unavailability of Reliable Building Energy Use Data

There are several cases whereby the unavailability of reliable building energy use data contributes to an increase in the gap between predicted and actual office building energy use. According to UK-GBC (2011), it is difficult to gain access to reliable data on UK building energy use and CO_2 emissions. Leaman et al. (2010) suggests that this is because people involved with poor building performance are embarrassed to publish the results, which indirectly helps perpetuate the same mistakes. This occurs during the design stage because when actual building data is unavailable, the prediction of building energy use is based more on assumptions and estimates (Demanuele et al., 2010). In the absence of actual building data, building energy use at the design stage may be based on assumptions and estimates from sources such as design guidelines and experience (De Wit, 1995; De Wit, 2004; and Demanuele et al., 2010). However, such assumptions and estimates do not equate to actual building energy use data (Demanuele et al., 2010; and Norford et al., 1994).

In the case of 30 St Mary Axe London, its presumed 23.56 percent gap increase is based on its predicted energy use in Foster and Partners (2006b), and its predicted-actual energy use data and assumptions in Buchanan (2007) and CTBUH (2009). However, the actual energy use data for 30 St Mary Axe London is unavailable. This makes it difficult to ascertain the actual energy use of 30 St Mary Axe London.

In the case of the Office of Energy Efficiency and Renewable Energy (EERE) Buildings Database (EERE, 2011), it is not possible to establish the gap between the predicted and actual energy use of 47 of its 50 office buildings, and this is because of the unavailability of either their predicted or actual energy use data.

These cases demonstrate the difficulty in gaining access to UK and U.S. building energy use data. This issue hinders the opportunity to learn from past mistakes, and contributes to the repetition of the same mistakes (Leaman et al., 2010). Pett et al. (2005) suggests a way to address this issue when it states that there is a need for a centralised registry of UK building energy use data or building energy-efficiency data. This study supports this suggestion. However, such a centralised registry or database should also be comprised of knowledge about the impacts of contributory factors in the gap between predicted and actual UK office building energy use. This will contribute to this study's development of a System for Bridging the Gap between Predicted and Actual Office Building Energy Use.

Therefore, the unavailability of reliable building energy use data contributes to the following impacts:

- A significant gap between predicted and actual building energy use due to a lack of knowledge about the factors affecting building energy use (Bordass et al., 2004; and Demanuele et al., 2010).
- A lack of opportunity for all those interested in building performance to learn from past mistakes, which indirectly helps perpetuate the same mistakes (Leaman et al., 2010).
- Prediction of building energy use that is based more on assumptions and estimates instead of actual building data (Demanuele et al., 2010).

3.4   Limitations of Building Energy Simulation Software

There are several cases whereby the limitations of building energy simulation software contributes to an increase in the gap between predicted and actual office building energy use. Bordass et al. (2004) and Demanuele et al. (2010) state that the significant gap between predicted and actual building energy use is partly attributable to the building energy simulation process used to predict building energy use at the design stage. Demanuele et al. (2010) suggests that the reason for this is because it is impossible for the simulation model to be an exact replica of the actual building, and such discrepancy affects the final result (De Wit, 2001; and MacDonald et al., 1999). A validation of building energy simulation programs undertaken by Lomas et al. (1997) suggests that the occurrence of variability is likely as a result of differences between the simulation programs, and not because of the way they were used. An investigation was undertaken by Raslan and Davies (2010) into prediction inconsistencies between accredited building energy simulation tools. This investigation revealed tool applicability limitations, and variability between produced compliance benchmarks (Raslan and Davies, 2010). However, De Wilde and Tian (2009) state that various factors, including weather variation, building use variation, electronic equipment and lighting trends, building refurbishment, and mechanical HVAC system upgrades introduce uncertainties into building simulation predictions.

In the case of City Hall London, a 50 percent gap increase occurred between its predicted and actual energy use (Tuffrey, 2005) despite Foster and Partners (2002) using building energy simulation for design studies to predict City Hall London's energy use and to develop its low energy use strategies.

In the case of the Caribou Weather Forecast Office, an 11.85 percent gap increase occurred between its predicted and actual energy use for 2003-2005 (Diamond et al., 2006; and EERE, 2004c). This occurred despite using the Multi-Criteria Decision Making Tool and DOE-2 building energy simulation software (EERE, 2004b).

In the case of the U.S. EPA Science and Technology Center, a 1.66 percent gap increase occurred between its predicted and actual energy use for 2003-2005 (Diamond et al., 2006; and EERE, 2006c). This occurred despite using the DOE-2 building energy simulation software (EERE, 2006c). In fact, it occurred despite the design team using energy modelling in order to undertake comparative studies between the U.S. EPA Science and Technology Center and a base case meeting the ASHRAE 90.1 standard (EERE, 2006c).

Therefore, the limitations of building energy simulation software contribute to the following impacts:

- The occurrence of a gap increase between predicted and actual building energy use because simulation models are not 100 percent exact replicas of actual buildings, and such discrepancy affects both the building energy simulation process used for predictions at the design stage, and the final result (Demanuele et al., 2010; De Wit, 2001; and MacDonald et al., 1999).

3.5   Level of Hours of Operation

There are several cases whereby the level of hours of operation contributes either to an increase or a decrease in the gap between predicted and actual office building energy use. Plant operating hours that may differ from assumptions made in initial predictions is one of the likely reasons for the gap between predicted and actual building energy use (Demanuele et al., 2010; and Mason, 2004). Furthermore, the longer hours of operation of equipment and appliances than originally predicted are another likely reason for the gap between predicted and actual building energy use (Demanuele et al., 2010).

In the case of City Hall London, the 50 percent gap increase between its predicted and actual energy use for 2004-2005 was partly attributable to an increase in annual operational hours than originally predicted (Tuffrey, 2005). In 2005, the Mayor of London explained that City Hall London's annual operational hours were over 50 percent greater than originally predicted (Tuffrey, 2005). He attributed this to City

Hall London's hosting of a range of evening and weekend events, and its high public access as a result of tourism (Tuffrey, 2005).

In the case of 30 St Mary Axe London, Maria Gillivan of 30 St Mary Axe (2011) states that the building is open to all its occupiers for 24 hours a day and seven days a week, that is, a total of 168 hours per week. This has contributed to its presumed 23.56 percent gap increase, which is based on its predicted energy use in Foster and Partners (2006b), and its predicted-actual energy use in Buchanan (2007) and CTBUH (2009).

In the case of the Caribou Weather Forecast Office, it is occupied 168 hours per week (EERE, 2004a), and had an 11.85 percent gap increase between its predicted and actual energy use for 2003-2005.

In the case of the U.S. EPA Science and Technology Center, it is occupied approximately 55 hours each week (EERE, 2006a), and had a 1.66 percent gap increase between its predicted and actual energy use for 2003-2005.

However, in the case of the Hawaii Gateway Energy Center, it is occupied 40 hours per week (EERE, 2009a), and had a 36 percent gap decrease between its predicted and actual energy use for 2006-2007.

Therefore, the level of hours of operation contributes to the following impacts:

- The occurrence of a gap increase between predicted and actual building energy use because of an increase in the level of hours of operation (Tuffrey, 2005).
- Discrepancy between shorter predicted hours of operation and longer actual hours of operation, which leads to the occurrence of a gap increase between predicted and actual building energy use (Tuffrey, 2005).
- The assumption that a gap decrease between predicted and actual building energy use will likely occur because of a decrease in the level of hours of operation.
- The assumption that a discrepancy between longer predicted hours of operation and shorter actual hours of operation will likely lead to a gap decrease between predicted and actual building energy use.

3.6   Level and Nature of Occupancy

There are several cases whereby the level and nature of occupancy, that is, values for number of people and their average heat output contributes either to an increase or a decrease in the gap between predicted and actual office building energy use. In fact, Azar and Menassa (2010) states that a contributory factor in the gap between predicted and actual building energy use is occupancy energy attributes, that is, the presence of occupants and their influence on energy use. Building energy performance is influenced by occupants because they exercise control over lighting, ventilation, internal temperature, and electrical/mechanical equipment (Demanuele et al., 2010; De Wit, 1995; and MacDonald et al., 1999). Occupancy patterns and behaviour are also unpredictable and this contributes to significant uncertainties in building energy use predictions (Demanuele et al., 2010; De Wit, 1995; and MacDonald et al., 1999).

In the case of City Hall London, the 50 percent gap increase between its predicted and actual energy use for 2004-2005 was partly attributable to an increase in its staff occupancy than originally envisaged (Tuffrey, 2005). In 2005, the Mayor of London explained that City Hall London had a 53 percent increase in the number of staff it originally planned to accommodate (Tuffrey, 2005). He attributed this to City Hall London accommodating 650 staff, that is, 224 more staff than the 426 staff originally envisaged (Tuffrey, 2005).

In the case of the 778m^2 Caribou Weather Forecast Office, it is typically occupied by 22 people at 168 hours per person per week, and its building operations room is usually occupied by two people during off-hours (EERE, 2004a). It also had an 11.85 percent gap increase between its predicted and actual energy use for 2003-2005 (Diamond et al., 2006; and EERE, 2004c).

In the case of the 334m^2 Hawaii Gateway Energy Center, it is typically occupied by four people at 40 hours per person per week, and 300 visitors per week at 2 hours per visitor per week (EERE, 2009a). It also had a 36 percent gap decrease between its predicted and actual energy use for 2006-2007 (EERE, 2009c).

In the case of the 6,680m^2 U.S. EPA Science and Technology Center, it is typically occupied by 110 people at approximately 55 hours per person per week (EERE, 2006a). It also had a 1.66 percent gap increase between its predicted and actual energy use for 2003-2005 (Diamond et al., 2006; and EERE, 2006c).

In the case of the 76,400m^2 of accommodation for 30 St Mary Axe London (Foster and Partners, 2006b), 30 St Mary Axe (2011) states that its predicted level of occupancy is 4,000 people. 30 St Mary Axe (2011) further states that 30 St Mary Axe London has design criteria of one person per 10m^2. However, Foster and Partners (2006a) states that 30 St Mary Axe London's predicted level of occupancy is 3,500 people. This appears to be a discrepancy in the predicted level of occupancy for 30 St Mary Axe London. It also had a presumed 23.56 percent gap increase, which is based on its predicted energy use in Foster and Partners (2006b), and its predicted-actual energy use in Buchanan (2007) and CTBUH (2009).

Therefore, the level and nature of occupancy contributes to the following impacts:

- The occurrence of a gap increase between predicted and actual building energy use because of an increase in the level and nature of occupancy (Tuffrey, 2005).
- An increase in occupancy heat output and significant uncertainties in building energy use predictions because of an increase in occupancy numbers, patterns and behaviour (Demanuele et al., 2010; De Wit, 1995; and MacDonald et al., 1999).
- The assumption that a gap decrease between predicted and actual building energy use will likely occur because of a decrease in the level and nature of occupancy.
- The assumption that both a decrease in occupancy heat output and a decrease in uncertainties in building energy use predictions will likely occur because of a decrease in occupancy numbers, patterns and behaviour.

4   Discussion

In several cases reviewed by this study, it was established that more instances of gap increases between predicted and actual office building energy use occurred compared to instances of gap decreases between predicted and actual office building energy use. For instance, gap increases between predicted and actual office building energy use occurred in five of the eight cases reviewed, and these are the: Caribou Weather Forecast Office; City Hall London; Gibbs Building; and U.S. EPA Science and Technology Center. A presumed gap increase between predicted and predicted-actual office building energy use occurred in one of the eight cases reviewed, and this is 30 St Mary Axe London. A gap decrease between predicted and actual office building energy use occurred in two of the eight cases reviewed, and these are the: Commerzbank Tower; and Hawaii Gateway Energy Center.

These suggest that it is more likely for actual office building energy use (in kWh/m^2) to be greater than predicted office building energy use (in kWh/m^2). However, this practice is unacceptable because it is preferable for actual office building energy use to be less than predicted office building energy use. Therefore, it is important to ensure that this becomes the practice because it signifies office building energy-efficiency and environmental best practice.

In the cases reviewed by this study, it was established that the gap increase and gap decrease between predicted and actual office building energy use are attributable to the impacts of several contributory factors such as:

- Nature of Environmental Design Measures Implemented: type of orientation; type of built form; type of HVAC strategy; type of lighting strategy; and type of services strategy.
- Weather Variation and Microclimates.
- Unavailability of Reliable Building Energy Use Data.
- Limitations of Building Energy Simulation Software.
- Level of Hours of Operation.
- Level and Nature of Occupancy.

There were also much more cases reviewed by this study that, for instance, demonstrated the difficulty in gaining access to UK and U.S. building energy use data. It is difficult to gain access to reliable data on office building energy use. This contributes to the perpetuation of past mistakes that are associated with building energy use. This study has had challenging experiences while attempting to gain access to both predicted and actual UK and U.S. office building energy use data. In addition to the gap between predicted and actual office building energy use, this study identified another type of gap, which is the gap between available and unavailable reliable office building energy use data. The latter type of gap occurs when only one of either predicted or actual reliable office building energy use data is available. Therefore, this study supports the establishment of a Centralised Database of Reliable UK Office Building Energy Use, which should also be comprised of knowledge of the impacts of contributory factors in the gap between predicted and actual UK office building energy use. This will contribute to this study's ideas for the development of a Building Management System for Bridging the Gap between Predicted and Actual Office Building Energy Use, otherwise known as 'BMS-Optimum'.

It is expected that such a Centralised Database of Reliable UK Office Building Energy Use – and comprised of knowledge of the impacts of contributory factors in the gap between predicted and actual UK office building energy use – will, for instance, contribute to reducing the limitations of building energy simulation software. It is expected that it will do this by facilitating the creation of simulation models that: are much better replicas of real office buildings and their energy use scenarios; and lead to more accurate office building energy use predictions. For instance, this is important in order to successfully forecast and pre-empt discrepancies in office building hours of operation so that office buildings achieve actual hours of operation that are shorter than their predicted hours of operation. It is also important in order to successfully forecast and pre-empt discrepancies in office building occupancy so that office buildings achieve actual occupancy levels that are shorter than their predicted occupancy levels. Perhaps this will prevent cases like that of City Hall London whereby the 50 percent gap increase between its predicted and actual energy use in 2004-2005 was attributable to an increase in its hours of operation and occupancy levels than originally envisaged.

However, the nature of environmental design measures implemented, including the type of HVAC strategy such as natural ventilation is important because it was shown by this study to contribute to: a gap decrease between predicted and actual office building energy use; and office building energy-efficiency. For instance, in the cases of 30 St Mary Axe London and City Hall London, the implementation of natural ventilation partly contributed to energy use that was below Energy Consumption Guide 19 typical energy use for office type 4. Furthermore, the implementation of natural ventilation contributed to: a 20 percent gap decrease in the Commerzbank Tower's electricity energy use than originally predicted; and the Philip Merrill Environmental Center's 66 percent less energy use than a typical office building of the same volume.

Although implementation of environmental design measures contributed to gap decreases between predicted and actual office building energy use and office building energy-efficiency, this study also showed that gap increases also occurred in several cases despite the implementation of environmental design measures. For instance, in nine cases reviewed, it was revealed that there were more instances of gap increases between predicted and actual office building energy use despite the implementation of environmental design measures. In fact, gap increases between predicted and actual office building energy use occurred in five of the nine cases reviewed, and these are: the Caribou Weather Forecast Office; City Hall London; the Gibbs Building; and the U.S. EPA Science and Technology Center. A presumed gap increase between predicted and predicted-actual office building energy use occurred in one of the nine cases reviewed, and this is 30 St Mary Axe London. However, even though these six cases experienced gap increases and a presumed gap increase despite the implementation of environmental design measures, they still achieved a level of actual or expected office building energy-efficiency. Furthermore, another one of the nine cases reviewed, that is, the Philip Merrill Environmental Center also achieved a level of actual office building energy-efficiency. This is good news because it shows that the implementation of environmental design measures does contribute to office building energy-efficiency. In fact, the implementation of environmental design measures contributed to gap decreases between predicted and actual office building energy use in two of the nine cases reviewed, and these are the Commerzbank Tower and the Hawaii Gateway Energy Center. This is also good news because it shows that the implementation of environmental design measures does contribute to a gap decrease between predicted and actual office building energy use.

The implementation of environmental design measures is also important in order to successfully counteract the impacts of weather variation and microclimates. For instance, this study showed that the implementation of a key environmental design measure occurred in the case of the Hawaii Gateway Energy Center, and this is the use of a photovoltaic system for the generation of additional electricity annually. In fact, the Hawaii Gateway Energy Center's use of a photovoltaic system, which takes advantage of high insolation to generate electrical power, contributes to its generation of more electrical energy than it uses. This is further good news because it shows that the implementation of a key environmental design measure such as the use of an appropriate photovoltaic system can counteract the impact(s) of weather by converting solar radiation into electrical energy. It also further demonstrates that the implementation of environmental design measures can contribute to a gap decrease between predicted and actual office building energy use.

5   Conclusion

This study used a combination of literature reviews, multiple case study research and comparative studies in order to build theory, which involved a desk based study in order to establish the impacts of contributory factors in the gap between predicted and actual office building energy use, and these are:

1. Type of orientation such as a north-south axis orientation – as part of the nature of environmental design measures implemented – contributes to the following impacts:
   - Office building energy-efficiency, for instance, energy use that is 34 percent below the Energy Consumption Guide 19 typical energy use for office type 4 (air conditioned, prestige office).

2. Type of built form such as the spheroid-like built forms of City Hall London and 30 St Mary Axe London – as part of the nature of environmental design measures implemented – contributes to the following impact:
   - Office building energy-efficiency, for instance, energy use that is expected to be 50 percent below the Energy Consumption Guide 19 typical energy use for office type 4 (air conditioned, prestige office).

3. Type of HVAC strategy such as energy-efficient HVAC – as part of the nature of environmental design measures implemented – contributes to the following impacts:
   - Office building energy-efficiency, for instance, energy use that is 10 percent below Energy Consumption Guide 19 good practice energy use for office type 4, and 45 percent below Energy Consumption Guide 19 typical energy use for office type 4.

- A gap decrease between predicted and actual office building energy use, for instance, a 20 percent gap decrease between predicted and actual office building energy use.

4. Type of lighting strategy such as energy-efficient lighting – as part of the nature of environmental design measures implemented – contributes to the following impacts:
   - Office building energy-efficiency, for instance, expected annual energy use that is 32 percent less than that of a conventional building designed to ASHRAE 90.1-1999 standards.
   - The assumption that a gap decrease between predicted and actual office building energy use will likely occur because of reduced demand for lighting.

5. Type of services strategy such as energy-efficient services – as part of the nature of environmental design measures implemented – contributes to the following impacts:
   - Office building energy-efficiency, for instance, energy use that is designed to be approximately 20 percent of that of a comparable building designed in minimal compliance with ASHRAE 90.1-1999 standards.
   - Office building energy-efficiency, for instance, energy use that is 5 percent below Energy Consumption Guide 19 good practice energy use for office type 4, and 41 percent below Energy Consumption Guide 19 typical energy use for office type 4.
   - A gap decrease between predicted and actual office building energy use, for instance, a 36 percent gap decrease between predicted and actual office building energy use.
   - Exploitation of high insolation for generation of electrical power, particularly the generation of more electricity annually than an office building uses.

6. Weather variation and microclimates contribute to the following impacts:
   - Variable building heating and cooling requirements and variable building energy performance.
   - A gap increase between predicted and actual office building energy use.
   - The assumption that a gap decrease between predicted and actual office building energy use will likely occur because of reduced exposure to weather variation and microclimates.

7. Unavailability of reliable building energy use data contributes to the following impacts:
   - A gap increase between predicted and actual building energy use because of limited knowledge of the factors affecting building energy use.
   - Perpetuation of the same mistakes because of limited opportunity for all those interested in building performance to learn from past mistakes.
   - Prediction of building energy use based more on assumptions and estimates instead of actual building data.

8. Limitations of building energy simulation software contribute to the following impacts:

- The assumption that gap increases between predicted and actual building energy use will likely occur because simulation models, which are not exact replicas of actual buildings will compromise both the building energy simulation process used for design stage predictions, and the final predictions.

9. Level of hours of operation contributes to the following impacts:

- A gap increase between predicted and actual building energy use because of a discrepancy between shorter predicted hours of operation and longer actual hours of operation.
- The assumption that a gap decrease between predicted and actual building energy use will likely occur because of a decrease in the level of hours of operation.
- The assumption that a discrepancy between longer predicted hours of operation and shorter actual hours of operation will likely lead to a gap decrease between predicted and actual building energy use.

10. Level and nature of occupancy contribute to the following impacts:

- A gap increase between predicted and actual building energy use.
- An increase in occupancy heat output and uncertainties in building energy use predictions.
- The assumption that a gap decrease between predicted and actual building energy use will likely occur because of a decrease in the level and nature of occupancy.
- The assumption that both a decrease in occupancy heat output and a decrease in uncertainties in building energy use predictions will likely occur because of a decrease in occupancy numbers, patterns and behaviour.

This study has also contributed to ideas for the development of a Building Management System for Bridging the Gap between Predicted and Actual Office Building Energy Use, otherwise known as 'BMS-Optimum'. The BMS-Optimum is comprised of optimum conditions and considerations such as: optimum environmental design principles; optimum weather and microclimate considerations; accessibility to reliable office building energy use data; optimum building energy and environmental assessment; optimum hours of operation; and optimum level and nature of occupancy.

6 Future Work

Future work will include further development of the BMS-Optimum, using methods such as: multiple case study research supported by building energy use audits, observations, questionnaire surveys, interviews, benchmarking and comparative studies; building energy simulation within multiple scenarios, parameters and variables, and supported by benchmarking and comparative studies; and peer reviews and focus group sessions. These will also help establish and validate a Framework for Improved Environmental Design and Energy Performance (FEDEP).

References

30 St Mary Axe, 30 St Mary Axe London (2011), `http://www.30stmaryaxe.co.uk` (retrieved October 7, 2011)

Action Energy, Energy Consumption Guide 19: Energy Use in Offices. Action Energy, Best Practice Programme, UK (2003)

Azar, E., Menassa, C.: A Conceptual Framework to Energy Estimation in Buildings using Agent Based Modeling. In: Proceedings of the 2010 Winter Simulation Conference (2010)

Beautyman, M.: HOK Earns Sustainable Leadership Award. Interior Design (2006), `http://www.interiordesign.net/article/484923-HOK_Earns_Sustainable_Leadership_Award.php` (retrieved March 25, 2012)

Bordass, B., Cohen, R., Field, J.: Energy Performance of Non-Domestic Buildings: Closing The Credibility Gap. In: Proceedings of IEECB 2004 Building Performance Congress, Frankfurt (2004)

Buchanan, P.: The Tower: An Anachronism Awaiting Rebirth. Harvard Design Magazine 10(26), 1–5 (2007)

C&A, Swiss Re Performance Story. Swiss Re Took a Walk with Us (2005)

Clark Construction Group, Philip Merrill Environmental Center (2011), `http://www.clarkconstruction.com/index.php/projects/feature_project/38` (retrieved May 23, 2011)

Cramer, J.P.: Sustainability Leadership: From Stagnation to Liberation. Design Intelligence (2011), `http://www.di.net/articles/archive/3616/` (retrieved March 25, 2012)

Crawley, D.B., Hand, J.W., Kummert, M., Griffith, B.T.: Contrasting the Capabilities of Building Energy Performance Simulation Programs. Building and Environment 43(4), 661–673 (2008)

CTBUH, Tall Buildings in Numbers. Tall Buildings and Embodied Energy. Council on Tall Buildings and Urban Habitat (CTBUH) Journal (3), 50–51 (2009)

Demanuele, C., Tweddell, T., Davies, M.: Bridging the Gap between Predicted and Actual Energy Performance in Schools. In: World Renewable Energy Congress XI, Abu Dhabi, September 25-30 (2010)

De Wilde, P., Tian, W.: Identification of Key Factors for Uncertainty in the Prediction of the Thermal Performance of an Office Building under Climate Change. Building and Simulation 2(3), 157–174 (2009)

De Wit, M.S.: Uncertainty Analysis in Building Thermal Modelling. In: Proceedings of International Building Performance Simulation Association, Wisconsin, August 14-16 (1995)

De Wit, M.S.: Uncertainty in Predictions of Thermal Comfort in Buildings. PhD Thesis. Delft University of Technology, Delft, Netherlands (2001)

De Wit, M.S.: Uncertainty in Building Simulation. In: Malkawi, A., Augenbroe, G. (eds.) Advanced Building Simulation, pp. 25–60. Spon Press, New York (2004)

Diamond, R., Opitz, M., Hicks, T.: Evaluating the Energy Performance of the First Generation of LEED-Certified Commercial Buildings. ACEEE Summer Study on Energy Efficiency in Buildings (2006)

EERE, Buildings Database. Caribou Weather Forecast Office (WFO) (2004a), `http://eere.buildinggreen.com/overview.cfm?ProjectID=334` (retrieved July 10, 2011)

EERE, Buildings Database. Caribou Weather Forecast Office (WFO) (2004b), `http://eere.buildinggreen.com/process.cfm?ProjectID=334` (retrieved July 10, 2011)

EERE, Buildings Database. Caribou Weather Forecast Office (WFO) (2004c), `http://eere.buildinggreen.com/energy.cfm?ProjectID=334` (retrieved July 10, 2011)

EERE, Buildings Database. U.S. EPA Science and Technology Center (2006a), `http://eere.buildinggreen.com/overview.cfm?projectid=323` (retrieved July 8, 2011)

EERE, Buildings Database. U.S. EPA Science and Technology Center (2006c), `http://eere.buildinggreen.com/energy.cfm?ProjectID=323` (retrieved July 8, 2011)

EERE, Buildings Database. Hawaii Gateway Energy Center (2009a), `http://eere.buildinggreen.com/overview.cfm?projectid=592` (retrieved July 10, 2011)

EERE, Buildings Database. Hawaii Gateway Energy Center (2009c), `http://eere.buildinggreen.com/energy.cfm?ProjectID=592` (retrieved July 10, 2011)

EERE, Buildings Database (2011), `http://eere.buildinggreen.com/mtxview.cfm?CFID=106521378&CFTOKEN=72666690` (retrieved July 8, 2011)

EIA, Commercial Buildings Energy Consumption Survey (CBECS) 2003 (2011), `http://www.eia.doe.gov/emeu/cbecs/` (retrieved March 17, 2011)

Foster and Partners, Foster and Partners City Hall. Foster and Partners Limited, London (2002)

Foster and Partners, City Hall London 1998 – 2002. Foster and Partners Limited, London (2003)

Foster and Partners, 30 St Mary Axe, London, Facts and Figures. Foster and Partners Limited, London (2006a)

Foster and Partners, 30 St Mary Axe, Swiss Re Headquarters, London, England 1997 – 2004. Foster and Partners Limited, London (2006b)

Foster and Partners, Norman Foster (2012), `http://www.fosterandpartners.com/Team/SeniorPartners/11/Default.aspx` (retrieved March 25, 2012)

Freiberger, M.: Perfect Buildings: The Maths of Modern Architecture. Plus Magazine (42) (2007), `http://plus.maths.org/content/perfect-buildings-maths-modern-architecture` (retrieved October 9, 2011)

Greener Buildings, Green Skyscraper: The Basis for Designing Sustainable Intensive Buildings (2008), `http://www.greenerbuildings.com` (retrieved March 25, 2012)

Holm, I.: Ideas and Beliefs in Architecture and Industrial design: How attitudes, orientations, and underlying assumptions shape the built environment. Oslo School of Architecture and Design (2006)

Hui, S.C.M.: Sustainable Architecture and Building Design (SABD). Building Energy Efficiency Research (BEER), Hong Kong University (2002), `http://www.arch.hku.hk/research/BEER/sustain.htm` (retrieved November 11, 2004)

Lane, M.: Modern Britain's Instant Icon. British Broadcasting Corporation (BBC) News Online Magazine (2004), `http://news.bbc.co.uk/2/hi/uk_news/magazine/3663971.stm` (retrieved June 15, 2011)

Leaman, A., Stevenson, F., Bordass, B.: Building Evaluation: Practice and Principles. Building Research and Information 38(5), 564–577 (2010)

Lomas, K.J., Eppel, H., Martin, C.J., Bloomfield, D.P.: Empirical Validation of Building Energy Simulation Programs. Energy and Buildings 26(3), 253–275 (1997)

MacDonald, I.A., Clarke, J.A., Strachan, P.A.: Assessing Uncertainty in Building Simulation. In: Proceedings of International Building Performance Simulation Association, Japan (September 1999)

Majumdar, M.: Energy Efficiency in Architecture: An Overview of Design Concepts and Architectural Interventions. In: Majumdar, M. (ed.) Energy Efficient Buildings in India. Ministry of Non-conventional Energy Sources (MNES) and Tata Energy Research Institute, TERI (2002)

Mason, M.: Where Legals Dare. Ecolibrium, 14–16 (2004)

McLennan, J.F.: The Philosophy of Sustainable Design (2004)

Menezes, A.: Bridging the Gap between Predicted and Actual Energy Consumption in Non-Domestic Buildings. Innovation and Research Focus (87), 7 (2011)

Mortimer, N.D., Elsayed, M.A., Grant, J.F.: Patterns of Energy Use in Non-Domestic Buildings. Environment and Planning B: Planning and Design 27(5), 709–720 (2000)

Norford, L.K., Socolow, R.H., Hsieh, E.S., Spadaro, G.V.: Two-to-one Discrepancy between Measured and Predicted Performance of a 'Low-Energy' Office Building: Insights from a Reconciliation Based on the DOE-2 Model. Energy and Buildings 21, 121–131 (1994)

Osaji, E.E.: The BEEHive (Building Energy-Efficient Hive). Bachelor of Architecture (BArch) Thesis, University of Lagos (2002)

Osaji, E.E., Hudson, J., Chynoweth, P.: Exploration of the Energy Efficiency of the Greater London Authority Building (GLA Building/City Hall). Journal of Applied Sciences and Environmental Management 11(2), 53–56 (2007)

Osaji, E.E., Price, A.D.F.: Parametric Environmental Design for Low Energy Healthcare Facility Performance. In: 6th International Conference on Innovation in Architecture, Engineering and Construction (AEC), Pennsylvania, June 9-11 (2010)

Pett, J., Guertler, P., Hugh, M., Kaplan, Z., Smith, W.: Asset Value Implications of Low Energy Offices. Phase 2 Report. Association for the Conservation of Energy (2005), `http://www.ukace.org/index.phpoption=com_content&task=view&id=372&Itemid=26` (retrieved July 4, 2011)

Raslan, R., Davies, M.: Results Variability in Accredited Building Energy Performance Compliance Demonstration Software in the UK: An Inter-Model Comparative Study. Journal of Building Performance Simulation 3(1), 63–85 (2010)

Shu-Yang, F., Freedman, B., Cote, R.: Principles and Practice of Ecological Design. Environmental Reviews 12, 97–112 (2004)

Spring, M.: City Hall revisit: Time has told. Building (42) (2005)

Strong, D.: Avoiding the Need for Active Cooling – The Systems/Whole Building Approach. A Future Buildings Forum: Cooling Buildings in a Warming Climate, Sophia Antipolis, June 21-22 (2004)

The Economist, The Rise of the Green Building. The Economist Newspaper Limited (2004), `http://www.chipwalter.com/articles/emerging/greenrise.pdf` (retrieved October 16, 2011)

Thomas, D.: Architecture and the Urban Environment – A Vision for the New Age. Architectural Press, Oxford (2002)

Tuffrey, M.: Mayor Answers to London: City Hall Energy Consumption. Liberal Democrat Group London Assembly Report (2005)

UK-GBC - UK Green Building Council, Existing Non-Domestic Buildings (2011), `http://www.ukgbc.org/site/info-centre/display-category?id=24` (retrieved April 17, 2011)

Wade, J., Pett, J., Ramsay, L.: Energy Efficiency in Offices: Assessing the Situation. Association for the Conservation of Energy, London (2003)

Wade, J., Pett, J., Ramsay, L.: Energy Efficiency in Offices: Motivating Action. Association for the Conservation of Energy, London (2003)

Wellcome Trust, Environmental Report 2006 (2011), `http://www.wellcome.ac.uk/About-us/Policy/Policy-and-position-statements/WTX022977.htm` (retrieved June 26, 2011)

Chapter 69
Improving Multiple Source Power Management Using State Flow Approach

Aziz Naamane and Nacer Msirdi

Laboratoire des Sciences de l'Information et des Systèmes
Avenue escadrille normandie Niemen, 13397, Marseille cedex 2O-France
`Aziz.naamane@lsis.org`

Abstract. The optimum design of a multiple power source supply system becomes complicated through uncertain renewable energy supplies and load demand. The optimum configuration and optimum control strategy of systems supplied by multiple power sources need to dispatch the power by matching the supply and demand in accordance with the power management tasks. Accurate load matching is especially critical for renewable energy sources such as photovoltaic panel and wind aero turbines, because it impacts on the available power utility. This paper advocates the use of the state flow approach as an alternative mean to manage the multi power source distribution for multiple supply systems by a load matching switch.

1 Introduction

Solar and wind energy systems are being considered as promising power generating sources due to their availability and topological advantages for local power generations in remote areas. Utilization of solar and wind energy has become increasingly significant, attractive and cost-effective, since the oil crises of early 1970s.However, a drawback, common to solar and wind options, is their unpredictable nature and dependence on weather and climatic changes, and the variations of solar and wind energy may not match with the time distribution of load demand. This short coming not only affects the system's energy performance, but also results in batteries being discarded too early. Generally, the independent use of both energy resources may result in considerable over-sizing, which in turn makes the design costly. It is prudent that neither a stand-alone solar energy system nor a wind energy system can provide a continuous power supply due to seasonal and periodical variations [1] for stand-alone systems.

Fortunately, the problems caused by the variable nature of these resources can be partially or wholly overcome by integrating these two energy resources in a proper combination, [2,3,4] using the strengths of one source to overcome the weakness of the other. The use of different energy sources allows improving the system efficiency and reliability of the energy supply and reduces the energy storage requirements compared to systems comprising only one single renewable energy source. Of course, with increased complexity in comparison with single energy systems, the optimum design of a hybrid system becomes complicated through uncertain renewable energy supplies and load demand, non-linear characteristics of the components, high number of variables and parameters that have to be considered for the optimum design, and the fact that the optimum configuration and optimum control strategy of the system are interdependent. This complexity makes the hybrid systems more difficult to be designed and analysed. So renewable energy sources are, and will be more and more, brought to cohabit on the same site. [11] However, they have not yet the subject of a real overall energy management strategy. As several systems rely on multiple energy sources, power distribution strategy must be implemented by matching the supply and the demand. The balance between production and consumption must be carefully conducted to ensure the availability of power. This paper addresses decentralized control strategies of multi-sources and multi-users energy systems. The objective is to describe, by using the stateflow approach, a decentralized multi-sources, multi-users energy.

2   Multi-energy Sources Structure

To illustrate the possibility to propose a power distribution switch that maximizes the total available power from the different power sources[5,7,8,9,10]., the example of a multi-energy system of figure 1 is taken. This hybrid system contains three energy sources: solar, wind and a battery. It is obvious that other energies sources could be added with the same principle.

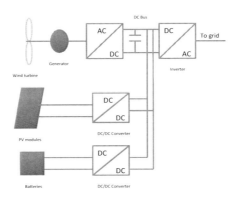

Fig. 1. Configuration of multi-sources hybrid energy systems

The amount of energy produced depends mainly on weather conditions, hence the importance of maximizing the amount of energy produced.

2.1 Switching Problem in a Hybrid Energy System

The main difficulty of such a distributed management of the energy problem is to use the most possible appropriate way all the possible sources of production in relation to all identified needs. For this, it should switch on an ad hoc manner some sources to some users, depending on the state's energy system at time t and on the forecasts of its operation in the near future.

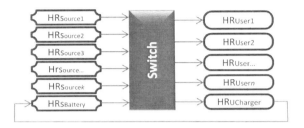

Fig. 2. Switching of multi-sources and multi-users energy system

The switch is then a connection function between users and power sources. The various connection configurations constitute solutions of such a multi-sources and multi-users energy system. Thus the control of an energy system consists in choosing among these connection configurations that will be the most appropriate at a given time allowing an optimum use of the majority of the produced energy

3 Stateflow Chart Approach

A Stateflow chart is an example of a finite state machine. A *finite state machine* is a representation of an event-driven (reactive) system. In an event-driven system, the system makes a transition from one state (mode) to another, if the condition defining the change is true. So Staeflow is based on finite State Machines concept that has been developed (Moore, Mealy) to account for the operation of discrete event system. In these systems the passage from one State to another is governed by discrete events.

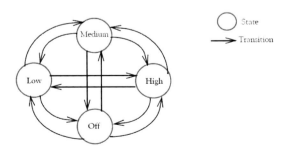

Fig. 3. State transition diagram

The example of the management of energy derived from several sources is represented by the Simulink model represented by figure 3. the Simulink model takes into account owing to stateflow, various modes of operation of the system associated with the changes of energy productions related to weather changes.

One can notice in effect the presence of a State logic block, built with stateflow, providing several signals intended precisely to modify the State of the relays.

Stateflow is based on finite State Machines concept that has been developed in order to deal with event driven systems. In these systems the passage from one State to another is governed by discrete events.Staeflow CAN manage the control part of the whole automated system by the sequence State- transition.

With stateflow, simulink it is possible to simulate the functionning of complex hybrid systems involving the continuous and the discrete event aspects.This allows to simulate continuous systems operating on several different modes sequenced by stateflow.

It is possible then using Simulink for the operative part and stateflow for the control one to simulate completely an automated hybrid system. Staeflow toolbox can even ensure the generation and implementation of executable code in a calculator.

3.1　Application Example

In order to explain and apply the stateflow approach the following case of managing power from three sources is described.

The used variables in stateflow chart are:

The system's inputs:
  P: Consumer Power Demand
  PPv: Power provided by the photovoltaic panels
  PEol: Power provided by the windturbine
  PB: Power supplied by the battery

The system's outputs :
  PV: switch On / Off - PV
  Wind: relay On / Off -windturbine
  Battery: relay On / Off –battery

Local variables:
  Test: the M function introduced in the chart
  i: the resultTest function

The conditions used are presented in the following table:

Table 1. Strategy Conditions

P<PPV	PV On
P<PEol	Eol on
P<PSt	Stockage on
P<(PPV+PEol)	PV et Eol on
P<(PPV+PSt)	PV et St On
P<(PEol+PSt)	Eol et St on
P<(PPV+PEol+PSt)	PV, Eol et St on
P>(PPV+PEol+PSt)	PV, Eol et St off shutdown

The machine that it has designed presented in figure 4 is composed of an initial state and of seven States of actions. These States are enabled if the condition of the junction is validated and the actions will be performed later. States are bounded between them at the beginning and at the end by connecting junctions that allow splitting a transition in several transitions and introduce points of decisions.

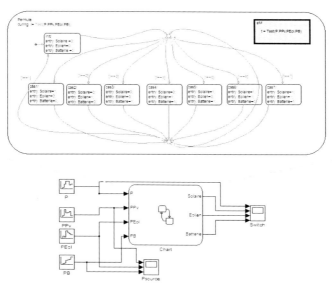

Fig. 4. Control Conditions diagram

4 Simulation Results and Interpretation

In these simulations, the different power sources production are generated randomly as well as the user's energy demand. According to this last figure, the general state of the switch is well defined, and the productions e power sources that supplies the desired power are well defined too.

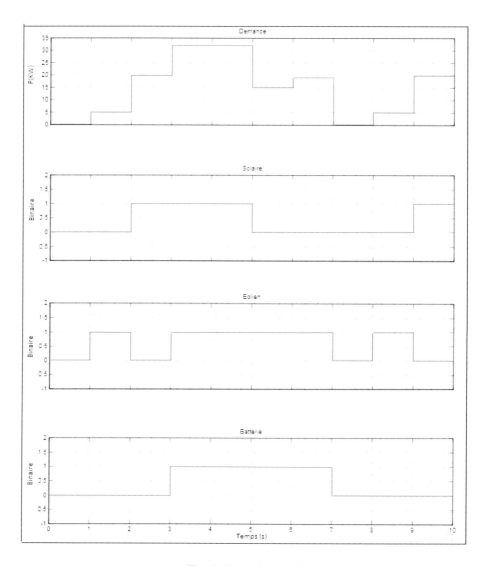

Fig. 5. Simulation results

5 Conclusions

Energy management is nowadays a subject of great importance and complexity. It consists in choosing among a set of sources able to produce energy that will give energy to a set of loads by minimising losses and costs. The sources and loads are heterogeneous, distributed and the reaction of the system, the choice of sources, must be done in real-time to avoid power outage.

For efficient management of hybrid renewable and classical energy systems, stateflow approach has been presented in this paper, it can be advantageously used to tackle the power management issues with the possibility to match the production and the demand.

The future developments of this study will focus on practical tests. First, a dSPACE card will be used to validate the principle in an experimental way. This work could be extended to develop an architecture proposal for an intelligent and autonomous demand-response energy management system, based on a fully interactive ICT infrastructure that meets specific requirements, the main purpose of which, by the cooperation between a house and the grid, is to help the end user to achieve energy savings.

References

1. Lagorse, J., Simoes, M.G., Miraoui, A.: A Multiagent Fuzzy-Logic-Based Energy Management of Hybrid Systems. IEEE Transactions on Industry Applications (6), 2123–2129 (2009)
2. Dawei, G., Zhenhua, J., Qingchun, L.: Energy management strategy based on fuzzy logic for a fuel cell hybrid bus. Journal of Power Sources 1(185) (2008)
3. Abbey, C., Joos, G.: Energy management strategies for optimization of energy storage in wind power hybrid system. In: PESC Record IEEE Annual Power Electronics Specialists Conference (2005)
4. Abras, S., Ploix, S., Pesty, S., Jacomino, M.: A multi- agent design for a home automation system dedicated to power management. In: Christos Boukis. Aristodemos Pnevmatikakis, and Lazaros Polymenakos,
5. Frik, R., Favre-Perrod, P.: Proposal for a multifunctional energy bus and its interlink with generation and consumption, Diploma thesis, High Voltage Laboratory, Swiss Federal Institute of Technology (ETH) Zurich (2004)
6. Klöckl, B., Favre-Perrod, P.: On the inuence of demanded power upon the performance of energy storage devices. In: Proc. of the 11th International Power Electronics and Motion Control Conference (EPEPEMC), Riga, Latvia (2004)
7. Manfren, M., Caputo, P., Costa, G.: Paradigm shift in urban energy systems through distributed generation: Methods and models. Applied Energy 88(4), 1032–1048 (2011)
8. Manwell, J.F.: Hybrid energy systems. In: Cleveland, C.J. (ed.) Encyclopedia of Energy, vol. 3, p. 215. Elsevier, London (2004)
9. Clarke, J.A.: Energy simulation in building design, 2nd edn., p. 362. Butterworth Heinemann, Oxford (2001) ISBN 0 7506 5082 6
10. Sontag, R., Lange, A.: Cost effectiveness of decentralized energy supply systems taking solar and wind utilization plants into account. Renewable Energy 28(12), 1865–1880 (2003), ISSN 0960-1481
11. Fouzia. OUNNAR- Aziz NAAMANE - Patrick PUJO - Kouider nacer M'SIRDI "Pilotage multicritère d'un système énergétique multi-sources et multi-utilisateurs", 1er Congrès International en Génie Industriel et Management des Systèmes, Fès (Maroc), 04

Chapter 70
Technical-Economic Analysis of Solar Water Heating Systems at Batna in Algeria

Aksas Mounir[1], Zouagri Rima[1], and Naamane Aziz[2]

[1] Laboratoire de Physique, Energétique Appliquée (LPEA),
El-Hadj-Lakhdar University of Batna, Algeria
[2] Laboratoire LSIS, Polytech Marseille, France
m_aksas@hotmail.com

Abstract. The solar water heater (SWH) is one of the most important applications of solar energy because it affects several major hot water consumer sectors, such as houses, hotels, hospitals, barracks, etc.., where it can satisfy up to 70% of the needs and contributes to reducing greenhouse gas (GHG) emissions and saving energy. To do this, Algeria has established a major program for the development of SWH for the different sectors. The aim of this work is to study the technical-economic feasibility of SWH integration in Hospital Centers (HC) in the province of Batna, and the possibility of reducing GHG emissions. SWH installations analysis was done by RETScreen software, a mathematical model for clean energy projects analysis. The analysis showed the possibility of significant energy savings with SWH installation in the HC (Total annual provided energy (MWh) = 1427,1) and a considerable reduction in GHG (Total annual net reduction of GHG = 905,84 tCO_2, which corresponds to 2106,1 barrels of not consumed crude oil). However, the main barriers to the development of such projects are the low cost of fuel in Algeria and, conversely, the exorbitant cost of SWH installations. This highlights the need for government subvention for this type of project.

Keywords: Solar Energy, Solar water heater, RETScreen, technical economic study.

1 Introduction

Energy, in all its forms, is undoubtedly one of the most prominent parameters in the development and economic growth of a country. Energy consumption during the last century has greatly increased. It's expected that it will grow from 12 billion tonnes of oil equivalent (TOE) in 2010 to 17 billion TOE in 2030 with an average increase of 1.8% per year, whereas during the period 1990-2000, the increase was 1.4% per year [1, 2].

Ninety per cent of the energy consumed in 2005 in the world comes from nonrenewable fields while the remainder is from renewable sources. The increase in energy consumption and fuel prices in the last years leads us to ask ourselves

A. Håkansson et al. (Eds.): *Sustainability in Energy and Buildings*, SIST 22, pp. 787–796.
DOI: 10.1007/978-3-642-36645-1_70   © Springer-Verlag Berlin Heidelberg 2013

increasingly on the necessity to focus on other types of energy, alternative energies that can reduce not only energy costs but also the rate of GHG emissions. Indeed, over the past 30 years, the solar thermal system has experienced strong growth in the world [3]. For this reason, Algeria has begun to adopt an energy strategy that allows responding to these needs [4]. In Algeria, fossil non-renewable energy resources and their exploitation represent 40% of the GDP (Gross domestic product) and 98% of the national exports [5]. The national potential of renewable energy is strongly dominated by the solar energy. With its ideal location, Algeria has the largest solar field in the Mediterranean Basin. The average duration of sunshine in Algerian territory exceeds 2,000 hours per year, and may reach nearly 3900 hours of sunshine in the Sahara desert. The daily energy received on a horizontal surface of 1 m^2 is about 5 kWh of most of the national territory and corresponds nearly to 1700kWh/m^2/year in the northern region and 2263 kWh/m^2/year in the south of the country. The total energy received is estimated at 169 400 TWh/year, or 5,000 times the annual electricity consumption of the country [6]. Fig. 1 shows the maximum solar radiation received by Algeria on the unmodified plane in July.

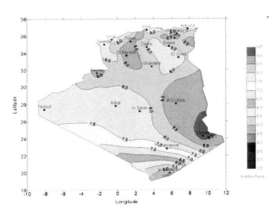

Fig. 1. Daily global irradiation received on normal plane in July [6]

Concerning wind, biomass and geothermal energy potentials they are much less important than that of solar energy. The hydroelectric potential is also very low.

Algeria has adopted a promotion program for SWH [7]. This program called "ALSOL" aims to promote the SWH, to prime the market, to encourage the creation of new industrial operators, and to develop networks of installers and energy establishment services. Ultimately, we expect the establishment of local manufacture for SWH production and of a sustainable solar thermal energy market in Algeria. This program is the first pilot program of this kind in Algeria. It predicts a direct financial support estimated at 45% of the cost of installed individual SWH and 35% of the cost of collective solar heating system through the National Fund for Energy Conservation (FNME). For 2010, the program "ALSOL" aimed at the promotion and dissemination, through the overall national territory, of 400 individual SWH, to produce hot water.

For 2011, the program "ALSOL" involves the installation of 2000 individual SWH and 3000 m² of solar collectors for the collective solar heating [7].

The objective of this paper is to show the technical and economic feasibility of SWH integration in Hospital Centers (HC) in the province of Batna, and the possibility of reducing GHG emissions

1.1 Technology of Solar Water Heaters

The principle of solar thermal energy is the recovery and utilization of heat from sunlight. SWH systems can be used for domestic hot water, space heating, and swimming pool heating. SWH systems comprise: a surface for solar radiation collection (collector or sensor), a heat transport system that transfers the energy extracted from the sensor to the storage element (storage of calories), a thermal storage, and a distribution network. The water is heated and stored inside the storage tank. There are two types of SWH systems: active (forced circulation) which have circulating pumps and controllers (Fig. 2), and passive where no pumping is required as the hot water naturally rises into the tank through thermosiphon flow.

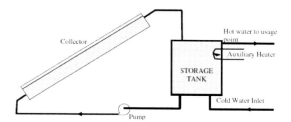

Fig. 2. Diagram of an active SWH system

There are several solar heat collector technologies whose performance and type of use differ. Unglazed solar collectors are simple sensors generally made of plastic polymer. They are adapted to cold temperatures and insensitive to the radiation incidence angle, they can be used for swimming pools heating. The glazed flat plate solar collectors are well suited to houses needs. Its operating temperatures correspond to temperatures of space heating and hot water production. An evacuated tube solar collector is composed of a series of transparent glass tubes of 5 to 15 cm in diameter. The tubes are evacuated to prevent convective heat loss from the absorber; this type of sensor is used for applications requiring greater levels of temperatures. It is located in industrial applications, but can be used for space heating and producing hot water for individual or collective housing.

Accurate sizing of performance can be done to achieve maximum performance of an installation, and this can be done for various parameters: daily consumption, area, orientation and angle of collectors, geographic location, possible masks, sensor type, as well as the solar tracking mode, the final temperature necessary for water use and the temperature of cold water entering.

1.2 The Software RETscreen

The RETScreen Clean Energy Project Analysis Software is a unique tool for decision support, developed in collaboration with many industry experts, the Canadian government and academia. Free, it can be used around the world to assess the production and energy savings, the cost over the life cycle, emission reductions, and the financial viability and risk for various types of Renewable-energy and Energy-efficient Technologies (RETs).

The RETScreen® International Solar Water Heating Model can be used to evaluate the energy production, life-cycle costs and greenhouse gas emissions reduction for three basic applications: domestic hot water, industrial process heat and swimming pools (indoor and outdoor), ranging in size from small residential systems to large scale commercial, institutional and industrial systems. It contains six worksheets: Energy Model, Solar Resource and Heating Load Calculation (Solar Resource and Heating Load), Cost Analysis, GHG Analysis, Financial Summary and Sensitivity and Risk Analysis (Sensitivity). The model also calculates the energy requirements corresponding to a given use. In the case of swimming pools heating, a cover can be used or not. In the case of SWH, we have the choice to use a storage tank of hot water or not. Three types of solar collectors can be used flat plate collectors glazed or not, and evacuated tube solar collectors.

This model calculates the energy produced by a solar system within a year using monthly data of sunshine, temperature, relative humidity and wind speed. The calculation method depends on the type of system and application. For SWH with storage tank, RETScreen uses the method known as the "f-chart". The method is explained in detail in [8]. This proven method calculates the percentage of hot water needs that can be met provided by solar energy, using the results of extensive simulations and laboratory models which helped to develop algorithms based on dimensionless numbers [9]. For SWH without storage, RETScreen uses the method of potential use. This method determines one threshold value level of solar radiation below which we cannot have solar gains when one takes into account the heat losses in solar temperature at which they must operate to meet the needs. The two methods are summarized in engineering manual and case studies RETScreen [9].

2 Case Study

To highlight the benefits of SWH installation and to understand the significance of their implementation in Algeria, the following cases studies are presented.

2.1 Selected Region and Its Parameters

The area chosen for the study is the commune of N'gaous, part of the province of Batna in Algeria, about 80 km west of Batna, which lies between 35 ° 33 'North and 06 ° 10' is in the junction of the Tell Atlas and the Saharan Atlas (Fig. 3).

Fig. 3. Location of N'Gaous commune in the province of Batna

The climate of the city is semi arid. To estimate the climatic data of Batna, we used the weather database that is in the RETScreen Software. The meteorological data used by RETScreen are given in Table 1.

Table 1. Characteristics the city of weather N'gaous

	Air temperature °C	Relative humidity %	Daily solar radiation - horizontal kWh/m²/d	Atmospheric pressure KPa	Wind speed m/s	Earth temperature C	Heating degree-days °C-d	Cooling degree-days °C-d
January	5,2	70,8%	2,50	94,4	3,3	7,0	397	0
February	6,3	66,9%	3,45	94,3	3,6	9,3	328	0
March	9,5	61,1%	4,51	94,1	3,6	13,7	264	0
April	12,9	58,0%	5,51	93,9	3,8	18,5	153	87
May	18,2	54,1%	6,38	93,9	3,8	25,1	0	254
Jun	23,3	45,1%	6,93	94,1	3,8	31,6	0	399
July	26,7	37,9%	7,13	94,1	3,8	34,7	0	518
August	26,1	41,7%	6,08	94,1	3,5	33,1	0	499
September	21,0	54,6%	4,86	94,2	3,1	26,9	0	330
October	16,7	60,1%	3,60	94,3	3,0	20,6	40	208
November	10,4	67,0%	2,65	94,2	3,2	13,3	228	12
December	6,6	74,2%	2,24	94,4	3,1	8,4	353	0
Annual	15,3	57,6%	4,66	94,2	3,5	20,2	1 763	2 307

The case study is a hospital located in the west of N'gaous commune, on the route to Setif province. The capacity of the hospital is 260 beds. It was built in 1985 over an area of 250 000 m². We used the RETScreen Software by choosing the energy model for a hospital. The parameters returned to the software for the basic scenario model 1 are given in Table 2.

Table 2. Input parameters for the RETscreen software (basic scenario model 1)

Parameter	Value	Parameter	Value
Beds number	260	Capacity (KW)	44,79
Occupancy rate (%)	60	Storage capacity / solar collector area (L\m2)	100
Daily hot water usage estimate (L/day)	30 707	Heat exchanger efficiency (%)	85
Hot water temperature (°C)	60	Miscellaneous losses (%)	5
Operating days per week	7	Seasonal efficiency for Base case	65
Supply water minimum temperature (°C)	11,8	Seasonal efficiency for Proposed case	75
Supply water maximum temperature (°C)	19,3	The transmission and distribution (T&D) losses (%)	8
Tilt angle (°)	35,8	Project life (yr)	25
Azimuth (°)	0	Fuel rate for Base case ($/kWh)	
The collector type	Glazed	Fuel rate for Proposed case ($/kWh)	0,004
Manufacturer	Agena SA énergies	Electricity rate ($/kWh)	0,052
Model	Azur 6	Fuel cost escalation rate (%)	2
Fr (tau alpha) coefficient	0,74	Inflation rate (%)	2
Fr UL coefficient (W/m²)/°C	4,28	Discount rate (%)	10
Gross area per solar collector (m²)	2,26	Total annual savings and income	($)
Aperture area per solar collector (m²)	2,06	Debt ratio (%)	0
Number of collectors	31	Incentives and grants (%)	35
		GHG emission factor (tCO$_2$/MWh)	0,169

The first scenario is for a recovery rate of the load equals to only 10%, and the provided heat is 41.1 MW. Other scenarios can also be developed for higher rates of recovery of the load.

2.2 Analysis of RETScreen Results for N'gaous Hospital

The results of the cumulative cash flows are shown in Fig. 4. They are produced by RETScreen for the baseline scenario, without financial support and considering that the fuel cost escalation rate is zero.

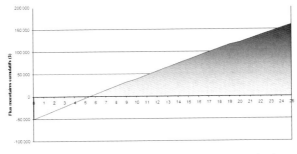

Fig. 4. Graph of cumulative cash flow for the baseline scenario (scenario N°1)

The Internal Rate of Return (IRR) before tax, which represents the actual performance of the project during its lifetime before tax, called also "return on equity invested", is in this case 17.8%. Simple payback and the return on investment (equity) (ROI) are of 5.3 and 5.4 years respectively.

The government will provide a grant equal to 35% of initial cost and the price per liter of diesel is currently 13.70 algerian dinars (AD), but could increase by 10% per year over the next 10 years and reach 45 AD in 2019 [10].

The calculation of the IRR after tax shows that the previous value becomes 42.5% and payback periods, and on equity of 3.4 and 2.8 years respectively.

The results of cumulative cash flow for 35% grant and 10% of fuel cost escalation rates are shown in Fig. 5.

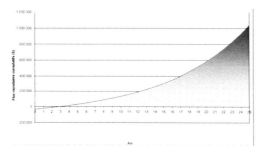

Fig. 5. Cumulative cash flow graph for 35% grant and 10% of fuel cost escalation rate

It is clear from these results that the SWH will be sufficiently profitable. The other part, which is as important, is the reduction of CO_2 emissions. In this study, net emissions of greenhouse gases are reduced by 79.8 tons of CO_2 equivalent (tCO_2) which corresponds to 186 barrels of not consumed crude oil. This quantity is important and may become more important when the rate of recovery of charge is increased, and when all hospitals in Patna will be considered.

This study was done for a recovery rate of the load equal to only 10%, so we studied other scenarios of SWH projects for different values of the load recovery rate that are higher. The results of this analysis are presented in Table 3.

Table 3. Results variation following load recovery rates for N'gaous hospital

Recovery rates (%)	Number of panels	Annual energy (MWh)	Annual GHG emission reduction (tCO_2)	Equivalent in oil barrels	Initial cost ($)	IRR (%)	Simple payback (yr)	Equity (yr)
10	31	41,1	79,8	186	49 724	42,5	3,4	2,8
20	62	79,2	89	207	86 148	28,4	6	4,5
30	100	122	101	235	122 572	22,4	8,5	5,9
40	140	162	109,6	256	140 524	20,3	9,9	6,6
46	165	181	113,8	265	184 485	16,7	13,2	8,1
60	234	237	130	302	246 444	13,8	17,2	9,7
70	310	283	142	330	328 770	10	26,1	12,5

794   A. Mounir, Z. Rima, and N. Aziz

Note that the higher the rate of recovery, the higher the SWH project is expensive with fewer benefits.

2.3   Analysis of the Results for the Whole Region of Batna

The province of Batna contains several hospitals that are different in size. They are distributed over several health sectors. These hospitals are distributed according to the size of the population covered by the health sector. Their number reaches 12 with a total capacity of 2121 beds in 2004. The province center contains three hospitals with a capacity to 953 beds, whereas the health sector of N'gaous contains a single hospital with a capacity of 260 beds. The number of beds in various hospitals in the province is in Table 4.

Table 4. Beds umber in each hospital of Batna province

Communes (municipalities)	Number of hospitals	Number of beds
Batna (center)	3	953
El Madher	1	75
Aïn Touta	1	146
Barika	2	260
Merouana	2	216
N'Gaous	1	260
Arris	2	211
Total	12	2121

Hospitals size in the province of Batna varies from one health sector to another, that's why it is important to analyze scenarios of SWH project for different sizes of hospitals, from the largest hospital in the province center with a capacity that reaches 636 beds to the small hospital that is located in El Madher with a capacity of 75 beds. The analysis was made for a recovery rate of 46%. The results of this analysis are presented in Table 5.

Table 5. Results variation according to the number of beds in the hospital

Number of Beds	Number of panels	Annual energy (MWh)	Annual GHG emission reduction (tCO2)	Equivalent in oil barrels	Initial cost ($)	IRR (%)	Simple payback (yr)	Equity payback (yr)
75	43	48,3	30,4	70,7	62 238	12,3	24	11,3
96	49	66,3	42,3	98,4	78 602	13,9	18,7	10
115	70	78,6	50	116	90 646	14,4	17,3	9,5
130	81	90,9	57,8	134	102 031	14,9	16,2	9,1
140	86	96,9	61,5	143	107 503	15,1	15,8	9
146	91	102,3	65,3	152	112 809	15,3	15,4	8,8
260	161	181	113,8	265	184 485	16,7	13,2	8,1
636	402	451,9	288,2	670	434 921	18	11,6	7,5

Note that the more the number of beds is important, the more the project of SWH is advantageous.

The results of the investigation of the benefits in using SWH in Batna province are summarized in Table 6.

Table 6. Summary of the results obtained with the case study of Batna province

Total annual provided energy (MWh)	Total annual net reduction of GHG (tCO$_2$)	Total oil barrels equivalent
1427,1	905,84	2106,1

The above results show that the proposed SWH in the province of Batna is very profitable, and there is a considerable saving in both energy savings and reducing GHG emissions. The benefits of this profitability are on the government and the environment.

3   Conclusion

This work helps to show that the annual energy saved is about 1427.1 MWh for the province of Batna. This energy is equivalent to 123 "toe".

Another parameter which is very important in studying the profitability of clean energy projects is to reduce total annual net greenhouse gas emissions. Our study shows that this parameter is equal to 905.84 tCO$_2$. This is equivalent to 2106,1 barrels of oil not consumed, and 0.02% reduces greenhouse gas emissions at national level, where GHG emissions due to final energy consumption reached 46 million tCO$_2$ These values become very important when we consider all national provinces.

All these data indicate that the replacement of fuel (diesel) water heating systems in hospitals by SWH systems will be possible, so solar hot water production facilities are cost-effective in Algeria.

Parameters that play a very important role in the profitability of SWH projects are fuel prices and government financial supports. Algeria is ranked fifth worldwide in terms of lowest price of fuel. The fact that the government subsidizes fuel prices and that the initial investment of SWHs is very high, hinders the development of SWH projects. For this, it is very important that the government finance enough such projects.

References

[1]   Rojey, A.: Energie et climat. Éditions technip, paris (2008)
[2]   Furfari, S.: Le monde et l'énergie, enjeux géopolitiques. Éditions Tchnip Paris (2007)
[3]   EUROBSERV'ER. Le baromètre Européen 2005 des énergies renouvelables. Paris. Systèmes solaires, p.32 (2005)
[4]   Boudghene Stambouli, A.: Algerian renewable energy assessment: The challenge of sustainability. Energy Policy 39(8), 4507–4519 (2011)

[5] Hamouda, C., Benamira, A.: et Malek. A. La formation de conseillers en maîtrise d'énergie et protection de l'environnement à l'université de Batna: Exemple d'un audit énergétique dans le secteur des matériaux de construction. Revue des Energies Renouvelables 10(2), 157–172 (2007)

[6] Ministry of Energy and Mining, Solar energy (2012),
http://www.mem-algeria.org

[7] CDER (Renewable Energy Development center), http://www.cder.dz

[8] Duffie, J.A., Beckman, W.A.: Solar engineering of thermal processes, 2nd edn. Wiley-Interscience, New York (1991)

[9] Clean Energy Project Analysis. RETScreen® Engineering and Case Studies Textbook, Solar Water Heating. Analysis Project Chapter, 3rd edn. (2005),
http://www.retscreen.net

[10] Bouzidi, B., Haddadi, M.: Implication financière des GES pour le développement des systèmes photovoltaïques de pompage de l'eau. Revue des Energies Renouvelables 10(4), 569–578 (2007)

Chapter 71
Design and Control of a Diode Clamped Multilevel Wind Energy System Using a Stand-Alone AC-DC-AC Converter

Mona F. Moussa and Yasser G. Dessouky

Arab Academy for Science and Technology and Maritime Transport
Miami, P.O. Box: 1029, Alexandria, Egypt
mona.moussa@yahoo.com

Abstract. The major application of the stand-alone power system is in remote areas where utility lines are uneconomical to install due to terrain, the right-of way difficulties or the environmental concerns. Villages that are not yet connected to utility lines are the largest potential market of the hybrid stand-alone systems using diesel generator with wind or PV for meeting their energy needs. The stand-alone hybrid system is technically more challenging and expensive to design than the grid-connected system that simply augments the existing utility system.

Multilevel inverter technology has emerged recently as a very important alternative in the area of high-power medium-voltage energy control. This paper presents the topology of the diode-clamped inverter, and also presents the relevant control and modulation method developed for this converter, which is: multilevel selective harmonic elimination, where additional notches are introduced in the multi-level output voltage. These notches eliminate harmonics at the low order/frequency and hence the filter size is reduced without increasing the switching losses and cost of the system. The proposed modulation method is verified through simulation using a five-level Diode-clamped inverter prototype. The system consists of a 690V wind-driven permanent magnet synchronous generator whose output is stepped down via a multiphase transformer, designed to eliminate lower order harmonics of the generator current. The transformer secondary voltages are rectified through an uncontrolled AC/DC converters to provide different input DC voltage levels of the diode clamp quazi phase multilevel inverter where the pulse widths are adjusted to eliminate low order harmonics of the output voltage whose magnitude is kept constant with different loading condition by controlling the inverter switching and maintaining low total harmonic distortion THD.

Keywords: Selective Harmonic Elimination, stand alone systems, converters, wind energy, renewable energy and Diode clamped Multilevel Inverter.

1   Introduction

In this paper a regulated AC-DC-AC converter is studied, where the AC-DC converter has lower THD while the elimination of harmonics using diode-clamped multilevel inverter (DCMLI) has been implemented. The problem of eliminating harmonics in switching inverters has been the focus of research for many years. The current trend of modulation control for multilevel inverters is to output high quality power with high efficiency. For this reason, popular traditional PWM modulation methods are not the best solution for multilevel inverter control due to their high switching frequency. The selective harmonic elimination method has emerged as a promising modulation control method for multilevel inverters. The major difficulty for the selective harmonic elimination method is to solve the equations characterizing harmonics; however, the solutions are not available for the whole modulation index range, and it does not eliminate any number of specified harmonics to satisfy the application requirements. The proposed harmonic elimination method is used to eliminate lower order harmonics and can be applied to DCMLI application requirements. The diode clamped inverter has drawn much interest because it needs only one common voltage source. Also, it is efficient, even if it has inherent unbalanced dc-link capacitor voltage problem [1]. However, it would be a limitation to applications beyond four-level diode clamped inverters for the reason of reliability and complexity considering dc-link balancing and the prohibitively high number of clamping diodes [2].

By increasing the number of levels in the inverter, the output voltages have more steps generating a staircase waveform, which has a reduced harmonic distortion. However, a high number of levels increases the control complexity and introduces voltage imbalance problems. Three different topologies have been proposed for multi-level inverters: diode-clamped (neutral-clamped) [8]; capacitor-clamped (flying capacitors) [13]; and cascaded multicell with separate dc sources [14]-[17].

The system configuration, as shown in Fig.1, is made-up of wind stand-alone system, multi-phase transformer connected to pulse series-type diode rectifier, dc link filter, diode clamped multilevel inverters, trap filters, and the load.

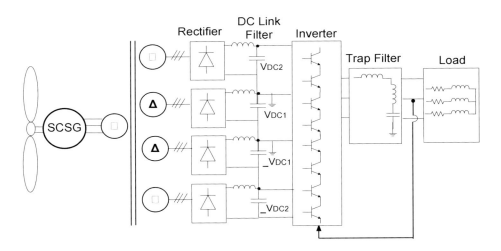

Fig. 1. Diode clamped multilevel wind energy system using a stand-alone AC-DC-AC converter system

An interior permanent magnet synchronous generator IPMSG is feeding a multi-phase transformer with four secondary windings. In order to reduce the line generator current THD, multipulse diode rectifiers powered by phase-shifting transformers are often employed. Consequently, each winding of the transformer is connected to 6-pulse series-type diode rectifier whose DC output is regulated by a DC link LC filter to feed a diode clamped inverters such that they are controlled independently in order to improve the performance under different load conditions. The output voltage of the inverter is supplying a three-phase 440V, 60kVA load with regulated voltage through a feedback signal from output load voltage to control the pulse width of the upper transistor.

2 Wind Stand-Alone System

The stand-alone wind system using a constant speed generator is has many features similar to the PV stand-alone system. For a small wind system supplying local loads, a permanent magnet IPMSG makes a wind system simple and easier to operate. The battery is charged by an AC to DC rectifier and discharged through a DC to AC inverter.

The wind stand-alone power system is often used for powering farms. In Germany, nearly half the wind systems installed on the farms are owned either by individual farmers or by an association. The generalized d-q axis model of the generator is used to model the synchronous generator [19].

3 Multi-phase Transformer Connected to Pulse Series-Type Diode Rectifier

Fig. 2 shows the typical configuration of the phase-shifting transformers for 12-pulse rectifiers. There are two identical six-pulse diode rectifiers powered by a phase-shifting transformer with two secondary windings. The dc outputs of the six-pulse rectifiers are connected in series. To eliminate low-order harmonics in the line current i_A, the line-to-line voltage v_{ab} of the wye-connected secondary winding is in phase with the primary voltage v_{AB} while the delta-connected secondary winding voltage $v_{\tilde{a}\sim b}$ leads v_{AB} by $\delta = 30°$. The rms line-to-line voltage of each secondary winding is $V_{ab} = V_{\tilde{a}\sim b} = K\ V_{AB}/2$. From which the turns ratio of the transformer can be determined by [20]:

$$\frac{N_1}{N_2} = 2K \text{ and } \frac{N_1}{N_3} = \frac{2}{\sqrt{3}}K \text{ , where } K \text{ is the step down ratio.}$$

The configuration of a Y/Z-2 phase-shifting transformer is shown in Fig. 2, where the primary winding remains the same as that in the Y/Z-1 transformer while the

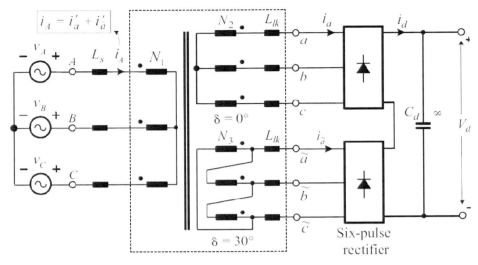

Fig. 2. 12-pulse diode rectifier

secondary delta-connected coils are connected in a reverse order. The transformer turns ratio can be found from:

$$\frac{N_3}{N_2 + N_3} = \frac{sin(30^o + \delta)}{sin(30^o - \delta)}$$

$$-30^o \leq \delta \leq 0$$

$$\frac{N_1}{N_2 + N_3} = \frac{1}{2sin(30^o + \delta)} \cdot \frac{V_{AB}}{V_{ab}}$$

The phase angle δ has a negative value for the Y/Z-2 transformer, indicating that V_{ab} lags V_{AB} by |δ|. The phase-shifting transformer is an indispensable device in multipulse diode rectifiers. It provides three main functions: (a) a required phase displacement between the primary and secondary line-to-line voltages for harmonic cancellation, (b) a proper secondary voltage, and (c) an electric isolation between the rectifier and the utility supply [20].

4 Trap Filters

To attenuate the penetration of harmonics into the a.c system from a rectifier load, harmonic filters can be connected to the neutral from each line. The manner in which the harmonics currents are by-passed is to provide harmonic filters as shown in Fig. 3. For a 6-pulse system, tuned harmonic filters are provided for the 13th and 17th harmonic components. For the higher order harmonics, a high pass filter is provided. Care must be taken to avoid excessive loss at the fundamental frequency. A practical problem is that of frequency drift, which may be as much as ±2% in a public supply system. Either the filters have to be automatically tuned or have a low Q-factor to be effective.

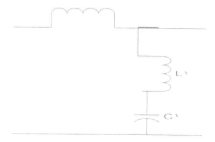

Fig. 3. Harmonic trap filter

When, trap filters are designed to eliminate the 11th harmonic, they do this by providing a low impedance path for that harmonic [21].

5   Diode-Clamped Multilevel Inverter

The diode-clamped inverter, or the neutral-point clamped (NPC) inverter, effectively doubles the device voltage level without requiring precise voltage matching [22].

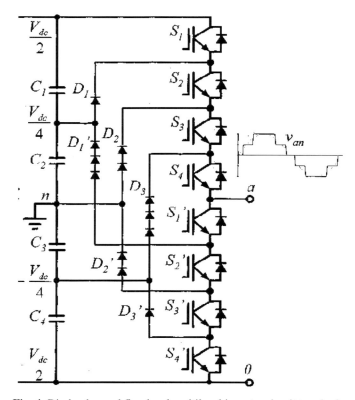

Fig. 4. Diode-clamped five-level multilevel inverter circuit topologies

Fig. 4 shows a five-level diode-clamped converter in which the dc bus consists of four capacitors, C_1, C_2, C_2, and C_1. For dc-bus voltage V_{dc}, the voltage across each capacitor is V_{dc1}, V_{dc2}, V_{dc2}, and V_{dc1} respectively, and each device voltage stress will be limited to one capacitor voltage level $V_{dc}/4$ through clamping diodes [23]-[25].

To explain how the staircase voltage is synthesized, the neutral point n is considered as the output phase voltage reference point. There are five switch combinations to synthesize five level voltages across a and n.

1. For voltage level $V_{an} = V_{dc1}$, turn on all upper switches S_1–S_4.
2. For voltage level $V_{an} = V_{dc2}$, turn on three upper switches S_2–S_4 and one lower switch S_1'.
3. For voltage level $V_{an} = 0$, turn on two upper switches S_3 and S_4 and two lower switches S_1' and S_2'.
4. For voltage level $V_{an} = -V_{dc1}$, turn on one upper switch S_4 and three lower switches $S_1' - S_3'$.
5. For voltage level $V_{an} = -V_{dc2}$, turn on all lower switches $S_1' - S_4'$.

6 Determination of Output Waveform Shape

The concept of the proposed technique is to combine the selective harmonic eliminated PWM method with the optimised harmonic step waveform method. The Selective Harmonic Elimination (SHE) method introduces additional notches in the basic voltage waveform of the square wave inverter. The inverter output voltage is "chopped" a number of times at an angle(s) to eliminate the selected harmonic(s) [26]-[29]. These angles are calculated in off-line correlating the selected harmonics to be eliminated in the inverter output voltage. In similar lines, for the multilevel inverter, the notches are optimised to eliminate the lower order harmonics in the output voltage of a multilevel inverter. In the Optimized Harmonic Stepped-Waveform Technique OHSW method the number of switching is limited to the number of level of the inverter [30].

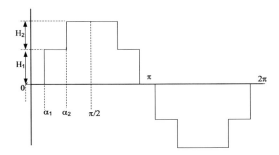

Fig. 5. Output voltage waveform of a diode clamped inverter

The output voltage waveform V(t) shown in Fig. 6 can be expressed in Fourier series as [31]:

$$V(t) = \sum_{n=1}^{\infty} V_n \, \sin n\alpha_n \qquad (1)$$

The amplitude of the nth harmonic is expressed only with the first quadrant switching angles α_1, α_2, as:-

$$V_n = \frac{4V_{dc}}{n\pi} [H_1(\cos n\alpha_1) + H_2(\cos n\alpha_2)] \qquad (2)$$

Where $0 < \alpha_1 < \alpha_2 < \frac{\pi}{2}$

V_n is equated to zero for the harmonics to be eliminated [31], as follows:

$$V_5 \ = 0 \ = H_1(\cos 5\alpha_1) \ + \ H_2(\cos 5\alpha_2) \qquad (3)$$

$$V_7 \ = 0 \ = H_1(\cos 7\alpha_1) \ + \ H_2(\cos 7\alpha_2) \qquad (4)$$

$$V_{11} = 0 \ = H_1(\cos 11\alpha_1) + H_2(\cos 11\alpha_2) \qquad (5)$$

These are three equations and also:

$$H_1 + H_2 = 1 \qquad (6)$$

Solving these four equations together using MATHCAD software, the value of H_1, H_2, α_1, and α_2 can be obtained as shown in table 1. Having got the values of the angles α and DC voltage heights H, the spectrum analysis using Fourier Transformation for the output voltage can be obtained.

Table 1.

$\alpha_1 = 10.97°$;	$\alpha_2 = 35.24°$;
$H_1 = 0.634$;	$H_2 = 0.365$;

7   Simulation

The stabilized AC-DC-AC power supply used is shown in Fig. 6 which consists of a step down transformer with one primary and four secondary and whose turns ratio are $(0.8:H_1)$, $(0.8:H_2)$, followed by an uncontrolled rectifier and then a dc link low pass filter, therefore, two different DC voltages are obtained. Each DC voltage is feeding a quasi-single inverter whose angles are α_1, and α_2 (given in table I) which are clamped to get the required wave form. This system consists of a load voltage feedback control signal in order to maintain the voltage of the load constant at 380V irrespective of the loads variations by controlling the DC voltage levels of through the controlled rectifiers. The inverters operate at 60Hz. The system has been tested from 10% to 20% of full load at time t = 0.9s, from 80% to 100% of full load at time t = 1.9s, and from 100% to 110% of full load at time t = 2.9s. The transient response results are shown in Fig. 7.

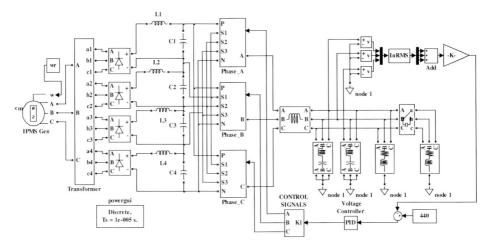

Fig. 6. Simulink block diagram

The proposed simulink system configuration is shown in Fig. 7, where a three-phase 50kVA, 690V step-up transformer has one primary coil and four secondary coils which feed three bridge rectifiers, which are cheap since they are uncontrolled devices in converting the AC to DC. Each bridge rectifier is connected to a DC/AC diode clamped inverter. The output voltages of the inverters are clamped together and connected to supply a three-phase 380V, 60kVA load with regulated voltage through a feedback signal from output load voltage to control the thyristor rectifiers.

The value of the inductance of the 1st trap filter is 0.0022 H, while that of the 2nd trap filter is 0.0013H. Both trap filters have a capacitance of 19.2µF. The value of the inductances of the four DC link filters are the same L_1, L_2, L_3, L_4 = 5mH, while the capacitances of the 1st and 4th DC link filters are $C_1 = C_4$ = 17600 µF. and the capacitances of the 2nd and 3rd DC link filters are $C_2 = C_3$ = 8800 µF.

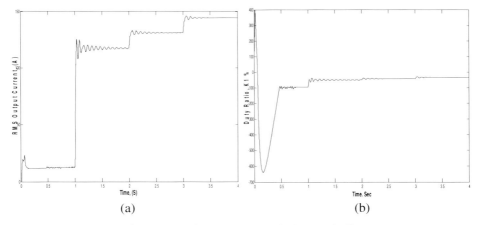

Fig. 7. (a) RMS output current, (b) Duty ratio K_1

71 Design and Control of a Diode Clamped Multilevel Wind Energy System

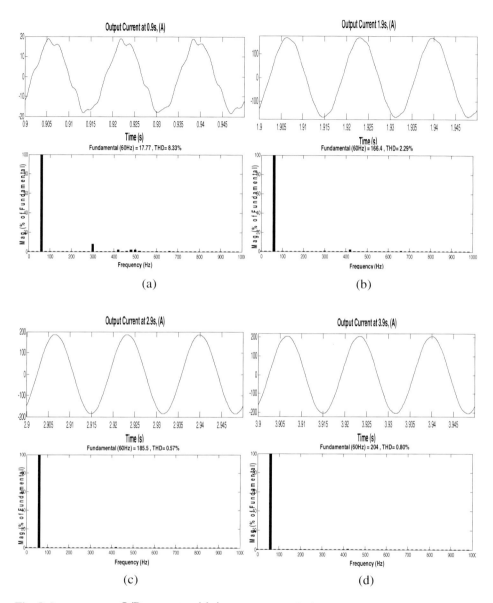

Fig. 8. Instantaneous O/P current, with its spectrum analysis at: (a) 10%, (b) 80%, (c) 100%, (d) 110% of the load

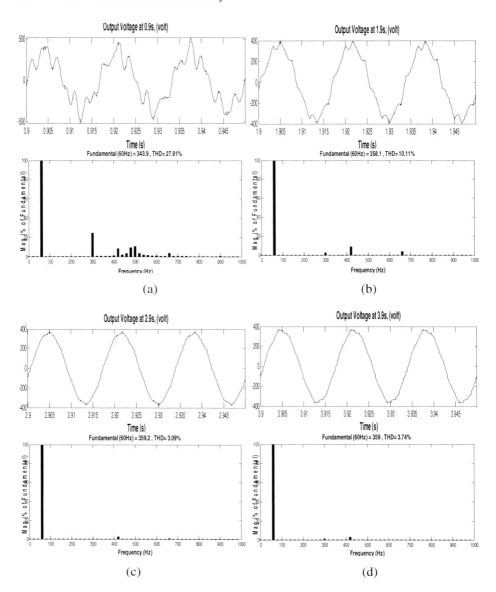

Fig. 9. Instantaneous O/P voltage, with its spectrum analysis at: (a) 10%, (b) 80%, (c) 100%, (d) 110% of load

71 Design and Control of a Diode Clamped Multilevel Wind Energy System 807

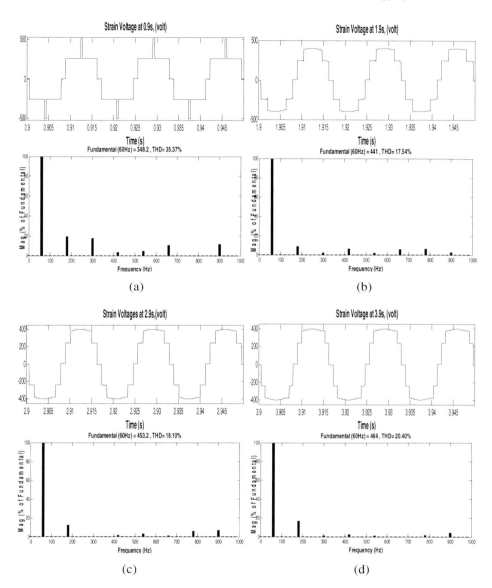

Fig. 10. Strain voltage, with its spectrum analysis at: (a) 10%, (b) 80%, (c) 100%, (d) 110% of the load

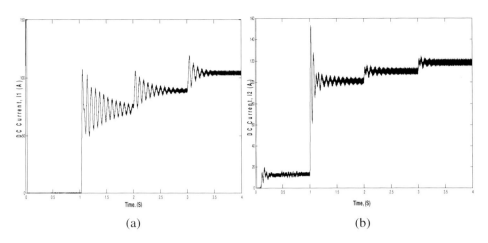

Fig. 11. (a) DC current I1, (b) DC current I_2

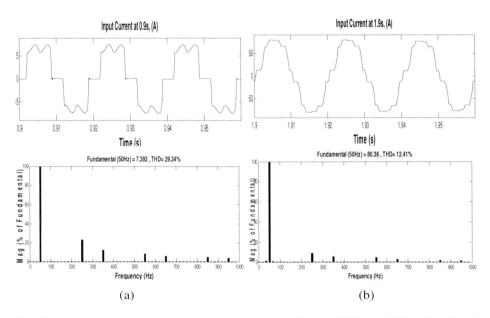

Fig. 12. Input current, with its spectrum analysis at: (a) 10%, (b) 80%, (c) 100%, (d) 110% of the load

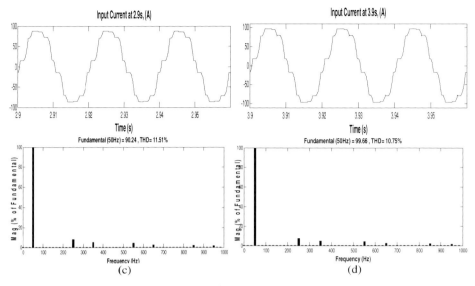

Fig. 12. (*continued*)

It should be noted that some power suppliers use the transformer after the DC to AC inverter to step up the voltage in the 60 Hz frequency level [32] while in this system the step up transformer is used before the AC to DC rectifier in the 50 Hz frequency level which has an advantage of lower iron core losses and less noise. The results of Fig. 8 to Fig. 13 show that the effectiveness of this AC/DC/AC converter to supply a regulated AC voltage regardless of the load changes with low THD.

From the simulation results, it can be noted that the phase-shifting transformer is an indispensable device in multipulse diode rectifiers. It provides three main functions: (a) a required phase displacement between the primary and secondary line-to-line voltages for harmonic cancellation, (b) a proper secondary voltage, and (c) an electric isolation between the rectifier and the utility supply. Also, it is clear that the trap filters are provided to eliminate the 13th and 17th harmonic components.

The inverters are diode clamped together to result in an output voltage free of 3rd, 5th, 7th, 9th and 11th harmonics. The result showed that the output voltage resulted in an almost sinusoidal current.

8   Conclusion

The selected harmonic elimination is a popular issue in multilevel inverter design. The proposed selective harmonic elimination method for DCMLI has been validated in simulation. The simulation results show that the proposed algorithm can be used to eliminate any number of specific lower order harmonics effectively and results in a dramatic decrease in the output voltage THD. In the proposed harmonic elimination method, the lower order harmonic distortion is largely reduced in fundamental switching. Multilevel inverters include an array of power semiconductors and capacitor voltage sources, the output of which generate voltages with stepped waveforms. The commutation of the switches permits the addition of the capacitor voltages, which reach high voltage at the output, while the power semiconductors must withstand only reduced voltages. Thus, by increasing the number of levels in the inverter, the output voltages have more steps generating a staircase waveform, which has a reduced harmonic distortion. However, a high number of levels increases the control complexity and introduces voltage imbalance problems. The most attractive features of multilevel inverters are that, not only, they draw input current with very low distortion, but also, they generate smaller common-mode (CM) voltage, thus reducing the stress in the motor bearings. In addition, using sophisticated modulation methods, CM voltages can be eliminated.

A high performance static AC-DC-AC converter is designed. The controller has a good control property. The system topology adopts two single-phase diode clamped inverters such that they are controlled independently in order to improve the performance under different load conditions. It is clear that, the phase-shifting transformer is an indispensable device in multipulse diode rectifiers, since it provides three main functions: (a) a required phase displacement between the primary and secondary line-to-line voltages for harmonic cancellation, (b) a proper secondary voltage, and (c) an electric isolation between the rectifier and the utility supply. With the help of the developed algorithm, the switching angles are computed from the non-linear equation characterizing the Selective Harmonic Elimination problem to contribute minimum THD in the output voltage waveform. Therefore, lower order harmonics like 3rd, 5th, 7th, 9th, 11th, 13th, and 17th are eliminated and higher-order harmonics are optimized in case of fundamental switching without using PWM. The selected harmonic elimination is a popular issue in multilevel inverter design. The proposed selective harmonic elimination method has been validated using Matlab Simulink. The simulation results show that the proposed algorithm can be used to eliminate any number of specific lower order harmonics effectively and results in a dramatic decrease in the output voltage THD. In the proposed harmonic elimination method, the lower order harmonic distortion is largely reduced in fundamental switching.

References

[1] Rodriguez, J., Lai, J.S., Peng, F.Z.: Multilevel inverters: a survey of topologies, controls, and applications. IEEE Transactions on Industrial Electronics 49(4), 724–738 (2002)

[2] Lai, J.S., Peng, F.Z.: Multilevel converters-a new breed of power converters. IEEE Transactions on Industry Applications 32(3), 509–517 (1996)

[3] Tolbert, L., Peng, F.-Z., Habetler, T.: Multilevel converters for large electric drives. IEEE Trans. Ind. Applicat. 35, 36–44 (1999)

[4] Teodorescu, R., Beaabjerg, F., Pedersen, J.K., Cengelci, E., Sulistijo, S., Woo, B., Enjeti, P.: Multilevel converters — A survey. In: Proc. European Power Electronics Conf. (EPE 1999), Lausanne, Switzerland, CD-ROM (1999)

[5] Hochgraf, C., Lasseter, R., Divan, D., Lipo, T.A.: Comparison of multilevel inverters for static var compensation. In: Conf. Rec. IEEE-IAS Annu. Meeting, pp. 921–928 (October 1994)

[6] Hammond, P.: A new approach to enhance power quality for medium voltage ac drives. IEEE Trans. Ind. Applicat. 33, 202–208 (1997)

[7] Cengelci, E., Sulistijo, S.U., Woom, B.O., Enjeti, P., Teodorescu, R., Blaabjerge, F.: A new medium voltage PWM inverter topology for adjustable speed drives. In: Conf. Rec. IEEE-IAS Annu. Meeting, St. Louis, MO, pp. 1416–1423 (October 1998)

[8] Chiasson, J.N., Tolbert, L.M., Mckenzie, K.J., Du, Z.: Elimination of harmonics in a multilevel converter using the theory of symmetric polynomials and resultants. IEEE Trans. Control Syst. Technol., 216–223 (2005)

[9] Ray, R.N., Chatterjee, D., Goswami, S.K.: Harmonics elimination in a multilevel inverter using the particle swarm optimisation technique. IET Power Electron 2, 646–652 (2009)

[10] Chiasson, J.N., Tolbert, L.M., McKenzie, K.J., Du, Z.: A unified approach to solving the harmonic elimination equations in multilevel converters. IEEE Transactions on Power Electronics 19(2), 478–490 (2004)

[11] Du, Z., Tolbert, L.M., Chiasson, J.N., Li, H.: Low switching frequency active harmonic elimination in multilevel converters with unequal DC voltages. In: Annual Meeting of the IEEE Industry Applications Society, vol. 1, pp. 92–98 (October 2005)

[12] Ozpinecib, B., Tolbert, L.M., Chaisson, J.N.: Harmonic optimization of multilevel converters using genetic algorithms. In: Proc. IEEE Power Electronic Spectrum Conf., pp. 3911–3916 (2004)

[13] Ozdemir, S., Ozdemir, E., Tolbert, L.M., Khomfoi, S.: Elimination of Harmonics in a Five-Level Diode-Clamped Multilevel Inverter Using Fundamental Modulation (2007)

[14] Pan, Z., Peng, F.Z., et al.: Voltage balancing control of diode clamped multilevel rectifier/inverter systems. IEEE Transactions on Volts Industry Applications 41, 1698–1706 (2005)

[15] Lezana, P., Rodriguez, J., Oyarzun, D.A.: Cascaded multilevel inverter with regeneration capability and reduced number of switches. IEEE Trans. on Industrial Electronics 55(3), 1059–1066 (2008)

[16] Dahidah, M.S.A., Agelidis, V.G.: Selective harmonic elimination PWM control for cascaded multilevel voltages source converters: A generalized formula. IEEE Transactions on Power Electronics 23(4), 1620–1630 (2008)

[17] Huang, A.Q., Sirisukprasert, S., Xu, Z., Zhang, B., Lai, J.S.: A high-frequency 1.5 MVA H-bridge building block for cascaded multilevel converters using emitter turn-off thyristor (ETO). In: Proc. IEEE APEC, Dallas, TX, pp. 25–32 (March 2002)

[18] Lai, J.S., Peng, F.Z.: Multilevel Inverters: A Survey of Topologies, Control and Applications. IEEE Trans. Ind. Elec. 49, 724–738 (2002)

[19] Patel, M.R.: Wind and solar power systems (1999)

[20] Wu, B.: High power converters and AC drives, 2nd edn. John Wiley & Sons, Inc., Publication

[21] Lander, C.W.: Power electronics, 2nd edn. Mc Graw-Hill Book Company

[22] Dell Aquila, A., Monopoli, V.G., Liserrre, M.: Control of H-bridge Based Multilevel converters. IEEE Trans. Power Electron, 766–771 (October 2002)

[23] Muthuramalingam, A., Balaji, M., Himavathi, S.: Selective Harmonic Elimination Modulation Method for Multilevel Inverters. In: India International Conference on Power Electronics (2006)

[24] Khomfoi, S., Tolbert, L.M.: Multilevel Power Converters. In: Power Electronics Handbook, 2nd edn., ch. 17, pp. 451–482. Elsevier (2007) ISBN 978-0-12-088479-7

[25] Chiasson, J.N., Tolbert, L.M., McKenzie, K.J., Du, Z.: A complete solution to the harmonic elimination problem. IEEE Transactions on Power Electronics 19(2), 491–499 (2004)

[26] Du, Z., Tolbert, L.M., Chiasson, J.N.: Active harmonic elimination for multilevel converters. IEEE Transactions on Power Electronics 21(2), 459–469 (2006)

[27] Moussa, M., Hussien, H., Dessouky, Y.: Regulated AC/DC/AC Power Supply Using Scott Transformer. In: Proc. IET Power Electronics Machines and Drives Conf., PEMD (2012)

[28] Lienhardt, A.M., Gateau, G., Meynard, T.A.: Zero-Steady-State-Error input-current controller for regenerative multilevel converters based on single-phase cells. IEEE Trans. on Industrial Electronics 54(2), 733–740 (2007)

[29] Hosseini, S.H., Sadigh, A.K., Sharifi, A.: Estimation of flying capacitors voltages in multicell converters. In: Proc. ECTI Conf., vol. 1, pp. 110–113 (2009)

[30] Massoud, A.M., Finney, S.J., Cruden, A., Williams, B.W.: Mapped phase-shifted space vector modulation for multilevel voltage source inverters. IET Electr. Power Appl., 622–636 (2007)

[31] Bina, M.T.: Generalised direct positioning approach for multilevel space vector modulation: theory and implementation. IET Electr. Power Appl., 915–925 (2007)

Chapter 72
Analyzing the Optical Performance of Intelligent Thin Films Applied to Architectural Glazing and Solar Collectors

Masoud Kamalisarvestani[1], Saad Mekhilef[2], and Rahman Saidur[1]

[1] Department of Mechanical Engineering, University of Malaya, Kuala Lumpur, Malaysia
[2] Department of Electrical Engineering, University of Malaya, Kuala Lumpur, Malaysia

Abstract. Windows provide us with natural light and fine-looking connections to the outdoors. However, a huge amount of energy is lost through these building envelopes. Similarly, the glazing is used in many other systems such as solar collector covers and photovoltaic cells. This study reviews the most common and contemporary coatings of the glass surface with thin films. It is analyzed how the smart windows operate and help us in enhancing energy efficiency, getting the most out of indoor comfort and improving the performance of solar collectors and PV cells. These intelligent coatings feature different and selective optical properties in different environmental conditions. The analysis emphasizes on the radiation equations and discusses the different scenarios based on these equations.

1 Introduction

A significant amount of energy is consumed for maintaining thermal comfort in buildings. This energy is mostly burnt off to keep vapor compression cycles and AC devices running. The building energy consumption is noticeably high in hot and humid regions, contributing to one third to half of the electricity produced in some countries [1-3]. Therefore, energy saving measures should be implemented in order to decrease energy losses [4]. It is highlighted as a major responsibility of designers of new buildings not only to cut down on electricity consumption in lighting and HVAC systems but also to choose building materials wisely [5]. There are two approaches in energy saving strategies, the active strategies and the passive ones. Improving HVAC systems and building lighting can actively increase the building's energy efficiency, whereas measures amending building envelopes are the passive methods for the above mentioned purpose. Any building element, such as wall, roof and fenestration which separates the indoor from outdoor is called building envelope [4]. Using cool coatings on roofs, adding thermal insulation to walls and coated window glazing are among the effective envelope-based passive techniques that ameliorate energy efficiency [6-8].

Windows are known as one of the most energy inefficient components of buildings [9]. Cooling and heating energy is noticeably wasted via windows. Particularly in commercial buildings, they are excessively exposed to solar radiation due to large area fenestration leading to thermal discomfort [10, 11]. According to the National

---

A. Håkansson et al. (Eds.): *Sustainability in Energy and Buildings*, SIST 22, pp. 813–826.
DOI: 10.1007/978-3-642-36645-1_72   © Springer-Verlag Berlin Heidelberg 2013

Renewable Energy Laboratory report 2009, windows are culpable of about 30 percent of building heating and cooling electrical loads and applying high-tech fenestration techniques can potentially save approximately 6 percent of the energy consumption nationwide [12].

Consequently, if we curtail these losses by improving the windows thermal performance less electricity costs and greenhouse gas emissions will be resulted. Therefore, controlling solar gain and loss by means of fenestration should be emphasized in building design. While reducing radiation transmitted, the window materials should be capable of sufficient transmission of visible light through windows [13]. Modern architecture lends a lot to the concept of residents' comfort [14]. From the aesthetic standpoint, these transparent facades are essential building elements that provide a comfortable indoor environment by creating eye-catching views as well as illuminating the interior space by inviting light inside [15]. Moreover, there are some goals, which cannot be achieved by conventional materials such as metals and plastics whereas glass will be suitable [16]. There is quite a vast number of parameters influencing the heat transfer through windows such as outdoor conditions, shading, building orientation, type and area of window, glass properties and glazing characteristics [15]. Improving glazing characteristics of windows such as thermal transmittance and solar parameters is the most important criterion to be considered in building windows standards [17].

In recent years, glazing technologies and materials have been the major focus of many studies. Aerogel glazing, vacuum glazing, switchable reflective glazing, suspended particle devices film, holographic optical elements [18, 19], low-e coatings, all-solid- state switchable mirror glass [20, 21], gas cavity fills and improvements in frame and spacer designs [22] are among the most common glazing technologies in terms of controlling solar heat gain, insulation and lighting. Electrochromic (EC), thermochromic (TC) and photochromic materials are employed in glass industry for different, sometimes odd, applications. The two most recent developments in the industry, "self-cleaning glass" and "smart glass", offer excellent energy efficient and environmentally friendly features in various applications [16].

1.1 Switchable Reflective Glazing (Intelligent Glazing)

This type of glass -also named "smart window"-is generally based on optical switching along with modulation in glass properties. These dynamic tintable windows are categorized in to passive and active systems.

In passive devices, the switching process is activated automatically in accordance with the environmental conditions. This environmental factor can be light in case of photochromic windows; or temperature and heat in thermochromic windows (TCW). Alternatively, the active systems require an external triggering mechanism to perform the modulation. For instance, electricity is the actuating signal in electrochromic windows (ECW). The active switchable glazing systems offer supplementary options compared to the passive systems whereas their dependency on power supply and wiring should be reckoned with as a drawback. Chromic materials, liquid crystals, and suspended particle windows are the three most common active-controlled intelligent windows [9]. The latter two share the disadvantage of their dependency on an electric field to be maintained when a transparent mode is desired; resulting in excessive electricity consumption. This is not the case in EC glazing that wants electricity only for transition [23]. However, chromic materials are classified into four types: electrochromic (EC), gasochromic, photochromic and thermochromic (TC). The first two belong to active glazing, responding to electricity and hydrogen gas respectively as a function of solar irradiation [9, 19].Smart windows are apt to glazing the cooling load demanding buildings with large solar gain [18], though providing a see-through mode is a must in any application.

The decisive factors, based on which the performance of intelligent windows can be evaluated, are ordered below with respect to their importance [9]:

1. Transmission modulation in the visible and outer visible spectrum
2. Anticipated life time and the number of cycles without degradation
3. Response time; the time required to switch between colored and bleached states, which depends on the size of the window
4. The resulted window size
5. Overall energy consumption
6. Operating voltage and temperature

We have brought different classes of switchable glazing in the following:

1.1.1 Electrochromic Windows (ECW)

The EC effect which was first explained in 1969 is a characteristic of a device which varies its optical properties when an external voltage triggers the EC material. The EC device modulates its transmittance in visible and near IR when a low DC potential is applied [24, 25].

It is usually consisted of several layers deposited on glass. The glass substrates are usually coated with transparent conducting films with natural colors-mostly tin oxide doped with either indium (ITO) or fluorine (FTO). The three major deposited layers cover the coated glass substrate as follows:

1-The Electrochromic film (cathodic electro-active layer with reversible transmittance modulation characteristic) which gets a darker color when the external circuit transfers electrons into the EC lattice to compensate for the positive ions injected from the adjacent ion storage layer, 2-Ion conductor (ion conducting electrolyte), and 3-Ion storage layer (anodic electro-active layer) that becomes darker while releasing positive ions [10, 25-28].

The electro-active layers (also named electrochromics) switch between their oxidized and reduced forms causing variations in their optical properties and colors as well. Ideally, it is desired that electrochromics act more reflective rather than absorptive in their colored state compared with their bleached mode [24].

EC windows should provide daylight while acting as a barrier to heat. Obviously, this type of window is not capable of providing both effects simultaneously [29]. The EC function can be controlled by thermal load, temperature and sunlight. The latter is stated to be the best governing parameter, especially from the comfort point of view [30-33]. All the more, self-powered EC windows are also developed using semitransparent PV cells, which provide the required activating electricity [34-42].

Figure.1a and 1b demonstrate the structure and function of EC windows.

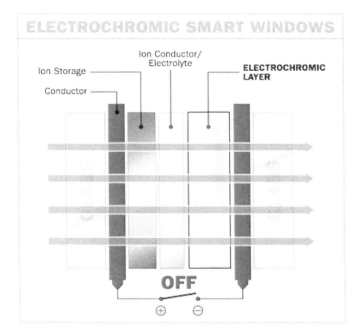

Fig. 1a. Electrochromic windows in bleached (transparent) mode [43]

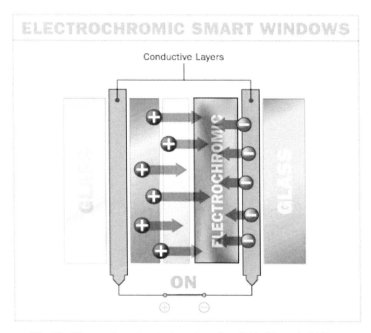

Fig. 1b. Electrochromic windows in colored (dark) mode [43]

1.1.2 Gasochromic Windows

The function of gasochromic devices is also based on electrochromism in EC windows. The main difference is that instead of DC voltage, a hydrogen gas (H2) is applied to switch between colored and bleached states. Compared to their counterpart, gasochromic devices are cheaper and simpler because only one EC layer is enough and the ion conductor and storage layers are not needed anymore. Although, gasochromic devices exhibit some merits such as better transmittance modulation, lower required voltage, staying lucid in the swap period, and adjustability of any middle state between transparent and entirely opaque; only a few numbers of EC materials can be darkened by hydrogen. Furthermore, strict control of the gas exchange process is another issue [44].

1.1.3 Liquid Crystals (LC)

Commonly used in wrist watches, LC technology is getting more popular as a means of protecting privacy in some interior applications such as bathrooms, conference halls and fitting rooms in stores. As it can be seen in Figure 2, two transparent conductor layers, on plastic films squeeze a thin liquid crystal layer, and the whole set is pressed between two layers of glass. Normally, the liquid crystal molecules are situated in random and unaligned orientations scattering light and cloaking the view to provide the interior space with privacy. When the power is switched on the two conductive layers provide an electric field via their electrodes. The field causes the

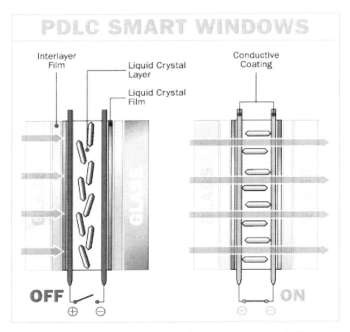

Fig. 2. PDLC technology used in smart windows [43]

crystal molecules to be positioned in an aligned direction causing a change in transmittance [45, 46]. The LC technology suffers from the disadvantage of high power demand in transparent mode, resulting in an electricity usage of 5-20 W/m2. These devices have problems, in long term UV stability and high cost disadvantages as well [9]. The technology using liquid crystals in intelligent windows is called Polymer dispersed liquid crystals (PDLC) which is illustrated in figure 2.

1.1.4 Electrophoretic or Suspended-Particle Devices (SPD)

SPDs have many things in common with LC devices: they are both fast in switching between phases, high electricity consumptive and dependent on an electric field. Figure 3 shows the construction and operation of SP windows. According to the figure, they consist of the liquid like active layer formed by adsorbing dipole needle-shaped or spherical particles (molecular particles), i.e. mostly polyhalide, suspended in an organic fluid or gel sandwiched between two sheets of glass coated with transparent conductive films. Normally, the device is in the dark reflective state because of the random pattern of the active layer's light absorbing particles. When the electric field is applied, the particles will align resulting in the clear transmissive state. As soon as the power turns off the device switches to its dark state. Typically, the transmission of SPDs varies between 0.79-0.49 and 0.50-0.04, with 100 to 200 ms switching time and 65 to 220 V AC requirements [9].

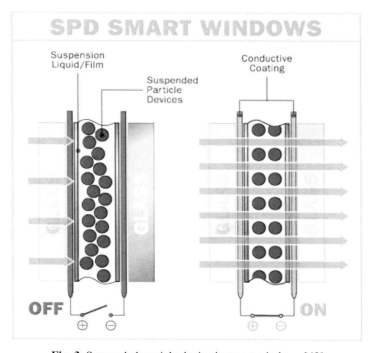

Fig. 3. Suspended particle device in smart windows [43]

1.1.5 Thermochromic windows (TCW)

A thermochromic material changes color as the temperature changes [47]. TCW is a device on which a thin film of TC materials is deposited. This can reduce the energy demand of buildings by changing the device's reflectance and transmission properties, reducing the solar energy gain [48, 49]. The TC thin film is initially in its monoclinic state at lower temperatures (usually room temperature). Monoclinic materials behave as semiconductors, less reflective, especially in near IR radiation. As the temperature rises, the TC material changes its nature from monoclinic to the rutile state. This effect is called metal to semiconductor transition (MST). In the rutile state, the material acts like a semi-metal, reflecting a wide range of the solar spectrum [50]. In high temperatures, it blocks near-IR (800-1200 nm) the wavelengths from which most of the heat is originated and far-IR (1200-2500 nm) while in low temperatures it allows those parts of the spectrum to pass [51].

2 Optical Analysis

In the previous sections, we reviewed the common intelligent glazing. The smart coatings are deposited on the glass surface and they amend the optical properties of the surface regarding wavelength and environmental conditions.

A typical sketch of the coating on a glass surface is demonstrated in figure 4. As it can be observed, the radiation will be reflected on two surfaces, first on the coating and second the coating-glass interface.

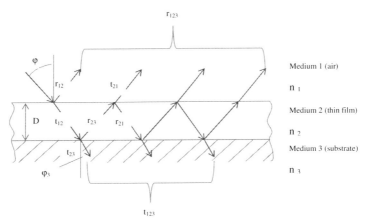

Fig. 4. The continuous reflections and transmissions through a thin film coated on the surface of substrate

As it is observed in the figure, the incident radiation first hits the coating surface (first surface). A fraction of the radiation gets reflected (r12) and the rest pass through the coating (t12) until it reaches the thin film- substrate interface (second surface). The second surface also reflects a portion of radiation (r23) and lets the remaining pass through (t23). The reflected part again reaches the first surface on which it can whether be reflected (r21) again or transmitted (t21).

Considering the phase difference (2δ) between r12 and t21 the following equations are obtained for the overall transmission and reflection [52, 53].

$$r_{123} = \frac{r_{12} + r_{23}e^{2i\delta}}{1 - r_{21}r_{23}e^{2i\delta}} \tag{1}$$

$$t_{123} = \frac{t_{12}t_{23}e^{i\delta}}{1 - r_{21}r_{23}e^{2i\delta}} \tag{2}$$

Where, for φ as incident angle, the phase gain can be calculated by equation (3).

$$\delta = 2\pi\nu D\sqrt{n_2^{\,2} - \sin^2\varphi} \tag{3}$$

Therefore, the reflected and transmitted energy of the film-substrate shown in figure 4 can be estimated using equations (4) and (5).

$$R = \mid r_{123}\mid^2 \tag{4}$$

$$T = \frac{\mathrm{Re}(n_3\cos\varphi_3)}{\mathrm{Re}(n_1\cos\varphi)}\mid t_{123}\mid^2 \tag{5}$$

Finally, the absorptance of the whole system will be

$$A = 1 - T - R \tag{6}$$

The absorbed incident energy of a solar collector plate with αc absorptivity can be computed by equation (7) [54].

$$A_c = \frac{\alpha_c T}{1 - (1 - \alpha_c)R} \tag{7}$$

If normal incidence is the going to be analyzed equation (8) can be introduced in which the Fresnel's formulae have been employed to simplify the equations using only refractive indices of the layers.

$$R = \frac{n_2^{\,2}(n_1 - n_3)^2 - (n_1^{\,2} - n_2^{\,2})(n_2^{\,2} - n_3^{\,2})\sin^2(2\pi n_2 D/\lambda_0)}{n_2^{\,2}(n_1 + n_3)^2 - (n_1^{\,2} - n_2^{\,2})(n_2^{\,2} - n_3^{\,2})\sin^2(2\pi n_2 D/\lambda_0)} \tag{8}$$

Where, λo is the wavelength in vacuum.

3   Discussion and Conclusions

The ideal windows are those, which let the light pass through in the visible range and block the unwanted radiation which cause heat loss or heat gain regarding the climate. The smart coatings such as electrochromic and thermochromic thin films feature spectrally selective behavior in the different wavelengths. Accordingly, in the visible range (400 nm – 700 nm) they are highly transmissive and in the near infrared (NIR) range (700 nm- 2500 nm) the reflectivity and transmissivity change in response to external triggers such as electricity, light and temperature. The infrared radiation is the greatest contributor to heat among the different sections of spectrum [55].

In order to compare different glazing technologies we should first define which goals take priority over the others. Do we want to block the IR radiation in the price of losing view or the view is more important? Is lighting energy is of more importance than cooling energy? Is the resulted color of the window important? How desirable is privacy?

Obviously, each coating has different effects on different parts of spectrum and results in diverse colors. Table 1 compares TC glazing, EC windows, LC technology and SPD in terms of their thermal effect, optical effect, visual performance, the activating factor and the challenges of these technologies. TC and EC windows show better upshot in reducing transmission and providing outside view.

Table 1. A comparison between Thermochromic, electrochromic, Liquid crystal and suspended particle devices [56]

	Thermal performance	Optical performance	View	Actuator	Challenge
TC	Reducing transmitted radiation UV transmissive in colored mode; operates best in the near IR	Low transmission in visible range	Transparent at high IR ; reduction in light intensity but still transparent	Heat (surface temperature)	Low visibility (can be solved by choosing the suitable dopant)
EC	reducing transmitted radiation	transparent in the short wavelength region coupled with opacity in the long wavelength region	Reduction in light intensity	Voltage or current	Electric field dependent; Wiring required
LC	Low reduction in transmitted radiation	Opaque in colored mode	Reduction in visibility; opaque	Voltage	High electricity consumption
SPD	Low reduction in transmitted radiation	Opaque in colored mode	Reduction in visibility; opaque	Current	High electricity consumption

As it is demonstrated in figure 5, electrochromic and thermochromic windows are the best glazing types in terms of reducing the required cooling load. As it is mentioned earlier in introduction, one of the most crucial parameters in evaluating the performance of smart windows is transmission modulation and their ability to pass through the visible light. Thermochromic and electrochromic windows fulfill this requirement. The overall energy consumption will also plunge considerably by using these two chromogenic smart windows. However, the necessity of wiring in electrochromic glazing and the better ability of thermochromic windows to maintain the visible transmission when it is properly doped [57] besides their simple structure [58] have given thermochromic windows a cutting edge compared to the other counterparts.

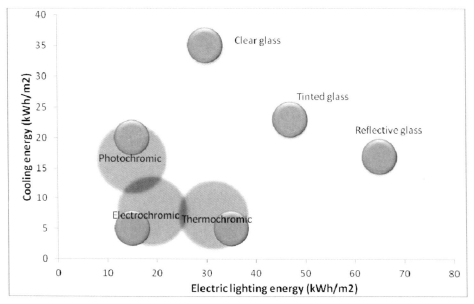

Fig. 5. Comparison of electric lighting energy and cooling energy between different glazing types adapted from Ref. [59]

Considering equation (8) it is clear that in order to acquire lower reflectivity, the refractive index of film should be lower. Hence, according to the application one should select the type of thin film and the value of refractive index. For fenestration in hot climates the desired window should be highly reflective in infrared while being transmissive enough to the visible light [60]. In this case, films should have higher refractive index in near infrared and should show lower refractive index in visible range. For instance, thermochromic thin films should have this characteristic in order to be effective in saving energy. According to experiments, at temperatures below the transition temperature, thermochromic films show higher reflectivity in visible range and lower reflectivity in NIR range. Whereas, at temperatures higher than transition point the optical properties are just reverse [50].

For the case of solar collectors, in order to absorb more energy, besides the high absorptivity of the plate itself the optical properties of the glass cover should show a desirable trend. Based on equation (7) it is concluded that the glass cover should have higher transmissivity and lower reflectance. Using smart thin films, which are more transmissive, especially in the visible range, can live up to this goal. However, it should be taken into consideration that beside the optical performance of the glass cover itself, there are some other omnipresent parameters, such as dust and humidity that can affect the level of solar absorption [61].

To recapitulate, it can be stated that until now, many experiments have been conducted on the optical performance of smart thin films but there is not enough attention to the optical equations available. Through elaborating the equations in this study, a systematic approach is constructed to evaluate performance of thin films coated on glass substrates.

Acknowledgments. The authors would like to thank the Ministry of Higher Education of Malaysia and University of Malaya for providing financial support under the research grant No.UM.C/HIR/MOHE/ENG/16001-00-D000024

References

1. Al-Rabghi, O.M., Hittle, D.C.: Energy simulation in buildings: overview and BLAST example. Energy Conversion and Management 42(13), 1623–1635 (2001)
2. de Wilde, P., van der Voorden, M.: Providing computational support for the selection of energy saving building components. Energy and Buildings 36(8), 749–758 (2004)
3. Kwak, S.Y., Yoo, S.H., Kwak, S.J.: Valuing energy-saving measures in residential buildings: A choice experiment study. Energy Policy 38(1), 673–677 (2010)
4. Sadineni, S.B., Madala, S., Boehm, R.F.: Passive building energy savings: A review of building envelope components. Renewable and Sustainable Energy Reviews 15(8), 3617–3631 (2011)
5. Thormark, C.: The effect of material choice on the total energy need and recycling potential of a building. Building and Environment 41(8), 1019–1026 (2006)
6. Bojic, M., Yik, F., Sat, P.: Influence of thermal insulation position in building envelope on the space cooling of high-rise residential buildings in Hong Kong. Energy and Buildings 33(6), 569–581 (2001)
7. Cheung, C.K., Fuller, R., Luther, M.: Energy-efficient envelope design for high-rise apartments. Energy and Buildings 37(1), 37–48 (2005)
8. Synnefa, A., Santamouris, M., Akbari, H.: Estimating the effect of using cool coatings on energy loads and thermal comfort in residential buildings in various climatic conditions. Energy and Buildings 39(11), 1167–1174 (2007)
9. Baetens, R., Jelle, B.P., Gustavsen, A.: Properties, requirements and possibilities of smart windows for dynamic daylight and solar energy control in buildings: A state-of-the-art review. Solar Energy Materials and Solar Cells 94(2), 87–105 (2010)
10. Zinzi, M.: Office worker preferences of electrochromic windows: a pilot study. Building and Environment 41(9), 1262–1273 (2006)
11. Arpino, F., Buonanno, G., Giovinco, G.: Thermal conductance measurement of windows: An innovative radiative method. Experimental Thermal and Fluid Science 32(8), 1731–1739 (2008)
12. (the US National Renewable Energy Laboratory) (November 29, 2009),
    `http://www.nrel.gov/buildings/windows.html`
    (access : September 17, 2011)

13. Correa, G., Almanza, R.: Copper based thin films to improve glazing for energy-savings in buildings. Solar Energy 76(1-3), 111–115 (2004)
14. de Wilde, P., Augenbroe, G., van der Voorden, M.: Design analysis integration: supporting the selection of energy saving building components. Building and Environment 37(8-9), 807–816
15. Hassouneh, K., Alshboul, A., Al-Salaymeh, A.: Influence of windows on the energy balance of apartment buildings in Amman. Energy Conversion and Management 51(8), 1583–1591 (2010)
16. Axinte, E.: Glasses as engineering materials: A review. Materials & Design 32(4), 1717–1732 (2011)
17. Tarantini, M., Loprieno, A.D., Porta, P.L.: A life cycle approach to Green Public Procurement of building materials and elements: A case study on windows. Energy 36(5), 2473–2482 (2011)
18. Sullivan, R., et al.: Energy performance of evacuated glazings in residential buildings, Lawrence Berkeley National Lab, CA, United States (1996)
19. Bahaj, A.B.S., James, P.A.B., Jentsch, M.F.: Potential of emerging glazing technologies for highly glazed buildings in hot arid climates. Energy and Buildings 40(5), 720–731 (2008)
20. Yamada, Y., et al.: Toward solid-state switchable mirror devices using magnesium-rich magnesium-nickel alloy thin films. Japanese Journal of Applied Physics Part 1 Regular Papers Short Notes ond Review Papers 46(8A), 5168 (2007)
21. Tajima, K., et al.: Optical switching properties of all-solid-state switchable mirror glass based on magnesium-nickel thin film for environmental temperature. Solar Energy Materials and Solar Cells 94(2), 227–231 (2010)
22. Robinson, P.: Advanced glazing technology for low energy buildings in the UK. Renewable Energy 5(1-4), 298–309 (1994)
23. Lampert, C.M.: Smart switchable glazing for solar energy and daylight control. Solar Energy Materials and Solar Cells 52(3-4), 207–221 (1998)
24. Syrrakou, E., Papaefthimiou, S., Yianoulis, P.: Eco-efficiency evaluation of a smart window prototype. Science of The Total Environment 359(1-3), 267–282 (2006)
25. Granqvist, C.G.: Handbook of inorganic electrochromic materials. Elsevier Science Ltd. (1995)
26. Papaefthimiou, S., Leftheriotis, G., Yianoulis, P.: Study of electrochromic cells incorporating WO3, MoO3, WO3-MoO3 and V2O5 coatings. Thin Solid Films 343, 183–186 (1999)
27. Papaefthimiou, S., Leftheriotis, G., Yianoulis, P.: Advanced electrochromic devices based on WO3 thin films. Electrochimica Acta 46(13-14), 2145–2150 (2001)
28. Papaefthimiou, S., Leftheriotis, G., Yianoulis, P.: Study of WO3 films with textured surfaces for improved electrochromic performance. Solid State Ionics 139(1-2), 135–144 (2001)
29. Lee, E.S., DiBartolomeo, D.: Application issues for large-area electrochromic windows in commercial buildings. Solar Energy Materials and Solar Cells 71(4), 465–491 (2002)
30. Karlsson, J., et al.: Control strategies and energy saving potentials for variable transmittance windows versus static windows. In: Proc. of Eurosun, Copenhagen, Denmark (2000)
31. Sullivan, R., et al.: Effect of switching control strategies on the energy performance of electrochromic windows (1994)
32. Sullivan, R., et al.: The energy performance of electrochromic windows in heating-dominated geographic locations (1996)

33. Sullivan, R., Rubin, M., Selkowitz, S.: Energy performance analysis of prototype electrochromic windows, Lawrence Berkeley National Lab, CA, United States (1996)
34. Bechinger, C., Gregg, B.: Development of a new self-powered electrochromic device for light modulation without external power supply. Solar Energy Materials and Solar Cells 54(1-4), 405–410 (1998)
35. Benson, D.K., et al.: Stand-alone photovoltaic (PV) powered electrochromic window, Google Patents (1995)
36. Huang, L.M., et al.: Photovoltaic electrochromic device for solar cell module and self-powered smart glass applications. Solar Energy Materials and Solar Cells (2011)
37. Deb, S.K., et al.: Stand-alone photovoltaic-powered electrochromic smart window. Electrochimica Acta 46(13-14), 2125–2130 (2001)
38. Deb, S.K.: Opportunities and challenges in science and technology of WO3 for electrochromic and related applications. Solar Energy Materials and Solar Cells 92(2), 245–258 (2008)
39. Gao, W., et al.: First a-SiCá: áH photovoltaic-powered monolithic tandem electrochromic smart window device. Solar Energy Materials and Solar Cells 59(3), 243–254 (1999)
40. Gao, W., et al.: Approaches for large-area a-SiC: H photovoltaic-powered electrochromic window coatings. Journal of Non-Crystalline Solids 266, 1140–1144 (2000)
41. Hauch, A., et al.: New photoelectrochromic device. Electrochimica Acta 46(13-14), 2131–2136 (2001)
42. Pichot, F., et al.: Flexible solid-state photoelectrochromic windows. Journal of the Electrochemical Society 146, 4324 (1999)
43. http://home.howstuffworks.com/home-improvement/construction/green/smart-window.htm (access : September 25, 2011)
44. Georg, A., et al.: Switchable glazing with a large dynamic range in total solar energy transmittance (TSET). Solar Energy 62(3), 215–228 (1998)
45. Doane, J.W., Chidichimo, G., Vaz, N.A.P.: Light modulating material comprising a liquid crystal dispersion in a plastic matrix, Google Patents (1987)
46. Fergason, J.L.: Encapsulated liquid crystal material, apparatus and method, Google Patents (1992)
47. Ritter, A.: Smart materials in architecture, interior architecture and design. Birkhauser (2007)
48. Greenberg, C.B.: Undoped and doped VO2 films grown from VO (OC3H7) 3. Thin Solid Films 110(1), 73–82 (1983)
49. Parkin, I.P., Manning, T.D.: Intelligent thermochromic windows. Journal of Chemical Education 83(3), 393 (2006)
50. Blackman, C.S., et al.: Atmospheric pressure chemical vapour deposition of thermochromic tungsten doped vanadium dioxide thin films for use in architectural glazing. Thin Solid Films 517(16), 4565–4570 (2009)
51. Lee, M.H., Cho, J.S.: Better thermochromic glazing of windows with anti-reflection coating. Thin Solid Films 365(1), 5–6 (2000)
52. Stenzel, O.: The physics of thin film optical spectra: an introduction, vol. 44. Springer (2005)
53. Duffie, J.A., Beckman, W.A., Worek, W.: Solar engineering of thermal processes, vol. 2. Wiley, New York (1991)
54. Siegel, R., Howell, J.R.: Thermal radiation heat transfer, vol. 1. Taylor & Francis (2002)
55. Saeli, M., et al.: Energy modelling studies of thermochromic glazing. Energy and Buildings 42(10), 1666–1673 (2010)

56. Addington, D.M., Schodek, D.L.: Smart materials and new technologies: for the architecture and design professions. Architectural Pr (2005)
57. Kanu, S.S., Binions, R.: Thin films for solar control applications. Proceedings of the Royal Society A: Mathematical. Physical and Engineering Science 466(2113), 19 (2010)
58. Granqvist, C.G., et al.: Progress in chromogenics: New results for electrochromic and thermochromic materials and devices. Solar Energy Materials and Solar Cells 93(12), 2032–2039 (2009)
59. Huovila, P.: Buildings and climate change: status, challenges, and opportunities, United Nations Envir Programme (2007)
60. Durrani, S., et al.: Dielectric/Ag/dielectric coated energy-efficient glass windows for warm climates. Energy and Buildings 36(9), 891–898 (2004)
61. Mekhilef, S., Saidur, R., Kamalisarvestani, M.: Effect of dust, humidity and air velocity on efficiency of photovoltaic cells. Renewable and Sustainable Energy Reviews 16(5), 2920–2925 (2012)

Chapter 73
A Sunspot Model for Energy Efficiency in Buildings

Yosr Boukhris, Leila Gharbi, and Nadia Ghrab-Morcos

National School for Engineers of Tunis,
Laboratory of Materials, Optimization and Energy for Durability,
BP 37, 1002 Tunis Belvédère, Tunisia

Abstract. Studying the distribution of beam solar radiation in buildings is becoming more important with the emphasis on curtailing energy consumption. This paper presents a model developed in order to evaluate sunspot position and area through a window on each wall of a parallelepiped room. The results of this developed model have been compared with numerical published data. An application of the model to study the sunspot progression on walls of a Tunisian building is also presented.

1 Introduction

The sun has influenced architectural design since primitive times. In the sixth century, the Greek philosopher Xenophanes wrote: "In houses with a south aspect, the sun's rays penetrate into the porticoes in winter, but in the summer the path of the sun is right over our heads and above the roof, so that there is shade. If, then, this is the best arrangement, we should build the south side loftier to get the winter sun, and the north side lower to keep out the cold winds".

This principle is still true; buildings must be oriented and designed to admit a minimum amount of direct sun rays in summer and a maximum amount of solar heat during the winter.

Furthermore, natural daylight is an abundant source of effective lighting. With shortages and increasing costs of available produced energy, the importance of daylighting in building design is being viewed with new emphasis.

To ensure human needs for natural light, buildings are designed with large window openings, making sun control very important. Solar radiation affects thermal comfort in hot season while it can supplement the heat source in winter. Thus it is increasingly important to know and understand the sun's effect on the design and the engineering of a building, hence the need for the knowledge of the distribution of beam solar radiation in buildings.

Beam solar radiation penetrating through a window is generally taken into account by numerical programs in an overly simplified way, considering that all the flux arrives on the floor. In other numerical codes, the percentage of beam solar radiation arriving on each wall must be defined by the user and kept constant during the entire simulation. A common practice is to consider that 60% arrives on the floor.

A. Håkansson et al. (Eds.): *Sustainability in Energy and Buildings*, SIST 22, pp. 827–836.
DOI: 10.1007/978-3-642-36645-1_73   © Springer-Verlag Berlin Heidelberg 2013

Wall [Wall, 1997] showed that the heating demand in a highly glazed building was underestimated in most simulation software because the beam solar radiation was insufficiently taken into account.

More recently, Tittelein [Tittelein and al., 2008] concluded that the difference in heating demand between the simulation where beam solar radiation is projected on the floor and the simulation where the sunspot is calculated for each time step is significant.

The first part of this paper describes the model that we have developed to calculate the beam solar radiation distribution in a parallelepiped room. Obtained results are then compared with published data [Bouia and al., 2002]. In the second part we present an application of our model to study the sunspot progression on the walls of a room located in Tunis.

2 Sunspot Distribution in a Parallelepiped Room

Several methods for calculating the position and the area of a sunspot on inside walls of a room have been developed.

The Swiss engineering consultant company Sorane has developed a program called SUNREP [Chuard, 1992], which can be used as a processor for programs such as TRNSYS. The principle is to break the window down into small rectangular elements. Then it is considered that only one sun's ray passes across this element through the center of the rectangle. The place where the ray arrives on walls of a parallelepiped room is then determined.

Serre [Serre, 1997] and Trombe [Trombe and al., 2000] suggested a geometrical method based on the projection of two rectangles that are inside and outside the bare part of the window. Each rectangle is projected on infinite horizontal and vertical planes corresponding to the walls and the floor of the room. Then the intersection between each parallelogram obtained and the real rectangular wall is considered. To determine the sunspot area, the intersection of both projections of rectangles (inside and outside the bare part of the window) is made on each wall.

3 Description of the Sunspot Model

Our goal is to develop a model that is both simple and requires only modest computation time on desktop computer to be integrated into the thermo-aeraulic building simulation tool ZAER [Gharbi and al., 2005; Boukhris and al., 2009].

The sunspot's position and area depend upon the sun's position and the geometry problem. Thus, to know how the rays will strike a building and how far the rays will penetrate through the opening; to shade certain areas and irradiate others; we must calculate, in the first stage, the following data for a particular surface at a specific time of the day studied:

- The Solar Time
- The Hour Angle
- The Declination Angle
- The Solar Altitude angle
- The Solar Azimuth angle

The relation between Solar and Standard Time

Solar Time is the time as measured by the sun. It is not uniform; it speeds up and slows down, because the earth moves slower and faster in its orbit around the sun, and because of different distances from the sun.

The following equation converts Standard Time to Sun Time:

$$SuT = StT - ET + 4 (SM - L)$$

where:

SuT : Sun Time (hours and minutes)

StT : Standard Time (hours and minutes)

ET : Equation of Time (minutes)

4 : Number of minutes required for the sun to pass over one degree of longitude

SM : Standard Meridian (longitude) for the local time zone

L : Longitude of Location

The Equation of Time denotes the factor for the non-uniformity of Sun Time. For a day of the year "d" (1 to 365), it is given by [Bouia and al., 2002]:

$$ET = 0.0002 - 0.4197 \cos(d) + 3.2265 \cos(2d) + 0.0903 \cos(3d) + 7.3509 \sin(d) + 9.3912 \sin(2d) + 0.3361 \sin(3d)$$

Sun Time is used to define the rotation of the earth relative to the sun. An expression to calculate the Hour Angle from Sun Time is [Al-Rawahi and al., 2011]:

$$HA = \left(\frac{360}{24}\right)(SuT - 12)$$

where HA is defined as the angle between the meridian parallel to the sun rays and the meridian containing the observer.

Calculation of the apparent sun position

When we observe the sun from an arbitrary position on the earth, we are interested in defining the sun position relative to a coordinate system based at the point of observation, not at the center of the earth. The conventional earth-surface based coordinates are a vertical line and a horizontal plane containing the north-south line and and the east-west line. The position of the sun relative to these coordinates can be described by two angles; the *solar altitude angle* α and the *solar azimuth angle Az* (figure 1).

The solar altitude angle (α) is the angular height of the sun in the sky measured from the horizontal. The solar altitude varies throughout the day. It also depends on the latitude of a particular location and the day of the year.

The other angle defining the position of the sun is the solar azimuth angle (Az). It is the angle, measured clockwise on the horizontal plane, from the projection of the sun's central ray to the south-pointing coordinate axis.

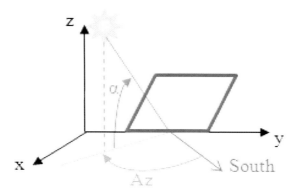

Fig. 1. Relative Solar angles and position of the surface

The solar altitude and the azimuth, expressed in terms of location's latitude (φ), the hour angle (HA), and the sun's declination (δ), are calculated by the equations:

$$\alpha = \sin^{-1}\left[(\sin\varphi \ \sin AH) + (\cos\varphi \ \cos\delta \ \cos AH)\right]$$

$$Az = \cos^{-1}\left[\frac{(\sin\varphi \ \cos\delta \ \cos AH) - (\cos\varphi \ \sin\delta)}{\cos\alpha}\right]$$

where the declination angle (δ) is the angle between the equator and a line drawn from the center of the earth to the center of the sun. It varies seasonally due to the tilt of the earth on its axis of rotation and the rotation of the earth around the sun.

One such approximation for the declination angle is [Al-Rawahi and al., 2011]:

$$\delta = 23.45 \sin\left[\frac{360}{365}(d+284)\right]$$

The developed method

We consider a room which is as a perfect parallelepiped with a window having a shape of a perfect rectangle as shown in figure 2.

A', B', C', D' are respectively the projection of the window's vertices A, B, C, D on the room's internal walls.

Twenty different cases are distinguished depending on the position of A', B', C', D' (sunspot on the floor; sunspot on the floor and the north wall, ...).

To distinguish the different cases, the vertices A, B, C and D are projected on the plane of the back wall (A_1, B_1, C_1, D_1) and on the plane of the floor (A_2, B_2, C_2, D_2). The comparison of the coordinates of these eight points on (O, x, y, z) to the room's dimensions is sufficient to determine which case we have to study.

When the case is determined, each wall is broken down into elementary rectangles. We consider that the rectangle is irradiated when its barycenter belongs to the polygon obtained by the projection of the window's vertices. The sunspot's area on each wall is then calculated by the sum of the irradiated rectangles.

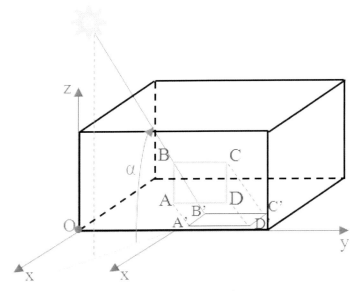

Fig. 2. Geometrical notations

4   Evaluation of the Sunspot Model

The developed model is tested by comparing its results achieved with predictions made by another numerical model [Bouia and al., 2002].

Comparisons are made for a north-south oriented room (5x3x3m) equipped by two glazed surfaces in the southern wall (figure 3).

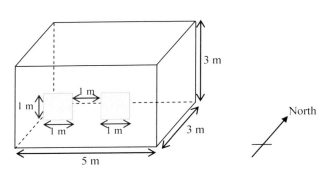

Fig. 3. Shape and dimensions of the room

This room is located in two different cities in France where the geographic coordinates are presented in Table 1.

Figure 4 shows the sunspot distribution obtained by our model (left column) and the one given by Bouia [Bouia and al., 2002] (right column). We observe that our results are in good agreement with the published data.

Table 1. Geographic coordinates of the studied cities

City	Latitude (φ)	Longitude (L)
Lyon	45.75°	4.85°
Marseille	43.30°	5.40°

12h StT 21/06/2002_ Lyon

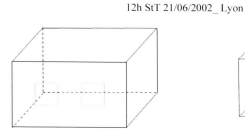
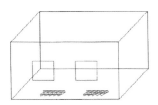

12h StT 21/12/2002_ Lyon

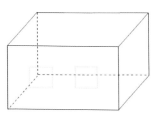
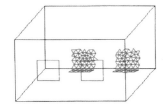

12h StT 21/12/2002_ Marseille

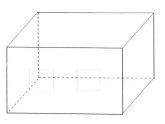
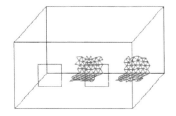

16h StT 21/12/2002_ Marseille

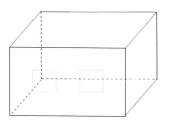
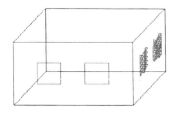

Fig. 4. Comparison of sunspot distributions (adapted from [Bouia and al., 2002])

5 Consideration of the Sunspot Distribution In Thermal Simulation

To evaluate accurately the distribution of the short-wave radiation, each wall of the building envelope was divided into NFT facets.

The internal distribution of short-wave radiation was computed with a method similar to that of the radiosities. This method introduces a total exchange factor $\hat{F}_{ij}$ which is defined as the fraction of the diffuse radiation leaving the facet i and striking the facet j after multiple reflections [Ghrab, 1991].

Optical characteristics of the glazing, dependant on incident angle, are calculated for each simulation time step [Gharbi and al., 2005].

The net radiative short-wave flux of an interior envelope facet i has the following expression:

$$\phi_{riSW} = -\alpha_{iSW} \left[B_i\, S_i + \sum_{k=1}^{NFT} \xi_{kSW}\, \hat{F}_{ki}\, S_k\, B_k + \sum_{v=1}^{Nv} \hat{F}_{vi}\, \tau_{bv}\, S_v\, E_{bv} \right]$$

α_{iSW} is the absorption coefficient for radiation, ξ_{kSW} the reflection coefficient for radiation and S_i the facet surface.

$B_i\, S_i$ is called primary beam solar radiation, where B_i represents the beam solar radiation that strikes the facet i from the direct solar radiation passing through the N_v glazings of the building. It is expressed as:

$$B_i = \frac{\sum_{v=1}^{N_v} \tau_{bv}\, S_v\, E_{bv}\, P_{vi}}{S_i}$$

P_{vi} is the fraction of incoming beam radiation that strikes the facet i, τ_{bv} the glazing transmittance for the direct solar radiation and E_{bv} the incident direct radiation upon the glazed facet.

6 Case Study: Application to a Passive Solar Tunisian Building

The simulated building is a parallelepiped room which has a floor area of 12 m^2, and a height of 3 m (Figure 5). It had a glazed surface of 2.4 m^2 in the middle of the southern wall.

The room was located in Tunis (Tunisia, latitude: 36.93°, longitude: 10.25°). We have applied the meteorological data (direct and global horizontal solar radiation) typical of the climate of Tunis.

We have chosen to study the sunspot's progression, on walls of the room, at each seasonal position of the Sun. The results are presented on figures 6 to 9.

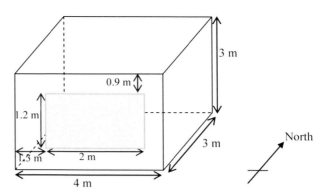

Fig. 5. Simulated room

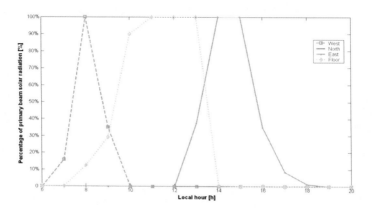

Fig. 6. Distribution of primary beam solar radiation on March 20

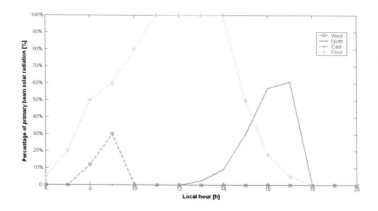

Fig. 7. Distribution of primary beam solar radiation on June 20

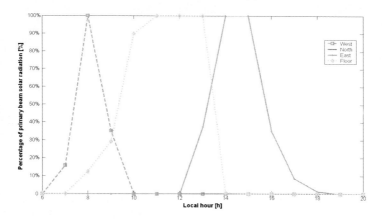

Fig. 8. Distribution of primary beam solar radiation on September 22

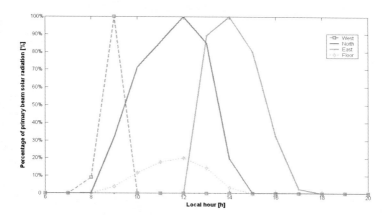

Fig. 9. Distribution of primary beam solar radiation on December 21

These results show that, at the latitude studied and for this geometry, assuming that the entire beam solar radiation is on the floor may be applicable only in summer.

Indeed, the sun reaches its maximum height on June 21. That's why the majority of the flux reaches the floor. However, during the winter solstice, when the sun is low, the projection of beam solar radiation is greater on the lateral walls than on the floor. This sunspot can even reach only the lateral walls.

On the vernal and autumnal equinoxes, we observe the same distribution of sunspot's on the walls.

Moreover, it seems that during the day, the sunspot can reach every wall in the room. Thus, using sunspot calculations makes it possible to take into account in a best way primary beam solar radiation and then an accurate evaluation of the internal distribution of the short-wave radiation.

7   Conclusion

A FORTRAN computer code has been developed to calculate the position and the area of the beam solar radiation reaching each wall of a parallelepiped room.

The comparison of the obtained results with the published data, for two cities and at different times, shows that our simulation tool is an accurate one.

An application of the model to study the sunspot's distribution at each seasonal position of the Sun proves that the commonly assumptions in thermal programs, considering that all the flux reaches the floor and it is independent of time, is far from reality.

The advantage of this model is that it permits an accurate calculation of the internal distribution of the short-wave radiation. Work in progress is coupling the developed sunspot model with the thermo-aeraulic building simulation tool ZAER. This will allow the realistic simulation of the thermal behaviour of large highly-glazed buildings.

References

Al-Rawahi, N.Z., Zurigat, Y.H., Al-Azri, N.A.: Prediction of Hourly Solar Radiation on Horizontal and Inclined Surfaces for Muscat/Oman. The Journal of Engineering Research 8(2), 19–31 (2011)

Bouia, H., Roux, J.J., Teodosiu, C.: Modélisation de la tache solaire dans une pièce équipée d'un vitrage utilisant un maillage en surface de Delaunay. In: Proceedings of IBPSA France, pp. 15–19 (2002) (in French)

Boukhris, Y., Gharbi, L., Ghrab-Morcos, N.: Modeling coupled heat transfer and air flow in a partitioned building with a zonal model: Application to the winter thermal comfort. Building Simulation: International Journal 2(1), 67–74 (2009)

Chuard, D.: Solar distribution computing program. IEA Task 12 Project A3 - ATRIA, SORANE SA (1992)

Gharbi, L., Ghrab-Morcos, N., Roux, J.J.: Thermal and airflow modelling of Mediterranean Buildings: Application to thermal comfort in summer. International Journal of Ventilation 4(1), 37–48 (2005)

Ghrab, N.: Analyse et simulation du comportement thermique des structures architecturales vis à vis des apports solaires. Thèse d'Etat. Faculté des Sciences de Tunis (1991) (in French)

Serres, L.: Etude de l'impact d'une perturbation thermique locale de type tache solaire. Influence sur le confort thermique. Thèse de doctorat. Institut National des Sciences Appliquées de Toulouse (1997) (in French)

Tittelein, P., Stephan, L., Wurtz, E., Achard, G.: Distribution of Beam Solar Radiation in Buildings. Effect on Heating Demand. In: Proceedings of COBEE, pp. 223–230 (2008)

Trombe, A., Serres, L., Moisson, M.: Solar radiation modelling in a complex enclosure. Solar Energy 67(4-6), 297–307 (2000)

Wall, M.: Distribution of solar radiation in glazed spaces and adjacent buildings. A comparison of simulation programs. Energy and Buildings 26(2), 129–135 (1997)

Chapter 74
Towards 24/7 Solar Energy Utilization: The Masdar Institute Campus as a Case Study

Mona Aal Ali and Mahieddine Emziane

Solar Energy Materials and Devices Lab.
Masdar Institute of Science and Technology, PO Box 54224, Masdar City, Abu Dhabi, UAE
{Maalali,memziane}@masdar.ac.ae

Abstract. In addition to the cost, storage of energy from solar generators remains a critical problem that is preventing the solar industry from reaching its full potential. Storage technologies are expensive and may not be suitable for large scale installations. Recent research suggests that electric vehicles (EVs) can be used for matching power demand with generation from distributed PV installations. This approach, called Vehicle-to-Grid (V2G), has never been demonstrated on a residential district to make it independent of the grid. In this study, we adopt the V2G concept and apply it to the Masdar Institute (MI) campus. Our aim is to investigate the feasibility of powering MI campus 24/7 from its own rooftop solar installations by using its EVs infrastructure for storage. The results showed that the generation and EV infrastructure already in place can power MI campus for 1 to 7 hrs of storage time, respectively. We have also conducted a feasibility and cost analysis to achieve 24/7 solar energy utilization by using different types of EVs.

1   Introduction

The UAE is blessed with high solar irradiation during the year with very few cloudy days [1-3]. The monthly average and maximum values of global horizontal irradiance (GHI) are shown in Figure 1 [4]. The use of rooftop PV, called technically Building Integrated Photovoltaics (BIPV) is increasing after launching the solar roof plan (SRP) which is part of the target of archiving 7% of the generation capacity of Abu Dhabi's energy from renewable resources by 2020. PV systems with a cumulative capacity of 500 MW are expected to be online in Abu Dhabi by the year 2020 according to this SRP.

With the rapid increase of the use of BIPV in Abu Dhabi, Dubai also started deploying BIPV technology [4,5]; however, if the UAE is to proceed with the installation of BIPV, one major problem arises. This problem is the management of electricity production. Due to the potential high solar penetration, the national grid may face difficulties in managing and distributing the extra electricity produced from the BIPV distributed generation systems because it cannot handle the fluctuation in electricity production and consumption. In other words, generation and transmission must be continuously managed to match the fluctuating load. Another limitation the

---

A. Håkansson et al. (Eds.): *Sustainability in Energy and Buildings*, SIST 22, pp. 837–845.
DOI: 10.1007/978-3-642-36645-1_74   © Springer-Verlag Berlin Heidelberg 2013

grid faces is that generation and consumption of electricity should be equal at all times or the grid will become unstable [6]. To minimize the load of excess electricity production on the grid, the apparent solution is to use batteries as a storage medium. However, storage like hydropower, hydrogen, ultra-capacitors and compressed air are expensive and may not be suitable for large scale installations in the UAE. Recent research suggests that electric vehicles can be used as a storage medium for matching power demand with generation from distributed PV installations. This approach, called Vehicle-to-Grid (V2G), has never been demonstrated on a residential district to make it independent of the grid.

In this study, we adopt the V2G concept [7] and apply it to the Masdar Institute campus. MI in Masdar City uses rooftop PV to supply the building with electricity during daytime hours. When no sunlight is available, MI campus is powered from the utility grid. The problem that needs to be investigated is how to store the excess solar energy using the transportation system available in MI campus, which is based on personal rapid transits (PRTs) and electric vehicles (EVs). The extra power generated from the solar installation during the day can be fed to the EVs and PRTs. During the night, when no sunlight is available, and when the PRTs and EVs are not used frequently, they can be plugged into the electric power system of the campus. By using this method, a 24/7 power generation may be achieved because power consumption during the night is minimum.

In recent years, interest in the V2G concept has increased. Ref. [8] reviewed the state of the art of plug-in-hybrid electric vehicles (PHEVs) and V2G concept. This paper investigates the feasibility of powering the MI campus 24/7 by using different scenarios to find the ultimate cost effective solution. Our main goal remains the optimization of the number of electric vehicles and PRTs to power the campus 24/7 by using the V2G concept. The contribution of this paper compared to previous works is that this study is more practical and based on the real performance of the MI campus, transportation infrastructure and PV installations. The following section reviews the published work on the V2G concept. Section 3 describes the research scope and methodology. A model of the infrastructure is described in section 4. Results and analysis are presented in section 5. Finally, section 6 provides conclusions.

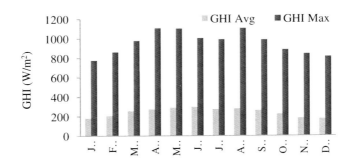

Fig. 1. Monthly average and maximum values of the GHI in Abu Dhabi [4]

2 Methodology

Masdar Institute campus in Masdar City is an interesting case study for this analysis due to its unique power architecture. The 1MW photovoltaic installed on the rooftop supplies the building with the electricity during the day only. When no sunlight is available, the building is powered from the utility grid. Figure 2 illustrates the system architecture of our case study in terms of its power generation, consumption, and transportation infrastructure. Figure 3 shows the total night hours needed for energy storage in Abu Dhabi. The figure shows that the maximum night hours needed for storage in most days is 14 hours. Based on this, the simulated day was selected. We used daily solar irradiation data to estimate the hourly solar energy output from the 10MW Masdar PV power plant and 1MW rooftop PV installation for January 2012. We developed hourly energy generation based on hourly global horizontal irradiance (GHI), and then we distributed the hourly energy generation of both PV installations over time. Figure 4 shows the hourly consumption of MI campus taking into account EVs average consumption and the 1MW energy profiles in kWh for the simulated day. The figure shows that during the day there is enough energy generated from the 1MW; however, the generation drops significantly at sunset. When the consumption fluctuates at night the building is being fed by the grid to meet the demand.

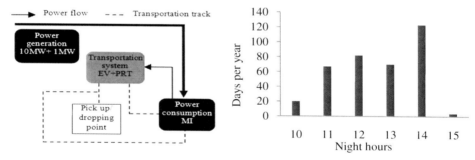

Fig. 2. System architecture of our case study

Fig. 3. Night hours needed to store energy for the year 2012 in Abu Dhabi

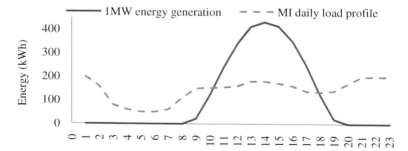

Fig. 4. Hourly demand load and 1MW energy profiles for the simulated day

To show the excess of energy from the 1MW, we calculated the difference between the energy generation of the 1MW and the energy consumption of the MI campus. The analysis showed that the total extra power output from the 1MW is around 1287 kWh (Figure 5). Based on the figures we obtained the hourly excess of energy from the 1MW ranges from 250 to 85 kWh with a total of 1287 kWh, while the total energy deficiency from 6 pm to 9 am of the MI campus is 2020 kWh. Those results show that the excess energy from the 1MW is not enough to power the MI campus during the night, thus we took the remaining 733 kWh from the 10MW PV power plant (Figure 6).

3 Model

A detailed hour by hour analysis was done to find the total number of vehicles required to store the excess amount of energy of 2020 kWh. In this study, analysis has been done for two types of vehicles. The first is the Mitsubishi innovative electric vehicle (i-MiEV). This electric vehicle is based on the gasoline-driven 660cc "i" mini-car. The i-MiEV has been lauded for its strong motor power, stability, quietness and its comfortable ride. Masdar in collaboration with Mitsubishi Heavy Industries (MHI) have launched this pilot project in January 2011. The EVs are used for intra-urban transportation and also for travel to Abu Dhabi international Airport and neighboring areas. Table 1 [9] shows the technical specification of the i-MiEV. The second type of vehicle is the Personal Rapid Transit system (PRT). The system is a transport method that offers personal, on-demand non-stop transportation between any two points on a network of specially built guide-ways. Table 2 shows the technical specifications of the PRT [10]. There are two EVs charging systems in MI campus. The first is a regular charge for a full charge in 6 hours from electrical outlets, and the second is a quick-charging connection for an 80% full charge in 30 minutes at two quick-charge stations powering up the vehicles. The i-MiEVs can generate or store electricity when parked, and with proper connections can feed power to the MI campus. For simplicity, we assume that by the end of the day, all vehicles are fully charged. So, a total of 208 and 173 kWh energy capacities are what the 13 EVs and 9 PRTs can provide, respectively. In this study, the effect of temperature, charging and discharging losses on the battery energy capacity are neglected. Also, in this model, each car will be recharged after discharging to be ready to use for the next day.

Fig. 5. Difference between the energy production from the 1MW and the energy consumption of the MI campus

74 Towards 24/7 Solar Energy Utilization: The Masdar Institute Campus as a Case Study

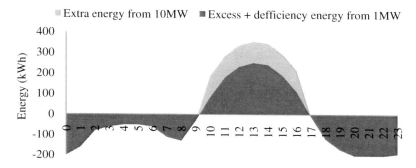

Fig. 6. Difference between the energy production from the 1MW and the energy consumption of the MI campus + extra energy needed from the 10MW

Table 1. Technical Specifications of EVs [9]

Dimensions and weights	Overall length	mm	3,475
	Overall width	mm	1,475
	Overall height	mm	1,610
	Kerb weight	kg	1,110
	Seating capacity	Persons	4
Performance	Electric range (NEDC)	km	150
	Maximum speed	Km/h	130
	Standing start 0-100km/h	sec	15.9
Power plant	Rated output	kW	35
Traction battery	Type		Lithium-ion
	Voltage	V	330
	Battery energy	kWh	16

Table 2. Technical Specifications of PRTs [10]

Capacity and weights	Seating capacity		4 adult + 2 children
	Payload weight	kg	1600
Performance	Maximum speed	Km/h	40
Traction battery	Type		Lithium-Phosphate
	Charging time	hour	1.5
	Battery energy	kWh	19.2

4 Results and Discussions

We conducted an analysis to find the number of vehicles and PRTs to provide power with a maximum energy capacity per vehicle. We focused on the following key variables: hourly energy consumption of the MI campus, hourly energy generation from the 1MW and the 10MW installations. The model results for the entire energy deficiency are shown in the following figures with one critical measure- the number of electric vehicles needed to feed the MI campus during the night. With an energy deficiency ranging from 200 to 9 kWh (Figure 7) we distributed the total number of EVs and PRTs which are 13 and 9 respectively, with their maximum storage capacity of 16 and 19.2 kWh on the energy deficiency chart to find the total discharge hours. We found that the existing 13 EVs will give 208 kWh of energy which is only enough to power the MI campus 1 hour during the night. For the PRTs, they can give 173 kWh of energy which is enough for 3 hours.

To compute the total number of the EVs and PRTs (n and m, respectively) needed to match the demand during the night, with the values in table 3, Eqs. (1), (2), and (3) are used:

$$16 \sum_{i=1}^{n} EV(i) + 19.2 \sum_{j=1}^{m} PRT(j) \geq C \quad (1)$$

$$n = \frac{A}{B} m \quad (2)$$

$$A + B = 1 \tag{3}$$

Where C is the total consumption of the MI campus during the night, n and m are the percentage of EVs and PRTs out of the total number of vehicles multiplied by their energy capacity, respectively. We considered 6 scenarios with different shares of EVs and PRTs to achieve 24/7 solar power generation. Figure 8 shows the total EVs and PRTs needed for discharging corresponding to each scenario. We also conducted a cost analysis to determine which scenario is best based on the forecast data of Lithium-ion battery. Note that in our module, the economy of scale factor was not considered. For instance, a study [11] showed the trend of reduction in Li-ion battery cost with an increase in production volume. As the production volume of Li-ion battery increases from 10 to 100,000 units/year, the battery cost decreases approximately from 3,000 to 400 $/kWh. Figure 9 shows the projected cost of i-MiEV EVs based on the future cost of the Li-ion battery [12]. It can be noticed that the cost of the vehicle decreases dramatically. The results of the cost analysis of the different scenarios are presented in Figure 10. The general trend resulting from these calculations shows a relative decrease in cost of these six scenarios. Notice that scenario 6 has the lowest cost, about $3M less than that of scenario 5 in 2012.

Table 3. Percentages of EVs to PRTs in scenarios

Scenarios	EV		PRT	
	A	n	B	m
Scenario1	0.7	84	0.3	36
Scenario2*	0.6	71	0.4	47
Scenario3	0.5	58	0.5	58
Scenario4	0.4	45	0.6	68
Scenario5	0.3	33	0.7	78
Scenario6	0.9	100	0.1	9

*Scenario2 corresponds to the infrastructure already in place.

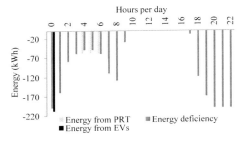

Fig. 7. Distribution of the total current EVs and PRTs matching the energy deficiency distribution

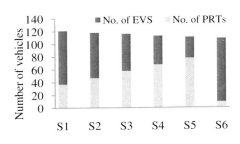

Fig. 8. The total number of EVs and PRTs needed for discharging in scenarios

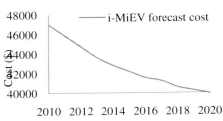

Fig. 9. The i-MiEV cost forecast based on the Li-ion cost

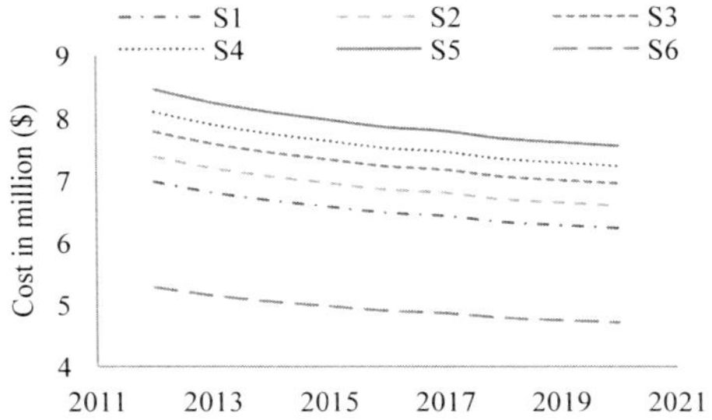

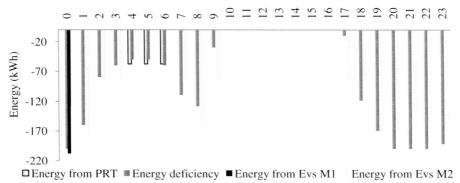

Fig. 10. Cost analysis of different scenarios

Table 4. Summary of the analysis for different models of EVs

Models	Battery Type	Battery capacity kWh	n	Cost $/unit	Cost/kWh	Total cost $
M1[9]	Li-ion	16	100	44,760	2,797	4,476,000
M2[13]	Li-ion	85	19	57,400	675	1,090,600
M3[14]	Li-ion	20	82	7,208	360	591,052
M4[15]	Li-ion	22	75	35,000	1,590	2,625,000
M5[16]	Li-ion	35	47	45,000	1,285	2,115,000
M6[17]	Li-FePO$_4$	48	34	35,000	729	1,190,000
M7[18]	Li-ion	24	68	35,000	1458	2,380,000

The best result obtained from the last analysis was 100 EVs and 9 PRTs; however, we noticed that the number of vehicles can be further minimized. So, we conducted a further analysis with different models of EVs. These models of vehicles have a battery energy capacity ranging between 20 and 85 kWh. These models were selected because they will have potential large share in the EV market (see table 4). Figure 11 shows the distribution of Model 2 because it has the largest battery energy capacity. This model resulted in only 19 vehicles needed for discharging during the night to power MI campus.

Fig. 11. The distribution of models 1, 2, and PRTs matching the distribution of the energy deficiency

5 Conclusions and Recommendations

This work has developed different scenarios in order to conduct a detailed hour by hour power generation for a specific case study by applying the V2G concept. The analysis has been applied for MI campus energy generation and consumption. The impact of three types of electric vehicles for night energy discharge was carefully studied. Altogether, the analysis shows that the storage capacity of the vehicles is a very critical factor in achieving 24/7 power. This was demonstrated when we compared two different models of EVs with their different capacity. The first model resulted in 100 vehicles, while the second model resulted in only 19 vehicles needed for night charging. Although initiating the SRP program is a promising solution to increase adoption of solar energy, it is still not sufficient since the buildings are powered by solar only during the day. Thus, it is important to introduce another program to adopt the penetration of electric vehicles. With the rapid development of EVs and the continuing decrease on the vehicle cost, this study showed that the number of EVs needed to power a campus can be reduced further.

Acknowledgments. Thanks are due to Masdar City for sharing data with us, and especially for Brian Fan, and Maki Nemoto.

References

[1] Isam, M., Alili, A., Kubo, I., Ohadi, M.: Measurement of solar-energy (direct beam radiation) in Abu Dhabi, UAE. Renewable Energy 35, 515–519 (2010)

[2] Chaar, L., Lamont, L.: Global solar radiation: Multiple on-site assessments in Abu Dhabi, UAE. Renewable Energy 35, 1596–1601 (2010)

[3] Isam, M., Kubo, I., Ohadi, M., Alili, A.: Measurement of solar energy radiation in Abu Dhabi, UAE. Applied Energy 86, 511–515 (2009)

[4] Al Ali, M., Mokri, A., Emziane, M.: On the current status of solar energy in the United Arab Emirates. In: Proceedings of the International Conference on Renewable Energy: Generation and Applications, Al-Ain, UAE, March 8-10 (2012)

[5] Al Ali, M., Mokri, A., Emziane, M.: Solar deployment projects at Masdar. In: Proceedings of the International Conference on Harnessing Technology, Oman, February 13-14 (2011)

[6] Kluza, J., Grose, J., Berenshteyn, Y., Holman, M.: Grid Storage: Show Me the Money. Luxresearch; LRPI-R-10-01 (March 2010), `http://luxresearch.web8.hubspot.com/Portals/86611/docs/research%20downloads/lux_research_grid_storage_show_me_the_money[1].pdf` (August 14, 2012)

[7] Camus, C., Esteves, J., Farias, T.: Integration of Electric Vehicles in the Electric Utility Systems, Electric Vehicles " The Benefits and Barriers," Dr. Seref Soylu (ed.), ISBN: 978-953-307-287-6, InTech (2011), `http://www.intechopen.com/books/electric-vehicles-the-benefits-and-barriers/integration-of-electric-vehicles-in-the-electric-utility-systems` (August 14, 2012)

[8] Sovacool, B., Hirsh, R.: Beyond batteries: An examination of the benefits and barriers to plug-in hybrid electric vehicles (PHEVs) and a vehicle-to-grid (V2G) transition. Energy Policy 37, 1095–1103 (2009)

[9] Mitsubishi Heavy Industries, Ltd. Presidential Administration Office Corporate Communication Dept. Japan, `http://www.mhi.co.jp/en/`, `http://www.mitsubishi-motors.com/special/ev/index.html` (August 14, 2012)

[10] 2getthere, sustainable Mobility Solutions (2012), `http://www.2getthere.eu/?page_id=10` (August 14, 2012)

[11] Amjad, S., Neelakrishnan, S., Rudramoorthy, R.: Review of design considerations and technological challenges for successful development and deployment of plug-in hybrid electric vehicles. Renewable and Sustainable Energy Reviews 14, 1104:10 (2010)

[12] Grose, J., Rutkowski, T., Wu, Y., Holman, M.: Unplugging the Hype around Electric Vehicles. Luxresearch; LPI-SMR-0905 (Septemper 2009), http://luxresearch.web8.hubspot.com/Portals/86611/docs/rese arch%20downloads/lux_research_unplugging_the_hype_around_ electric_vehicles[1].pdf?hsCtaTracking=cd003f11-0721-4c75-b45d-1c53bbf89295%7Ce0ed2fdf-dd35-4e3e-976b-f741d29b27d0 (August 14, 2012)

[13] Tesla motors Ltd. Kings Chase, King Street, Maidenhead, SL6 1DP, UK (2012), http://www.teslamotors.com/models/features#/battery (August 14, 2012)

[14] Chevrolet Chevrolet ChevroletChevrolet Motors, http://www.chevrolet.com/#volt (August 14, 2012)

[15] BMW i, http://www.bmw-i.com/en_ww/bmw-i3/ (August 14, 2012)

[16] Phoenix Cars LLC (April 23, 2012), http://www.phoenixmotorcars.com/ (August 14, 2012)

[17] BYD AUTO CO., LTD (2012), http://www.byd.com/ (August 14, 2012)

[18] Nissan Motors, http://www.nissanusa.com/leaf-electric-car/index#/ leaf-electric-car/index (August 14, 2012)

Chapter 75
Effect of Selective Emitter Temperature on the Performance of Thermophotovoltaic Devices

Mahieddine Emziane and Yao-Tsung Hsieh

Solar Energy Materials and Devices Laboratory
Masdar Institute of Science and Technology
Masdar City, P.O. Box 54224, Abu Dhabi, UAE
memziane@masdar.ac.ae

Abstract. We investigated the performance of thermophotovoltaic (TPV) systems that are composed of different cells and selective emitters. We selected three cells which are the Si conventional PV cell and the low bandgap Ge and InGaAs TPV cells. The three cells operating with Yb based selective emitters were simulated. The effects of the emission spectra on the the performance of these cells are presented and discussed, especially with regard to the emitter temperature. By comparing the overall cell performances, the best combinations of cells and selective emitters were determined.

Keywords: Thermophotovoltaics, Selective emitters, simulation, PV cells.

1 Introduction

The TPV technology is based on the use of the PV effect to convert infrared radiations (i.e. radiation from a thermal source) into electricity. A typical TPV system consists of three components: a heat source, a radiation emitter and a PV cell. The radiation emitter absorbs the thermal energy from the heat source, and then it radiates photons with various energies. These photons are incident on the PV cell, and only photons with suitable energies can be utilized and converted by the PV cell. High cost of fabrication and low energy conversion efficiency are the main reasons for the slow commercialization of TPV systems. Over the past decades, the development of materials science and fabrication technologies led to further progresses in PV cells and spectral control designs for TPV systems [1][2]. Nowadays, many significant improvements have been made in the designs of PV cells, and various techniques are available for controlling the spectra.

In order to enhance the TPV efficiencies considerably, a good match between the emitted spectrum and the sensitivity of the PV cells is essential. The use of selective emitters is considered in this research. The role of a selective emitter is to convert the incoming heat (i.e. a broad spectrum) into a narrower emission spectrum that is adapted to the sensitivity of the PV cell considered [3]. Once the optimum condition is reached by combining the suitable selective emitters and PV cells, the efficiency of

A. Håkansson et al. (Eds.): *Sustainability in Energy and Buildings*, SIST 22, pp. 847–857.
DOI: 10.1007/978-3-642-36645-1_75   © Springer-Verlag Berlin Heidelberg 2013

the TPV system can be improved significantly [4]. Because the shape and major peak of the radiation spectrum are controlled by the selective emitters, conventional PV cells and low bandgap TPV cells can be used as both of them have the potential to improve their efficiency by delivering high power output [5][6].

2 Methodology

2.1 Simulation Model

Through the PC1D simulation program, we can predict the performance of PV and TPV devices under different conditions, and optimize the TPV systems that integrate selective emitters [7]. Using PC1D, we were able to calculate the efficiency of PV and TPV cells by adjusting the cell parameters that strongly depend on the incident spectrum. In this research, selective emitters are considered as the photon source for PV and TPV cells. By using the emittance spectra of different selective emitters, we studied the effect of the key emitter parameters on the overall performance of the resulting systems.

In the simulation model, we plugged in different input data of emission spectra for the three cells considered: Si, Ge and InGaAs. The input intensity from the selectiv emitter is derived from their emission spectra. The results obtained from the progra include open-circuit voltage (V_{OC}), short-circuit current (I_{SC}), and maximum pow. (P_{max}). Based on these data, we calculate the fill factor (FF), current density (J_{SC}) an. cell efficiency (η).

In this paper, Yb based selective emitters are studied and employed in the modeling and simulations of TPV systems. The reason why we focus on this rare-earth selective emitters is that they can be used on either commercial PV cells like Si or lower bandgap TPV cells. Yb based selective emitters have an emission peak at about 1000 nm, which is especially suitable for conventional Si solar cells. W. extracted emission spectra of many selective emitters from experimental dat available in the literature and plugged them into our simulation models [6][8][9]. Fo these Yb based selective emitters, we particularly investigated the effect of their temperature on the overall system performance.

2.2 The Cells Used in This Study

In this research, we use three different cells as our simulation models. They are based on Si, Ge and InGaAs. The energy bandgap of Si is 1.12 eV and Ge is 0.67 eV. For the InGaAs cell, we use the one which was oprtimized and reported recently [10]. The energy bandgap of this InGaAs cell is 0.74 eV. We chose the Si and Ge cells which typically represent PV cell and TPV cell, respectively [7][11]. The InGaAs cell has a bandgap that is between those of Si and Ge, and thus it can have more flexibility and potential for TPV applications using selective emitters. Furthermore, these three cells have different energy bandgaps but also different quantum efficiencies which can affect the cell performance when different selective emitters are applied.

3 Results and Discussions

3.1 Yb$_3$Al$_5$O$_{12}$ Emitter

When the cells are combined with Yb$_3$Al$_5$O$_{12}$ as the emitter, their performance at temperatures above and below 1473 K shows a clear change. The performances of the three cells are presented in Figures 1 to 5. The open circuit voltage (V$_{oc}$) however increases slightly when the temperature rises, and its increase for the three cells is no more than 0.1 V when the temperature increases from 1273 K to 1773K. For the short-circuit current density (J$_{sc}$), the Si cell shows a value that is around nine times higher at 1673 K compared with 1473 K. The Ge and InGaAs cells have J$_{sc}$ that is about 5.5 times higher as the emitter temperature goes from 1473 K to 1673 K. For the max power density (P$_{max}$), the Si cell also has the largest increase among the three cells. It increases 9.76 times at 1673 K compared with 1473 K.

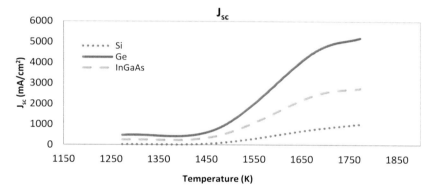

Fig. 1. Jsc versus the temperature of the Yb$_3$Al$_5$O$_{12}$ emitter

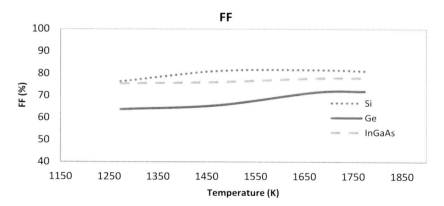

Fig. 2. FF versus the temperature of the Yb$_3$Al$_5$O$_{12}$ emitter

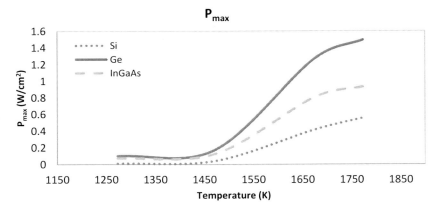

Fig. 3. Pmax versus the temperature of the $Yb_3Al_5O_{12}$ emitter

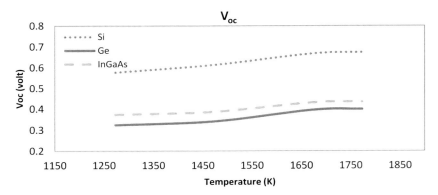

Fig. 4. Voc versus the temperature of the $Yb_3Al_5O_{12}$ emitter

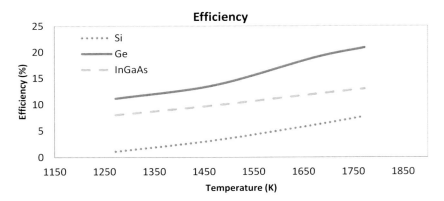

Fig. 5. Efficiency of the cells versus the temperature of the $Yb_3Al_5O_{12}$ emitter

For the Ge and InGaAs cells, they have values that are 7.1 and 6.2 times higher, respectively, when the emitter temperature increases in the same range. When the emitter temperature increases from 1273 K to 1773 K, the P_{max} the Si cell shows the highest increase (58.3 times), but the Ge and InGaAs cells have their power density values increased by 15.1 and 13.1 times, respectively, within the same temperature range. For the fill factor (FF), the Si cell also has the best performance. Its value is improved from 76.43 % to 81.11 % when the emitter temperature increases from 1273 K to 1773 K, and for the InGaAs cell, its FF is enhanced 2.48% (77.97 % at 1773 K) with the same increase of the emitter temperature. The Ge cell has the lowest fill factor compared to Si and InGaAs cells, which changes from 63.75 % to 72.01 % following the emitter temperature increase.

In terms of cell efficiency, the Si cell sees its value improved around seven times as the emitter temperature increases from 1273 K to 1773 K. The Ge cell has the highest efficiency (20.84 %) at all four temperatures compared with Si and InGaAs cells, even though it only doubles when the emitter temperature increases from 1273 K to 1773 K. The Si cell shows the lowest cell efficiency, but it is improved to 7.1 times its initial value as the emitter temperature increases to 1773 K. The InGaAs cell improves its efficiency to 1.62 times (12.97 %) at 1773 K. From Table 1, we can conclude that the Ge cell has the best performance when it is combined with the $Yb_3Al_5O_{12}$ emitter in a TPV system.

Table 1. Comparison of the cells performance at the $Yb_3Al_5O_{12}$ emitter temperature of 1773 K

Emitter temperature of 1773 K	Cells		
	Si	Ge	InGaAs
P_{max} (W/cm^2)	0.555	1.49	0.93
J_{sc} (mA/cm^2)	1016	5184	2740
η (%)	7.75	20.84	12.97

3.2 Yb$_2$O$_3$ Emitter

For all the cells with a Yb_2O_3 selective emitter, we see that the cell performance is enhanced when the emitter temperature increases from 1273 K to 1773 K (see Figures 6 to 10). The V_{oc} is only enhanced by about 0.07 V for all three cells. However, for the J_{sc} we can see different extents of the emitter temperature effect on the three cells. The increase of emitter temperature makes the Si cell J_{sc} increase around 14 times, but the Ge and the InGaAs cells J_{sc} both increase only 4.6 times.

The Si cell has the highest FF compared with the other two cells, and it also keeps its FF above 80 % at all temperatures. Moreover, the P_{max} of the Si cell has the highest increase (16 times). It increases 5.6 times for the Ge cell and 5.2 times for the InGaAs cell. As a result, we can conclude that increasing the Yb_2O_3 emitter temperature is more beneficial for the Si cell. If we compare the best performance of each cell, we can see from Table 2 that the Ge cell has the highest values of P_{max}, J_{sc} and cell efficiency at the Yb_2O_3 emitter temperature of 1773 K.

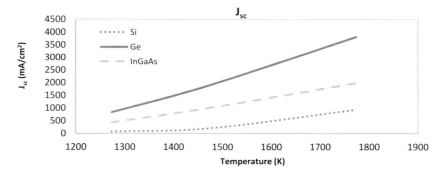

Fig. 6. Figure 6: J_{sc} as a function of the Yb_2O_3 emitter temperature

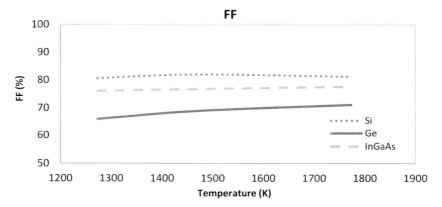

Fig. 7. FF as a function of the Yb_2O_3 emitter temperature

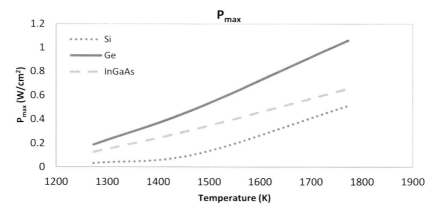

Fig. 8. Pmax as a function of the Yb_2O_3 emitter temperature

75 Effect of Selective Emitter Temperature on the Performance of TPV Devices   853

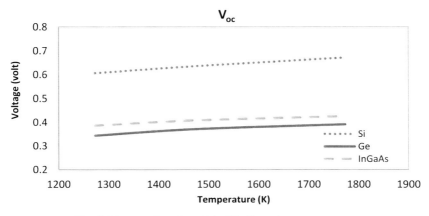

Fig. 9. Voc as a function of the Yb$_2$O$_3$ emitter temperature

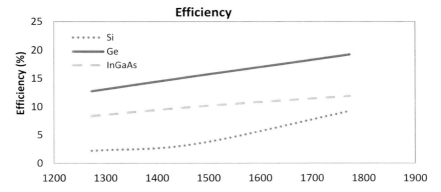

Fig. 10. Efficiency of the cells as a function of the Yb$_2$O$_3$ emitter temperature

Table 2. Comparison of the cell performance at the Yb$_2$O$_3$ emitter temperature of 1773 K

Emitter temperature of 1773 K	Cells		
	Si	Ge	InGaAs
Power density (W/cm^2)	0.511	1.06	0.66
J$_{sc}$ (mA/cm^2)	935.3	3810	1980
η (%)	9.24	19.22	11.85

3.3   Yb:Y$_2$O$_3$ Emitter

When we consider the emitter made of Yb:Y$_2$O$_3$, we observe that the efficiencies of the three cells are improved with the same trend as the previous two emitters; however, the extent of temperature effect becomes much less. The performances of the three cells are shown in Figures 11 to 15.

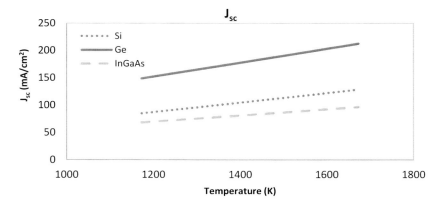

Fig. 11. Jsc versus the Yb:Y$_2$O$_3$ emitter temperature

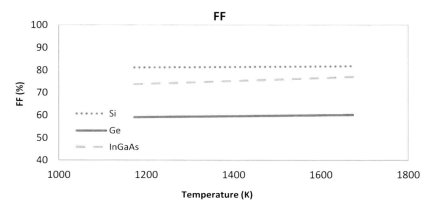

Fig. 12. FF versus the Yb:Y$_2$O$_3$ emitter temperature

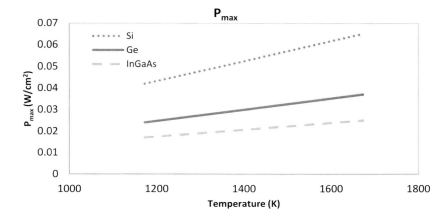

Fig. 13. Pmax versus the Yb:Y$_2$O$_3$ emitter temperature

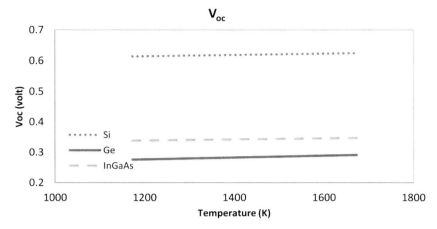

Fig. 14. Voc versus the Yb:Y$_2$O$_3$ emitter temperature

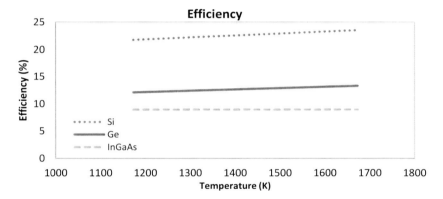

Fig. 15. Efficiency of the cells versus the Yb:Y$_2$O$_3$ emitter temperature

V_{oc} increases only 0.01 V for the three cells, which is much smaller than for Yb$_2$O$_3$ and Yb$_3$Al$_5$O$_{12}$ when they undergo the same temperature increase. For J_{sc}, the Si cell still has the highest increase (1.5 times), but the other two cells also increase with about the same amount (1.4 times). The same behavior is seen for the P_{max} of the three cells.

The Si cell has the largest FF (81.29 %) and increase of the cell efficiency (1.75%). Moreover, it can be observed that increasing this emitter temperature only produces a small effect on the fill factor (3.37% increase) and cell efficiency (0.01% decrease) for the InGaAs cell.

At 1673 K, the performances of the three cells are summarized in Table 3. It is noted that the Si cell has the highest power density and cell efficiency therefore. But in terms of the J_{sc}, the Ge cell has the highest one.

Table 3. Comparison of the cell performance at the Yb:Y_2O_3 emitter temperature of 1673 K

Emitter temperature Of 1673 K	Cells		
	Si	Ge	InGaAs
P_{max} (W/cm^2)	0.065	0.04	0.025
J_{sc} (mA/cm^2)	128.2	212.4	96.3
η (%)	23.54	13.29	8.93

4 Conclusions

When selective emitters are used to generate the input energy for the cells in TPV systems, the input intensity and the illumination conditions for the cells vary depending on the selective emitters. The emission behavior of these selective emitters is affected by their temperature and this important parameter was therefore studied in detail. Through our modeling and simulations, the way in which the emitter temperature influences the cell performance was predicted. Based on our results, it can be concluded that the Ge cell has the best performance and the highest cell efficiency when working in TPV systems with Yb based selective emitters. The results also indicated that the InGaAs cell shows a better performance in TPV than the Si cell. The InGaAs TPV cell technology is emerging and it has a large potential for further developments in TPV applications.

References

[1] Nelson, R.E.: A brief history of thermophotovoltaic development. Semiconductor Science and Technology 18(5), S141–S143 (2003)

[2] Fraas, L., Minkin, L.: TPV History from 1990 to Present & Future Trends. In: AIP Conference Proceedings, vol. 890(1), pp. 17–23 (February 2007)

[3] Nelson, R.E.: Rare Earth Oxide TPV Emitters. In: Proceedings of the 32nd International Power Sources Symposium, pp. 95–101. Electrochemical Society, Pennington (1986)

[4] Chubb, D.L.: Reappraisal of solid selective emitters. In: Photovoltaic Specialists Conference, Record of the Twenty First IEEE (May 1990)

[5] Chubb, D.L., Lowe, R.A.: Thin-film selective emitter. Journal of Applied Physics 74(9), 5687–5698 (1993)

[6] Licciulli, A., et al.: The challenge of high-performance selective emitters for thermophotovoltaic applications. Semiconductor Science and Technology 18(5), S174–S183 (2003)

[7] PC1D, School of Photovoltaic and Renewable Energy Engineering at the University of New South Wales, Australia

[8] Ghanashyam Krishna, M., Rajendran, M., Pyke, D.R., Bhattacharya, A.K.: Spectral emissivity of ytterbium oxide-based materials for application as selective emitters in thermophotovoltaic devices. Solar Energy Materials and Solar Cells 59(4), 337–348 (1999)

[9]   Panitz, J.-C., Schubnell, M., Durisch, W., Geiger, F.: Influence of Ytterbium Concentration on the Emissive Properties of Yb:YAG and Yb:Y2O3. In: CP401, Thermophotovoltaic Generation of Electricity: Third NREL Conference, pp. 265–276 (1997)

[10]   Emziane, M., Nicholas, R.J.: Optimization of InGaAs(P) photovoltaic cells lattice matched to InP. Journal of Applied Physics 101(5), 054503 (2007)

[11]   van der Heide, J., Posthuma, N.E., Flamand, G., Geens, W., Poortmans, J.: Cost-efficient thermophotovoltaic cells based on germanium substrates. Solar Energy Materials and Solar Cells 93(10), 1810–1816 (2009)

Chapter 76
New Tandem Device Designs
for Various Photovoltaic Applications

Mahieddine Emziane

Solar Energy Materials and Devices Laboratory
Masdar Institute of Science and Technology
Masdar City, PO Box 54224, Abu Dhabi, UAE
memziane@masdar.ac.ae

Abstract. We report on new tandem device designs based on IV or III-V semi-conductors for the top cell, and group IV materials for the bottom cell. In addition to the extended spectral coverage leading to more photons being converted, three and four-terminal device configurations were considered in order to avoid the current matching and the associated tunnel junctions between the two sub-cells. A comprehensive modeling analysis is presented where device structures were designed and optimized, and the behavior of the sub-cells studied. Optimal cell characteristics were obtained with the quantum efficiency. The applications of these devices were assessed and the output parameters were predicted as a function of the device simulated operating conditions.

Keywords: Tandem solar cells, device design, three terminals, four terminals, CPV, PV.

1   Introduction

Solar cells enable the use of abundant, sustainable and clean solar energy as a source of electricity generation. Due to their very high efficiency, multi-junction solar cells can potentially easily compete with the conventional sources of energy in terms of cost. However, they are not yet deployed commercially on a large scale due essentially to their high fabrication costs. The processes used and the materials involved are determined by the device design and the electrical configuration of the cell [1], showing the importance and relevance of design work in PV technologies.

In a recent paper, we showed that tandem solar cell devices can be designed [2]. The semiconductor materials used were GaAs and Ge for the top and bottom cell, respectively, in a monolithic device structure that is almost lattice-matched to Ge substrate. Assessment of these devices for conventional photovoltaics (PV) as well as concentrated PV (CPV) applications was presented.

This paper considers two novel designs of tandem PV devices. The objective here is to enhance the conversion efficiency by extending the spectral coverage and removing the constraints on the device structure.

A. Håkansson et al. (Eds.): *Sustainability in Energy and Buildings*, SIST 22, pp. 859–864.
DOI: 10.1007/978-3-642-36645-1_76   © Springer-Verlag Berlin Heidelberg 2013

2 Methodology

We have considered two tandem PV devices. The first one consists of a Si cell at the top together with a Ge bottom cell. We have adopted a three-terminal device configuration which allows an independent operation of the two individual cells without the need for current matching and the corresponding tunnel junctions. This configuration also provides an improved output compared to the same sub-cells used in a two-terminal device as it has been shown for structures based on III-V semiconductors.

The second tandem device is a four-terminal mechanically stacked double-junction based on GaAs as a top subcell and Si as a bottom subcell. A lattice mismatch of about 4% exists between the materials used, but this lattice mismatch does not contribute to any losses in this novel device configuration.

In both cases, by adding a bottom cell with a lower bandgap to the top single-junction cell, the spectral coverage of the resulting tandem device is extended in the near-infrared region leading to an enhanced overall performance through the conversion of more photons. Additionally, we adopted electrical connections that do not need any tunnel junctions or current matching unlike for the two-terminal device configuration.

Simulations of both tandem devices were carried out by modeling using the software PC1D [3]. Physical materials characteristics were used as input parameters for the device structure layers. PC1D is a commercial program which can solve the fully coupled nonlinear equations for the quasi-one-dimensional transport of electrons and holes in crystalline semiconductor photovoltaic devices. It is particularly useful for simulating PV device performance. **INPUTS**

3 Results and Discussion

In this section, design results are shown regarding the two tandem devices that were investigated, namely Si/Ge in a three-terminal device configuration and GaAs/Si in a four-terminal configuration.

3.1 Three-Terminal Devices

Fig.1 shows a typical quantum efficiency versus wavelength from these Si/Ge tandem PV devices (under AM1.5G) indicating an extended spectral coverage compared to a single-junction Si cell. The advantage of having an additional Ge cell at the bottom is further highlighted by the current-voltage (I-V) characteristics of these Si/Ge tandem PV devices. Fig. 2 shows typical I-V curves of top (Si) and bottom (Ge) cells simulated under the same illumination conditions.

For the purpose of implementing such Si/Ge tandem devices, a structure optimization was undertaken. Fig. 3 shows the total efficiency expected from these devices as a function of the individual thickness of Si and Ge cells. Fig. 3 indicates that close to optimal overall performance can be reached for the tandem device within the constraint of keeping it reasonably thin.

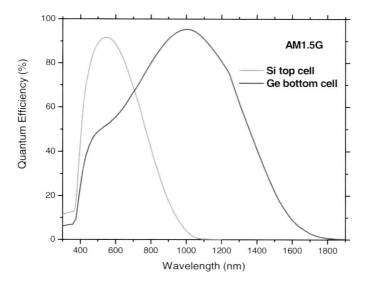

Fig. 1. Typical quantum efficiency of Si/Ge tandem PV devices with three terminals

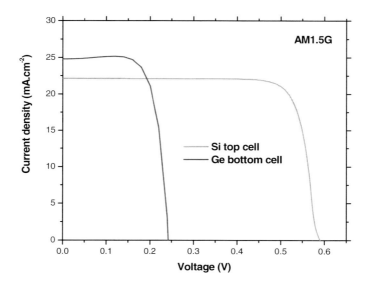

Fig. 2. Typical I-V curves of Si/Ge tandem PV devices with three terminals

Also important for the implementation of such devices are the optimal doping levels in both n and p sides of the Si and Ge junctions. An analysis was carried out for this purpose and the doping levels leading to the highest tandem device efficiency obtained under AM1.5G are summarized in Table 1.

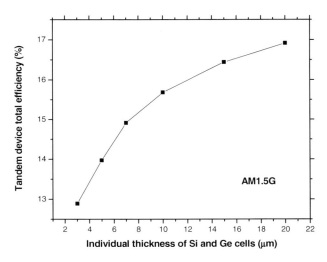

Fig. 3. Total efficiency as a function of the individual thickness of Si and Ge cells

Table 1. Optimal doping levels for a Si/Ge tandem PV device

	Si Top Cell	**Ge Bottom Cell**
p doping level (cm^{-3})	1×10^{15}	5×10^{15}
n doping level (cm^{-3})	1×10^{19}	5×10^{19}

3.2 Four-Terminal Devices

The optimization of the thicknesses of the top cell active layers was carried out based on a trade-off between the two cells, and the optimal values obtained were therefore for the entire tandem device rather than the individual cells.

A thickness of 0.1µm for the emitter layer of the GaAs top cell is found to be the optimal. Also, a thickness of 10µm, i.e. 2µm for the emitter and 8µm for the base, is chosen for the bottom the Si cell.

Optimal doping concentrations of 10^{18} cm^{-3} and 10^{16} cm^{-3} were predicted, respectively, for the emitter and base of the Si cell.

The I-V curves of the optimized device under standard conditions (i.e. one sun AM1.5G illumination and 25°C) are plotted in Fig. 4. The current-voltage characteristics generated by the proposed device exhibit the advantage of the four-terminal configuration, where the current mismatch between top and bottom cells is obvious and would lead to substantial losses if the traditional two-terminal device configuration was adopted. Instead, the two cells being operated independently, the maximum solar power converted will be collected through the individual cell circuits.

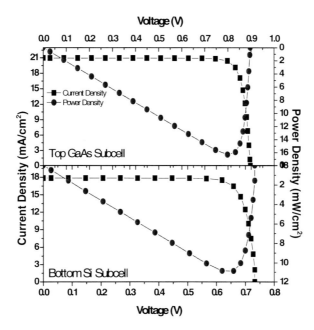

Fig. 4. I-V curves of optimal GaAs top and Si bottom cells with four terminals

4 Conclusions and Future Work

We have shown that efficient and relatively thin tandem PV devices can be designed, optimized and simulated. The devices reported have three terminals (with Si and Ge cells) and four terminals (with GaAs and Si cells).

Optimal device structure doping levels and thicknesses were determined for each configuration. The device operation under AM 1.5G conditions was simulated and the advantages of using the three- and four-terminal device configurations were also highlighted. An extended spectral coverage is shown by these double-junction devices compared to cells made of single junctions.

The I-V curves of the optimized device were generated, and the proposed device configurations were shown to benefit from the combination of appropriate materials without the conventional constraints. The electrical independence of the individual cells through the device designs proposed allows the carriers to be independently collected without losses that exist in the conventional two-terminal configuration.

Based on encouraging preliminary results, the operation of these novel device designs under concentrated light for CPV applications will be further investigated taking into account other crucial operating parameters such as the temperature. The purpose is to show the relevance and flexibility in the application of the proposed devices depending on the operation conditions.

Acknowledgements. The author would like to thank Masdar Institute of Science and Technology for funding this research. The contribution of A. Sleiman is also acknowledged.

References

[1] Dimroth, F., Kurtz, S.: High-efficiency multijunction solar cells. MRS Bulletin 32, 230–235 (2007)
[2] Emziane, M.: Two-Junction GaAs/Ge Cells with Three Terminals for PV and CPV Applications. In: Mater. Res. Soc. Symp. Proc., 1165, pp. M08-13 (2009)
[3] PC1D, version 5.9, School of Photovoltaic and Renewable Energy Engineering at the University of New South Wales, Australia

Chapter 77
GIS-Based Decision Support for Solar Photovoltaic Planning in Urban Environment

Antonio Gagliano, Francesco Patania, Francesco Nocera,
Alfonso Capizzi, and Aldo Galesi

Department of Industrial and Mechanics Engineering, University of Catania,
Viale A. Doria n.6 95125 - Catania, Italy
agagliano@diim.unict.it

Abstract. In 2007, the European Council decided a fixing goal of 20% contribution of the renewable energy sources (RES) to the total European electric energy production in 2020. Micro-generation systems integrated in urban environment are an interesting opportunity, in terms of research and development of RES. The development of a solar energy planning system to predict the potential of solar energy photovoltaic, solar water heating and passive solar gain is necessary for the optimization of energy efficiency strategies and integration of renewable energy systems in urban areas. The work discussed here relates to solar photovoltaic (PV), technology which has matured to become a technically viable large-scale source of renewable energy sources. This paper illustrates the capabilities of Geographic Information Systems (GIS) to determine the available rooftop area for PV deployment for an urban area and how the methodology may enable planners to consider the urban-scale application of solar energy with greatly increased confidence.

1 Introduction

Common problems in European towns and city are:

- discrepancy between high urban population densities and the low level of renewable energy source availability.
- deterioration of environmental conditions and increase of resource consumption levels

The sustainable renovation of urban area is a very complex task for both the necessity of preserving the original architectural and urban characteristics and for the difficulty related to the renewable energy sources integration.

For structural and aesthetic reasons, the basic layout and design of urban blocks and buildings can usually not be altered/adapted to suit extended exploitation of renewable energy uses. Nevertheless, there is great potential for the utilization of renewable energy which so far has remained largely untapped.

However, in order to maximize the use of Renewable Energy Source (RES) on a regular basis and in a large scale manner, a decisive pre-requisite is that renewable energy application is integrated in the urban planning process at the beginning.

A. Håkansson et al. (Eds.): *Sustainability in Energy and Buildings*, SIST 22, pp. 865–874.
DOI: 10.1007/978-3-642-36645-1_77   © Springer-Verlag Berlin Heidelberg 2013

Especially in regions like Sicily, with a predominantly warm and sunny climate, but also in Northern Italy regions, renewable energy use for heating, cooling and lighting can play an important role in municipal rehabilitation planning. The development of micro-generation systems from renewable energy are one of the frontiers of technological innovation, in alternative to macro-generation systems. The relocation of the wind and solar plants in urban area, where the human presence is already established, reduces the environmental impact and maintains the conditions present in the countryside areas.

Among the electrical energy production systems from renewable sources, the photovoltaic system represents today one of the most efficient and technically tested choices. In this context, an interesting perspective is the develop of the so called photovoltaic thermal hybrid panels (PV/T) [1]. Understanding the rooftop PV potential is critical for utility planning, accommodating grid capacity, deploying financing schemes and formulating future adaptive energy policies. Many studies have investigate the potentiality of photovoltaic and wind power integrated in built area [2,3].

The aim of this work is the evaluation of the potential energy production by solar photovoltaic (PV) within an urban residential area.

Different forms of financing for PV systems have been put into effect in the last decade: capital subsidies, VAT reduction, taxes credits, green tags, net-metering, Feed-in Tariffs (FiTs), etc.

Feed-in tariffs have proven to be the most effective government incentive program for renewable technologies: countries who have adopted FITs have been shown to have the largest growth rates in renewable energy technology deployment [4].

1.1 Italy PV Incentives

In 2010, Italy installed at least 1800 MW of new PV, bringing the total installed capacity to 2903 MW across more than 144,000 installations, according to GSE (national electricity service agency) [5]. Italy's PV FiTs were introduced in 2005, replacing an earlier incentive scheme that had been judged a moderate success. The Decree of the Ministry for Productive Activities (2005) defined a support System (Conto Energia) that allows the producers to obtain both FIT and net-metering for PV installations with rated power not over 20 kWp. For rated power over 20 kWp, the customer could choose to sell the whole electric energy produced by the PV system to the local Utility or to use part of this energy for its own consumption. The incentive duration is 20 years, like in Germany, with a constant remuneration.

A new Decree of the Ministry for Economical Development (2007) simplified the procedure to obtain the incentive and changed the FITs values distinguishing among FIPV, Partially Integrated in Building (PIPV) and Building Integrated PV Systems (BIPV). The same Decree established that the FITs values are decreased for every year after 2008 of 2% and are increased in case of energetic certification of the building. In Mid-2010, following a surge of activity in the PV sector, the Italian government announced another review, and the 3th "Conto Energia" was introduced from the start of 2011. On May 2011 the decree defining the 4th Conto Energia" was approved. This decree introduced a range of tariff reductions during the first four months of 2011 with further cuts foreseen at four-monthly intervals during the year. The new tariffs will apply to photovoltaic plants that will start to generate power between June 1, 2011 and December 31, 2016.For rooftop installations, the decreases ranged from 4.75% to 13.28%, depending on system size, although integrated BIPV was treated more generously. A further annual drop of 6% was planned for 2012 and 2013.

Regarding PV on agricultural land, only systems up to 1 megawatt will be facilitated. Such policy choice is based on the assumption that previous Government subsidies has adequately enabled the Italian photovoltaic industry to grow and that the market of such renewable energy is now ripe to their developing without the need for public support. The cost of Italy's, as in many countries, is passed on to electricity consumers by their regular bills. This translates into an increase for Italian consumers from 0.25 euro cents/kWh in 2009 to 1.42 euro cents/kWh in 2010, representing 6% of the electricity bill [6].

2 GIS Methodology

The energy provided by RES depends on climate, geographical conditions and morphological features of the site (height of buildings, typology of roof exposure, presence of obstacles, and so on). There are several hundreds of ground meteorological stations directly or indirectly measuring solar radiation throughout Europe. To derive spatial databases from these measurements different interpolation techniques are used, such as spline functions, weighted average procedures or kriging [7]. Spatially continuous irradiance values can be also derived directly from meteorological geostationary satellites (e.g. METEOSAT). Processing of satellite data provides less accurate values (compared to ground measurements), but the advantage is a coverage over vast territories at temporal resolution of 0.5-12 hours [8]. Other techniques of generating spatial databases are solar radiation models integrated within geographical information systems (GIS). They provide rapid, cost-efficient and accurate estimations of radiation over large territories, considering surface inclination, aspect and shadowing effects. Coupling radiation models with GIS and image processing systems improves their ability to process different environmental data and cooperate with other models.

The methodologies that use interpolation techniques of data measured from weather stations or satellite images, for mapping solar and wind energy usually allow to obtain reliable estimates only on large-scale [9,10].

The assessment of the energy potential needs to calculate the solar radiation on the available area for the installation of photovoltaic or solar thermal plants.

For this aim, it is necessary develop a tool that permits to implement the geographical conditions and morphological features of the site (height of buildings, typology of roof exposure, presence of obstacles, and so on) and then calculate the potential available energy resources. These goals have been achieved thanks to the capabilities of Geographical Information Systems (GIS), that permit to obtain the urban radiation mapping useful for both the quantitative assessment of the solar potential available and for planning resource management

2.1 Urban Mapping

Geographic and morphological data can be obtained from large scale vector mapping (1:2000), that is a digital representation of topographic map features with attributed points, lines and polygons.

The 3D digital models, normally utilized for GIS applications, use square mesh (DEM or DTM) or triangular mesh (TIN). The Triangular Irregular Network (TIN) is a vector-based representation (a digital data structure) used for the representation of a surface made up of irregularly distributed nodes and lines with three-dimensional coordinates (x, y, and z) that are arranged in a network of non overlapping triangles. An advantage of using a TIN over a raster DEM in mapping and analysis is that the points of a TIN are distributed variably based on an algorithm that determines which points are most necessary to an accurate representation of the terrain. ArcGIS uses Delaunay's triangulation method to construct these triangles.

On the other hand, TIN meshes are appropriate for contexts where morphology is gentle and continuous without sudden altitude changes, but they do not supply a good representation of complex morphology that is characteristic of urban areas (presence of buildings).

For these reason, a TIN modified model suitable for urban space morphology has been developed considering in the terrain map also the buildings elevations.

The ESRI software ArcGIS Desktop v.9.2 [11] with Spatial Analyst and 3D Analyst extensions has been used. Respect to the "classical" approach, for which each building is represented with a polygon that represents the building (gutter surface), the proposed procedure requires to add another polygon at ground height (ground surface) used as a "break line" in 3D modeling.

The first polygon, that represent the gutter surface, is "shrinked" with an offset of 5 cm, that is a minimum value to distinguish the two polygons, respect to the second polygon that represent the ground surface.

All this geo-data are used for the representation of spatial information through a geo-referenced cells matrix (GRID). The size meshes of the GRID are: in the XY plane 1 m x 1 m and each cell has its own altimetric quote (GRID-quote). The size of the cells establishes the level of spatial resolution.

In this way is possible to obtain the rasterized digital elevation model (DEM) of the urban area.

2.2 Solar Mapping

The 3D digital terrain model is useful for evaluating both of the incident solar energy on the buildings and the shades on each building roof. The solar radiation incident on any surface is calculated with the "Solar Radiation Area" function that is a sub-application of "Solar Analyst" tools.

The Solar Analysis Tools of ArcGIS, calculate solar insulation (Wh/m^2) at any location on the Earth's surface. This tool uses point-based imagery of local level elevation, slope, and aspect to determine the amount of energy available. Optimized algorithms account for variations in surface orientation and atmospheric weather data.

Total global radiation (Global$_{tot}$) is calculated from the sum of the direct and diffuse radiation of all sectors on the topographic surface. These are calculated separately for each location and the total produces an irradiance map for the whole study area. Detailed models and algorithms used to calculate the direct and diffuse solar radiation can be found in the Solar Analyst design document [12]. The output of the Solar Radiation Area function is the incident radiation measured in watt hours per square meter (Wh/m^2).

In this way a raster file is generated, where each cell contains the information about the annual average global solar radiation.

The calculation model permits the use of different configurations of time: a specific day, a full year, a range of days.

The result of the developed procedure is useful to estimate the solar potential everywhere inside the investigated area.

3   Pilot Study

To test the proposed methodology, this study presents a preliminary pilot study for an urban area in the municipality of San Cataldo, a small city located in the centre of Sicily (latitude: 37° 29' 0" N, longitude: 13° 59' 0" E) . Figure 1 shows the DEM of the studied urban area.

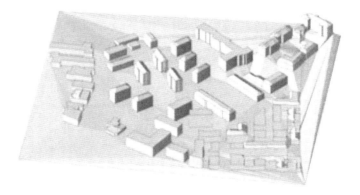

Fig. 1. DEM of a neighborhood of the city center of San Cataldo (CL)

In the figure 2 both the ortho-photo of the investigated area and the DEM are depicted. The overlapping allows noticing the good accuracy of the DEM representation respect to the real geometry of the built up area.

Following the procedure describes above, a raster file has been generated with a spatial resolution of 1,0 x 1,0 m (grid size), where each cell contains the information about the annual average global solar radiation. Figure 3 shows the yearly solar radiation in the investigated urban area.

Fig. 2. DEM with overlapping ortho-photos

Fig. 3. Map of yearly solar radiation (kWh/m^2)

The yearly solar radiation was calculated using the data and algorithms available in the Solar Analyst tools [12]. The values of the global solar radiation calculated was compared with the data calculated using the Enea Solar tool [13] that takes into account the orientations and the inclination of the surfaces.

The Enea solar tool gives a value of 1624,72 kWh/m^2 for a not shaded horizontal surface that is about the same value (1640 kWh/m^2) calculated by the GIS tools for the horizontal roofs of the higher buildings. This comparison confirms the reliability of data obtained by the Solar Analyst tools.

It is also possible observe that the façade surfaces are in many cases characterized by low values of incident solar radiation for the effect of the shadows of other buildings or for the effect of their orientation.

3.1 Assessment of Photovoltaic Energy Production

The procedure for evaluation of solar potential was implemented in GIS software using a plug-in for ArcGIS®, called "Spatial Analyst" and, in particular, the "Raster Calculator"

The Raster calculation allows making mathematical operations using the value of each single raster cell as an operator within a formula. In this way, a new raster file composed by cell whose value derives from mathematical operations performed on the previous raster cells (start raster), is generated.

The start raster file is the raster whose cells contain the average annual value of irradiation (Fig. 4). From the start raster file it was calculated the actual energy extracted from PV solar panels.

In order to derive the actual energy extracted from PV solar panels, a "fixed conversion efficiency of 16% is assumed, which is a fairly conservative considering that the current efficiency of the mono-crystalline cells is above 20% and that of amorphous cells is near 12%, with multi-crystalline cells falling in between.

A "standard" installation of photovoltaic modules was considered with a single inverter with a nominal input power of 80% of the installed PV peak power [14]. The estimated electrical energy production of PV plant was calculated by the following expression:

$$EPV = I \times \eta_{nom} \times \eta_{BOS} \times KPV \times S \qquad (1)$$

where:

EPV is the annual electricity production by the plant (kWh/year),

I is the annual solar irradiance (kWh/m^2 year),

η_{nom} is nominal module efficiency

η_{BOS} is the efficiency of BOS (inverter, wiring, etc.) assumed as 0.95,

KPV is the reduction coefficient due to the effect of real operational conditions respect to standard test conditions (STC), considered as 0.8,

S is Fraction of available roof area for solar PV application

Regarding the fraction of available roof area for solar PV application, the technique used assumes, in accord with other research [15, 16], that each raster cell represents potentially 0.50 active surface of PV modules.

The product ($\eta_{BOS} \times KPV$) corresponds to performance ratio (PR), which quantifies the overall effect of losses on the rated output due to inverter inefficiency, wiring, mismatch, module temperature, incomplete use of irradiance by reflection from the module front surface, soiling or snow and component failure.

Normally PR falls within the range of 0.6÷0.8 [17]. The value assumed in the present calculations is 0.76 in agreement with operational performance results collected in several IEA countries and in similar climate conditions [18].

A customer tool of the GIS software creates the raster file of potential PV production by implementing the previous following formula.

The figure 4 shows the map of the potential PV production in the urban area of San Cataldo.

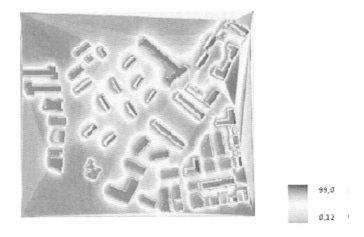

Fig. 4. Map of PV potential production (kWh/m^2 of roof surface)

The proposed procedure permits to obtained detailed information, with a very high level of spatial resolution. In fact, it is possible to know, for each building, the energy that can be produced daily, monthly or yearly by the photovoltaic modules. These information are useful to find the mismatch between the produced and consumed electric energy. In fact, the GIS platform can also contain the data on the population density, energy demand intensity, in this way it will be possible to create a data base which gives both the solar energy production and the electricity consumption for each building. In this context it is possible to express some preliminary consideration: the average consumes of Italian domestic household [19, 20] varying between 2,700 – 3.300 kWh per year, (including consumption for cooling and heating) consequently at least about 25,0-35,0 square meters of roof surfaces are necessary to satisfy the electric energy consumption of each family nuclei. Therefore in many cases, (i.e. multi level buildings) the available potential of PV energy production may be not sufficient to guarantee the energy demand for domestic use.

4 Conclusion

Urban districts represent an optimal scale for combining energy conservation programs with promising energy strategies implementation, like local RES generation. This research investigated the use of GIS-based methodology for planning solar PV energy potential. The results of this GIS-based methodology are thematic maps of PV energy production which are certainly a powerful tool to support the renewable energy planning.

The innovative aspect of this approach is mainly linked to its high productivity, which allows to automatically extend the evaluation of the potential production from a single site of specific interest to an entire urban area. This work contributes to the diversification of energy sources and maximization of on-site RES penetration in urban areas in accordance with the objectives set by the EU. The possibility to locate PV systems in the most advantageous position allows to make a quickly and efficient assessments that can properly guide decision-making

Future develops of the research will be the comparison between the hourly household electrical load and the hourly power from the PV plant, to create a stronger connection between energy demand and supply, with reference to dispatch and control strategies. The results can also give a substantial support to guide energy policy formulation.

References

[1] Rosa-Clot, M., Rosa-Clot, P., Tina, G.M.: TESPI: Thermal Electric Solar Panel Integration. Solar Energy 85, 2433–2442 (2011)
[2] Gadsden, S., Rylatt, M., Lomas, K., Robinson, D.: Predicting the urban solar fraction: A methodology for energy advisers and planners based on GIS. Energy and Buildings 35(1), 37–48 (2003)
[3] Gagliano, A., Patania, F., Capizzi, A., Nocera, F., Galesi, A.: A proposed methodology for estimating the performance of small wind turbines in urban areas. In: Proceedings of International Conference on Sustainability in energy and Buildings, Marseille (June 2011)
[4] Pietruszko, Renewable Energy Policy Network for the 21st Century (REN21) (2006)
[5] http://ww.gse.it
[6] http://www.renewableenergyworld.com
[7] http://re.jrc.ec.europa.eu/pvgis/
[8] Hofierka, J., Súri, M.: The solar radiation model for Open source GIS: implementation and applications. In: Proceedings of the Open source GIS - GRASS users conference 2002, Trento, Italy, September 11-13 (2002)
[9] Noia, M., Ratto, C.F., Festa, R.: Solar irradiance estimation from geostationary satellite data: I. Statistical models, Solar Energy 51, 449–456 (1993)
[10] Sorensen, B.: GIS management of solar resource data. Solar Energy Materials & Solar Cells 67, 503–509 (2001)
[11] http://www.esri.com/software/arcgis/index.html
[12] The Solar Analyst 1.0 user manual. Helios Environmental Modeling Institute, http://www.hemisoft.com
[13] http://www.solaritaly.enea.it/
[14] http://www.esat.kuleuven.be/electa/publications/fulltexts/pub_1057.pdf
[15] Wiginton, L.K., Nguyen, H.T., Pearce, J.M.: Quantifying rooftop solar photovoltaic potential for regional renewable energy policy Computers. Environment and Urban Systems 34, 345–357 (2010)
[16] Izquierdo, S., Rodrigues, M., Fueyo, N.: A method for estimating the geographical distribution of the available roof surface area for large-scale photovoltaic energy-potential evaluations. Solar Energy 82, 929–939 (2008)
[17] Marion, B., Adelstein, J., Boyle, K.: Performance parameters for grid-connected PV systems. In: Proceedings of the 31st IEEE Photovoltaics Specialists Conference and Exhibition, Lake Buena Vista, Florida (2005)

[18] Jahn, U., Nasse, W.: Performance analysis and reliability of grid-connected PV systems in IEA countries. In: Proceedings of the 3rd World Conference on Photovoltaic Energy Conversion, Osaka, Japan (2003)

[19] Cicero, F., Di Gaetano, N., Speziale, L.: I comportamenti di consumo elettrico delle famiglie italiane L'Energia Elettrica 57 (2010)

[20] Ruggieri, G:, Alcune note sui consumi elettrici nel settore domestico in Italia, http://www.aspoitalia.it/attachments/220_Gianluca%20Ruggieri%20-%20Consumi%20elettrici%20nel%20domestico.pdf

Chapter 78
Infrared Thermography Study of the Temperature Effect on the Performance of Photovoltaic Cells and Panels

Zaoui Fares[1], Mohamed Becherif[2], Mahieddine Emziane[3], Abdennacer Aboubou[1], and Soufiane Mebarek Azzem[4]

[1] MSE Laboratory, Biskra University, Biskra, 07000, Algeria
[2] FCLab FR CNRS 3539, FEMTO-ST UMR CNRS 6174 and UTBM University, 90010 Belfort Cedex, France
[3] Solar Energy Materials & Devices Lab., Masdar Institute of Science and Technology, Abu Dhabi, UAE
[4] Ferhat Abbas University, Setif,19000, Algeria

Abstract. Silicon solar cells are widely used in the Photovoltaic (PV) industry. The silicon PV electrical performance is described by its current–voltage (I–V) characteristic, which is a function of the device used and material properties. The PV cell efficiency is strongly temperature dependent. This work studies the electric performance of a polycrystalline Si solar panel under different atmospheric conditions by using thermographic images. The PV performance study is carried out as function of the junction temperature and solar insolation. An infrared analysis as close to the junction temperature has allowed measurements of the cells surface temperature in order to increase the measurement accuracy and to make a reliable assessment of the PV module performance.

1 Introduction

Specific issues about PV power plants can affect the PV modules or the inverters. Some effects regarding the PV modules are reported in [1]-[2], while specific models of defects implemented in Finite Element Method (FEM)-based software are reported in [3]. Reliability issues about several parts of PV plants are listed in [4]. Many PV devices exhibit poor performance under real conditions of mainly due to the internal parameters such as junction temperature and external parameters such as ambient temperature, insolation level, wind speed, wind direction, tilt angle, dust and shadow [5].

The current–voltage (I–V) characteristic of the solar cell describes its electrical performance. These I–V characteristics are determined by parameters such as diode saturation current, diode ideality factor, photo generated current and the presence of parasitic resistances (series and shunt resistances). These parameters depend. in turn, on the solar cell structure, material properties and operating conditions.

PV modules, which generally consist of a set of series-connected solar cells, are rated at the maximum power output under standard test conditions (STC):

A. Håkansson et al. (Eds.): *Sustainability in Energy and Buildings*, SIST 22, pp. 875–886.
DOI: 10.1007/978-3-642-36645-1_78   © Springer-Verlag Berlin Heidelberg 2013

25°C, 1 kW/m² insolation, Air Mass (AM) 1.5 global spectrum. The analysis of I–V characteristics at STC allows the determination of additional electrical parameters and also gives an indication of the presence of parasitic resistances [6]. The dark I–V characteristics enable the extraction of the device parameters and parasitic resistances. The presence of shunt paths in the solar cells of a PV module leads to excessive power loss at low insolation levels. The performance of PV modules/cells can be fully characterized using a suite of electrical, optical and mechanical evaluation tools [7], to detect any degradation and possible failure, or by using the effect of outdoor conditions such as dust and shadow on the performance [8]. Sometimes, the reliability measurements could be done by laser irradiation of the cell [9].

PV power systems may be subject to unexpected ground faults, like any other electrical system. Installed PV systems always have invisible elements other than those indicated by their schematics. Capacitance, resistance and stray inductance are distributed throughout the system [10].

This work focused on the electrical characterization of a polycrystalline silicon solar panel under outdoor conditions to compare the practical results with the performance given by the manufacturer and to determine and analyze the temperature effect on the efficiency of PV cells using thermography.

2   PV Cell Modelling

The mathematical model associated with a cell is deduced from that of a PN junction. It consists of the sum of the PV current I_{ph} (which is proportional to the illumination), and a term modeling the internal phenomena. The electrical equivalent circuit is depicted in Fig. 1. The current I in the output of the cell is then [11]:

$$I = I_{ph} - I_s \cdot \left(e^{\frac{q(U - R_s I)}{KT}} - 1 \right) - \frac{U + R_s \cdot I}{R_{sh}} \qquad (1)$$

I_{ph}: photocurrent, or current generated by the illumination (A)
I_s: saturation current of the diode (A) (about 100 nA)
R_s: serial resistor (Ω)
R_{sh}: shunt resistor (Ω)
k: Boltzmann constant, k= $8.62.10^{-5}$
q: charge of the electron e: $1.602.10^{-19}$
T: cell temperature (K)
   The solution of equation (1) is reported in [12].

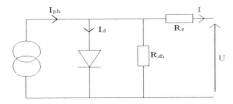

Fig. 1. Equivalent circuit of a solar cell

3 Infrared Thermography

Infrared thermography is the process of acquisition and analysis of the emitted radiation without direct contact with an object and converting the data to an image. All bodies emit infrared radiation when their temperature is higher than 0K. The determination of the panel temperature is obtained by using an infrared camera "Irisys 4000" which has a

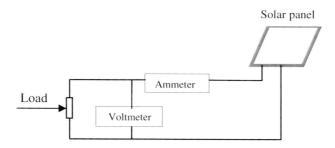

Fig. 2. PV schematic connection for I-V measurements

Fig. 3. Outdoor experimental setup

20° × 15° field of view lens whilst and a 160 × 120 (19200) pixel detector and provided with its own software. Among the options of this camera recording thermographs, is the possibility to measure the temperature at each point of the panel. The operating principle is based on the collection of the infrared radiation emitted by the solar panel. The thermal images stored on the supplied memory card can be transferred to a PC, and a software is supplied to display and analyse the recorded thermal images.

4   Experimental Setup

Using a multimeter, the short circuit current (Isc) is measured together with the open circuit voltage (Voc). The maximum current (Imax) and the maximum voltage generated by the panel (Vmax) are also measured. A variable resistor (rheostat) from 0 to 100Ω, considered as a load resistor is used, the variation of this resistor implies a variation of the values of Imax and Vmax. The infrared camera aim is to follow and to observe the cells temperature variation. Figs. 2 and 3 illustrate the PV schematic connection and experimental setup used to measure the I-V characteristics.

As depicted in Fig. 4, the thermograph camera is 1 meter distant from the PV panel in order to clearly distinguish each PV cell separately. The inclination is chosen in order to avoid the insolation reflection.

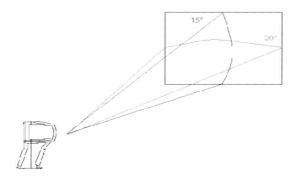

Fig. 4. Thermograph camera use

5   Results and Discussion

The outdoor tests have started in May 2010 in the Setif region (Algeria) where the maximum illumination during the day (at 12h30) is approximately (1000W/m^2). Fig. 5 shows the change in the insolation of a typical half day in May.

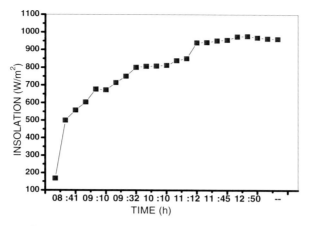

Fig. 5. The variation of insolation as a function of time

To plot the I-V characteristics and to ensure that the ambient conditions remain the same for a short time (time change implies a change in insolation and junction temperature), measurements of Imax and Vmax were taken in the same day at 12h30 pm with an ambient temperature of 28 °C and a panel temperature of 50 °C measured by an infrared camera. The simulation data of the I-V characteristics, performed using Matlab and the manufacturer data, are shown in Fig. 6 together with the measured characteristics.

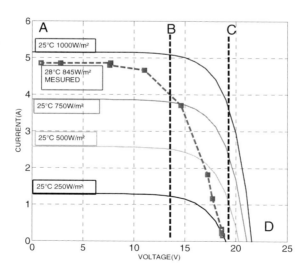

Fig. 6. Experimental and simulated I-V characteristics

880   Z. Fares et al.

According to the I-V characteristics simulated with the values given by the PV manufacturer and experimental measurements:

- The I-V curves have almost the same shape. According to the operating zones of the panel, the difference between measured and simulated I-V characteristics are divided into three zones:

  The first zone [AB]: the panel behaves like a current generator, i.e. the current is relatively constant and the maximum voltage is variable. It can be noted that the maximum value of the short circuit current Isc in simulation is higher_ than the measured one. Indeed, the short circuit current given by the manufacturer is in the range of 5.15A under STC. On the other hand, the real test conditions are not the same, the measured Isc cannot reach 5.15A but is about 4.84 A. The ambient temperature was 28 °C, the insolation was 845 W/m^2 and the panel temperature was 50 °C

- The second zone [CD]: the panel behaves like a voltage generator, that is to say the current is low and the voltage is relatively constant. The difference between the simulated and experimental voltages is below 2.9 V (18.7 V measured and 21.6 V simulated). This difference is due to the increase in the panel temperature that is about 50 °C (measured).

- The third zone [BC]: the functioning panel is optimal, i.e. the power is maximum. It is observed that the slope of the two curves (measured and simulated) is not the same. Indeed, the starting point of this area of the simulated I-V characteristics is about 16 V. On the other hand, in the measured characteristics, the starting point is only at 10 V. Therefore, the simulated value is greater than the measured one. Concerning the power-voltage characteristics presented in Fig. 7, the difference between the two curves is significant. In fact, there is a power loss of about 20 W (80 W simulated and 60 W measured). This loss is due to the increase of the PV temperature (a voltage decreased is observed: 21.6 V simulated and 18.7 V measured). The power decrease is proportional to the voltage decrease. In addition, the insolation is 845W/m^2 and the ambient temperature is 31°C.

To determine the temperature effect on the panel, the PV characteristic was simulated using Matlab for different temperatures (25 °C, 50 °C, 75 °C and 100 °C) [12]. Fig. 7 shows the simulation results of the temperature effect on the power characteristics. There is an increase of the photocurrent in particular because of the decrease of the band gap. This increase is of 2.7 mA/°C, i.e. a relative variation of 0.053%/°C. At the same time, there is a net decrease in the open circuit voltage (approximately -75.6 mV/°C), that is a relative variation of -0.35%/°C. The increase in temperature resulting in a reduction of the output power of -0.36W/°C that is a relative variation of -0.4%/°C.

Similarly, the variation in power as a function of the junction temperature is shown in Fig. 8. According to these simulated curves, when the temperature increases from 25 °C to 100 °C, the power decreases by about 50%, that is to say from 80 W to 45W. It is noted that there is an inverse relationship between the temperature and the power as increased temperature leads to decreased voltage and power.

78 Infrared Thermography Study of the Temperature Effect on the Performance

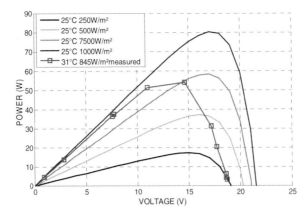

Fig. 7. Measured and simulated power-voltage characteristics

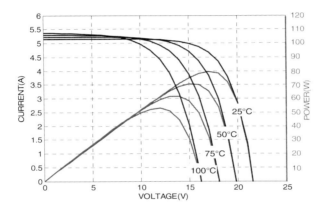

Fig. 8. The temperature effect on the I-V, power-voltage characteristics

According to the curves of Figs. 9 and 10, the PV panel power and temperature are proportional to the insolation.

The increase in panel temperature is due to:

- The increase of insolation received at the panel surface.
- The part of the solar spectrum that penetrates the junction and that is not converted into electricity.

As shown in these Figs., the increase in panel temperature produces a decrease in the power output, the panel temperature reached 50.5 °C and the power cannot exceed 62.20 W, which implies that an increase of 25 °C in temperature produces a power decrease of about 18 W. So the increase in temperature affects the panel yield.

The open circuit voltage will decrease with increasing temperature, in contrast to the short circuit current.

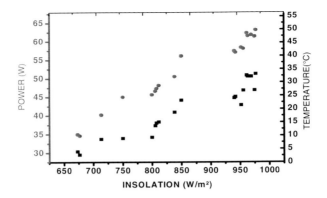

Fig. 9. The power and temperature variation as a function of insolation

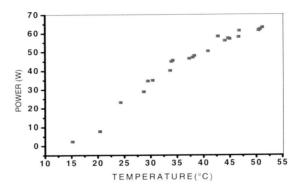

Fig. 10. The variation of the power as a function of temperature

The aim of this test is to determine the influence of temperature on the performance of the PV panel by measuring the panel characteristics (current, voltage and power). The measurement of the panel temperature (cells) is done by thermography. The experimental procedure consists of changing the panel temperature by using a cooling water jet.

The test is performed within a very short period of time to avoid the effect of variation of insolation and therefore the change in the PV efficiency η:

$$\eta = \frac{P_m}{P_{absorbed}}$$

where P_m is the maximum power and $P_{absorbed}$ is the power absorbed by the PV cells.

According to Fig. 11, it is noted that when the panel temperature increases (from 31 °C to 49.6 °C), the voltage decreases and the current remains almost constant (a slight increase). At the same time, the power decreases, and it is concluded that the panel temperature affects the performance negatively.

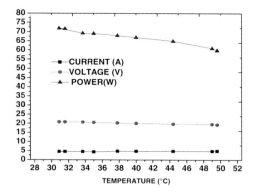

Fig. 11. Effect of temperature on the current, voltage and power output

Table 1 shows the variation of efficiency as a function of temperature.

Table 1. Panel efficiency and junction cells temperature

Temperature (°C)	30	35	40	45	50
Efficiency (%)	10.9	10.3	10.1	9.6	8.9

The efficiency of the panel decreases when the cells temperature increases. The thermographs in Fig. 12 show the variation of the temperature of the PV panel exposed to the sun from 9:18 am to 10:49 am. These thermographs are recorded by IR camera one after the other with a variable time interval.

The changes in the panel temperature according to the thermograph may be explained as follows:

During the exposure of the solar panel to the sun, the panel temperature increases with increasing cell temperature.

According to the results, there is a range of critical temperature of the panel (in this case from 37 °C to 45 °C). At these temperatures the operation of the panel is bad because there is no uniformity of temperature in the panel (the white cell temperature attained 50.2 °C). This means that the cells do not generate the same power and therefore the panel cannot generate a maximum power. This is due to the fact that the cells that generate less power act as resistance and the wind is playing a very important role. It can be seen that the cells on the left side are cooler that those on the right side as the wind direction at the time of the experiment was from left to right.

Fig. 13 shows the temperature distribution in the thermograph, taken according to the evolutionary change of temperature with time. The hot spot (cursor 2) characterized by a higher temperature compared to the average is due to the presence of the connection box on the back of the panel. This limits the cooling by convection and therefore the corresponding surface is warmer.

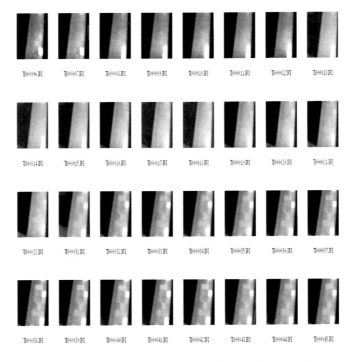

Fig. 12. Temperature of the panel surface by thermography

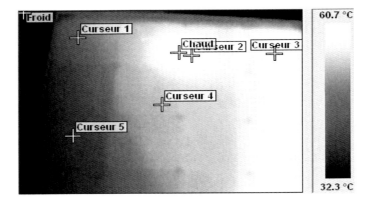

Fig. 13. The PV panel surface temperature distribution

This technique can be used in fault detection as a fault diagnosis or prognosis method. In fact, detecting the hot spot (or the hot cell) can help cool or change the right cell. Fault possibilities for PV panels are reported in [13].

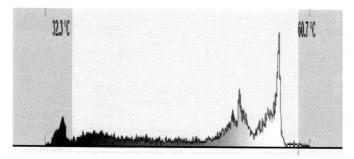

Fig. 14. The temperature density histograms

Fig. 14 is the histogram of the temperature density of Fig. 13, showing that the measured points are within a wide range.

6 Conclusions

This work shows that infrared analysis can be of great importance when used for the efficiency analysis of PV panels as it depends strongly on temperature. We studied the influence of temperature on the PV panel performance and therefore on its characteristics (i.e. power, current and voltage). The tests were performed on a standard PV panel with a rated output power of 80 W. The temperature was measured by thermographs recorded by an infrared camera. The electrical characteristics were measured under different insolation, time periods and temperature conditions.

The simulation, using Matlab, was implemented by using the panel parameters given by the manufacturer in order to compare the simulated values and the experimental ones.

From experimental and simulation, it is concluded that:

The panel temperature and the power are related to the insolation that varies gradually as the sun moves over the day hours. By observing the thermographs, the PV cells of the solar panel do not generate the same power throughout the panel and the temperature in the panel tends to stabilize over time. The measured temperature from 31 °C to 49.6 °C has an effect on the voltage, where the current remains almost constant and the power tends to decrease. Therefore, the variation of the temperature causes a variation of the PV panel efficiency and on the amount of the generated energy. The proposed technique can be of a great help in the PV cell diagnosis or prognosis by locating the hot spot (or the hot cell) and helps to remedy to the power decrease by cooling or changing the right cell.

References

1. Breitenstein, O., Rakotoniaina, J.P., Al Rifai, M.H., Werner, M.: Shunt type in crystalline solar cells. Progress in Photovoltaics Research and Application 12, 532 (2004)
2. Breitenstein, O., Langenkamp, M., Lang, O.: A. Shunts due to laser scribing of solar cell evaluated by highly sensitive lock-in thermography. Solar Energy Materials and Solar Cells, 55–6 (2001)

3. Vergura, S., Acciani, G., Falcone, O.: Modeling defects of PV-cells by means of FEM. In: IEEE-ICCEP 2009, 9-11/06/2009, Capri, Italy, pp. 52–56 ISBN 978-1-4244-2544-0
4. Petrone, G., Spagnuolo, G., Teodorescu, R., Veerachary, M., Vitelli, M.: Reliability Issues in Photovoltaic Power Processing Systems. IEEE-Trans. on Industrial Electronics 55(7), 2569–2580 (2008)
5. Ibrahim, A.: Analysis of Electrical Characteristics of Photovoltaic Single Crystal Silicon Solar Cells at Outdoor Measurements. Smart Grid and Renewable Energy 2, 169–175 (2011), doi:10.4236/sgre.2011.22020
6. Van Dyk, E.E., Meyer, E.L.: Analysis of the Effect of Parasitic Resistances on the Performance of Photovotaic Modules. Renewable Energy 29(3), 333–344 (2004), doi:10.1016/S0960-1481(03)00250-7
7. Meyer, E.L.: Assessing the Reliability and Degradation of Photovoltaic Module Performance Parameters. IEEE Transactions on Reliability 53(1), 83–92 (2004), doi:10.1109/TR.2004.824831
8. Ibrahim, A.: Effect of Shadow and Dust on the Perform-ance of Silicon Solar Cell. Journal of Basic and Applied Sciences Research 1(3), 222–230 (2011) ISSN 2090-424X
9. Ibrahim, A.: LBIC Measurements Scan as a Diagnostic Tool for Silicon Solar Cell. Journal of Applied Sciences Research 1(3), 215–221 (2011) ISSN 2090-424X
10. Bower, W., Wiles, J.: Photovoltaics Syst. Applications, Sandia Nat.Labs, Albuquerque, NM, Photovoltaic Specialists. In: Conference Record of the Twenty-Eighth IEEE, Anchorage, AK, USA, September 15-22, pp. 1378–1383 (2000)
11. Becherif, M., Ayad, M.Y., Henni, A., Aboubou, A.: Hybridization of Solar Panel and Batteries for Street lighting by Passivity Based Control. In: IEEE-ENERGYCON 2010, Kingdom of Bahrain (2010)
12. Walker, G.R.: Evaluating MPPT converter topologies using a MATLAB PVmodel. In: Australasian Universities Power Engineering Conference, AUPEC 2000, Brisbane (2000)
13. Ancuta, F., Cepisca, C.: Fault Analysis Possibilities for PV Panels" (IYCE). In: Proceedings of the 2011 3rd International Youth Conference on Energetics, Leira Portugal, July 7-9 (2011)

Chapter 79
Integrating Solar Heating and PV Cooling into the Building Envelope

Sleiman Farah, Wasim Saman, and Martin Belusko

Barbara Hardy Institute, School of Advanced Manufacturing & Mechanical Engineering,
University of South Australia, Mawson Lakes, South Australia 5095, Australia

Abstract. Photovoltaic/thermal (PVT) systems generate electrical and thermal energy. In summer, the usage of the collected heat is limited to domestic hot water heating. By contrast in winter, more useful heat collection is favorable, however, the PVT collectors require less cooling; therefore, the improved electrical output is limited. In this paper a new one-dimensional steady-state building integrated solar collector model is presented and examined, incorporating PVT and thermal (PVTT) collectors connected in series. In summer, the PVT collector is air-cooled, and the collected heat is discarded to the surroundings while the thermal collector heats the water for domestic use. In winter, both the PVT and thermal collectors are water-cooled generating domestic hot water. The efficiencies of the new collector are compared to that of a PVT collector, with both collectors having the same total area and characteristics. Both collectors are able to meet the summer thermal load and to provide useful thermal energy in winter. The PVTT collector reduces the collector thermal stresses and provides slight additional electrical power output.

List of Symbols and Abbreviations

$(\tau\alpha)$	transmittance-absorptance product
A_c	collector area (m^2)
C_p	specific heat (J/kg.°C)
F'	collector efficiency factor
F_R	heat removal factor
h_c	convective heat transfer coefficient between the absorber and glazing ($W/m^2.°C$)
h_{cd}	conductive heat transfer coefficient ($W/m^2.°C$)
h_r	radiative heat transfer coefficient between absorber and glazing
h_s	radiative heat transfer coefficient to sky
h_w	wind convective heat transfer coefficient ($W/m^2.°C$)
I	incident solar radiation (W/m^2)
k	conductivity (W/m.°C)
m	mass flow rate (kg/s)
M	air mass modifier
PV	photovoltaic

A. Håkansson et al. (Eds.): *Sustainability in Energy and Buildings*, SIST 22, pp. 887–901.
DOI: 10.1007/978-3-642-36645-1_79   © Springer-Verlag Berlin Heidelberg 2013

PVT	photovoltaic/thermal
PVTT	photovoltaic/thermal - thermal
Q_u	useful thermal energy (W)
Ra	Rayleigh number
Ra"	modified Rayleigh number
R_b	ration of beam radiation on tilted surface to that on horizontal surface
r	ratio of the top to the bottom heat flux
S	absorbed solar energy (W)
St	Stanton number
T	temperature (°C)
T_r	photovoltaic cells reference operating temperature (°C)
Th	thermal
U	heat transfer coefficient (W/m^2.°C)
u	free stream air velocity (m/s)
U_L	total heat transfer coefficient to the ambient (W/m^2.°C)
β	collector slope (°)
δ	thickness (m)
ε	emissivity
η_r	photovoltaic cells electrical reference efficiency
η_{PV}	photovoltaic cells electrical efficiency
ξ	photovoltaic cells efficiency temperature coefficient
ρ	density (kg/m^3)
a	ambient air
ab	absorber
b	beam radiation
d	diffuse radiation
g	glazing
h	horizontal
i	inlet
ins	insulation
o	outlet
til	tilted

1   Introduction

Solar energy is a promising distributed source of sustainable energy being used increasingly in building applications. Typical residential utilisation of solar systems use separate thermal and photovoltaic (PV) systems which generate heat and electricity respectively and independently. A more efficient energy generation is achieved by the integration of the photovoltaic and thermal collectors into one photovoltaic/thermal (PVT) collector [1]. The enhanced energy generation from PVT collectors is attained by the cooling effect of the thermal collector which extracts heat from the photovoltaic cells. This heat extraction reduces the operating temperature of the PV cells which normally have higher efficiencies at lower operating temperatures [2]. The PVT systems have additional features compared to the independent thermal and photovoltaic systems. The following features of PVT systems make them more suitable for residential applications:

- Higher total efficiency per unit area
- Higher electrical efficiency
- Better architectural integration
- Lower installation cost [3].

In order to maximize the performance of PVT systems, researchers have modelled, analyzed, and tested several systems using different configurations and different cooling fluids. In a comparative study, Zondag et al. [4] developed a dynamic 3D model for a PVT water system, and three steady state 3D, 2D, and 1D models. All the models were able to predict the experimental values with an error of 5%. The 1D model was able to evaluate the daily yield as accurately as the 3D dynamic model. However, the multi-dimensional models offered more flexibility, and detailed information for further improvement of the system.

Chow [5] developed a one-dimensional transient model for a glazed flat-plate PVT water system to analyse the effect of irradiation fluctuation. The model was able to calculate the instantaneous temperatures for the different components of the model as well as the electrical and thermal efficiencies. While some researchers developed dynamic and transient models, the literature is full of successful research based on steady state models as found by Daghigh et al. [6], Kumar and Rosen [7], Anderson et al. [8,9] and Solanki et al. [10], where the models predicted the performance of the systems with relatively high accuracy.

Tiwari and Sodha [11] studied a PVT water system. They concluded that the variation of the water flow rate had a slight effect on the temperature of the water in the system, and that the increase in water temperature with the increase of the length of the collector was not significant after 4 m. In addition, they studied four different configurations of PVT air/water collectors, glazed with tedlar, unglazed with tedlar, glazed without tedlar, and unglazed without tedlar. They concluded that the collectors using water performed better than those using air except for the glazed without tedlar system [12].

Tripanagnostopoulos et al. [13] compared water and air as cooling fluids for glazed and unglazed collectors. From the experimental results, they concluded that water is better than air for cooling for all the studied collectors. The glazing cover had a beneficial effect on the thermal efficiency up to about 30% with a decrease of the electrical efficiency by about 16%. This effect makes the unglazed collectors preferable when electricity production is a priority. They also concluded that electrical efficiency increased with the increase of thermal efficiency.

Zondag et al. [3] conducted a numerical analysis on different PVT collectors, and showed that PVT collectors performed better than two separate electrical and thermal collectors did under the same conditions, and for the same collector area. In addition, the "one-cover sheet and tube" collector was the second best, with just 2% efficiency reduction in comparison with "channel-below-transparent-PV". This difference of performance is made up for by the simplicity of construction of the "One cover sheet and tube" collector making it the most promising one.

Anderson et al. [8] studied a building integrated PVT collector (BIPVT) made of low cost pre-coated steel. They emphasized the need to maximize the ratio of the cooling channel width to the spacing between channels, and enhance the heat transfer between the PV cells and the absorber by applying thermally conductive adhesives. In addition, they mentioned the possibility of doing without the insulation at the rear of the BIPVT collector, since the low natural convective heat transfer in the roof attic would act as a heat barrier. The usage of the attic as heat barrier decreases the cost of the system by reducing the material and installation costs. These cost reductions by integration minimize the payback period of the system.

Chow et al. [14] simulated a facade BIPVT water system using numerical models that have been experimentally verified. They concluded that the natural water circulation system was thermally 5% more efficient than the forced water circulation system. Similarly and significantly, the electrical efficiency of the natural circulation system was 43% higher than that of the forced circulation system by saving the power consumed by the circulating pump. Both, natural and forced circulation systems were found to be significantly more economical than the conventional PV system, and they reduced the heat transmission through that facade by about 70%. The cost study showed that the payback period of the natural circulation system was about 14 years.

Dubey and Tiwari [15,16] studied different series and parallel arrangements of PVT collectors which were partially and fully covered by PV cells. They evaluated the annual thermal and electrical performances of these different arrangements. Their studies showed that the partially and fully covered collectors are recommended when the main objective is the production of both hot water and electricity.

2 Design Concept

The presented building integrated collector incorporates an unglazed PVT collector and a glazed thermal collector, connected in series to form one photovoltaic thermal/thermal (PVTT) collector (see Fig. 1). The PVTT collector is cooled by one fluid (water) or simultaneously by two fluids (air and water). The cooling fluid tube or channel is bounded by stiffened corrugated roof metal sheet and the absorber. The roof metal sheets are available in different profiles and they are typically made from aluminum or coated steel. These profiles can be easily modified to allow for the installation of the absorber over a trough space, and the installation of the PV laminate and the glazing of the PVT and thermal sections respectively (see Figs. 2, 3). The absorber is made from polymers that are thermally conductive with an extended surface to improve the heat transfer to the cooling fluid, and electrically nonconductive to avoid electrical short-circuiting. The back of the fluid channel is thermally insulated to reduce the collector thermal heat losses.

Under solar radiation, the PV cells of the PVT section convert the absorbed radiation to electricity at relatively low efficiency, and the remaining absorbed energy is converted to heat [6]. Simultaneously, the thermal section converts the solar radiation to heat [17]. The collected heat from both sections is transferred then to the cooling fluid.

In summer, the PVT collector absorbs an excess of thermal energy, which increases extensively the collector temperature and therefore deteriorates the electrical output of the PV cells. The temperature increase would be more significant in a closed domestic hot water system, especially as less hot water withdrawal is required. In order to avoid the high PV cells operating temperatures, the PVTT collector uses both air and water as cooling fluids. The PVT section is cooled by forced convection with air to maintain the PV cells at lower operating temperatures. The air collected thermal energy is discarded to the ambient surroundings to avoid the temperature increase of the collector PVT section. Water is heated for domestic use in the thermal section without the need of preheating in the PVT section.

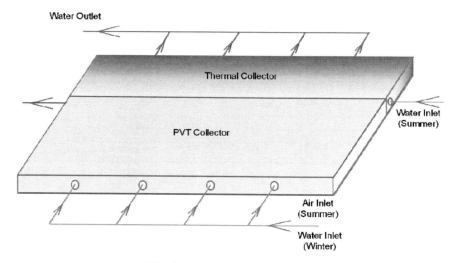

Fig. 1. Schematic PVTT Collector

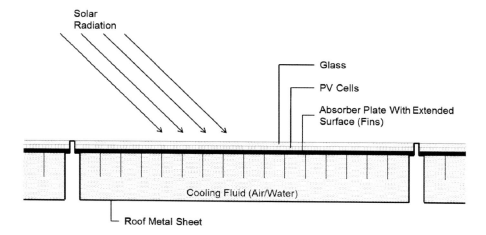

Fig. 2. Schematic of the PVT Cross-Section of the PVTT Collector

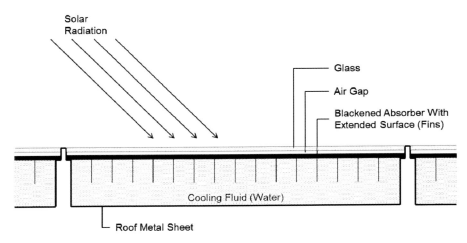

Fig. 3. Schematic of the Thermal Cross-Section of the PVTT Collector

In winter, the PVTT collector is water cooled in both the PVT and thermal sections. The PVT section transfers the absorbed heat to the water flowing through the collector. This heat transfer increases the temperature of the water, decreases the temperature of the PV cells, and consequently increases the electrical output. The water leaving the PVT section enters the thermal section as preheated water where it is heated up to a higher temperature by receiving an additional amount of thermal energy.

3   Collector Model

To analyse the electrical and thermal performances of the PVTT collector, a mathematical model is developed with the following major assumptions:

- Heat flow is one-dimensional
- Performance is steady state
- Collector thermal capacity is negligible
- Edge losses are negligible
- Ohmic losses are negligible.

The modeling of the PVTT collector is based on Hottel and Whillier thermal energy balance Eq. [18], with a modification in the modeling of the PVT section to account for the PV cells, where a part of the absorbed radiation is converted to electricity and the remaining radiation is converted to heat. The thermal energy balance of the PVT and thermal sections is

$$Q_{u,PVT} = A_{c,PVT} F_{R,PVT} [(1-\eta_{PV})S_{PVT} - U_{L,PVT}(T_{i,PVT} - T_a)] \quad (1)$$

and

$$Q_{u,Th} = A_{c,Th} F_{R,Th} [S_{Th} - U_{L,Th}(T_{i,Th} - T_a)] \quad (2)$$

with the calculations of F_R, S, η_{PV} and U_L being detailed in appendix A.

The total collector useful thermal output depends on the cooling fluids. In summer, as the PV section is air-cooled, the total collector useful output consists of the thermal output of the water-cooled section only, since the thermal energy carried by air is discarded to the surroundings. In winter, both collector sections are water-cooled, and the total collector useful thermal output is the sum of the PVT and thermal sections outputs. To facilitate performance comparison, both the thermal and electrical efficiencies are based on the total collector area with

$$\eta_{Th} = \frac{Q_u}{A_c I_{til}} \qquad (3)$$

and

$$\eta_{Electrical} = \frac{A_{PV} S_{PV} \eta_{PV}}{A_c I_{til}}. \qquad (4)$$

Table 1. PVTT Collector Parameters

	Summer Values	Winter Values	Units
A_c	24	24	m^2
Collector Width	6	6	m
PVT Section Length	3.670	3.670	m
Thermal Section Length	0.330	0.330	m
F'_{PVT}	0.729	0.912	
F'_{Th}	0.944	0.959	
$F_{R,PVT}$	0.564	0.856	
$F_{R,Th}$	0.903	0.956	
S_{PVT}	738.4	473.1	W/m^2
S_{Th}	625.6	387.9	W/m^2
u	3	3	m/s
$U_{L,PVT}$	14.825	12.809	$W/m^2.°C$
$U_{L,Th}$	7.899	6.432	$W/m^2.°C$
T_a	29.1	15.7	°C
η_r	0.180	0.180	
I_{til}	800	500	W/m^2
k_{ins}	0.050	0.050	$W/m.°C$
k_g	0.800	0.800	$W/m.°C$
m	0.020/0.020 (air/water)	0.020 (water)	$kg/s.m^2$
M	0.983	1.008	
β	20	20	°
δ_{ins}	0.050	0.050	m
δ_g	0.002	0.002	m
ε_g	0.880	0.880	m
ε_{ab}	0.880	0.880	m
ξ	0.0045	0.0045	$°C^{-1}$

The performance evaluation of the proposed system is illustrated using the collector parameters defined in Table 1. The selected values are based on typical domestic system characteristics and climate data for summer and winter encountered in Adelaide, Australia.

4   Results and Discussion

The performance of the PVT collector is evaluated for three different cases in summer and for two cases in winter. The difference of the studied cases between summer and winter is discussed in Section 4.2. A comparison of the summer and winter performances of the full length PVT collector are compared to that of the PVTT collector, taking into account the electrical consumptions of the fan (60 W) and pump (34 W). These power consumption values are estimates based on anticipated pressure drops using commercially available fan and pump data.

4.1   Summer Performance

The summer performance results for the cases being considered are summarised in Table 2 and Fig. 4. The first case to be considered is when the PVT collector water inlet is from the cold water mains. In this case, the collector inlet temperature is taken as 20 °C, which is the lowest average collector water inlet temperature, and consequently the system has the highest electrical and thermal performances. However, the water low outlet temperature (27.2 °C) makes the use of the extracted thermal energy limited if not useless, and more than 1.7 m^3/h of valuable water has to be discarded just to cool the collector.

The second case is when the PVT collector water inlet temperature is 40 °C, which is the arithmetical average of the cold water mains temperature (20 °C) and the typical domestic hot water tank set temperature (60 °C) to comply with the Australian standards [19]. For night time electrically heated systems, the collector inlet temperature starts at (60 °C) and gradually decreases with hot water withdrawals from the hot water tank. When all the available thermal energy is used, the water in the tank will be totally replaced by the cold water mains (20 °C). Consequently, this temperature (40 °C) represents the average inlet temperature. In this case, the collector thermal and electrical efficiencies are lower than the collector efficiencies in the first case, however, they represent the highest efficiencies of an actual domestic application where no water is discarded.

With this vast collector area, it is anticipated that the PVT collector will operate at near stagnation temperatures in summer. The third case is based on the anticipated summer temperature of a system supplied by a collector area of 24 m^2. In this case, the collector water inlet temperature is estimated as the average between two limits of 40 °C, when all the hot water is domestically used (case 2) and the collector stagnation temperature (82.8 °C) when no hot water is used. To estimate this average temperature, the ratio of the daily thermal energy load to the daily average solar radiation on 24 m^2 collector area is used as the interpolation point between the 40 °C and 82.8 °C. The Adelaide daily average global solar radiation is 26.7 MJ/m^2 [20], while the maximum daily thermal energy load is 32.2 MJ [21]. For the 24 m^2 PVT collector, the ratio of the daily thermal energy load to the daily average horizontal solar radiation is 0.05%. Therefore, the estimated actual average inlet temperature is 80.6 °C. At this inlet temperature, the PVT collector thermal and electrical efficiencies are reduced to 2.6% and 12.3% respectively.

Table 2. Performances of water-cooled PVT and air/water-cooled PVTT collectors in summer

	Inlet Water Temp. (°C)	Outlet Water Temp. (°C)	Thermal Eff. %	Thermal Energy Output (W)	Electrical Eff. %	Electrical Power Output (W)
PVT Case 1	20.0	27.2	75.2	14438	18.6	3579
PVT Case 2	40.0	44.5	46.7	8958	14.9	2869
PVT Case 3	80.6	80.8	2.6	497	12.3	2354
PVTT	40.0	45.8	5.0	965	13.2	2539
PVTT*	40.0	45.8	5.0	963	11.3	2164

* Natural Ventilation

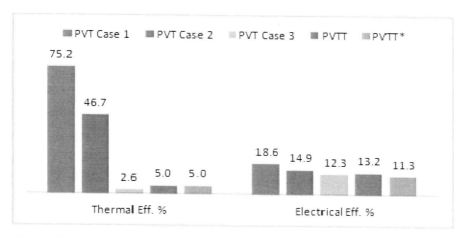

Fig. 4. Efficiencies of water-cooled PVT and air/water-cooled PVTT collectors in summer

Considering the case when the PVTT collector is cooled by air and water simultaneously, the air inlet temperature of the PVT section is 29.1 °C, which is the average of the mean maximum temperature for the months of December, January and February for the city of Adelaide [20]. The water inlet temperature of the thermal section is the average water inlet temperature when all the hot water is domestically used (40 °C). At these inlet temperatures, the thermal and electrical efficiencies are 5.0% and 13.2% respectively. The thermal efficiency of the PVTT collector is higher than that of the PVT collector due to the lower inlet temperature, however, the higher inlet temperature of the PVT collector indicates that the thermal potential of the PVT collector is higher than that of the PVTT collector. This potential is obvious in the difference of thermal efficiencies between the PVT (case 2) and PVTT collectors at the same inlet temperature (40 °C). The thermal section of the 24 m^2 PVTT collector provides 32.7 MJ of thermal energy, which is sufficient for providing the daily hot water thermal load (32.2 MJ). The higher electrical efficiency of the PVTT collector (13.2%) compared to that of the PVT collector (12.3%) provides 185 W of additional electrical power, with the fan and pump power consumptions being taken into account. The naturally ventilated PVTT collector is less efficient than both the water-cooled PVT and forced ventilated PVTT collectors.

4.2 Winter Performance

The winter performance results for the cases being considered are summarised in Table 3 and Fig. 5. The first case is similar to that in summer, when the PVT collector water inlet is from the cold water mains. In this case, the water outlet temperature (14.3 °C) is low and unsuitable for domestic use. This case is therefore associated with wasting valuable water at a rate of 1.7 m^3/h for cell cooling.

The second case is also similar to that in summer, when all the thermal energy is domestically used, and the PVT collector water inlet temperature (34.5 °C) is the arithmetical average of the cold water mains temperature (9 °C) and the typical domestic hot water tank set temperature (60 °C). At this inlet temperature, the PVT collector thermal and electrical efficiencies are 26.8% and 15.9% respectively. Unlike case 3 in summer, all the collected thermal energy in winter is useful, therefore, these efficiencies are the collector winter average thermal and electrical efficiencies, and they are to be compared to the efficiencies of the PVTT collector.

The PVTT collector is water-cooled in both sections. The water inlet temperature of the PVT section is the average water inlet temperature when all the hot water is domestically used (34.5 °C), and the inlet temperature of the thermal section is the outlet temperature of the PVT section. At these inlet temperatures, the thermal and electrical efficiencies are 27.9% and 14.6% respectively. Unlike summer, the thermal efficiency of the PVTT collector (27.9%) is higher than that of the PVT collector (26.8%), with thermal power outputs equal to 3349 W and 3211 W respectively. The PVTT collector electrical efficiency (14.6%) is lower than that of the PVT collector (15.9%) with 159 W of electrical power output difference.

Table 3. Performances of water-cooled PVT and PVTT collectors in winter

	Inlet Water Temp. (°C)	Outlet Water Temp. (°C)	Thermal Eff. %	Thermal Energy Output (W)	Electrical Eff. %	Electrical Energy Output (W)
PVT Case 1	9.0	14.3	88.2	10587	24.0	2881
PVT Case 2	34.5	36.1	26.8	3211	15.9	1907
PVTT	34.5	36.2	27.9	3349	14.6	1748

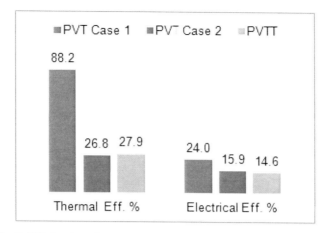

Fig. 5. Efficiencies of water-cooled PVT and PVTT collectors in winter

These results demonstrate that in summer, both collector arrangements provide the required domestic thermal load. In winter, the thermal efficiency of the PVTT collector (27.9%) is higher than that of the PVT collector (26.8%), when the thermal energy outputs of both collectors are totally used in the cold season. In addition, the PVTT collector electrical efficiency (13.2%) in summer is higher than that of the PVT collector (12.3%). Contrarily to the summer performances, in winter, the PVT collector electrical efficiency (15.9%) is higher than that of the PVTT collector (14.6%). Based on these results, the PVTT collector electrical power output improvement in summer is 185 W, while the winter electrical output decline is 159 W. Additional electrical output improvement could be expected in an annual hourly analysis, for the reason that the summer solar radiation is more available than that of winter.

5 Conclusion

This paper presents a new building integrated solar collector design concept (PVTT) and one-dimensional steady-state analysis of the proposed collector. The collector is cooled by air and water in summer and by water in winter. The change of the cooling fluid maximizes the electrical efficiency, without compromising the thermal output to meet the required domestic thermal load. The electrical and thermal efficiencies of the PVTT collector are calculated, and compared to the electrical and thermal efficiencies of PVT collector. For the same total collector area, and for the same PVT section characteristics, the PVTT collector is found to be better than the PVT collector by reducing the summer operating temperature. This temperature reduction avoids the collector thermal stresses and provides modest electrical power output improvement, however, this small improvement is associated with a more complex collector.

The results of this research are based on a steady state analysis using typical values of the different parameters. Further research is under way to continue this investigation in four main areas:

- Detailed hourly annual transient performance analysis
- Experimental validation of the PVTT model performance
- Identifying the optimal dates to switch the PV cooling mode from air to water and vice versa
- Cost analysis in comparison with separate collectors (PV panels and PVT collectors).

Appendix A. Collector Parameters Calculation

This appendix provides the calculations of the PVTT collector parameters present in Eqs. (1) and (2). From Duffie and Beckman [18] the heat removal factor (F_R) is

$$F_R = \frac{mC_P(T_o - T_i)}{A_c[S - U_L(T_i - T_a)]} \tag{5}$$

and the outlet temperature (T_o) in Eq. (3) is

$$T_o = T_a + \frac{S}{U_L} + \left(T_i - T_a - \frac{S}{U_L}\right)e^{\frac{-A_c U_L F'}{mC_P}}. \tag{6}$$

The absorbed solar radiation calculations of the PVT and thermal sections (S_{PVT} and S_{Th}) are based on the isotropic diffuse sky model, with negligible ground reflected radiation:

$$S_{PVT} = M\left[I_b R_b(\tau\alpha)_{b,PVT} + I_d(\tau\alpha)_{d,PVT}\left(\frac{1+\cos\beta}{2}\right)\right] \tag{7}$$

$$S_{Th} = I_b R_b(\tau\alpha)_{b,Th} + I_d(\tau\alpha)_{d,Th}\left(\frac{1+\cos\beta}{2}\right). \tag{8}$$

The PV cells electrical efficiency (η_{PV}) at the temperature (T_{PV}) is

$$\eta_{PV} = \eta_r\left[1 - \xi(T_{PV} - T_r)\right]. \tag{9}$$

The total heat loss coefficient (U_L) is the sum of the top and bottom heat losses coefficients (U_{top}) and (U_{bottom}). The top heat losses coefficients of the PVT and thermal sections are

$$U_{top,PVT} = \frac{h_{cd}(h_w + h_s)}{h_{cd} + h_w + h_s} \tag{10}$$

and

$$U_{top,Th} = \frac{(h_c + h_r)(h_w + h_s)}{h_c + h_r + h_w + h_s}$$ (11)

where

$$h_{cd} = \frac{k_g}{\delta_g},$$ (12)

$$h_w = St\rho_a C_{p,a} u_a,$$ (13)

$$h_s = \frac{\varepsilon_g \sigma(T_g^4 - T_s^4)}{T_{PV} - T_a},$$ (14)

$$h_c = \frac{k}{\delta_a} \left[1 + 1.44 \left(1 - \frac{1708(\sin(1.8\beta))^{1.6}}{Ra\cos\beta} \right) \left(1 - \frac{1708}{Ra\cos\beta} \right)^+ + \left(\left(\frac{Ra\cos\beta}{5830} \right)^{\frac{1}{3}} - 1 \right)^+ \right]$$ (15)

and

$$h_r = \frac{\sigma(T_{ab}^4 - T_g^4)}{\frac{1}{\varepsilon_{ab}} + \frac{1}{\varepsilon_g} - 1}.$$ (16)

For the bottom heat loss coefficient (U_{bottom}), the convective and radiative heat resistances are negligible compared to the conductive heat resistance through the back insulation, therefore, the bottom heat loss coefficient is

$$U_{bottom} = \frac{k_{ins}}{\delta_{ins}} \left(\frac{T_{ab} - T_{bottom}}{T_{ab} - T_a} \right).$$ (17)

The Nusselt number for the PVTT collector natural ventilation is adopted from the study of Mittelman et al. [22] and assumed to be valid at 20° collector inclination

$$N_u = \left[\frac{6.25(1+r)}{Ra''\sin\beta} + \frac{1.64}{(Ra''\sin\beta)^{\frac{2}{5}}} \right]^{-\frac{1}{2}}.$$ (18)

References

1. Chow, T.T., He, W., Ji, J., Chan, A.L.S.: Performance evaluation of photovoltaic-thermosyphon system for subtropical climate application. Solar Energy 81(1), 123–130 (2006)
2. Skoplaki, E., Palyvos, J.A.: On the temperature dependence of photovoltaic module electrical performance: A review of efficiency/power correlations. Solar Energy 83(5), 614–624 (2009), doi:10.1016/j.solener.2008.10.008
3. Zondag, H.A., de Vries, D.W., van Helden, W.G.J., van Zolingen, R.J.C., van Steenhoven, A.A.: The yield of different combined PV-thermal collector designs. Solar Energy 74(3), 253–269 (2003), doi:10.1016/s0038-092x(03)00121-x
4. Zondag, H.A., de Vries, D.W., van Helden, W.G.J., van Zolingen, R.J.C., van Steenhoven, A.A.: The thermal and electrical yield of a PV-thermal collector. Solar Energy 72(2), 113–128 (2002), doi:10.1016/s0038-092x(01)00094-9
5. Chow, T.T.: Performance analysis of photovoltaic-thermal collector by explicit dynamic model. Solar Energy 75(2), 143–152 (2003)
6. Daghigh, R., Ibrahim, A., Jin, G.L., Ruslan, M.H., Sopian, K.: Predicting the performance of amorphous and crystalline silicon based photovoltaic solar thermal collectors. Energy Conversion and Management 52(3), 1741–1747 (2011)
7. Kumar, R., Rosen, M.A.: Performance evaluation of a double pass PV/T solar air heater with and without fins. Applied Thermal Engineering 31(8), 1402–1410 (2011)
8. Anderson, T.N., Duke, M., Morrison, G.L., Carson, J.K.: Performance of a building integrated photovoltaic/thermal (BIPVT) solar collector. Solar Energy 83(4), 445–455 (2009), doi:10.1016/j.solener.2008.08.013
9. Anderson, T.N., Duke, M., Carson, J.K.: The effect of colour on the thermal performance of building integrated solar collectors. Solar Energy Materials and Solar Cells 94(2), 350–354 (2010), doi:10.1016/j.solmat.2009.10.012
10. Solanki, S.C., Dubey, S., Tiwari, A.: Indoor simulation and testing of photovoltaic thermal (PV/T) air collectors. Applied Energy 86(11), 2421–2428 (2009)
11. Tiwari, A., Sodha, M.S.: Performance evaluation of solar PV/T system: An experimental validation. Solar Energy 80(7), 751–759 (2006), doi:10.1016/j.solener.2005.07.006
12. Tiwari, A., Sodha, M.S.: Performance evaluation of hybrid PV/thermal water/air heating system: A parametric study. Renewable Energy 31(15), 2460–2474 (2006), doi:10.1016/j.renene.2005.12.002
13. Tripanagnostopoulos, Y., Nousia, T., Souliotis, M., Yianoulis, P.: Hybrid photovoltaic/thermal solar systems. Solar Energy 72(3), 217–234 (2002), doi:10.1016/s0038-092x(01)00096-2
14. Chow, T.T., Chan, A.L.S., Fong, K.F., Lin, Z., He, W., Ji, J.: Annual performance of building-integrated photovoltaic/water-heating system for warm climate application. Applied Energy 86(5), 689–696 (2009), doi:10.1016/j.apenergy.2008.09.014
15. Dubey, S., Tiwari, G.N.: Thermal modeling of a combined system of photovoltaic thermal (PV/T) solar water heater. Solar Energy 82(7), 602–612 (2008), doi:10.1016/j.solener.2008.02.005
16. Dubey, S., Tiwari, G.N.: Analysis of PV/T flat plate water collectors connected in series. Solar Energy 83(9), 1485–1498 (2009), doi:10.1016/j.solener.2009.04.002
17. Soteris, A.K.: Solar thermal collectors and applications. Progress in Energy and Combustion Science 30(3), 231–295 (2004), doi:10.1016/j.pecs.2004.02.001
18. Duffie, J.A., Beckman, W.: Solar engineering of thermal processes, 3rd edn. Wiley, Hoboken (2006)

19. Australian and New Zealand Standards, Plumbing and drainage Part 4: Heated water services. AS/NZS 3500.4:2003. SAI Global, Australia/New Zealand (2003)
20. Bureau of Meteorology, Climate statistics for Australian locations (2012), http://www.bom.gov.au/climate/averages/tables/cw_023037_All.shtml, http://www.bom.gov.au/climate/averages/tables/cw_023037_All.shtml (accessed June 2012)
21. Standard ANZ, Heated water systems-Calculation of energy consumption. AS/NZS 4234:2008. SAI Global, Australia/New Zealand (2008)
22. Mittelman, G., Alshare, A., Davidson, J.H.: Composite relation for laminar free convection in inclined channels with uniform heat flux boundaries. International Journal of Heat and Mass Transfer 52(21–22), 4689–4694 (2009), doi:10.1016/j.ijheatmasstransfer.2009.06.003

Chapter 80
Risk and Uncertainty in Sustainable Building Performance

Seyed Masoud Sajjadian, John Lewis, and Stephen Sharples

School of Architecture, University of Liverpool, Liverpool, UK

Abstract. Decision-making in the design of sustainable building envelopes will mostly consider the trade-off between initial cost and energy savings. However, this leads to an insufficiently holistic approach to the assessment of the sustainable performance of the building envelope. Moreover, the decisions that designers face are subject to uncertainties and risks with regards to design variations. This research examines a range of concepts and definitions of risk, uncertainty and sustainability in the context of climate, building construction and overheating. These concepts are then combined to objectify a range of risks and uncertainties affecting the decision. A simple computer model was used to analyze different building cladding constructions in terms of an overheating risk inside a building. The paper concludes by considering how the cladding materials may be chosen to optimize a model that will aid decision-making in design. The research suggests that none of the cladding systems would completely eliminate the risk of overheating for a range of climate change scenarios.

Keywords: Risk, Uncertainty, Climate Change, Overheating, Environment.

1 Introduction

For researchers in science and engineering the terms uncertainty and risk are explored and reviewed within a large number of academic articles and reports. Essentially, no general definitions are observed for these terms, although many constraints and context-dependent definitions exist. Almost every definition is problem-specific, implying that every time a decision problem is stated particular definitions for risk and uncertainty are presented for the decision problem. A consensus within these definitions, however, is that risk and uncertainty are frequently related. Every definition considered consists of three comprehensive areas - Economics and Finance, Operations Research, and Engineering (based on the affiliation of the author to the specific field). The challenge can be extended into a design problem by introducing design parameters in construction and demonstrating how climate change models uncertainty and quantifies risk in the availability of feasible options for the decision-maker.

2 Background

Increasingly, there is recognition that potential changes in UK climate are likely to have impacts on the built environment. Perhaps the most significant of these changes

---

A. Håkansson et al. (Eds.): *Sustainability in Energy and Buildings*, SIST 22, pp. 903–912.
DOI: 10.1007/978-3-642-36645-1_80   © Springer-Verlag Berlin Heidelberg 2013

concerns the influence of higher temperatures on thermal performance. Bill Dunster Architects and Arup R&D (2005) demonstrated the significance of mitigating climate change effects by designing homes with passive features to offset the expected increases in air temperatures. This research also identified that thermally lightweight homes would result in levels of discomfort by creating considerably higher room air temperatures. The study stated that masonry houses with inherent thermal mass can save more energy over their lifetime compared to a lightweight timber frame house.

The risk of overheating in highly insulated houses happens not only in the summer but also in other seasons. The risk of overheating exists as long as there is solar penetration into the building (Athienitis and Santamouris 2002). Orme et al. presented research work which illustrated that in a lightweight well-insulated house, external temperatures of 29°C may result in internal temperatures of more than 39°C (Orme, Palmer and Irving 2003).

Trying to calculate how a range of design variables will perform over time is fraught with uncertainty. Obviously, heating and cooling loads are influenced by the thermal properties of the building envelope, which are likely to be sensitive to future conditions. Therefore, the efficiency of decisions made due to the thermal characteristics of the built environment need to be considered in the light of climate change scenarios.

3   Sustainability and High Performance

Currently, in lean construction thinking, sustainability and high performance seem to be gaining significant momentum. The ASHRAE Standard 189.1 defines the high performance green building as a "building designed, constructed and capable of being operated in a manner that increases environmental performance and economic value over time". Consequently, the challenges observed from this definition are i) that it casts the problem as being one of definition; ii) in fact, it is more a question of prediction of which features will meet the criteria and of achieving consensus on which of those features would be deemed appropriate for inclusion; and iii) it does not account for the range of interrelated time and space scales. One of the essential challenges in the long term is uncertainty, and sustainability by any definition refers to the long term. The question arises of how to describe and manage sustainability under uncertainty. In this interaction the following must be answered:

- What are the factors that need to be sustained?
- At what level and for how long, should the factors last?
- What degree of uncertainty is acceptable?

Sustainability is about thoughtful choices, without spending more on non-essential options but with confidence of earning more return on investment. In a general sense, it is about dealing with nature – not ignoring it. Additionally, it is not about constructions that appear to be environmentally-responsible but which eventually sacrifice occupant comfort. It becomes clear therefore that "Sustainability in Buildings" is a multi-criteria subject, which includes interlinked parameters of economics, environmental issues, and social parameters (Vesilind, et al. 2006). Therefore, this paper tries to explore the interaction of each feature by using a set of criteria to optimize the thermal performance of a variety of construction types in dealing with uncertainty and risks.

4 Climate Change

Adaptation of buildings to climate change is becoming increasingly necessary. Adaptation, a responsive adjustment to decrease or remove risk, will be critically important since, in even the most optimistic projection of climate-change scenarios, temperatures will increase considerably around the world. It is very unlikely that the mean summer temperature increase will be less than 1.5°C by the year 2080 (IPCC 2010). Figures 1 and 2 illustrate a range of increasing temperatures in the UK in summer and winter with different probabilities. The 'worst-case scenario' is thought to be essential when considering change for construction types. The construction adapted for the extreme case should be the most robust design - a design that is durable in both the current climatic condition and in response to the maximum envisaged change in future climate.

Fig. 1. Summer mean temperature in 2020, 2050, 2080; 90% probability level, very unlikely to be less than the degrees shown on maps [Source:http://ukclimateprojections.defra.gov.uk/content/view/1293/499/]

Fig. 2. Winter mean temperature in 2020, 2050, 2080; 90% probability level, very unlikely to be less than degrees shown on maps [Source: http://ukclimateprojections.defra.gov.uk/content/view/1284/499/]

As can be seen from Figure 1, around an 8°C increase in the summer is 90% likely to happen in most of the UK. Obviously, the rate of increase would be less in winter but is still quite considerable. Revealing the potential impacts of climate change scenarios in dealing with construction types demonstrates the need for an optimization in the decision-making process to optimize both the thermal comfort of occupants and future energy consumption regardless of active design impacts. These are essential determinants when attempting to establish the vulnerability of occupants during a heat-wave, and the potential change in energy usage and CO_2 emissions are a consequence of changing climatic conditions. In essence, this approach causes effective and practical adaptation strategies for decreasing the potential for overheating in the homes. Currently, it has been observed that a large number of dwellings in the UK have no mechanical cooling systems. Therefore, an increase in summer temperatures has considerable potential to increase occupant vulnerability to overheating as well as the potential to considerably raise energy consumption (Collins et al 2010).

5   Thermal Comfort

ASHRAE defines thermal comfort as "that condition of mind which expresses satisfaction with the thermal environment". This symbolizes the complexity and uncertainty of the issue of thermal comfort and likewise overheating or discomfort levels (ASHRAE 2010). Figure 3 demonstrates the potential comfort adaption to a climate change.

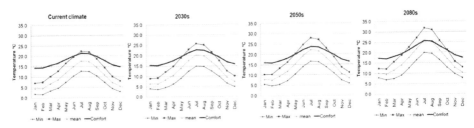

Fig. 3. Potential comfort adaption to a change in climate [Source: Gupta and Gregg 2012]

6   Methodology

For quantification purposes and simplification of decision-making, a building model using the Ecotect thermal software was used. The indoor air temperature at which overheating occurred was taken to be when the average interior home air temperature was 26°C or greater. This condition was used to simplify the image of overheating hours in the home. What is most important here is the relative change in 'overheating' hours between projections when different construction types are tested. Basically, in an optimization process concerned with risk and uncertainty, decisions are made on certain quantitative measures to determine the best course of action possible for a decision complexity. As such, three main elements are required to be considered before reaching a decision (Al-Homoud 1994):

- Selection options from which a selection is created (variables)
- Precise and quantitative information of the system variables' interface (constraints)
- Particular measure of system efficiency (objective function)

In this study the variables are a range of five typical cladding systems. Constraints are likely to be wall-thickness, environmental and economic performance. The objective function is established as the decrement factor (the ability to decrease the amplitude of temperature from outside to the inside), time constant (the time takes the maximum outside temperature makes its way to a maximum inside temperature), admittance (building fabric response to a swing in temperature) and U-value (overall heat transfer coefficient). The research considered five construction techniques, all of which are appropriate for use in house walls. This number was considered satisfactory to make useful comparisons but not to be excessive to consider in detail. The selection criteria were:

- Recent use for housing in the UK so the availability of detailed information is met
- Method appropriate for the UK housing use
- The potential of achieving Part L of UK thermal building regulations (U-value set to 0.12 W/m^{2}K)

The typical cladding systems examined were:

- Brick and block wall (BB)

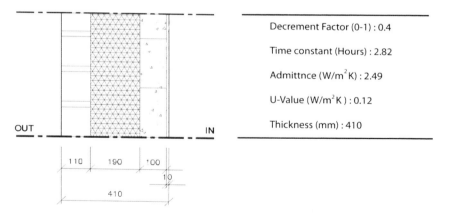

Fig. 4. From Out to in: 110mm Brick Outer Leaf, 190mm Phenolic Insulation, 100mm Aerated Concrete Block, 10mm Lightweight Plaster

- Timber frame wall (TF)

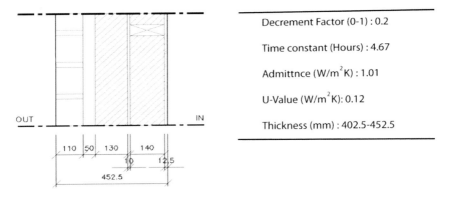

Fig. 5. From Out to in: 110mm Brick Outer Leaf, 50mm Air Gap, 130mm Rockwool, 10 mm Plywood, 140mm Rockwool, 12.5mm Plasterboard

- Insulating concrete formwork (ICF)

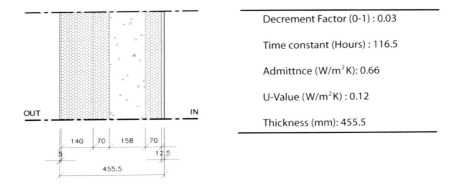

Fig. 6. From out to in: 5mm Rendering, 140mm Extruded Polystyrene (EPS), 70mm Extruded Polystyrene (EPS), 158mm Heavyweight concrete, 70mm Extruded Polystyrene (EPS), 12.5mm Plasterboard

- Structural insulated panel (SIPs)

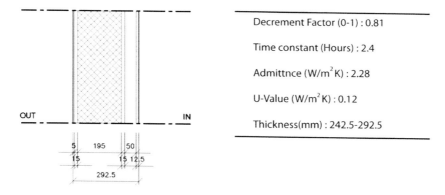

Fig. 7. From out to in: 5mm Rendering, 15mm Softwood board, 195mm Extruded Polyurethane (PUR), 15mm Softwood board, 50mm Air Gap, 12.5mm Plasterboard

- Steel frame wall (SF)

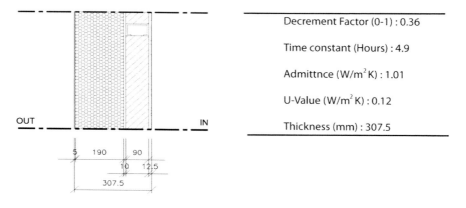

Fig. 8. From out to in: 5mm Rendering, 190mm Extruded Polystyrene (EPS), 10mm Plywood, 90mm Rockwool, 12.5mm Plasterboard

The environmental modeling software Ecotect was used to analyze the thermal performance of the building model shown in Figure 9. The weather data used were based on Manchester, UK climate data from the year 2011 with no heating period (1st of May to the 30th of September), and without any internal gains. The infiltration was assumed as 0.05 air change per hour (ACH) and no ventilation was considered. A U value of 0.1 W/m^{2}K for the roof and floor and 0.8 W/m^{2}K for triple glazed windows were assumed.

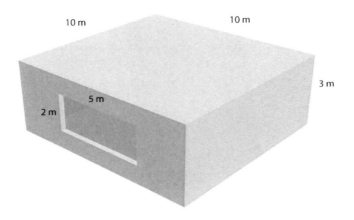

Fig. 9. Model building examined in Ecotect

7 Discussion and Results

Figure 10 compares the thermal properties of the examined wall system systems. As can be seen, a higher wall thickness means a lower admittance and decrement factor in most cases. It seems that ICF has the lowest decrement factor and admittance rate with the highest thickness. It could be observed that SF, with considerably less thickness, shows an acceptable level of performance.

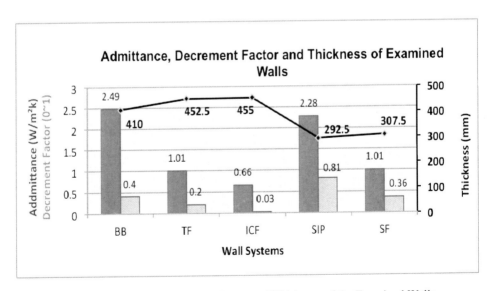

Fig. 10. Admittance, Decrement Factor and Thickness of the Examined Walls

With regard to overheating, Figure 11 illustrates that TF had the worst performance among the construction systems. The SF performance was not much better. As was expected, ICF seems to have the best performance, with the lowest percentage of overheating, although the maximum thickness does not seem to be ideal.

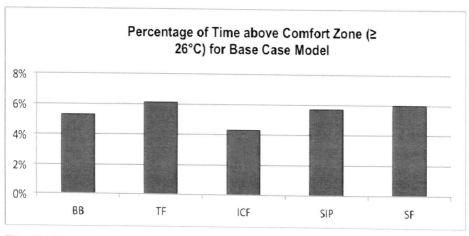

Fig. 11. Percentage of time above comfort zone (≥26°C) for base case model – current Manchester weather data

8 Conclusion

In this paper, the proposed approach models uncertainty in current climate and quantifies overheating risk. The wall construction options tested in this study have only a small difference in their performance in dealing with this risk. However, further works in his area should consider future climate change scenarios in assessing risk. This leads to a decision problem offering the decision maker an opportunity to arrive at a decision influenced by their knowledge. Clearly, one "correct" decision is not given to decision-makers but rather a small collection of choices to reduce or eliminate the negative impacts on comfort and energy consumption. It has been assumed that dealing with the uncertainty and risk proposed in this paper will make the decision-making more dynamic and environment- specific.

Unfortunately, the climate observations show that the right choice comes after it is needed. Climate models cannot deliver what is the present-day decision-makers necessary framework; the only answer is to modify design frameworks to enable them to take a range of uncertainties into account. Regarding this, a classification of "no-regret strategies" and "flexible strategies" in decisions is proposed for consideration. 'No-regret'' decisions represent the ability to deal with climate uncertainty. These strategies produce paybacks even if climate change does not happen. Improving building insulation is the most appropriate example of this strategy in construction systems, since this energy saving can frequently pay back the additional cost in the short term. Secondly, it seems wise to add external passive design strategies such as shadings and louvers, which are reversible, over permanent choices. Clearly, the aim is to minimize as far as possible the cost of being wrong about prospective climate change. Eventually, the research found that none of the construction types optimization common strategies could entirely remove the risk of overheating in the homes for current weather conditions in Manchester.

9   Further Research

A major risk for sustainable design is the uncertainty in future climate. Preferably, climate models would be able to produce more accurate climate statistics; clearly, this is the evidence that researchers in engineering and science need to optimize future investments. Basically, two major issues remain that make a precise model difficult for future scenarios. Initially, there is a scale misfit between what decision-makers need and what climate models can deliver. Secondly, the epistemic uncertainty of climate change is important. However, the initial issue can be alleviated by downscaling techniques such as using regional models with limited domains. But the second issue seems to be more difficult to overcome, at least in the short-term future; clearly, there is a real risk of misperception between old data and model output.

References

Al-Homoud, M.S.: Design optimization of energy conserving building envelopes. PhD thesis. Texas A&M University, Texas (1994)
ASHRAE. Standard for the design of high – performance green buildings except low-rise residential buildings. Atlanta: American Society of Heating. Refrigerating and Air-conditioning Engineers organization (2010)
Athienitis, A., Santamouris, M.: Thermal Analysis and Design of Passive Solar Buildings. Jame & James (Science Publishers) Ltd., London (2002)
Collins, L., Natarajan, S., Levermore, G.: Climate change and future energy consumption in UK housing stock. Building Services Engineering Research & Technology 1, 80–86 (2010)
Dunster, B.: Housing and Climate Change: Heavyweight vs. Lightweight construction. RIBA (January 2005),
http://www.architecture.com/Files/RIBAProfessionalServices/Practice/UKHousingandclimatechange.pdf (accessed June 28, 2012)
Gupta, R., Gregg, M.: Using UK climate change projections to adapt existing English homes for a warming climate. Building and Environment 55, 8–17 (2012)
IPCC. contribution of working groups I, II and III to the fourth assessment report of the intergovernmental panel on climate change. Geneva: Intergovernmental Panel on Climate Change (2010)
Orme, M., Palmer, J., Irving, S.: Control of overheating in well-insulated housing. In: Building sustainability: value and profit. (2003),
http://www.cibse.org/pdfs/7borme.pdf (accessed March 7, 2012)
Vesilind, P.A., Lauren, H., Jamie, R., Hendry, C., Susan, A.: Sustainability Science And Engineering: Defining Principles, vol. 1. Elsevier, Oxford (2006)

Chapter 81
Potential Savings in Buildings Using Stand-alone PV Systems

Eva Maleviti[1] and Christos Tsitsiriggos[2]

[1] University of Central Greece, Levadia, 32100, Greece
[2] Visiontask Consultancy, Athens, 15562 Greece

Abstract. This research analyses the electricity consumption and CO2 emissions of 3 hotels in Greece using energy auditing. Following this analysis, the energy generation, at the same month, from a stand alone PV system will be presented for each case, showing its contribution in electricity consumption and its possible share on the total electricity consumption. The PV systems are installed from the same company, and they have different power capacity of 75, 100 and 200kW. This research aims at showing potential energy and emissions savings from the selected cases, showing the necessity to promote the application of renewable energy sources and in particular PV stand alone systems in Greek hotels considering the current socio-economic situation of the country.

Keywords: Energy audit, buildings, hotel, stand alone pv systems.

1 Introduction

Tourism is a fast growing sector affecting the environment and the natural resources. Its continuous expansion is not compatible with sustainable development and possibly harms the local societies and traditional cultures. Additionally, uncontrolled tourism expansion has led to degradation of many ecosystems, particularly in coastal and mountainous areas. The premise of tourism involves people to commute from their homes to different destinations with various means of transportation. Furthermore, the construction of new accommodation and hotels, along with the existing hotels, which offer a high number of facilities to their customers and the large energy consumption in these facilities, signifies their severe contribution to environmental effects in a global scale but also in regional level, depending on where these facilities are located (UNEP,2003). However, tourism contributes 7% to the pollution in the Mediterranean region. Simultaneously, the growing pollution in Mediterranean countries is likely to have a negative effect to the tourism sector as well. It is estimated that every tourist in Europe generates at least 1 Kg of solid waste per day. (French Institute for the Environment, 2000). Nevertheless, the tourism industry is amongst the most profitable sectors within the commercial sector, especially in the Southern Europe, France, Greece, Italy, Spain, and Portugal. The tourism sector in these countries is an

A. Håkansson et al. (Eds.): *Sustainability in Energy and Buildings*, SIST 22, pp. 913–929.
DOI: 10.1007/978-3-642-36645-1_81   © Springer-Verlag Berlin Heidelberg 2013

essential development tool for the country's economy. Specifically, in Greece the year 2000 the tourism industry represents the 16.3% of the Greek Domestic Product, standing for the most important service industry. (World Travel and Tourism Council, 2006) In addition, the World Tourism Organization (WTO) forecasts that Greek tourism will be developed by 4.1% by 2016. (World Tourism Organization, 2005).Based on WTO's forecasts, in 2015, 20.8% of local employment will be linked to tourism (from 18% currently), while investments of the sector will represent 10.7% of total Greek investments. (World Tourism Organization, 2005). On the other side, energy consumption in hotels is among the highest in the non-residential building sector, not only in Greece but also across Europe; for instance, 215 kWh/(m^2a) in Italy, 287kWh/(m^2a) in Spain, 420 kWh/(m^2a) in France and 280kWh/(m2a) in Greece. (Argiriou, 2002) Hotels in Greece represent about 0.26% of the total Hellenic building stock. (Santamouris, 1996). Despite the fact, that hotels' percentage is small comparing to other buildings, they are indicating a high-energy consumption. The total annual energy demand of the Greek Hotel sector in estimated to 42TWh representing the 28% of the total energy demand of the tertiary sector. (Dascalaki, 2004) In particular, the high-energy consumption in hotels is due to the different and multiple operations and facilities offered to their customers. Specifically, in order a hotel to operate, various types of energy, such as electricity, gas, diesel fuel, natural gas and others are required. Still, the main energy source used is electricity, generally for air-conditioning, heating, lighting, lifts, kitchen equipment and many static and portable appliances (Onut,2006). Consequently, an effort to reduce energy consumption in the hotel sector would be significant.

2 Research Methodology

This study will demonstrate the Business as Usual Scenario (BaU), developed under the collected data from the 3 hotels. Energy audits in each hotel and interviews to hotel managers and engineers have been accomplished, in order to collect the appropriate data for this research. The customers/tourists needs, habits and behavior were considered in order to define electricity consumption, supporting this research to be as accurate as possible. The energy auditing included inspection in all areas of the hotels. Information about the lighting of each area such as power capacity per lamp in each room, the daily use of the equipment in the kitchen, the laundry facilities and the rooms, along with the technical characteristics of each device were collected. The following graphs will exhibit future projections on the electricity consumption of the 3 hotels, since from the energy auditing has been observed that the main source of energy consumed is electricity. The data analysis will demonstrate the current trend in electricity consumption in the selected case studies. This analysis represents the baseline scenario, without policy interventions, giving approximations and estimates on energy consumption in these 3 case studies. The next part of the analysis demonstrates three possible installations of PV systems in the selected cases according to their needs. Economic viability of the investments is also examined.

3 Data Analysis

3.1 Three-Star-Hotel in Corfu (Island)

The case study assessed for the installation of the PV system is a three star hotel located in Corfu island. It has a full season operation with 60 rooms in total. The building has four floors, with 450 m^2 per floor. The total floor area of the facility is almost 6700m^2. The occupancy rates vary during the year according to the period; 80% in summer, 50% in autumn, 30% in winter and 40% in spring. Oil and LPG are used to cover heating and hot water needs and the rest are covered from electricity.

Table 1. Fuel consumption (2010 prices)

Fuels	Amount	Cost(€)
Oil	25,000 lt	17,000
LPG	20,000 lt	11,600
Electricity	120,000kWh	10,624

Table 2. Electricity consumption per service

Service	MWh
Lighting	11.879
Cooling	15.552
Cooking	2.835
Refrigeration	28.518
Laundry	39.000
Other	21.000
Total	**119.024**

As Table 1 depicts, oil consumption has the highest share of total energy consumption in the building. Oil is used for heating and LPG for cooking purposes. Energy efficiency measures and alterations in three-star-hotels are examined in this case in order to evaluate the cost of the proposed equipment. Table 2 displays a breakdown of electricity use for a variety of services. The high amount of electricity consumption in each service needs to be reduced. The first immediate and easy method for that is the application of efficient electrical equipment. The necessity to reduce the electricity consumption as well as to alter the fuels used arises from the increase of the price per fuel used in the facility. The projections in increase of oil, LPG and electricity prices are based on the Energy Information Administration forecasts (2010). According to the Ministry of Development, the estimated increases in electricity, oil and LPG consumption as well as assumptions of energy use, occupancy rates and operations for three-star-hotels, have been taken into consideration for the calculation of the energy cost projections. The increases in the fuel prices and consequently in the energy costs are presented in Figure 1. In particular by 2020 oil costs would increase 26% by 2020, electricity costs 28% and LPG costs 25%.

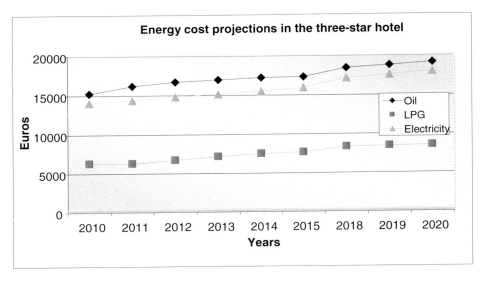

Fig. 1. Energy cost projections in the three-star-hotel

Thus, considering the uncertain and difficult economic situation of Greece along with the increase of fuel prices, hoteliers need to consider methods that would reduce their facilities' energy costs. The two options considered in this section are the use of electrical efficient equipment and the installation of PV systems. Table 3 represents the changes in the total electricity consumption under the use of efficient equipment.

Table 3. Electricity Consumption using efficient equipment

Service	MWh
Lighting	2970
Cooling	11664
Cooking	2126
Refrigeration	24240
Laundry	33150
Other	15750
Total	89900

Between the total electricity consumption values in Tables 2 and 3, there is a reduction in energy almost 29MWh. This proves the effectiveness of this type of equipment in reducing the hotel's electricity consumption. Despite its efficacy in reduction of electricity, a determining factor for their installation is their cost. In particular, the economic viability of the efficient equipment is examined, with the net present values of each equipment, and according to the Ministry of Finance, interest rate to be 5% as it is the current interest rate today in Greece (2010). These values are presented in the following section. The second option considered for reduction of the building's electricity consumption is the PV system. The Greek Energy Efficiency Action Plan proposes that at least 10% of the buildings' energy requirements be covered from renewable energy sources. The average annual solar activity in Greece is 1350 kWh (Institute of Environment and Sustainability,2010). For the material used in the panels there is a small loss from the amount of energy generated. The losses in photovoltaic panels are 0.05% per year (Conergy AG,2010). The transmit losses which are the losses of the energy generated occurring while transmitting are 5% annually. In this plan, the repair and maintenance costs are 0.5% of the total investment cost and other expenses for the operation of the panels are 0.2% of the total investment cost per year. In the following case the possibility of generating electricity from a 100kW PV panel and use it for the buildings' energy needs, is examined. The panels have a lifespan of 20-25 years where the energy output is the highest, under the appropriate maintenance, as indicated from panels' provider (Conergy AG,2010). In addition, the components of the PV require maintenance so as to operate better and longer adding a cost to the total of the investment. The interest rate of return –or alternatively the return of the investment, is calculated by finding the discount rate that causes the net present value of the project to be equal to zero. If the IRR of the project is equal or greater than the required IRR of the investor, then the project would be acceptable (Pratt, 2001). The installation of a PV system in a building is more successful when the energy efficient equipment or standards are applied, prior to the PV system.

A.   Net Present Values (NPV) for energy efficient equipment

This section presents the net present values for the equipment considered for replacement in this case. The interest rate considered is 5%. The net present value method examines the financial viability of the energy efficient equipment in this hotel. When the calculated values of NPV are positive, then this type of investment is acceptable.

Table 4. Net Present values of energy efficient equipment with 5% interest rate by 2020

Services	NPV(5%)
WASHING MACHINES	-3.359,1 €
REFRIGERATORS	-1.236,1 €
MINI-BARS	-6.840,2 €
LIGHTING	3.942,0 €
A/C	-5.470,3 €
FREEZERS	4.790,5 €
DRYERS	202 €
SOLAR THERMAL (LPG)	11324 €

Table 4 displays that not all the existing equipment should be replaced with efficient ones. In specific, for washing machines in the three different values of interest rate, the NPV is negative, thus this replacement would not be beneficial to the hotel. The same stands for the replacement of refrigerators, mini-bars and air-conditioning equipment.

The rest of the equipment considered could add value to the hotel. This analysis examines the investment for the installation of a 75kW photovoltaic panel. The panel is made of crystalline silicon, the most commonly used, with 74.63kW capacity including 5% system losses. The total cost of the investment of 75kW capacity is 290.000€. 40% of the total cost is covered from the National Development Law 3299/2004 (Ministry of Development, 2008). In the following case the possibility of generating energy and using it for the buildings' energy needs, is examined. Electricity consumption is responsible approximately 21% of the total energy consumption in this building. The electricity generation from the PV is approximately 96MWh on an annual base. The monthly electricity consumption of the building and the monthly electricity generation from the PV system is displayed in Figure 8.5 after the possible implementation of the efficient equipment. The final electricity generation from the PV could cover the building's electricity needs. One kWh of electricity consumption, costs 0.083€/kWh in 2010, therefore for 89 MWh consumed in the building, there are 89,000kWh x 0.083€/kWh = 7387 € spent, which is 2,500€ less than the cost occurring without the use of efficient equipment. From 96MWh generated from the PV, 89MWh are used to cover the buildings' energy needs. The remaining 7MWh could be sold to the main grid. In that case, the kWh generated from the PV and sold to the main grid is 0.45€/kWh (Ministry of Development, 2008); thus this hotel could have a profit of 0.45€/kWh * 7,000kWh=3,150€ approximately per year. For this unit, with this specific characteristics and fuel consumption the installation of a PV system is economically and environmentally profitable, since not only it covers the electricity needs of the building and offers an extra 3150€ profit, but it also prevents from emissions being released.

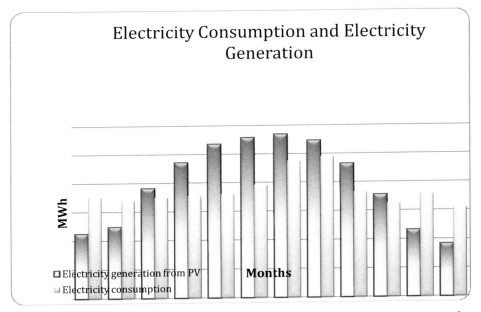

Fig. 2. Electricity generation from PV and electricity consumption in the building per month

For this project the IRR factor is calculated from the annual cash flows and it is -9% and the NPV is -32,342€.

B. Marginal abatement carbon cost (MACC)

With this method the volume of abatement in emissions is compared to a given cost per tonne in a specific time period. The graph displayed below, has been developed considering the NPV of each equipment, divided by the CO_2 emissions saved per year. This simple method has been examined assuming that the carbon savings for each equipment are constant for a ten-year period. The Marginal Abatement Carbon Cost has been developed for this case considering the suggested template proposed by the Carbon Trust (Low Carbon Cities Website, 2008). Figure 8.7 demonstrates the graph of MACC carbon discounted (€/tonne) and the Cumulative CO2 savings (tonnes/year). Figure 3 represents the graph for the case of 5%, since this is the interest rate today. In specific, the width of each column represents the amount of potential abatement from each measure. The height of each bar represents the unit cost of each measure. The bars that are below the X axis are the most cost effective. This means that they would save money to the facility, as well as CO_2. For this facility, the most cost effective equipment and CO2 emissions savings occur from the washing machines, the refrigerators, mini-bars and air-conditioning equipment. These are also represented below the X-axis. The lighting, freezers, solar thermal heaters, PV system and dryers are the most expensive and they are found above the X-axis. In addition, the area of each bar stands for the total cost to deliver all CO_2 emissions savings from the measures below X-axis. Following to that, the sum of the area of all the bars represent the total cost to deliver the total CO_2 emissions savings from all the proposed equipment. For this hotel the MACC value is 113€/year for reduction of 1 tonne of CO2 emissions for this facility

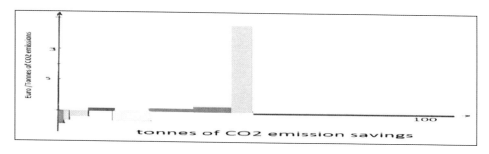

Fig. 3. MACC for the three-star-hotel

3.2 Four-Star-Hotel in Rio- Patras (Mainland)

This economic assessment is for a four-star-hotel located in the Rio region of Patras, in South-Western Greece. It has 255 rooms and it is operating 12 months per year.

The occupancy rates vary throughout the year, with 69% in summer, 54% in autumn, 31% in winter and 36% in spring. The building has three floors with a total area of 30,240m^2. Electricity is the main source used. Oil and LPG are used for heating and hot water facilities hence the fuel usage per year is displayed in Table 5. It is important to note that this hotel has no efficiency standards for any of the equipment used. Thus, before continuing with the economic assessment of a possible installation in PV system, it is crucial to observe the possible reduction in electricity consumption if energy efficiency standards are applied.

Table 5. Final end-use by fuel

Fuels	Amount	Cost (€)
Oil	30,200 lt	20,838
Electricity	1,828MWh	151,724
LPG	168,950 lt	97,991
Total		270,553

Table 6. Electricity consumption by end-use

Service	MWh
Lighting	221
Cooling	273
Cooking	110
Refrigeration	317
Launderette	290
A/C	397
Others	220
Total	1,828

As it is observed from Table 5, electricity consumption is very high and there are no energy efficiency standards applied. Thus, it is important firstly to examine what reduction would occur in final electricity consumption before continuing with the PV system installation. The assumptions used, concern the efficiency measures proposed from the policy framework presented in Chapter 3 and used for the development of the scenarios for four-star hotels in Chapter 6. Table 7 presents the electricity consumption in the current state, without energy efficiency standards. The electricity consumption is presented for each service in Table 6. The increase in the electricity costs only are expected to rise by 20%, oil 22% and LPG 50% by 2020. In order to prevent further increases in the hotel's energy costs, the use of efficient equipment is essential. As observed in Table 8, the final electricity consumption could be reduced by 695MWh. In energy costs this is 695,000 kWh x 0.083€/kWh = 57,686 €.Efficient lighting is not used in this facility, thus the use of LED lighting is proposed instead of the incandescent lights. The average 30W capacity applied in the building could be replaced with 3W LED lights. The price between new and old lights is almost 4 times higher; as LED prices vary from 12 to 18 € for these considered in this case, while the conventional ones vary from 2.5 to 4.5€ (Phillips Website, 2010). For LED lights, the reduction in electricity consumption is significant with reductions up to 75%.

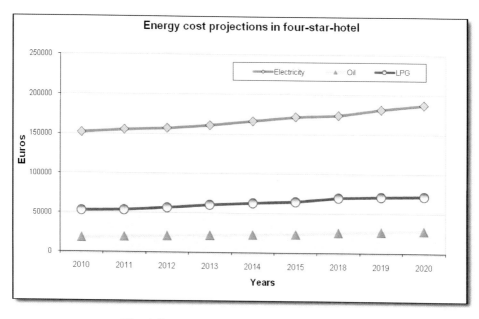

Fig. 4. Energy cost projections in four-star-hotel

Table 7. Electricity consumption by end-use using efficient equipment

Service	MWh
Lighting	65
Cooling	177.45
Cooking	99
Refrigeration	206
Launderette	188.5
A/C	256
Others	143
Total	**1,134.95**

The existing equipment could be replaced with efficient ones and according to the type of the equipment, the electricity consumption reductions would vary from 5 to 15% depending on the equipment opposed to the existing equipment of the hotel.

C. *Net Present Values (NPV)*

Table 8 displays the net present values with 5% interest rate. The net present values are positive for laundry services, air-conditioning, solar water heaters and PV systems, thus this type of investment in this specific equipment for this hotel is economically viable.

Table 8. Net Present Values of energy efficient equipment

Services	NPV(5%)
MINI-BARS	-8302
LAUNDERETTE/DRYERS	108552
REFRIGERATION	-127055
LIGHTING	-33761
A/C	13790
SOLAR THERMAL (OIL)	39226

Therefore, the above Table demonstrates that the use of the equipment with the positive NPVs is beneficial not only in terms of the reduction of environmental effects but also in terms of the economic benefits offered in the building. Even though the equipment considered for minibars, refrigeration and lighting is significant for the reduction in energy consumption, the energy savings and the prevention of further emissions, it is not in fact an economically viable investment for this hotel. This indicates that certain types of equipment may be beneficial in terms of reduction in environmental effects but the capital cost of the equipment is not beneficial for the facility in terms of economics. For that reason, the marginal abatement cost (MAC) is presented in the following section, in order to demonstrate a clearer picture of the economic benefits, the environmental reductions in the buildings and the relation between these two factors. Before that, the IRR for a PV installation of 100kW is also considered in this building, since it is one of the measures for reduction in energy consumption and emissions. This technology is also included in the MACC calculations.

D. Internal rate of return for PV project (IRR)

The installation of PV systems could significantly change the hotel's energy profile. The installation of 100kW plant is examined for this hotel. The economic information required for the PV plan is presented below. The electricity production from the PV system could only cover approximately 10% of the total electricity consumption in the building. It would be more efficient if the PV system were applied in a facility that has already got energy efficient standards. Thus this 10% share of electricity that is generated from the PV system, may seem a very low share comparing to the total, but in terms of reduction in electricity use and energy costs, is very important. The electricity consumption and electricity generation from the PV system is provided in Figure 5. Considering the results of the previous section, after the application of efficient equipment, the total electricity is reduced to 1,135 MWh. On condition that 127 MWh are covered from the PV system then, the electricity would greatly reduced. This means that with efficiency standards in the facility the energy costs are reduced to 1,135,000kWh * 0,083€/kWh = 94,205€ and this amount could be reduced even more if another 11% is covered from the PV system; thus 127,000kWh*0,083€/kWh=10,540€ per year. Examining the economic viability of the PV investment, the interest rate of the loan defines the values of the NPV.

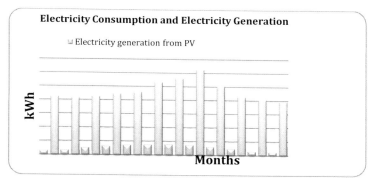

Fig. 5. Electricity generation from PV and electricity consumption in the building per month

The NPV is positive for the 5% interest rate and this indicates the viability of this investment. The payback period is after the year 2012. The installation of the 100kW PV system is an economically viable investment for this hotel. Based on these values, the margin abatement carbon cost (MACC) is calculated in the following section, along with the MACC of the energy efficient equipment.

E. Marginal Abatement Carbon Cost (MACC)

Figure 6 represents the graph for the MACC for four-star-hotel. The values calculated for the 5% interest rate are presented since it is the current one used now in Greece. As the figure displays, refrigeration, lighting and mini-bars have the lowest margin carbon abatement cost. The most expensive technologies for CO_2 emissions savings are the laundry facilities, air-conditioning equipments, solar thermal heaters and the PV system. The total sum of the bars is calculated in -83€/year in 1 tonne of CO_2 emissions savings.

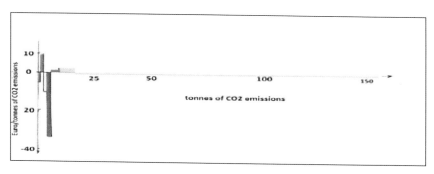

Fig. 6. MACC in four-star-hotel

3.3 Five-Star-Hotel in Santorini (Island)

The economic assessment refers to a five-star-hotel in Santorini Island in the Aegean Sea. This hotel is operating from March to November and it has 17 rooms- suits. The total floor area is approximately 2800m². The occupancy rates vary during the operational period with 90% of occupancy rates in summer and 50% in autumn and in spring. Oil is used for hot water needs by using a boiler and electricity for heating. The rest of the services are covered from electricity. The following Table displays the electricity consumption per service in the building.

Table 9. Final end-use by fuel

Service	MWh
Lighting	14
Cooling	67
Cooking	44
Refrigeration	12
Launderette	49
Others	27
Total	213

Table 10. Electricity consumption by end-use

Fuels	Amount	Cost (€)
Electricity	213 MWh	37690
Oil	5200 (lt)	3536

This unit uses oil and electricity to cover the required energy needs. Table 10 displays the final energy consumption by fuel. The energy cost of this facility is already very high; up to 41000 € per year. Considering the projections in increase of fuel prices, it is necessary to reflect on alternative fuel sources and changes in the final energy use.

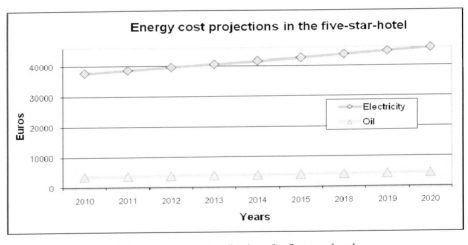

Fig. 7. Energy cost projections for five-star-hotel

Thus, it is essential to consider improvements in energy efficiency and use renewable energy projects that would replace a share of the total electricity consumption. It is thoroughly examined which of the electrical equipment is considered economically viable for this facility and which contributes to a significant reduction in energy consumption as Table 11 displays.

81 Potential Savings in Buildings Using Stand-alone PV Systems   925

Table 11. Electricity consumption by end-use using efficient equipment

Service	MWh
Lighting	3,4
Cooling	47
Cooking	31
Refrigeration	8,5
Launderette	34
Others	18,6
Total	**142.5**

As it is depicted in Table 11 the replacement of existing equipment with efficient ones could reduce final electricity consumption approximately by 70MWh. This indicates that this type of equipment could be beneficial in terms of reduction of electricity costs. However, it is essential to consider the cost of the equipment and how economically viable is the suggested equipment for this facility. This analysis is provided in the following section examining the net present values of each equipment considered.

F. Net Present Values of Energy Efficient Equipment

Table 12 displays the net present values for different interest rates. It is observed that refrigerators and freezers would not be economically viable for this facility, due to their high capital cost. All other equipment considered, are economically viable for this facility since the net present values are positive and thus they are acceptable as an investment.

Table 12. Net Present values of equipment in the five-star-hotel

Service	NPV(5%)
WASHING MACHINES	18831
REFRIGERATORS	-1425
MINI-BARS	5264
LIGHTING	6005
A/C	26536
FREEZERS	-669
DRYERS	19869
SOLAR THERMAL (OIL)	9214

The above presented values in Table 12 are used in the following section of MACC calculations. In these calculations, the net present values of the PV system and the internal rate of return are included. Electricity consumption covers 82% of the total energy consumption in the building and the rest is from oil consumption. In contradiction to the previous case, in this one the installation of a PV system with 200kW capacity is considered. The reason is that the final electricity consumption is 213MWh and the electricity production from a 100kW PV installation would not be enough for covering the hotel's electricity needs. One kWh of electricity consumption costs 0.13€/kWh, therefore the total cost for electricity consumption is 278,500kWh x 0.13€/kWh= 36,205€. Since the island is not connected to the main grid, the price is different to the one in the connected grid, thus the price per kWh is 0.13€/kWh. In the following section, the installation of the PV system with capacity of 200kW is thoroughly examined, since this capacity would cover more than 60% of the building's electricity demand, a lot more than the proposed share of the Action Plan which is 20%. As Figure 8 depicts, the electricity consumption in the building could not be fully covered from the PV system. The difference from the previous case is that this hotel is operating only from March to November; therefore, the profit is occurring from a three-month-operation without electricity consumption and consequently electricity costs, since the hotel is not operating. As previously mentioned, the PV system generates 255MWh per year and the building would consume 142MWh per year. Thus, the total electricity consumption could be covered from the electricity generated from the PV. That is a very significant amount since not only the costs but the emissions are decreased as well. From the consumption of 1kWh the amount of emissions released is 1,1kg.

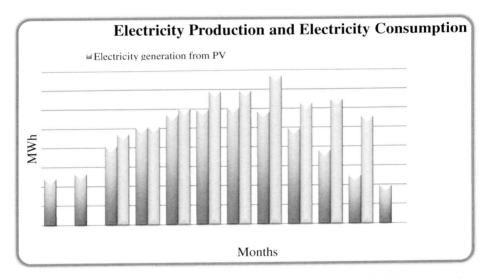

Fig. 8. Electricity generation from PV and electricity consumption in the building per month

Thus 142,000kWh x 1.1kg/kWh = 156,200kg = 156.2 tonnes of CO_2 are prevented from being emitted. The detail of the investment plan that should be followed is shortly presented in the following Table 8.16. For this investment, the IRR factor is 1%, which makes the investment acceptable and profitable. The following Figure displays the capital cost of the PV system assuming that electricity prices could remain the same over the 10-year period.

G. Margin Abatement Carbon Cost (MACC)

The MACC values are presented in figure 9 which depicts that the most cost effective technologies are the all the proposed ones except for the refrigerators and freezers considered for replacement in this facility. Based on these values. As Figure 8.15, demonstrates the technologies with values above X-axis are the most expensive ones and these are the refrigerators and freezers, as previously mentioned. The rest of the technologies considered are the most effective since they save money as well as CO_2 emissions. The total sum of the areas of all the bars presented in the graph, represents the overall emissions savings available and it is calculated in -50 € per year for the saving of 1 tonne of CO_2 emissions.

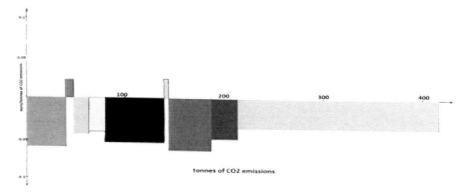

Fig. 9. MACC for the five-star-hotel

Table 13. Results of Economic assessment in the three cases

	MACC	IRR	NPV
Three-Star-Hotel	113€ of 1 tonne of CO_2	-9%	-49,503€
Four-Star-Hotel	-83€ of 1 tonne of CO_2 emission savings	77%	10,495€
Five-Star-Hotel	-50€ of 1 tonne of CO_2 emissions savings	1%	387,300

4 Conclusion

All the above values have been generated considering the 5% interest rate that currently is applied to this type of investments in Greece. The IRR values are for the PV installation and the MACC and NPV values are for both efficient equipment and PV system. The NPV values presented in Table 8.18 have been calculated for all the equipment together and for not each one separately as displayed in the previous sections.

The above results represent that the above considered efficient equipment and PV system are economically viable for the four and five-star hotel. Only for the three-star hotel the calculations for the NPV and IRR displayed that this type of investment could not be economically viable for this case. It is noteworthy that in this case the MACC value is translated to 113€ to be spent in order to prevent 1 tonne of CO2 emissions from being released. The results displayed that from the three cases, the most inefficient is the first case. Despite the fact that the PV considered requires less own capital than other cases since it is only 75kW, it is the less economically viable. Further than that, this hotel unit is the smallest one compared to the other two, thus the energy costs would be expected to be more easily reduced, but this is not the case. If the proposed investments are not economically viable then the hotelier would know follow these options. It is important to note as well, that this case could have a profit of 3150€ per year from the electricity sold to the main grid, considering that its electricity costs are covered from the PV system. The four-star-hotel on the other side, shows that even if there is not electricity sold to the main grid, but all the building's electricity needs are covered from the PV system (100kW), still it is economically viable. It is also noteworthy that these two cases are totally different between them in size and types of operations and consequently in energy costs. Moreover, the five-star-hotel follows the same trend as the four-star-hotel, having both efficient equipment and PV system (200kW) to be economically viable.

References

Diakoulaki, D., et al.: A bottom-up decomposition analysis of energy-related CO2 emissions in Greece. Energy 31(14), 2638–51 (2006)

European Commission (in cooperation with Eurostat), EU energy and transport in figures, Statistical pocketbook, Luxembourg (2003)

European Commission. Priorities for energy conservation in buildings. 48. European Commission. Ref Type: Report (2006)

European Commission. European Directive on the Energy Performance in Buildings, 2002 Directive 2002/91/EC, European Parliament (December 2002)

Hellenic Republic. National Allocation Plan (2005)

IEA, 30 Years of Energy Use in IEA Countries, International Energy Agency (2004)

IEA, Energy Technology Perspectives, OECD/IEA, Paris (2006a)

IEA, World Energy Outlook 2005, OECD/IEA, Paris (2006b)

Michaelis, K., et al.: HOTRES: renewable energies in the hotels. An extensive technical tool for the hotel industry. Renewable and Sustainable Energy Reviews 10(3), 198–224 (2006)

Onut, S., Soner, S.: Energy efficiency assessment for the Antalya Region hotels in Turkey. Energy and Buildings 38(8), 964–971 (2006)

Santamouris, M., et al.: Energy characteristics and savings potential in office buildings. Solar Energy 52(1), 59–66 (1994)

Santamouris, M., et al.: Energy conservation and retrofitting potential in Hellenic hotels. Energy and Buildings 24(1), 65–75 (1996)

Santamouris, M., et al.: On the impact of urban climate on the energy consumption of buildings. Solar Energy 70(3), 201–216 (2001)

Santamouris, M., et al.: Investigating and analysing the energy and environmental performance of an experimental green roof system installed in a nursery school building in Athens, Greece. Energy 32(9), 1781–188 (2007)

United Nations Framework Convention on Climate Change, Article I. United Nations Framework Convention on Climate Change (retrieved on January 15, 2007)

World Travel and Tourism Council, Greece: The Impact of Travel and Tourism on Jobs and the Economy 200

Chapter 82
The Application of LCCA toward Industrialized Building Retrofitting – Case Studies of Swedish Residential Building Stock

Qian Wang[*] and Ivo Martinac

Civil and Architectural Engineering Department,
Division of Building Service and Energy System
KTH Royal Institute of Technology, Brinellvägen 23, Stockholm, Sweden
qianwang@byv.kth.se, qianwang@kth.se

Abstract. This study analyzed how industrialized building retrofitting measures contribute better decision supports for building retrofitting strategy to the energy saving potential from a Swedish building typology approach. Contributions to cost-effectiveness retrofitting from one distinguishing but major type of Swedish building stock in Stockholm, Sweden, one of which case was studied from a life cycle perspective as demonstrations for the introduced renovation alternatives. A basic life cycle costing tool coupled with building energy demand calculation was applied. The study focus on the relative costing impact mainly from retrofitting materials and building energy consumptions, as well as corresponding importance of the cost contribution from four life cycle stages.

The tool analyzes the retrofitting material costs and embodied energy consumption after undergoing proposed retrofitting work packages as regards as the relevant payback time simulation. For the case type of building stock, a retrofitting measure compounds in terms of various energy saving and architectural service refurbishments were introduced, the most costly measures could be the most cost-effectiveness alternatives in different life cycle stages based on the typology of the target building.

Every building is unique and represents its own contexts, the proposed approach addresses getting an efficient general knowledge for the whole retrofitting and future building performance costs by life cycle thinking, aims at finding the similarities in Swedish building stock for providing greater resource-efficient, lower life cycle costing, simpler decision making and higher profitable building retrofitting strategy.

Keywords: Building retrofitting, building typology, life cycle costing, energy demand, material cost, decision making.

---

[*] Corresponding author.

A. Håkansson et al. (Eds.): *Sustainability in Energy and Buildings*, SIST 22, pp. 931–946.
DOI: 10.1007/978-3-642-36645-1_82 © Springer-Verlag Berlin Heidelberg 2013

1 Introduction

Building stock is one of the largest energy consumers worldwide; approximately 20-40% energy was consumed by the building sectors. According to the EU energy and environment target, 20% share of renewable energy in total energy consumption, 20% reduction in the use of primary energy through energy efficiency should be met by 2020, as regards as climate targets to decrease the emission of greenhouse gas by 20% and 50% by 2050 below the level of 1990[1-2]. In Sweden, 68%-75% of major energy consuming buildings are considered as residential buildings, more importantly, Swedish building sectors emit 15 Mton CO_2 eq/yr, the amounts are approximately 20% of contributions regarding to the greenhouse gas emissions.

Currently, a number of studies [3-5] have focused on residential buildings, which were considered as the major civil energy consumer compare with other type of building stock, Fig. 1 shows the statistical results toward the percentage of residential building energy consumption in different countries.

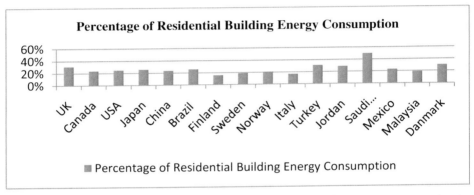

Fig. 1. Comparison results of the percentage of residential building energy consumption in different countries

In Sweden, new building construction rate is among the lowest in Europe, according to the European building production review [6-7], Swedish new building rate is around 8% compare with average level 10%-14% in Europe, since 2006, all buildings and construction value index has declined 70.1%, with the residential value index declining by a massive 87.8%. Moreover, home building rates in Sweden collapsed from around 65,000 units annually in the early 1990s to around half that level throughout the 20th century [6]. In another word, Swedish new buildings only contribute 10%-20% additional energy consumption by 2050, more than 80% of the building energy consumption will be influenced by the existing building stock. According to the market report of K. Mjörnell[8], if most of the building construction companies put efforts on making extensive renovation that includes both additional insulation and replacement of the ventilation and heating systems, we can save over 2 TWh per year in energy use which corresponds to almost ten percent of the total energy use for buildings in Sweden, obviously, building retrofitting has a striking energy and market potential.

The retrofitting of building stock in Sweden has a long historical development accompanied with the building design, architectural technology, energy planning and the development of urban civilization. Currently, we have paid more attentions to the systematic retrofitting operations based on not only fulfillment of functional satisfactory and replacement for the housing maintenances, but also offer optimal renovation strategy compounds of lower energy consumption, cost, life-cycle environmental impacts and higher housing comforts, which can be massively imitated to similar building types. Y. Juan [9] developed several hybrid decision making methodologies combined genetic algorithm with searching algorithms for optimal renovation selections, G. Mark etc[10-11] has combined LCC and savings-to-investment ration into the approach for selecting optimal renovation solutions from economic perspectives, other studies from O. Jinlong etc [12-15] also performed LCC based hybrid methodologies and cases studies toward different cities in Belgium, Germany, Greece and China, aims at finding feasibility of renovation measures and optimization-based approaches or multi-criteria decision-making process for producing relatively suitable retrofitting measures to in general or to the specific case buildings.

2 Aim

The aim of this study was to examine how energy use and retrofitting costs can be reduced by strategies planning in renovation phases. This was achieved by exploring a compound of retrofitting work packages on one of the major types of Swedish building stock classifications with the economic life cycle costing analysis approach. Specific attention was paid to the life cycle costs of building embodied operational energy consumption and the material-related investment costs in relation to the overall retrofitting life cycle costs and payback time.

3 Methodology

3.1 The Selected Typology of Swedish Building Stock

Sweden building constructions has over 100 years of development. The intensive construction of the Swedish building was firstly started at major cities in Stockholm area and the south part of Sweden. Around 50% of the buildings are constructed in Stockholm area, 17% in North area and 35% in South parts mainly Gothenburg and Malmö[3], Fig. 2 shows a survey of Swedish building stock distributions among four building periods, before 1931, 1931-1945, 1946-1960, 1961-1975. Accordingly, the modern building construction booming periods are 1940-1960, 1965-1975, in which large amounts of residential buildings were designed and constructed with a rapid demanding of foreign migration and domestic trends toward metropolitan cities, currently, the buildings constructed in "miljönprogrammet" are facing a challenging stage of energy retrofitting and envelope refurbishment.

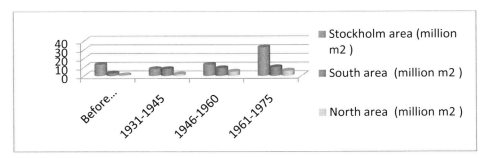

Fig. 2. Swedish building stock distributions and construction periods

The typology of the Swedish building is diverse and developed with the civilization and social demands. Generally, multi-dwelling houses (flerbostadsvillor), closed apartment block (sluten kvartersbebyggelse), slab house (lamellhus), tower house (punkthus) are the main constructions, in which "Lamellhus" plays the major role in the whole construction history [3] with a relatively large amounts compare with other building stock. According to statistics from Boverket [5], 800 000 apartments were constructed during 1961-1975 [8] [13], of which 300 000 were slab houses (lamellhus) and only few of the buildings were retrofitted, industrialized building retrofitting strategies are necessarily demanding both practically and academically, Fig 3 shows the statistic results for the construction types of Swedish building stock including the building storeys in four duration classifications.

In the study, one case Skivhus building, which is considered as one major type of large slab house (lamellhus), constructed in the miljönprogrammet is selected as a demonstration case for the LCCA of industrialized building retrofitting. The features

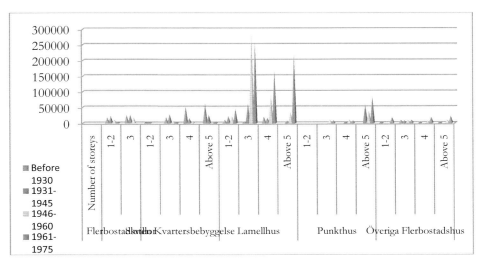

Fig. 3. Building types and construction amounts of Swedish building stock in four duration classifications

of the Skivhus are summarized similar with slab house, but higher and more floors, usually with 8-9 floors and polished concrete shed façade elements with balcony access. The insulation material of the external wall is mineral wools and double glazing systems, the ventilation system is usually exhausted air ventilation, as regards as heating radiator as the space heating equipment with central distribution.

3.2 Basic Energy Demand Calculation Tool

The energy demand simulation is calculated in the Consolis Energy tool, which is a dynamic Excel based software initially developed at KTH [16][17], by building material data inventory and material selections, the tool is capable of calculating two types of zone cases, which can be applied as the energy simulation of the base case and building performance after retrofitting. The results of the energy demand calculation are illustrated based on the effective thermal capacity and ISO 13790 calculation method, energy for heating, cooling, household electricity and domestic hot water can be separately presented by the tool. According to the developer of the software [16], the results are in the rage of 0-8% error, which is acceptable for this case study.

3.3 LCCA Calculation Tool for the Case Buildings

Several studies have developed life cycle costing techniques for building and energy systems to help decision makers have an integrated economic understanding of different building design strategies, B.S.Dhillion [18] introduced the classical approach of LCC for buildings as equ. (3.1), as well as energy LCC in building sectors as equ.(3.3), which can be applied in the early design phase for the building constructions.

$$LCC_b = CC + OC + RMC + DC \qquad (3.1)$$

Where CC is the capital cost, which is composed of land and construction costs, OC is operation cost associated with items such as energy, insurance and wages, RMC is repair and maintenance cost, DC is demolition cost.

Other building institutions as Stanford University [19] proposed similar guidelines and comprehensive calculation components for LCCA toward the building and housing performances as equ. (3.2),

$$LCC_d = PC + UC + RMC + SC + ELC \qquad (3.2)$$

Where PC is the project cost including labor, material costs, UC is the utility cost including energy costs and non-energy costs such as domestic water and sewer costs, RMC is the replacement and maintenance costs, SC is the building service costs such as janitorial services, pest control and elevator maintenance, ELC is the end of life costs.

$$LCC_e = EC + IC + SV + NPOMC + NRC \qquad (3.3)$$

Where EC is the present value of energy cost, IC is the present value of investment cost, SV is the present value of salvage cost, $NPOMC$ is the present value of the annually recurring nonfuel operation and maintenance cost, NRC is present value of the nonrecurring nonfuel operation and maintenance cost.

3.3.1 Cost Components of Building Retrofitting LCCA

In the study, the full life cycle costing components are defined as equ. (3.4), four calculating components are selected as the main stages for current calculation metrics.

$$LCC_r = IC + EEC(+UC) + RMC + ELC \tag{3.4}$$

Where LCC_r is the retrofitting life cycle cost expressed by present value, IC is the investment value expressed by present value, including the material costs, labor costs in the construction duration as well as the relevant tax costs, EEC is the embodied energy costs expressed by present value, UC is the utility costs including the domestic water and sewer costs, which are not included in current calculations, RMC is the replacement and maintenance costs, and ELC is the end of life costs expressed by the present value.

3.3.2 LCCA Techniques and Payback Time Calculation

The money of today and the money spend in the future are different, net present value is one of the common approaches to calculate the capital flows as equ. (3.5),

$$NPV = \sum_{n=0}^{L} \frac{C_n}{(1+r)^n} \tag{3.5}$$

Where NPV is the life-cycle cost expressed as a present value, n is the year considered, C_n is the sum of all cash flows in year n, r is the discount rate, and L is the service life-span

The net present value for a future cash flow C_0 expected to fall due every year during the service life-span L, the annual energy and utility costs can be calculated by equ. (3.6),

$$NPV = C_0 \times \frac{1-(1+r)^{-L}}{r} \tag{3.6}$$

$$d_{real} = \frac{1+d_{nominal}}{1+d_{inflation}} - 1 \tag{3.7}$$

Where d_{real} is the real rate, $d_{nominal}$ is the nominal rate and $d_{inflation}$ is the inflation rate;

$$LCC = C + PV_{recurring} - PV_{residual\ value} \quad [19] \tag{3.8}$$

Where LCC is the life cycle costs, C is the year 0 investment costs (IC), $PV_{recurring}$ is the present value of all recurring costs (embodied energy costs, maintenance and replacement costs, if any utility costs), $PV_{residual\ value}$ is the present value at the end of the study life.

4 Description of the Case Buildings

The residential building studied is located in Upplands Väsby, Stockholm, Sweden (Fig. 4, 5and 6, Table 1). The building is one of the eight similar apartment clusters. The apartment is a typical high Skivhus with slab in 7 floors. It was initially designed in "Miljonprogrammet" between 1965 and 1974 with a parallel set connection with other apartment buildings. The apartment block contains 6600 m² of residential space (total heated area 6860 m²) with heating radiator facilities, it's a 7 storey building with a load bearing structure of reinforced concrete and steel, slab materials are the major elements of the constructions. The ceiling is flat designed concrete structure covered with asphalt felt and unpainted sheet metals. The detailed building components can be found at Table 1, including the energy data inventory of the apartment.

The average U value of the building envelope before and after retrofitting can be found at Table 2, the energy consumption is approximately 972043 kWh/yr, 141, 7 kWh/m² yr, which including the building and user electricity, space heating and domestic hot water energy demands, which exceeds a bit the latest requirement for new building constructions in the Swedish building code of 130 kWh/m² yr for non-electric resistance heated buildings of Swedish climate zone II in Stockholm area. The apartment block is facing the stage of both energy and cost-effectiveness retrofitting.

The strategy planning of the retrofitting work packages are mainly based on the typology of the building stock and the relevant energy system, the initial U-value of the insulation as well as the potential of electricity savings of the buildings and users, the retrofitting work packages are listed in Tab. 2.

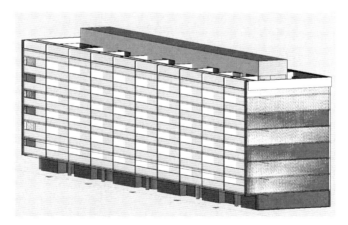

Fig. 4. 3D photographs of the apartment block before retrofitting (South-East)

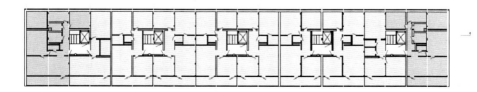

Fig. 5. Floor plan from 1-7th floor

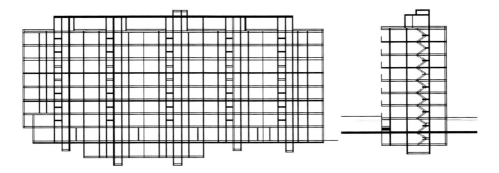

Fig. 6. Façade toward north and west (basement included)

5 Results

Table 2 shows the calculated results for the original building case and retrofitting investment costs with different improvements, including envelope insulation adding, installation of heat exchanger unit and relevant low energy facilities installations,. Figure 7 and 8 shows the differences of monthly energy flow for the base case and the case after retrofitting. The results illustrate a distinguished annual operational energy demand reduction 28.7% after retrofitting from 141. 7 kWh/m^2 yr to 101 Kwh/kWh/m^2 yr, which well meet the requirement of Swedish residential building energy code 130 kWh/m^2 yr, The improvements regarding the envelope aims at reaching the Swedish U-value regulations with a relative cost-effectiveness way. Figure 9 and Figure 10 shows the conclusion of energy reductions of different energy consuming elements, in which space heating and energy system have a distinguished decreasing after retrofitting due to a large efforts are made on the insulations on the envelope system based on the existing features of the target building.

The economic analysis based on the LCC calculated the cumulative operational embodied energy consumption after and before the retrofitting mainly focus on the material and energy perspectives, Figure 11shows a cumulative embodied operational energy costs in 30 years study from a life cycle perspective, the result illustrates the decreasing rates of energy costs are changing with time due to the variable of currency and energy prices, in this study, the energy prices are assumed with a constant rate and the inflation of currency are calculated by the discounting rate 4.4%, escalations rate 0,5% in the first five years and 1% in the following years.

82 The Application of LCCA toward Industrialized Building Retrofitting – Case Studies 939

Table 1. Building data inventory for features of building type and energy profile

Building Data Inventory		Energy Data Inventory	
Location	Upplands Väsby, Stockholm	Averaged daily Use Time	10h/person
Typology	Residential skivhus (High Slab House)	Indoor temperature winter	$22\,^{0}C$
No. of storeys	7	Air change rate	0.5
Room No. each floor	10	Heat exchange efficiency	0,6
External wall area (S)	$910\ m^{2}$	Fuel Type	Swedish el mix
Building ventilation system	Exhausted Air Ventilation	Heat emission system	Water radiator
Building ventilation volume	$18865\ m^{3}$		
Life time, years	50		
Averaged outdoor temperature	$6.3\ ^{o}C$	Heat generation	District heating
Set indoor temperature	$22\ ^{o}C$		
Roof	Asphalt felt insulation 200 mm Concrete 300 mm Steel sheet 20 mm	Energy for heating and ventilation	594,743kWh/yr
External Walls	Tile 200 mm Concrete 250 mm	Building and user electricity	205,800kWh/yr
Ground Floor	Gypsum 0.13 mm Concrete 300 mm		
Internal Walls	Concrete 250 mm Plaster 5mm Gypsum plasterboard 13 mm	Energy use for domestic hot water	171,500kWh/yr

Table 1. (*continued*)

Building Data Inventory		Energy Data Inventory	
Windows	Double glazing timber frame (South) Triple glazing aluminum frame (North)	Total Energy use (incl users energy and domestic hot water)	972,043kWh/yr
Doors	Wood Veneer with plastic film		

Table 2. Investment costs of retrofitting work packages for Skivhus and the corresponding U-value

Building elements	Retrofitting measure	Unit (W/m^2 K)	Initial Value	Reno- vated Value	Invest- ment (SEK)
Exterior wall	Exterior extra insulation	(W/m^2 K)	0,91	0,89	3 567 160,91
Roof	Ceiling insulation	(W/m^2 K)	1,7	0,18	772 705,50
Window	Replacements and re-design the area of the Southern side and North side	(W/m^2 K)	1,9 1,5	0,8 0,8	1 704 687,60
Internal wall	Interior extra 13mm gypsum insulation	-	-	-	948 499,86
Heat ex-changer	Heat exchanger unit installation	-	0	0,87	960 000,0
Public lighting	Low energy lighting and white wares	Each Floor	-	-	135 000,0
Facade	Re-paint the façade	m^2	-	-	535 074,14
Others	Household energy saving facilities recommenda-tions	Each Room	-	-	-

5.1 Energy Results

For the case building type, the retrofitting improvements explored related to changes in the building envelope and energy system generate relatively large building energy consumptions as well as the LCC of the building performance after retrofitting. The building envelope improvements giving the largest energy savings are increasing the external insulation in the exterior walls, ceiling extra insulations, windows re-design and replacements with a better U-value as well as the heat exchanger unit insulations. The retrofitting of the ground floor and the renewable energy sources applications are not considered due to the complexity of construction toward the case building and the climate condition of Stockholm. However, more low-electricity facilities can be further installed in the public and private areas if possible. The retrofitting packages with

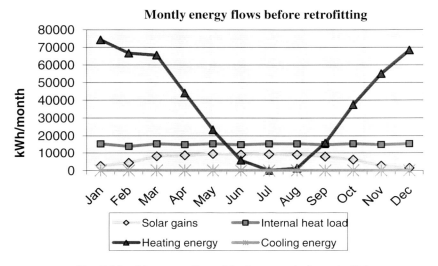

Fig. 7. Monthly energy flow of the base case before retrofitting

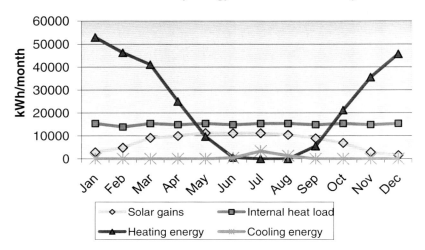

Fig. 8. Monthly energy flow of the apartment block after retrofitting

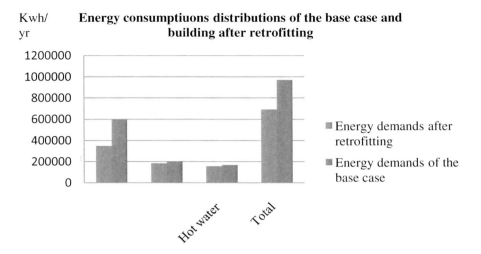

Fig. 9. Comparisons of energy consumption before and after implementing retrofitting work packages

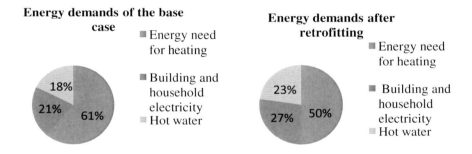

Fig. 10. Comparisons of the energy demands and consumer percentage for the base case and after retrofitting

the highest effect were changing the windows with the original one to high performance triple windows partial in the South and North area, as well as the insulation of the heat exchanger units due to the large façade area of the building and the large amounts of tenants, however, these results may different according to the typology of the building stock.

Figure 10 shows the proportional contribution changes to energy consumption from space heating, building and household electricity and domestic hot water respectively before and after retrofitting. As shown in the pie chart, space heating energy consumption was reduced from 61% to 50% of the total, which illustrates a relatively high efficiency of envelope insulation retrofitting compare with other retrofitting measures.

5.2 LCCA Results

Figure 11 shows the comparison of the cumulative embodied operational energy costs in a study years of 15, which are largely depending on the energy production and sources locally. The case study performed the price of the energy in Stockholm area, in which the heating are district heating with industrial waste and biofuels, which has a relatively high price compare with the electricity but lower CO_2 emission, the electricity is Swedish electricity mix with hydro and nuclear resources. In the performed 15 years study, the cumulative operation energy costs have a distinguished decreasing in the first five years and a following decreasing rate in the left years, due to the inflation of the currency has an impact on the energy price, however, if taken into considerations of the energy shortage crisis and escalations of the energy price, the results will be different, which are not considered in this study.

Figure 12 illustrates the portions of the cost elements from a 15 year life cycle perspective; compare with major investment costs of constructing a new building, the retrofitting project performs main costs of operational embodied energy costs 68%, 14% material costs, 11% labor costs and 4% maintenance costs and 3% other services costs, which shows energy saving measures not only contributes the annual building energy consumptions, but also to the whole life cycle costs of the target case building.

Figure 13 shows the calculated payback time of the retrofitting work packages, an 11.2 years payback was concluded for the case building. Due to a relatively large investment for the retrofitting and the typology of the apartment block, the payback time is longer than a multiple family house, however, with the increasing of the energy price and a partial adaption of the retrofitting work packages to other case buildings, the payback time may be shorten accordingly.

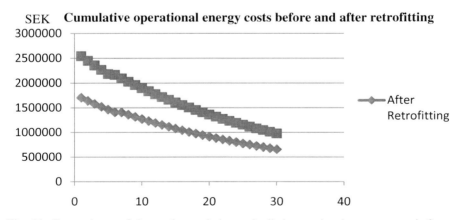

Fig. 11. Comparisons of the total cumulative embodied operational energy costs before and after retrofitting

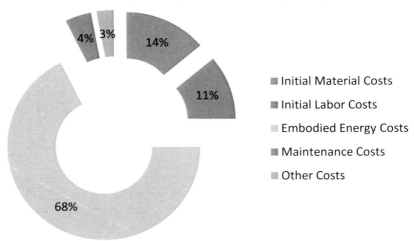

Fig. 12. Results of the cumulative life cycle cost percentage analysis in 15 years study

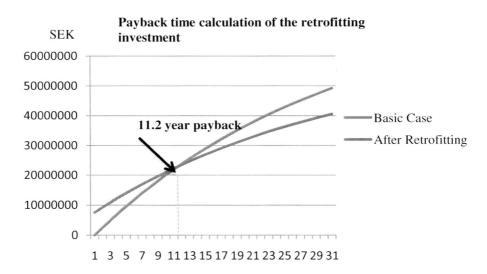

Fig. 13. Payback time calculation of the retrofitting work packages in 30 years study

6 Discussion

The life cycle costs analysis which used in this study are mainly focus on the material costs and energy costs in the phase of building retrofitting and performance after the proposed renovation work packages, which are listed in Table 2, with the updated values and investments including raw material, labor as well as relevant construction costs during implementing the retrofitting strategy, the retrofitting measures are collected based on the features of the case building type [3] and common prefabricated and isolated retrofitting alternatives [8][20]. The proposed building changes in the study demonstrate a possibility to reduce the operational energy in a life cycle perspective of 15 years by nearly around 49%. However, it is not concluded that all the similar Swedish apartment block should take the same work packages due to the fact that every building is unique, none less, the study give a rough general knowledge for the LCC based thinking of retrofitting decision making reference toward Swedish apartment block constructed in 1965-1971.

With the increasing interests for industrialized building retrofitting and the generation of efficient energy saving and high cost-effectiveness building retrofitting strategies, the study gave a simplified decision making approach to optimal the retrofitting alternatives for the similar building types. The chosen service life span is crucial to the cumulative operational embodied energy costs, additionally; the price of the energy will naturally influence the life cycle calculation of the energy costs, which are expected to be studied.

Externally, this study are based on some assumptions in the life cycle stages in terms of the maintenance costs, which are not detailed calculated due to the difficulties of the costs data for maintenance are hard to get and complexity, the energy calculations from CONSOLIS gave a rough quantified energy guidelines for comparison of the building performances before and after retrofitting, however, which don't precisely reflect the energy demands of the building due to the accuracy limitation of the tool, which needed to be further developed.

The study focused on one type of building case and its relevant economic analysis for the industrialized building retrofitting, this is a starting study of the amounts of variable Swedish building stock and their corresponding retrofitting strategies, in addition, optimal retrofitting work packages to one type of building stock are also expected to the future studies.

References

[1] The Building Performance Institute Europe. Europe's Buildings under the Microscope (2010)
[2] European Commission.EU action against climate changes; 2010 (March 28, 2010)
[3] Björk, C., Kallstenius, P., Reppen, L.: Så Byggdes husen 1880-2000
[4] Sarafoglou, N.: Regional efficiencies in building sectors research in Sweden. Comput. Environ. and Urban Systems 44, 117–132 (1990)
[5] Teknisk status i den svenska bebyggelsen-resultat från projektet BETSI, Boverket (2010)
[6] 2010 Swedish Production in Building and Construction Index
[7] RICS 2011 European Housing Review

[8]   Mjörnell, K.: Rationell isolering av klimatskärmenbpå befintliga flerbostadshus. Förstudie inför Teknikupphandling (2010)

[9]   Juan, Y., Gao, P., Wang, J.: A hybrid decision support system for sustainable office building renovation and energy performance improvement. Energy and Buildings 42, 290–297 (2010)

[10] Mark, G.: Optimizing renovation strategies for energy conservation in housing. Building and Environment, 583–389 (1995)

[11] Zhao, J., Zhu, N.: Technology line and case analysis of heat metering and energy efficiency retrofit of existing residential buildings in northern heating areas of China. Energy Policy 37, 2106–2112 (2009)

[12] Jinlong, O.: Economic analysis of energy-savings renovation measures for urban existing residential buildings in China based on thermal simulation and site investigation. Energy Policy 37 (2009)

[13] Verbreeck, Hens, G.H.: Energy saving in retrofitted dwellings: economically viable? Energy and Buildings 37 (2005)

[14] Jinlong, O.: A methodology for energy-efficient renovation of existing residential buildings in China and case study. Energy and Buildings 42 (2011)

[15] P Agis M., Feasibility of energy saving renovation measures in urban buildings. The impact of energy prices and the acceptable payback time criterion. Energy and Buildings, 455–466 (2002)

[16] Gudni, J.: Building Energy-A pratical design tool meeting the requirement of the EPBD, KTH-IBRI. In: 7th Nordic Building Physics Symposium (2005)

[17] Gudni, J.: Building Energy-A design tool meeting the requirements for energy performances standards and early design-validation. In: International Building Physics/Science Conference, IBPC3 (2006)

[18] Dhillion, B.S.: Life cycle costing for engineers

[19] Stanford University Land and Buildings Department, Guidelines for life cycle cost analysis (2005)

[20] Levin, P., Larsson, A.: Projektengagemang, Rekorderlig Renovering-demonstrationsprojekt för energieffektivisering i befintliga flerbostadshus från miljonprogramstiden (2011)

[21] Wikells, Sektionnsfakta –Rot, Teknisk-ekonomisk sammanställning av rot.byggdelar (2012)

Chapter 83
A Proposal of Urban District Carbon Budgets for Sustainable Urban Development Projects

Aumnad Phdungsilp[*] and Ivo Martinac

Division of Building Service and Energy Systems
Royal Institute of Technology, Stockholm 100 44, Sweden
aumnadp@kth.se

Abstract. Energy security and carbon emissions are key issues for policy-makers and research communities worldwide. Climate change mitigation poses many challenges for all levels of society. Energy-related carbon emissions in urban areas have received a great deal of attention. This paper builds on the principle that urban areas are major sources of emissions and play an important role in the carbon cycle. Urban development can serve as a cornerstone for achieving transition towards a sustainable city. This paper proposes and describes a framework for carbon budgets with a focus on urban district level. The urban district carbon budget is a mechanism for embedding long-term total emission restrictions into the urban economy. This paper proposes a proposal of urban district carbon budgets in an effort to provide the figure for emission allowances that can be emitted in a given amount of time. The paper presents a design framework of urban district carbon budgets and discusses the scope and scale of carbon budget allocation approaches. It also examines the emission reduction potential and co-benefits of the proposal.

Keywords: urban districts, carbon budgets, climate change, sustainability.

1 Introduction

Interest in the sustainability of urban areas has increased in recent years. The World Bank reported that cities are responsible for as much as 80 percent of global green-house gas (GHG) emissions, while they are also significantly impacted by climate change (World Bank 2011). Urban areas are having an impact on the local and global environment that is mobilising political representatives in all countries.

Urban sustainability is growing in importance in energy and climate research. Consequently, academics, consultants, national and international bodies have developed a large number of assessment tools (Devuyst 2001; Jensen and Elle 2007; Iverot and Brandt 2011). In order to reduce emissions as well as the flows of urban metabolism (energy and materials), it is important to facilitate the integration of technical and policy innovations into urban systems. In future urban districts, previous study

---

[*] Corresponding author.

A. Håkansson et al. (Eds.): *Sustainability in Energy and Buildings*, SIST 22, pp. 947–954.
DOI: 10.1007/978-3-642-36645-1_83   © Springer-Verlag Berlin Heidelberg 2013

suggests to include a clear structure of the assessment process, which would ensure the quality of gathered data and facilitate the development of sustainable urban districts (Iverot and Brandt 2011).

Urban districts can be a testing ground for new solutions. Urban areas are assigned an emissions budget and must keep local emissions (i.e., transport and building) within this figure. Previous research (Salon et al. 2010) has proposed making local government responsible for deciding which strategies to use and how to implement them. Such strategies could include building codes for new construction, use of renewable technologies, congestion zones and car-share schemes. Two suitable methods of budget allocation, based on current emissions in each locality, have also been identified. The first suggested that the budgets be adjusted over time to arrive at a single per capita emission level across all localities. The second suggestion would be to reduce budgets each year by a given percentage according to a predetermined schedule.

While urban researchers have long emphasised the importance of planning decisions, a shift has also emerged towards market-based approaches and the carbon economy. Fig. 1 shows that a growing number of cities around the world have taken actions and initiatives to reduce their GHGs or CO_2 emissions. Some cities have already adapted the concept of carbon budgets under different targets and base years. According to Fig. 1, it shows emission reduction targets in relative emission amount compared to the base year (that is, 1990, 2000 and 2005).

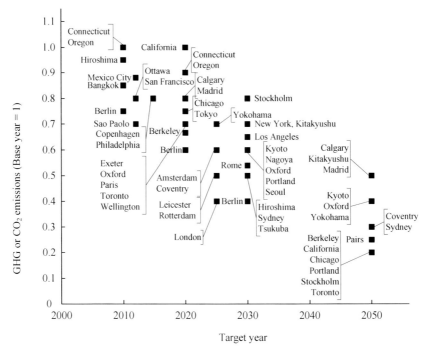

Fig. 1. Emission targets of the cities across the world (Source: Gomi et al. 2010; Global Carbon Project 2011; C40 Cities 2011)

This paper proposes and describes a framework for carbon budgets with a focus on urban district level. The urban district carbon budget is a mechanism for embedding long-term total emission restrictions into the urban economy. The paper aims to contribute to the emerging literature on urban sustainability in the context of climate change. The objectives are to propose the concept of urban carbon budgets and to highlight the application to urban development projects. The paper describes the emissions base year, methodological approach for carbon allocating, and co-benefits of carbon budgets.

This work was inspired by the path towards a low-carbon society and the call for environmental focus at the urban scale to become a sustainable urban district. There are various supporting reasons for developing a carbon budget approach for urban districts. Urban districts are differences, both in terms of urban scale and characteristics. Local government could differentiate and target its politics in relation to the various characteristics of urban development projects. There are also some earlier urban district initiatives, which aim to reduce the district's resource consumption and emissions in order to create a sustainable urban district. Examples of this are the Sydney Olympics in 2000 (Newman 1999) and Hammarby Sjöstad (Iverot and Brandt 2011). The carbon budget approach can be an effective policy instrument for sustainable urban development projects.

2 Methodology

2.1 The Sustainable Urban District

Sustainability means different things to different people in different situations. Consequently, sustainable urban district results in different definitions. In this paper, the sustainability of urban district only relates to the environmental aspects of sustainability. The social and economic aspects are co-benefits from the emission reductions of the urban systems. This is because the objective of this work is to propose a framework in the planning and development of the urban district in order to mitigate and manage carbon emissions.

Girardet (1992) used the metabolism approach for a study of sustainable city in which he argued for the circular metabolism of sustainable cities from the linear metabolism of modern cities. In fact, the linear metabolism has led to increased consumption of natural resources and the disposal of emissions and waste in the environment. Girardet used a biological perspective to argue that outputs from the city (i.e., exhaust gases, waste and sewage) must be used to generate energy, new materials and plant nutrients that should be imported to the city again. This perspective has attracted increasing attention in the scientific community. As the urban metabolism defined by Girardet was based on biological, thereafter Newman (1999) extended the model (Fig. 2) and stated that *"it is possible to define the goal of sustainability in a city as the reduction of the city's use of natural resources and production of wastes while simultaneously improving its liveability, so that it can better fit within the capacities of the local, regional, and global systems"* (Newman 1999).

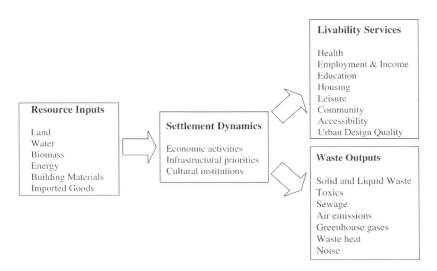

Fig. 2. The metabolism model of the city (Adapted from Newman 1999)

2.2 Carbon Budget Approaches

The urban carbon emissions can be broadly separated into four categories: mobile sources, static local sources, semi-static sinks, and remote sources. In this study, the carbon budgets approach focuses on energy-related CO_2 emissions; carbon stocks and sequestration in the natural materials are not included. The energy-related CO_2 emissions can be further sub-divided into several energy services, including heating, cooling, mobility and lighting. Across these energy services, the energy sources are differentiated, including electricity, natural gas and fossil fuels.

According to Fig. 1, cities used a variety of methodologies to set their targets. Many cities utilized existing international methodologies, such as the ICLEI Local Government Operations Protocol or the Intergovernmental Panel on Climate Change (IPCC). Some cities applied a combination of approaches with their own proprietary methodologies that fit local circumstances.

The Carbon Disclosure Project (CDP) argued that although cities recognize the value of international protocols, they still find them lacking. There is no preferred methodology. Many cities have adopted a number of software tools, with varying levels of sophistication. The most common involve spreadsheets (CDP 2011).

The various sizes, populations and development patterns of the urban development projects means it is difficult to decide on a standard mechanism for carbon budgeting that is fair to all. There is no correct way of allocating carbon budgets; several methods are possible. In any case, the carbon budget allocation method should satisfy two criteria: (i) it should have a clear and predetermined schedule for what the carbon budgets will be in the future; and (ii) it should encourage economic growth (Salon et al. 2010). Based on previous works by Dhakal (2008), Fong et al. (2008), Salon et al. (2010) and Kennedy et al. (2010), the present study proposes the five following alternative carbon budget allocation methods.

(1) Allowance auctioning. In carbon cap-and-trade policy regimes, auctions are often promoted as being economically efficient mechanisms to allocate responsibility for reducing emissions (Burtaw et al. 2001). However, devolution of a portion of emissions reduction responsibility to lower levels of government is fundamentally different from allocating emissions reduction responsibility to polluters.

(2) The equal share approach. This approach is based on the assumption that urban districts have the same degree of responsibility for emission reductions towards achieving national targets. For example, if a country's CO_2 emissions target is 20 percent below the 1990 level, then all urban districts should be responsible for reducing CO_2 emissions to 20 percent below the 1990 level by 2050.

(3) The per capita approach. The carbon budgets (CB) of each urban district are decided based on the proportion of population (POP) of the district compared to the national population. This can be expressed in the following equation:

$$CB_{district, \ per \ capita} = \frac{CB_{country}}{POP_{country,2010}} \times POP_{district,2010} \tag{1}$$

This equation is assumed using the 2010 population as a baseline. This approach can be assigned uniform allocation on a per capita basis, with a predetermined schedule for reducing the allocation over time so that everyone is allowed to emit the same level. It can also use baseline per capita emissions based on current emissions as a starting point and transition to a uniform allowance allocation. It can be implemented by setting budgets equal to baseline or current emissions and reducing them by a given percentage each year according to a predetermined schedule.

(4) The economic approach. This is based on a country's gross domestic product (GDP) and urban district's gross product (GP) values. The carbon budgets (CB) are decided by the subject urban district's GP as a proportion of national GDP. However, the available data and data collection might be questionable in this approach. The economic approach can be expressed in the following equation:

$$CB_{district, \ GP} = \frac{CB_{country}}{GDP_{country,2010}} \times GP_{district,2010} \tag{2}$$

(5) The consumption-based approach. This approach allocates all upstream emissions to goods and services consumption in the urban district (that is, energy and material flows along the whole production chain, whether they took place inside or outside the boundary). This approach contrasts with the production-based approach, which accounts for the emissions produced within the urban boundaries.

2.3 Emission Coverage

The carbon budgets approach is an urban responsibility concept that reduces GHG or CO_2 emissions at a local level. The portion of responsibility for emission reductions corresponds either to direct or indirect sources. In urban districts, large portions of these emissions include those from transport and building sectors. It is important to develop a monitoring, reporting and verification (MRV) system for emissions to the locality.

An accurate emission inventory is the key to the success of the carbon budgets approach. Decision-makers will have to decide which emissions (e.g., GHG, CO_2, CO_2-equivalent) will be included in the inventory. Apart from the energy systems and transportation, other emission sources have high potential reductions, such as waste management, water system, deforestation and changes in urban vegetation.

3 Findings and Discussion

In a proposal of urban carbon budgets, urban districts are the points of action for emission reductions. Urban district carbon budgets can complement a city's efforts to combat with climate change, as well as the development of energy infrastructures. The co-benefits of action are substantial, including energy savings, security of energy supply systems, improved public health, and cost savings. City and national governments can serve an important function in formulating policy and providing information, technical and financial support.

The co-benefits approach addresses both global and local environmental problems while contributing to local development needs. In developing countries in particular, energy and climate co-benefits make it possible to overcome many environment and development challenges with limited capacity and resources within a rapidly growing urban area.

While urban areas are often identified as sources of emissions, they are also part of the solution. Local decision-makers have good access to relevant stakeholders and have the ability to identify or develop projects with high local co-benefits, other than emission reductions. Wilbanks and Kates (1999) argued that analysis at the local level provides useful insights into ways of maximizing the cost-effectiveness of carbon management strategies. Previous studies have shown that addressing climate change issues often reveals climate policy linkages through co-benefits related to activities in energy efficiency, air pollution and transport management (Cifuentes et al. 2001; Dhakal 2006).

Many efforts are now underway to reduce carbon emissions from urban areas. New tools, such as GHG emissions standard for cities, Urban Risk Assessment, and Global City Indicator Facility, are being launched in an effort to manage the efficient resource use and energy-related carbon emissions of cities. New financing options, such as Green Bonds, a city-wide approach to carbon finance (now proposed in Amman, Jordan), and Emissions Trading Systems such as the one launched in Tokyo, are also being implemented (World Bank 2011). However, new urban development projects should be responsible for deciding which set of emission reduction strategies to pursue, for selecting and implementing those strategies and actions. There are some interesting examples of sustainable urban development projects that help to reduce emissions and are also models for urban development measures; these include Norra Djurgårdsstaden and New Albano, both in the city of Stockholm (The Delegation for Sustainable Cities 2012).

Moreover, carbon markets have become an important new source of funding. Urban development projects can generate their revenues by selling of carbon credits; for example, the Certified Emission Reductions (CERs) through Clean Development Mechanism (CDM) project. Many urban projects, such as energy efficiency improvements, energy conservation, fuel switching, and waste management, are provided such promises (Dhakal 2008). The implementation of carbon budgets will provide a blueprint for future urban development projects striving for sustainability and will serve as a model for how all future urban projects should be built.

4   Conclusion

Urban areas will be challenged with a twofold task. On the one hand, they must provide distinctive and high-quality places that can compete on a regional or global scale. On the other hand, they must develop responsive solutions so that their effective functioning will be secured and the needs of their citizens satisfied.

The proposed methodological framework of this study is laid out in order to fulfill the criteria of the national inventories; that is, so that they are transparent, consistent, comparable, complete and accurate. A proposal of urban carbon budgets can be used as a policy instrument for embedding long-term total emission restrictions into both existing and new urban development projects. Urban carbon budgets will include political and regulatory negotiations, research and development of technological innovations, and coordinate between citizens, project developers and local government. It is highly recommended that urban development projects are free to decide which set of emission reduction strategies to pursue and implement.

Since climate change has already happened, urban areas will have to adapt by limiting the impacts of climate change. Urban districts should pursue both climate change mitigation and adaptation. Therefore, the future development of carbon budgets approach could be extended to address adaptation issue as well.

References

[1]   Burtraw, D., Palmer, K., Bharvirkar, R., Paul, A.: The effect of allowance allocation on the cost of carbon emission trading. Resources for the Future Discussion Paper 01-30 (2001), http://www.rff.org (as accessed on April 8, 2011)

[2]   C40 Cities–Climate Leadership Group, HP, http://www.c40cities.org (as accessed on April 8, 2011)

[3]   Carbon Disclosure Project(CDP) CDP Cities 2011: Global Report on C40 Cities. Report (2011), http://www.cdproject.net (as accessed on May 21, 2012)

[4]   Cifuentes, L., Borja-Aburto, V.H., Gouveia, N., Thurston, G., Davis, D.L.: Assessing the Health Benefits of Urban Air Pollution Reductions Associated with Climate Change Mitigation (2000-2020): Santiago, São Paulo, Mexico City, and New York City. Environmental Health Perspectives 103(3), 419–425 (2001)

[5]   Devuyst, D., Hens, L., De Lannoy, W. (eds.): How Green Is the City?: Sustainability Assessment and the Management of Urban Environments. Columbia University Press (2001)

[6] Dhakal, S.: Urban Transport and Environment in Kathmandu Valley Nepal: Integrating Global Carbon Concerns into Local Air Pollution Management. Institute for Global Environmental Strategies, Hayama, Japan (2006)

[7] Dhakal, S.: Climate Change and Cities: The Making of a Climate Friendly Future. In: Droege, P. (ed.) Urban Energy Transition – From Fossil Fuels to Renewable Power. Elsevier Publications, Oxford (2008)

[8] Fong, W.K., Matsumoto, H., Lun, Y.: Establishment of City Level Carbon Dioxide Emission Baseline Database and Carbon Budgets for Developing Countries with Data Constraints. Journal of Asian Architecture and Building Engineering 7(2), 403–410 (2008)

[9] Girardet, H.: Cities: New directions for sustainable urban living. Gaia Books, London (1992)

[10] Global Carbon Project, HP,
http://www.gcp-urcm.org/Resources/CityActionPlans
(as accessed on April 8, 2011)

[11] Gomi, K., Shimada, K., Matsuoka, Y.: A low-carbon scenario creation method for a local-scale economy and its application in Kyoto city. Energy Policy 38(9), 4783–4796 (2010)

[12] Iverot, S.P., Brandt, N.: The development of a sustainable urban district in Hammarby Sjöstad, Stockholm, Sweden? Environment, Development and Sustainability 13, 1043–1064 (2011)

[13] Jensen, J.O., Elle, M.: Exploring the Use of Tools for Urban Sustainability in European Cities. Indoor and Built Environment 16(3), 235–247 (2007)

[14] Kennedy, C., Steinberger, J., Gasson, B., Hansen, Y., Hillman, T., Havranek, M., Pataki, D.E., Phdungsilp, A., Ramaswami, A., Mendez, G.V.: Methodology for inventorying greenhouse gas emissions from global cities. Energy Policy 38(9), 4828–4837 (2010)

[15] Newman, P.W.G.: Sustainability and cities: Extending the metabolism model. Landscape and Urban Planning 44, 219–226 (1999)

[16] Salon, D., Sperling, D., Meier, A., et al.: City carbon budgets: A proposal to align incentives for climate-friendly communities. Energy Policy 38(4), 2032–2041 (2010)

[17] The Delegation for Sustainable Cities, HP,
http://www.hallbarastader.gov.se (as accessed on April 30, 2012)

[18] Wilbanks, T.J., Kates, R.W.: Global change in local places: how scale matters. Climate Change 43, 601–628 (1999)

[19] World Bank, HP, http://www.worldbank.org/climatechange
(as accessed on January 24, 2011)

Chapter 84
A Study of the Design Criteria Affecting Energy Demand in New Building Clusters Using Fuzzy AHP

Hai Lu[*], Aumnad Phdungsilp, and Ivo Martinac

Division of Building Service and Energy Systems
Royal Institute of Technology, Stockholm, Sweden
hai.lu@byv.kth.se

Abstract. The level of concern regarding the total energy consumption in new building clusters/urban districts (BCDs) has increased recently. Rising living standards have led to a significant increase in building energy consumption over the past few decades. Therefore, along with sustainability requirements, it is essential to establish an effective and precise energy demand model for new building clusters/districts. In principle, energy demand in building clusters is hard to plan and pre-calculate because a number of design criteria influence energy performance. Establishing such a model would require a decision-making base, and the present study proposes two methods for achieving this objective. The study uses general survey aims to collect and identify the design criteria that affect the energy demand model and to evaluate the priorities of each criterion using the fuzzy analytical hierarchy process (AHP) method. Four main criteria – location, building design, government and cluster design – are established, along with a total of 13 secondary criteria. The results show that the use of the AHP method can accurately guide the energy demand model and automatically rank significant criteria. The method can provide the weighting value for each criterion as well as the relative ranking for the energy demand building model. According to the sustainability concept, one crucial benefit is an improvement in the energy performance of building clusters/urban districts and a reduction in energy consumption. Another advantage of this methodology is that it can provide accurate energy input for future energy supply system optimisation.

Keywords: Energy demand, New building clusters/urban districts, Design criteria, Fuzzy AHP.

1 Introduction

Increasing demand for sustainability has meant that concerns about energy quality management (EQM) have become the new focus for energy systems design. EQM represents the management of overall condition of whole energy chain, from energy

---
[*] Corresponding author.

A. Håkansson et al. (Eds.): *Sustainability in Energy and Buildings*, SIST 22, pp. 955–963.
DOI: 10.1007/978-3-642-36645-1_84 © Springer-Verlag Berlin Heidelberg 2013

generation to end-use and covers a wide range of issues, from energy demand to energy supply systems. EQM must also be highlighted when developing the design and planning procedures of new building clusters/urban districts (BCDs). BCDs are more complicated than individual buildings because they cover more parameters, such as surrounding conditions and relationship of individual buildings. Trias energetica is a simple and logical concept created by Delatter (2006) that helps achieve energy savings, reduce fossil fuel dependence, and decrease the environment burdens. The initial step is to minimise energy demand from the Trias energetica view in the 4th Annex 44 forum (2006). To avoid extra energy waste and keep indoor climate comfortable, it is critical to calculate energy demand as accurately as possible.

2   Objectives

The studies on energy performances for BCDs are very limited, especially those that focus on the energy issue. The energy consumption is influenced by many criteria, from individual building characteristics to cluster characteristics. That means it is critical to clarify the relation between energy demand and BCD design criteria. This paper aims to describe and understand the following issues for new BCDs:

- Reasonable methodology for helping designers select the most appropriate criteria for BCDs energy demand
- Comparison of the importance different design criteria using fuzzy AHP.

This paper is organised as follows. Section 3 briefly introduces the fuzzy AHP method. Section 4 discusses the establishment of a hierarchical structure and identifies the design criteria. Sections 5 and 6 provide the results and concluding remarks of this study.

3   Methodology Analysis

3.1   Fuzzy Analytic Hierarchy Process

In order to distinguish the important criteria affecting energy demands in new BCDs, multi-criteria decision-making (MCDM) has been applied widely in energy studies, such as those by Hobbs and Meier (2000), Phdungsilp and Martinac (2004), Zhou et al. (2006), Lee and Hwang (2010). Comprehensive reviews of MCDM application in the area of energy studies can be found in Pohekar and Ramachandran (2004) and Wang et al. (2009). To compare with these MCDM approaches, the most important benefit of Analytic Hierarchy Process (AHP) is relative simple mathematical analysis process. This benefit makes it easy to control and be proficient, especially for policy-maker and designer without enough mathematics knowledge. Therefore, AHP has been broadly used in the MCDM approach. This method can be used to specify numerical weights representing the relative rankings of different design criteria as well as design factors. Numerical weight is a comparative index of a specific criterion and can stand for the relative importance of the criterion in the whole system. The higher the numerical weight, the more important the criterion is.

Earlier AHP for basic management and decision-making studies were reported included by Cheung et al. (2001), Cheng and Li (2002) and Cheung and Suen (2002). Johnny et al. (2008) applied AHP methodology for intelligent building systems. The methodology was conducted to prioritise and assign the important weightings for the perceived intelligent building criteria that had been selected and identified in the literature review process.

However, in applying AHP, it is easier and more user-friendly for evaluators to assess "criterion A is much more important than criterion B" than it is to consider "the importance of principle A and principle B is seven to one", as proposed by Hsieh (2004). This comparison means it is easier for designers and developers to use fuzzy concepts, such as more importance, less importance and equal importance, than a certain scale like the nine-point scale proposed by Saaty (1980). Therefore, Buckley (1985) extended traditional AHP to the case where the evaluators were allowed to employ fuzzy ratios in place of exact ratios in order to handle the difficulty of assigning exact ratios when comparing two criteria and deriving the fuzzy weights of criteria by the geometric mean method. For this reason, the fuzzy AHP methods were introduced into decision-making process and systematically studied by researchers, such as Deng (1999), Sheu (2004) and Eunnyuong et al. (2010). Owing to the fuzzy concept, the results covered a relatively wide range, from technological, market-related and economic, to environmental and policy-related.

3.2 Fuzzy AHP Structure Model

AHP structures the problems by decomposing them into a hierarchy and influencing a system by incorporating different levels: objectives, sub-objectives, criteria and sub-criteria according to Crowe's research (1998). The first step is to design the pairwise comparison matrix. In this step, the relative importance of each criterion should be investigated from a subjective viewpoint. The relative importance of the criteria and sub-criteria is assigned using fuzzy numbers. Fuzzy numbers are a fuzzy subset of real numbers, representing the expansion of the idea of the confidence interval. According to the definition of Laarhoven and Pedrycz (1983), a triangular fuzzy number (TFN) should possess basic features listed in Table 1, which indicates the level of relative importance using TFN.

Table 1. TFN of pairwise comparison scale (Chiou, 2001)

Intensity of weight	Definition	TFN
$\tilde{1}$	Equal importance	(1,1,3)
$\tilde{3}$	Weak/moderate importance of one over another	(1,3,5)
$\tilde{5}$	Essential or strong importance	(3,5,7)
$\tilde{7}$	Very strong or demonstrated importance	(5,7,9)
$\tilde{9}$	Absolute importance	(7,9,9)
$\tilde{2}\,\tilde{4}\,\tilde{6}\,\tilde{8}$	Intermediate values between the two adjacent scale values	

The subsequent step is to calculate the weight value of design criteria. The root method is utilised for this fuzzy AHP analysis. The comparison matrix was established by the pairwise comparisons and expressed as $\widetilde{A}$:

$$\widetilde{A}=(\tilde{a}_{ij})_{n*n} \tag{1}$$

$\tilde{a}_{ij}$ represents the pairwise fuzzy comparison value of criterion i to criterion j.

The geometric mean technique can be used to define the fuzzy geometric mean and fuzzy weights of each criterion as below:

$$\widetilde{M}_i=\tilde{a}_{i1}\otimes \tilde{a}_{i2}\otimes \cdots \otimes \tilde{a}_{in},\ i=1,2,\cdots,n \tag{2}$$

$$\overline{\overline{M}}_i=\sqrt[n]{\widetilde{M}_i},\ i=1,2,\cdots,n \tag{3}$$

$$\widetilde{\omega}_i=\overline{\overline{M}}_i\otimes(\overline{\overline{M}}_1\otimes \overline{\overline{M}}_2\oplus \cdots \oplus \overline{\overline{M}}_n)^{-1},\quad i=1,2,\cdots,n \tag{4}$$

$\widetilde{M}_i$ is the geometric mean of the fuzzy comparison value of criterion i to each criterion. $\widetilde{\omega}_i$ is the fuzzy weight of the *i*th criterion and can be indicated as below:

$$\widetilde{\omega}_i =(L\omega_i,\ M\omega_i, U\omega_i) \tag{5}$$

Here, $L\omega_i$, $M\omega_i$ and $U\omega_i$ stand for the lower, middle and upper values of the fuzzy weight of *i*th criterion. After the fuzzy weight confirmation, the next step is the procedure of defuzzification. This procedure is to locate the Best Non-fuzzy Performance value (BNP):

$$BNP_i=[(U\omega_i- L\omega_i)+(M\omega_i- L\omega_i)]/3+ L\omega_i \tag{6}$$

The weight values can be expressed as below:

$$W= (BNP_1, BNP_2,\ldots, BNP_n)^T\ (\textstyle\sum_{j=1}^{m} BNP_j=1) \tag{7}$$

The BNP of every element could show its own relative prioritisation for the objective or sub-objective.

4 Identifying Criteria and Sub-criteria

4.1 Main Criteria and Sub-criteria

The first step in identifying and collecting the criteria and sub-criteria is to review the papers and reports about building energy/power demands simulation. We can then choose the significant design criteria that affect energy demand. Obviously, BCDs' energy demands are influenced by many factors, such as their location. The initial step is to make sure the location of BCD, which could affect the outdoor climatic conditions.

Apart from this criterion, the energy demand in new BCDs is also affected by building architectural design. Many crucial elements related to building energy performance are determined by architectural design, including the orientation of buildings, building glazing and building envelope (Rolsfman, 2002; Fesanghary, 2012). Building use patterns also influence individual building energy demand. Many authors (e.g., Chow, 2004; Kari, 2007; Sofia 2010; Webber, 2011) have concluded that individual building energy demand is influenced by building users and occupants, building using time or opening time and building types (functions). As a relative new research focus, indoor climate comfort has been considered in energy/power demand simulation process (Kari, 2007; Sofia, 2010).

The next step is a critical one that complements certain criteria for the identifying process, especially for cluster characteristics and regulations and rules. Little research has highlighted cluster characteristics influence for energy/power demand. The work (Chow, 2004) showed that BCDs could simplify the estimation of thermal load demands to multiply the number of building models. It is relatively simple and inaccurate for the BCDs energy/power model. Therefore, future work should highlight criteria regarding outdoor surroundings and the relation between individual buildings.

The final criterion that should be emphasised relates to regulations and rules. Building code is one type of regulation and it provides a maximum energy consumption limitation for BCDs.

4.2 Description of Main Criteria and Factors

The literature survey identified four main criteria: location of BCDs, building characteristics, regulations and standards, and cluster characteristics. Table 2 presents the main criteria and sub-criteria.

Table 2. Criteria and sub-criteria effecting energy/power demands

Main criteria	Sub-criteria
S1. Location of BCDs	SS1. Climatic conditions
	SS2. Topography
S2.Building characteristics	SS3. Building functions
	SS4. Indoor climate requirements
	SS5. Building operation patterns
	SS6. Users & occupants
	SS7. Building architectural design (Building envelops and building HVAC systems)
S3.Regulations & standards	SS8. Building standards
	SS9. Energy saving regulations
S4. Cluster characteristics	SS10. Outdoor area design
	SS11. Cluster density and size
	SS12. Outdoor transportation
	SS13. Cluster configuration

5 Results

The results can be divided into local and global results. The local results focus on the ranking of main criteria for different energy demand objectives, while the global results show the priories of each sub-criterion for holistic energy demand.

5.1 Local Results

Four energy demand sub-objectives –space heating, electricity, domestic hot water, and cooling – were selected for local results analysis. BNPs were used to subjectively prioritise all criteria and sub-criteria. The criterion with the highest BNP is the most important criterion for specific energy demand. Figure 1 shows the BNPs of the four criteria.

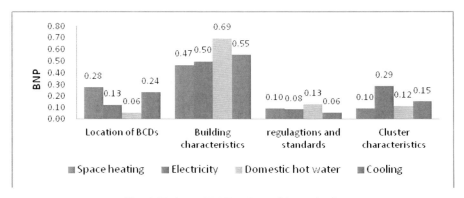

Fig. 1. Estimated BNP values of four criteria

All sub-objective groups estimated S3 to have the lowest importance and S2 to have the highest. For other two criteria (S1 and S4), the estimated BNPs were different in different sub-objective groups. S1 is the second-highest BNP in space heating and cooling groups, but it had no impact on the domestic hot water group. S4 was estimated to have the second-highest BNP for electricity groups but very limited influence on the other three groups. S2 and S1 were the most important for the space heating and cooling group. S2 and S4 had the highest importance in the electricity group. S2 was dominant in the domestic hot water group and the location of BCDs criterion had little impact in this group.

5.2 Global Results

Table 3 presents the results of the global priorities calculated by multiplying the BNPs of the criteria (parents) by the BNPs of factors (siblings).

84 A Study of the Design Criteria Affecting Energy Demand in New Building Clusters 961

Table 3. Results of global priorities

Factors	Overall energy	Space heating	Electricity	Domestic hot water	Cooling
SS1	0.154	0.243	0.106	0.019	0.196
SS3	0.136	0.114	0.114	0.198	0.137
SS7	0.132	0.188	0.093	0.031	0.180
SS6	0.103	0.064	0.093	0.208	0.079
SS5	0.095	0.037	0.093	0.208	0.079
SS4	0.067	0.064	0.087	0.033	0.079
SS8	0.057	0.134	0.011	0.013	0.043
SS9	0.053	0.027	0.074	0.119	0.014
SS12	0.051	0.033	0.092	0.020	0.058
SS11	0.040	0.033	0.042	0.026	0.058
SS10	0.037	0.010	0.113	0.016	0.012
SS2	0.033	0.035	0.021	0.039	0.039
SS13	0.033	0.018	0.042	0.053	0.027

The climatic conditions factor (SS1) was estimated to have the highest BNP among all the factors in the calculation of the overall energy demand (the highest in space heating and cooling group, but little impact for the domestic group). The building functions factor (SS3) and the building architectural design factor (SS7) tied for the second place in overall energy demand. The factors with the third-highest BNPs in total energy demand were the user and occupant factor (SS6) and the building operation pattern factor (SS5). The indoor climate requirement (SS4), building standards (SS8) and energy saving regulations (SS9) ranked equal fourth in overall energy demand. Interestingly, three factors (SS12, SS11, SS10) belonging to S4 had almost the same importance in terms of overall energy demand.

6 Conclusion

We established four criteria (location of BCDs, building characteristics, regulations and standards, and cluster characteristics) and a total of 13 sub-criteria using fuzzy AHP in order to evaluate the priorities of different factors. Several conclusions can be derived from the findings.

- Firstly, climatic conditions, building architectural design and building functions (including outdoor areas) were considered the most important of all the factors. The reason is that these elements are basic and essential components for BCD design.
- Secondly, factors such as building operation patterns, users and occupants, and indoor climate requirements, are of critical importance because more and more attention had been paid to social impacts. Apart from these factors, the building

standards factor and the energy savings regulation factor also played an important role in energy demand. As for BCDs, it is not sufficient to have only building characteristics criterion; consideration from cluster characteristics criterion is necessary. With increasing concern about BCDs energy/power demand, factors belonging to cluster characteristics will become more important.

- Thirdly, there were significant differences between the estimated priorities for different groups. The most disparate factor was the location of BCDs. The space heating group estimated the factor to be nearly 12.8 times more than the domestic hot water demand.

- At last, the analysis results could be described as following: electricity demand is mainly affected by building functions, users & occupants and using times as well as outdoor area electricity consumption. Thermal energy (space heating and cooling) are influenced by building envelopes, building HVAC systems, indoor & outdoor climatic conditions as well as heat generated by users & occupants. Domestic hot water demand is determined by building occupants and using times. All energy demands should obey specific standards and regulations.

References

1. Christof, D.: Regional and local strategies of development of use of renewable energy. Presented, Wroclaw, Poland, March 23-24 (2006)
2. Ad V.D.A. The Trias energetic – An integrated design approach. In: 4th Annex 44 Forum (2006)
3. Hobbs, B.F., Meier, P.: Energy Decisions and the Environment: A Guide to the Use of Multicriteria Methods. Kluwer Academic Publishers, Boston (2000)
4. Phdungsilp, A., Martinac, I.: A Multi-Criteria Decision-Making Method for Retrofitting of Designated Buildings in Thailand. In: Proceedings of the 21st Conference on Passive and Low Energy Architecture, Eindhoven, Netherlands, September 19-22 (2004)
5. Zhou, P., Ang, B.W., Poh, K.L.: Decision analysis in energy and environmental modeling: An update. Energy 31, 2604–2622 (2006)
6. Lee, D.J., Hwang, J.: Decision support for selecting exportable nuclear technology using the analytic hierarchy process: A Korean case. Energy Policy 38, 161–167 (2010)
7. Pohekar, S.D., Ramachandran, M.: Application of multi-criteria decision making to sustainable energy planning – A review. Renewable and Sustainable Energy Reviews 8, 365–381 (2004)
8. Wang, J.J., Jing, Y.Y., Zhang, C.F., et al.: Review on multi-criteria decision analysis aid in sustainable energy decision-making. Renewable and Sustainable Energy Reviews 13, 2263–2278 (2009)
9. Cheung, S.O., Lam, T.I., Leung, M.Y., et al.: An analytical hierarchy process based procurement selection method. Construction Management and Economics 19, 427–437 (2001)
10. Cheung, S.A., Suen, H.C.H.: A multi-attribute utility model for dispute resolution strategy selection. Construction Management and Economics 20, 557–568 (2002)
11. Johnny, K.W., Wong, H.L.: Application of the analytic hierarchy process (AHP) in multi-criteria analysis of the selection of intelligent building systems. Building and Environment 43, 108–125 (2008)

12. Cheng, E.W.L., Li, H.: Construction partnering process and associated critical success factors: quantitative investigation. Journal of Management in Engineerin 10, 194–202 (2002)
13. Ting-Ya, H., Shih-Tong, L., Gwo-Hshiung, T.: Fuzzy MCDM approach for planning and design tenders selection in public office buildings. International Journal of Project Management 22, 573–584 (2004)
14. Saaty, T.L.: The analytic hierarchy process: planning, priority setting, resource allocation. McGraw-Hill, USA (1980)
15. Buckley, J.J.: Fuzzy hierarchical analysis. Fuzzy Sets System 17(1), 233–247 (1985)
16. Deng, H.: Multicriteria analysis with fuzzy pairwise comparison. International Journal of Approximate Reasoning 21, 215–231 (1999)
17. Sheu, J.B.: A hybrid fuzzy-based approach for identifying global logistics strategies. Logistics and Transportation Review 40, 39–61 (2004)
18. Eunnyeong, H., Jinsoo, K., Kyung-Jin, B.: Analysis of the assessment factors for renewable energy dissemination program evaluation using fuzzy AHP. Renewable and Sustainable Energy Reviews 14, 2214–2220 (2010)
19. Crowe, T.J., Noble, J.S.: Multi-attribute analysis of ISO 9000 registration using AHP. International Journal of Quality and Reliability Management 15(2), 205–222 (1998)
20. Laarhoven, P.J.M., Pedrycz, W.: A fuzzy extension of Saaty's priority theory. Fuzzy Sets System 11(3), 229–241 (1983)
21. Chiou, H.K., Tzeng, G.H.: Fuzzy hierarchical evaluation with grey relation model of green engineering for industry. International Journal Fuzzy System 3(3), 466–475 (2001)
22. Rolfsman, B.: CO2 emission consequences of energy measures in buildings. Building and Environment 37, 1421–1430 (2002)
23. Fesanghary, M., Asadi, S., Zong, W.G.: Design of low-emission and energy-efficient residential buildings using a multi-objective optimization algorithm. Building and Environment 49, 245–250 (2012)
24. Chow, T.T., Fong, K.F., Chan, A.L.S.: Energy modelling of district cooling system for new urban development. Energy and Buildings 36, 1153–1162 (2004)
25. Kari, A.: Applications of decision analysis in the assessment of energy technologies for buildings. PhD thesis. Helsinki University of technology, Helsinki, Finland (2007)
26. Sofia, S.: Energy efficiency in shopping mall- energy use and indoor climate. Licentiate thesis. Chalmers University, Lund, Sweden (2010)
27. Weber, C., Shah, N.: Optimisation based design of a district energy system for an eco-town in the United Kingdom. Energy 36, 1292–1308 (2011)

Chapter 85
Cooling Coil Design Improvement for HVAC Energy Savings and Comfort Enhancement

Vahid Vakiloroaya and Jafar Madadnia

Centre for Built Infrastructure Research, School of Electrical,
Mechanical and Mechatronic Systems, University of Technology, Sydney, Australia
Vahid.vakiloroaya@engineer.com

Abstract. In designing an energy-efficient HVAC system, several factors being to play an important role. Among several others, the performance of cooling coil which is embodied through its configuration, directly influence the performance of HVAC systems and should be considered to be crucial. This paper investigates and recommends design improvements of cooling coil geometry contributes for a central cooling system by using a simulation-optimisation approach. An actual central cooling plant of a commercial building in the hot and dry climate condition is used for experimentation and data collection. An algorithm was created in a transient system simulation program to predict the best design. Available experimental results were compared to predicted results to validate the model. Then different models of several new designs for cooling coil were constructed to evaluate the potential of design improvements. Afterwards, the computer model was used to predict how changes in cooling coil geometry would affect the building environment conditions and the energy consumption of the HVAC components.

Keywords: Cooling coil, Design Optimization, Energy Saving, Comfort enhancement.

1 Introduction

The increasing consumption of energy in buildings on heating, ventilating and air-conditioning (HVAC) systems has initiated a great deal of research aiming at energy savings. With the consolidation of the demand for human comfort, HVAC systems have become an unavoidable asset, accounting for almost 50% energy consumed in building and around 10-20% of total energy consumption in developed countries (Perez-Lombard et al. 2008) Because building cooling load varies with the time of the day, an HVAC system must be complemented with an optimum design scheme to reduce the energy consumption by keeping the process variables to their required set-point efficiently in order to maintain comfort under any load conditions. One of the effective ways of achieving energy efficiency is to design cooling coil configurations properly that has motivated us to propose a design procedure which significantly lead to an overall reduction in HVAC energy consumption. Therefore, it is not surprising

A. Håkansson et al. (Eds.): *Sustainability in Energy and Buildings*, SIST 22, pp. 965–974.
DOI: 10.1007/978-3-642-36645-1_85   © Springer-Verlag Berlin Heidelberg 2013

that the design of energy efficient HVAC components is receiving a lot of attention. (Jabardo et al. 2006) presented results from an investigation carried out with commercial air coils of 12.7 mm of tube diameter. They tested coils with different fin pitch and tube rows in order to determine their effect over the thermal performance. (Sekhar and Tan 2009) investigated the performance of an oversized coil at different conditions during the operation stage.the results showed that the humidifying performance of the oversized coil at the reduced loads during normal operation can be considerably enhanced by changing the effective surface area of the coil through a simple mainpulation of the effective number of rows. (Cai et al. 2004) derived a model for a cooling coil based on energyconservation and heat transfer principles. Catalogue fittings of published coil data and experiments on a centralized HVAC pilot plant were conducted and the results showed that the model can achieve good and accurate estimation over the entire operating ranges and thus the model can be used to handle real time information.However, no work has been mentioned to optimize the cooling coil geometry by using combined simulation of building dynamic behaviour with a detailed operational data of a real tested central HVAC system.

The objective of this paper is to minimize the energy consumption of building cooling system by using the design improvements of cooling coil geometry contributes while satisfying human comfort and system dynamics. For this purpose, a real-world commercial building, located in a hot and dry climate region, together with its central cooling plant (CCP) is used for experimentation and data collection. The existing central cooling plant was tested continuously to obtain the operation parameters of system components under different conditions. In order to take into account the nonlinear, time varying and building-dependent dynamics of the CCP, a transient simulation software package, TRNSYS 16, is used to predict the CCP energy usage. The cooling coil model was developed and coded within the TRNSYS environment.On the basis of the TRNSYS codes and using the real test data, a simulation module for the central cooling plant is developed and embedded in the software. An optimization algorithm which uses an iterative redesign procedureis developed and implemented in the cooling coil module in order to calculate and select its optimum configuration.The simulation results are compared with the monitored data in order to analyze the performance and feasibility of the proposed method. To show the effect of proposed approach, the comfort condition index, predicted mean vote (PMV), is studied.

2   Methodology

2.1   Cooling Coil Model

The central cooling plant which is installed in the building consist of one water cooled chiller, one cooling tower, one air handling unit (AHU), two chilled water pumps and two condenser water pumps.In this section, a mathematical model is developed for the cooling coil of AHU in order to truly simulate the effects of its operation on the whole system performance:

In this paper, it is studied the sustainable integration of photovoltaic systems in the Island of Pantelleria. In this case, the integration of RES is strategic due to the strong dependence of the existing energy system, based on fossil fuel, on the mainland. Other benefits concern the reduction of CO_2 emissions and the increase of the energy efficiency, with possible economic side effects such as the attainment and selling of white or green certificates. In the same way, it is also quite important, in the light of the singularity and of the value of the cultural and environmental heritage of the Island of Pantelleria, that the technical analysis is coupled with an equally deep analysis about the entire heritage of the Island. The particular value of the landscape and of the architecture in the Island of Pantelleria is widely recognized and it is also a resource, infact tourism, together with agriculture, are the most important economic factors of the Island. It is thus necessary to start with an integrated feasibility analysis of the existing energy system that takes into account both the production and management potential as well as morphological and natural constraints and the architectural and regulatory features that the considered context holds. In this view, in this work, it is proposed a feasibility study about the integration of the Energy resources sized and located in [6] with special attention to the solar energy (photovoltaic and solar thermal) trying to limit the depletion of the heritage of the Island. Such study goes through the understanding of the original constructions, alterations, actual conditions, qualities, material and immaterial values, lacks [7]. The methodological approach followed in this paper is structured in different phases that can be summarized in the flowchart in figure 1. Some of the steps of the proposed methodology must be carried out by experts from different areas but use the same indicators. As an example, the morphological analysis of the territory is part of the preliminary analysis needed both for architectural and for energetic considerations.

What is most important of the proposed methodological approach is the possibility to define sensitive indicators characterizing territories for sustainable energy and territorial planning. The final aim of the study is that to define such indicators; in this view, the analysis proposed in this paper gives a first insight to the problem.

2 Natural and Historical Analysis of the Territory, Urban Site and Heritage Buildings

The Island of Pantelleria is situated in Canale di Sicilia at 70 km from the coast of Africa and at about 100 km from the south eastern coast of Sicily (Italy). Its area is of about 83 km^2 and the inhabitants are around 7.800.

The Island is located close to a submerged rift, 2000 m deep, and is the highest part of a submarine volcano. The shape of the Island, extended towards the direction NW-SE, follows the general course of the rift in Canale di Sicilia. The territory is shaped by the activity of the volcano in different ways and at different times. From all this derives the varied morphology of the Island and its surface manifestations of geothermal activity. The morphological structure of the island has been strongly affected in the centuries the human settlement [8]. In the Island, there are three main urban centers (the main center called Pantelleria and two smaller centers, Khamma-Tracino and Scauri-Rekhale) that have developed along the centuries around two historical nucleus below the mountains and close to the plain lands that are still cultivated. Agriculture is indeed the main economical fact in the Island of Pantelleria, strongly characterizing the landscape. The impervious morphology of the Island has been modified through contention walls made with local rocks (*Terrazzamenti*), in this way, more than 50% of the Island is dedicated to agriculture; also the typical architecture of the island *Dammuso,* was originally for agricultural uses. Such typical edification is made of the local volcanic rock with small openings and a dome roof and has characterizing features that allows to classify it as an example of bioclimatic architecture. The large thermal inertia, given by the thick perimeter walls, guarantees a good thermal insulation inside; the limited number of small openings, located in a repaired position with reference to the dominating winds, activates a natural ventilation; the height of the dome lets the hot air to go up keeping fresh the internal rooms during summer. The use of lime protects from solar radiation as well as the shape of the covering which allows the rain water harvesting and its conveyance in underground tanks (quite important for the lack of hydro resources of the Island). All these are elements saying that the existing housing typology is feasible for the territorial needs. The housing typology all over the Island is indeed the Dammuso. An exception is the main center of Pantelleria (rebuilt after the second world war with three floor buildings).

Another important issue for the Island is the presence of many natural reserves [8]. About 80% of the Island, for its features, resides in areas identified through the Directives 79/409/CEE and 92/43/CEE as SIC (Sites of communitary interest) and ZPS (Special protection Areas). Following such classification, the Island has been included into the 'Rete Natura 2000' identifying the areas devoted to the conservation of biodiversity within the European Union.

From what said before, it is evident that the territory of the Island of Pantelleria, such as many other islands in the Mediterranean sea, is quite rich and delicate, and particular attention must be devoted to any possible new installation to efficiently integrate/substitute the energy supply system.

3 Analysis of Functions, Performance and Needs of Users and Current Energy Supply System

The energy system supplying the Island of Pantelleria, such as all Sicilian islands and most Mediterranean islands, is disconnected from the main grid but serves a touristic attraction. Such condition is quite critical from the energetic point of view. In the Island of Pantelleria the growth of the population concentrated in high season periods (June – September), almost doubling the number of inhabitants, from 7.800 to 15.000 in August, is certainly a source of richness but also creates a problem of management of resources, among which also the energy resources. The current delivery of fossil fuel, essential to the energy production system (as well as to the internal mobility system) is carried out through maritime transportation. Such management is certainly not environmentally sustainable and also not efficient since in critical times of the year the fuel cannot be delivered to the Island due to adverse weather conditions. The following graph in figure 2 [9] shows how the consumption per inhabitant of fossil fuel in Pantelleria is much larger than the regional and provincial consumptions.

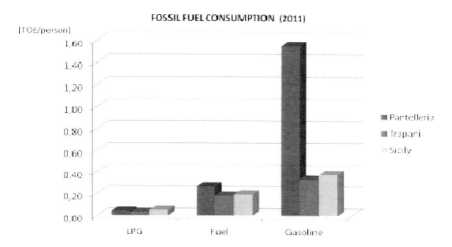

Fig. 2. Comparison in the consumption of fossil fuel in Tons of Oil Equivalent per person (TOE/person) between the Island of Pantelleria, the province to which it belongs, Trapani, and the Region to which it belongs, Sicily – 2011

It must be underlined that at provincial and regional level the use of LPG is mostly connected to mobility, while in Pantelleria its consumption is only for domestic use (stored in gas bottles). Gasoline is instead used in all cases for agriculture and for mobility and heating; moreover in Pantelleria the same fuel is used for the production of electrical energy (the gasoline employed for this reason covers 81% of the total consumption). The energy distribution system in the Island is made of a thermal electric central generation system with a Medium Voltage radial distribution system composed of 4 main feeders with rated voltage 10,5 kV. The network is made of 150 MV/LV substations from which the LV system supplies all the utilizations. The main generation system is made of 8 diesel groups with total rated power 20 MW. The entire network is automated with a DCS (Distributed Control System) for the control of the central station and of some of the secondary substations. The entire system allows the regulation of frequency and voltage at the central station and can execute a monitoring of the lines to detect possible outages and isolate portions of the network. The functions of the same system can be easily extended to cope with RES integration.

The island electrical energy request is entirely covered by the central station and is of about 43 GWh/year. As it can be observed from figure 3, half of the yearly consumption is concentrated between June and September with absorption peaks in August during the evening hours.

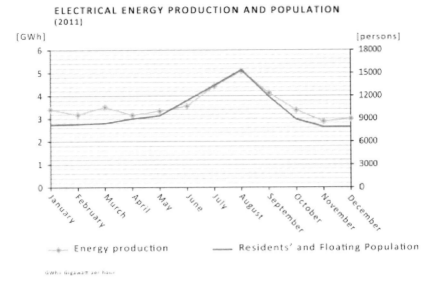

Fig. 3. Electrical energy production (blue line) and inhabitants (red line) in Pantelleria - 2011

The fluctuation of the electrical energy consumption has indeed required an oversized fossil fuel generation system as compared to the winter demand, such condition implies high management costs, due to the discontinuous usage of the machines.

The electrical energy production, 4.462 kWh per inhabitant in 2011, is much higher than the regional average of 3.783 kWh per inhabitant in the same year, this is due to the fact that about 16% of the produced energy is lost in generation and distribution inefficiencies (losses and maintenance). The non barycentric position of the generation system compared to the loads, requires the transportation of energy along long lines giving rise to high voltage drops and large Joule losses. Of the remaining produced energy, 31% supplies the two water desalination plants. As the following graph in figure 4 shows, only 53% of the produced energy goes to public and private uses and mainly for the domestic production of hot water (attained excluding 'desalinator' and 'others', namely generation and distribution losses).

As far as the evaluation of CO_2 emissions is concerned, the calculation executed on the basis of the number of kWh produced per year and on the emissions coefficient connected to the entire life cycle of energy for standard plants of thermal electric production based on gasoline [10], shows how the emissions connected to the energy production are for the Island equal to 5,23 tons per inhabitant. Such value is much higher than that registered for the province (Trapani) equal to 1,78 tons per inhabitant. To the just outlined issues also economical figures must be added, infact the production of electrical energy through gasoline is even more costly due to the maritime transportation costs, for this reason, there is a public contribution from the government to the utility. In this way, it is possible for the same utility to sell energy at the fixed national price.

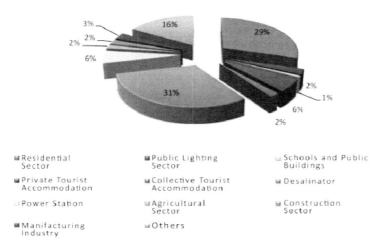

Fig. 4. Final use of electrical energy in Pantelleria- 2011

In order to limit the use of fossil fuel, an analysis of the available local natural resources has been carried out. Based on the existing literature on the topic and on the preliminary analysis (natural resources and economy) of the Island of Pantelleria [11], in this paper the sustainability analysis of the designed solar energy system has been analyzed, considering the impact on the built environment and landscape. The study uses climate data, considers the regulatory frame and the local technical directives ruling the use of such Energy generation installations and the cultural heritage and the landscape, in order to mitigate as much as possible the impact of the new solar energy systems in the territory. All data on the Island of Pantelleria presented in this section are derived by processing data from the Municipality of the Island of Pantelleria and from the fuel storage station of the Island (D'Aietti Petroli S.r.l).

4   Evaluation of the Compatibility and Integration of the Solar Energy System

The climate data show that in the island there is a solar irradiation of about 1,69 MWh/m2/year (from 1,90 kWh/m2/day in january to 7,2 kWh/m2/day in july). The yearly course of the irradiation is analogous to that of the population, and thus more or less to that of the electrical loading. The solar energy can be used for the production of electrical energy through photovoltaic modules and also for the hot water production which is the largest part of the consumption for domestic use. In the aim of identifying the optimal location of such plants, a deep analysis of the existing heritage has been carried out with a special attention to the constraints put by the Technical Actuation Norms and the existing plans. Based on these norms and on what said about the valuable landscape of the Island the ground installations of the modules was excluded also to avoid the land depletion in a territory with strong agricultural vocation. Thus all installations have been located on roofs. To do so, the architectural value of the existing built edifications has been considered. As it was said before, the built environment is mostly composed of Dammusi. In the main center, instead, the recently built three floor buildings did not show a consistent architectural value and have been considered suitable for PV installations also due to their flat roofs. These buildings have uniform height limiting the phenomenon of mutual shading and consequent strong reduction of the performances of the solar plants, especially photovoltaic systems. Figure 5 shows a study carried out for a representative building, the city hall. The figure shows the visibility study carried out. If the building on the other side of street would be higher (more than 26,75 m) there would be a shading effect of the modules. Figure 6 shows how the bulwark of one of the two volumes composing the building can hide the solar panels. From the energetic point of view, it must be considered that 29% of the energy consumption is connected to residential buildings, with a large use for hot water production. Since 40% of families live in the considered

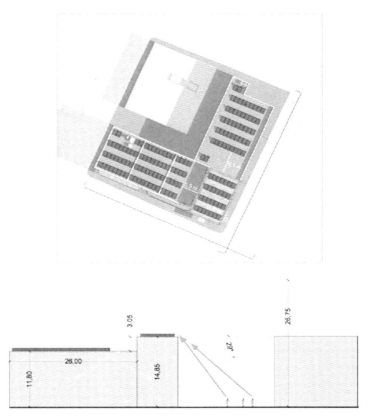

Fig. 5. City hall floor plan and visibility study for solar modules (via Concezione)

area, there is no doubt that it is convenient to propose the installation of both PV modules for electrical energy production and of solar thermal modules for the production of hot water in the same area. The overall available surface for solar modules is of about 205.358 m2, considering an average solar irradiation on the horizontal plan of 4,55 kWh/m2, the solar energy harvested from such surface is of about 935 MWh/day. The solar thermal modules contribution has been considered translating the thermal contribution into electrical energy. Such surface cannot be employed entirely, since the surface considered does not account for the shadings due to bulwarks or other obstacles. Based on a study carried out on one of the quarters of the main center of Pantelleria, a reduction of 40% of the flat surfaces was considered.

Such value has then been decreased further of 20% as an effect of the presence of small obstacles on the buildings roofs (tv antennas, tanks, ecc..), while the reduction related to the use of some roofs as terraces was estimated of about 5%.

Besides it was also considered that there would not be a full acceptance from citizens for the installation of such modules, and thus the surface has been still reduced of 25%. Based on what was said before the surface of flat roofs available for the installation of solar thermal plants and of PV modules for the main center of Pantelleria is of 23103 m^2, to such surface also the area attainable through the use of curved roofs has to be added. Such type of roofs is used in the industrial area close to the city center and considering the reduction due to the mutual shadings, the available surface is equal to 6.750 m^2. It was decided to install modules with an inclination of 30° so as to combine the maximum production efficiency with a reduced visibility from the streets. Figure 7 shows a part of the city center area, the quarter surrounding the city hall, suitable for the installation of solar modules. The shaded area cannot be used because it is subjected to constraints.

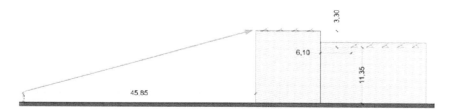

Fig. 6. Visibility study for solar modules in the city hall (piazza Cavour)

Considering the technical data about the types of commercially available modules, both PV and solar thermal, the inclination of the solar rays in southern Italy (28°,5'; latitude 38°) and the methodology for the calculation of the distance between the rows of modules in order to avoid the mutual shading and to optimize the efficiency of the overall system [12], it was calculated that the overall surface needed for the installation (panel and shaded area) of 1 kWp is of 12 m2. It was indeed chosen to use silica monocristalline modules (rated power 240 Wp) at 30° on flat roofs and amorphous silica thin film modules on curved roofs (rated power 120 Wp) of the industrial area. The latter choice was mandatory due to the geometry of the roofs.

Fig. 7. Rendering of the available areas for the installation of solar systems in the quarter surrounding the city hall

As a results, the electrical energy that can be produced from all the installations in the main center of Pantelleria (urban center and industrial area) is of about 3.260 MWh/year. Such value was calculated considering the number of equivalent hours of peak functioning for PV systems in the Island of Pantelleria (2007 hours/year). The production of energy takes into account the average conventional efficiency of an on-grid plant (connected to the MV network) considered equal to 0,8 (in agreement with the Italian norm CEI 82-25).

The energy produced by the solar plants should be injected in the LV network and through secondary substations in the MV system in the high loading hours. In this way, the entire distributed production system relieves the main fuel based generation system.

The sizing of the solar thermal system was calculated considering the number of families in the town of Pantelleria (1.314 in year 2011, municipal data). It was considered for each family a hot water production system with a 300 lt boiler and two solar thermal panels installed with an inclination of 30°.

The calculated value of the energy produced from PV panels has been summed up with the thermal energy produced with the solar thermal systems and converted in the equivalent of electrical energy for about 4.022 MWh/year.

The entire balance of the installations brings a coverage of the entire consumptions in the Island from solar source to 17% with a consequent 16% reduction in CO_2 emissions. The reduction of the emissions was calculated as missed emissions of CO_2, considering that the installation of PV plants and solar thermal systems would substitute a total of 7.282 MWh/year of energy from fossil fuel and would turn out that the avoided emissions are 6.773 tons/year of CO_2.

5   Conclusions

In this paper, a complete feasibility study for the sustainable installation of PV systems and solar thermal modules for the production of electrical energy and heat for hot water in a site with particular value of the landscape and of the architecture was considered. The study carried out will be completed with the analysis of the integration of other energy sources that are available on the Island, such as wind energy and geothermal energy, as well as biomass, trying to harmonize the technical installations with what already exists and trying to use adequate shapes and sizes of the plants. The aim of the entire work is that to create guidelines for an energy aware and thus sustainable planning of the territory using integrated knowledge. The problem is particularly relevant in isolated area (islands, small isolated centers), where the availability of supply (fossil fuel in the studied case) is also connected to not sustainable transportation systems. The idea is thus that to create methodological tools for local administrators for the future territorial planning.

References

[1] Projects "microgrids" and "more microgrids" website,
     `http://www.microgrids.eu`
[2] Tselepis, S.: Greek Experience with Microgrids Results from the Gaidouromantra site, Kythnos Island. In: Vancouver 2010 Symposium on Microgrids Fairmont Pacific Rim, Vancouver, Canada Thursday (July 2010)
[3] Lucchi, E.: Energy Efficiency in Historic Buildings: a Tool for Analysing the Compatibility, Integration and Reversibility of Renewable Energy Technologies. World renewable energy congress, Sweden (2011)
[4] Project "UP-RES" website, `http://aaltopro2.aalto.fi/projects/up-res/`
[5] Project "Regions for RES" website, `http://www.resregions.info/fileadmin/res_e_regions/WP_5/RES-e_Folder_01.pdf`
[6] Cosentino, V., Favuzza, S., Graditi, G., Ippolito, M.G., Massaro, F., Riva_Sanseverino, E., Zizzo, G.: Smart renewable generation for an islanded system. Technical and economic issues of future scenarios. Energy Elsevier 39(1), 196–204 (2012)
[7] Riva_Sanseverino R.: Atlante sulla forma dell'insediamento: le isole minori della Sicilia – Analisi e studi sul territorio delle microisole, progetto e strategie di pianificazione, Palermo, Flaccovio Libreria Dante (2002)
[8] Costantino, D.: Piano territoriale Paesistico dell'isola di Pantelleria, Regione Sicilia (1996-1997)
[9] Regional Dept of Energy "Rapporto Energia- Dati sull'energia in Sicilia", Regione Sicilia (2011)
[10] Ala, G., Cosentino, V., Di Stefano, A., Fiscelli, G., Genduso, F., Giaconia, C., Ippolito, M.G., La Cascia, D., Massaro, F., Miceli, R., Romano, P., Spataro, C., Viola, F., Zizzo, G.: Energy Management via Connected Household Appliances. McGraw-Hill (2008)
[11] Graditi, G., Bertini, I., Cosentino, V., Favuzza, S., Ippolito, M.G., Massaro, F., Riva Sanseverino, E., Zizzo, G.: Studio di fattibilità e progettazione preliminare di dimostratori di reti elettriche di distribuzione per la transizione verso reti attive. Report 1 – Caratterizzazione delle reti attuali e analisi di possibili scenari di sviluppo
[12] TuttoNormel - Le Guide Blu – Fotovoltaico

Chapter 87
Mobile Motion Sensor-Based Human Activity Recognition and Energy Expenditure Estimation in Building Environments

Tae-Seong Kim[1], Jin-Ho Cho[1], and Jeong Tai Kim[2]

[1] Department of Biomedical Engineering, Kyung Hee University, Yongin, South Korea
tskim@khu.ac.kr
[2] Department of Architectural Engineering, Kyung Hee University, Yongin, South Korea

Abstract. This paper presents a work on human activity recognition (HAR) using motion sensors embedded in a smart phone in building environments. Our HAR system recognizes general human activities including walking, going-upstairs, going-downstairs, running, and motionless, using statistical and orientation features from signals of motion sensors and a hierarchical Support Vector Machine classifier. Upon activity recognition, our system also generates energy expenditures of the recognized physical activities: energy expenditures are computed based on Metabolic Equivalents (METS) values, step count, distance, speed, and duration of activities. By testing our system in building environments, we have obtained an average recognition rate of 98.26% with physically consumed energy information. With the presented system, different building designs and environments can be evaluated in terms of energy consumptions of residents for their physical activities.

1 Introduction

Humans spend most of their daily life in indoor environments such as homes and buildings. Since living environments affect physical activities of residents, research works are underway to investigate the effects of indoor environments on physical activities that are closely related to the health of residents. For instance, Rassia et al. investigated the activity energy expenditure of the occupants in office buildings [1]. Also, Stamatina et al. proposed a simple quantitative model through which daily energy expenditures within an office layout could be estimated without the need for specialized medical equipment [2].

Recently, motion sensors including accelerometers and gyroscopes are widely utilized to measure physical activities. For instance, Lee et al. presented a novel human activity recognition (HAR) system using accelerometers in [3]. Also in [4], Nanami et al. proposed a calorie count application using an accelerometer which is embedded in a mobile phone. As shown, motion sensors built in a mobile phone offers an opportunity to be used in HAR. However, an effective HAR system based on motion sensors of a mobile phone has not been realized that recognizes physical activities and generates energy expenditures in real-time.

A. Håkansson et al. (Eds.): *Sustainability in Energy and Buildings*, SIST 22, pp. 987–993.
DOI: 10.1007/978-3-642-36645-1_87   © Springer-Verlag Berlin Heidelberg 2013

In this work, we present mobile motion sensors-based HAR and energy expenditure generation system. In the system, statistical and orientation features of motion sensor signals are utilized in hierarchical recognition of human daily activities, including walking, going-upstairs, going-downstairs, running, and motionless. Upon activity recognition, our system also computes energy expenditures based on the information collected on the device. The presented system has been tested in building environments. The system should be useful as a tool to investigate the effects of different building designs and environments on physical activities and health status of residents.

2   Motion Sensor-Based HAR System

Our system is based on several motion sensors embedded in a smart phone including accelerometers, gyroscopes, and magnetometer. To acquire both the acceleration and orientation information from the sensors, Android APIs are utilized as shown in figure 1(a). Our system consists of three main parts as shown in figure 1(b): a feature extraction module, a human activity recognition module, and an energy expenditure generation module. In the first module, signal features are extracted from accelerometer and orientation signals. In the second module, human activities are recognized using hierarchically-structured classifiers. In the third module, energy expenditure of each physical activity gets computed from the recognition results on the device.

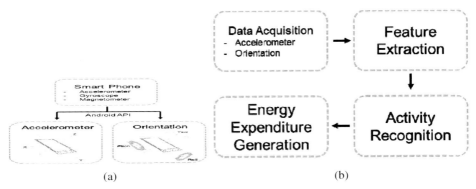

Fig. 1. (a) Motion sensors on a smart phone and (b) key components of our motion sensor-based HAR system

2.1   Sensor Information and Features

From the motion sensors on a mobile phone, we have obtained 3-axis accelerometer and orientation signals as shown in figure 2. Then from the accelerometer signals, we extract standard deviation of each axis, correlation between y and z axis, autoregressive coefficients of y axis, and signal magnitude area (SMA) as features. From the pitch angle of the orientation information, the features of mean, standard deviation and skewness are extracted [5, 6].

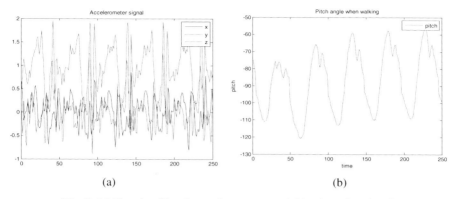

Fig. 2. (a) Signals of 3-axis accelerometers and (b) orientation signal

2.2 Activity Recognition Module

Our activity recognition module is implemented in a hierarchical structure as shown in figure 3(a). In the 1st level recognition stage, activities are recognized as one of motionless, walking-like, and running. In the 2nd level recognition stage, walking-like activities are further recognized as one of walking, going-upstairs, and going-downstairs. For recognition, two support vector machines (SVMs) are hierarchically implemented as a classifier: LibSVM [7] has been adopted. In training of the 1st SVM, the set of 1st features as shown in figure 3(b) are used. In the similar way, the 2nd SVM is trained with the 2nd features listed in figure 3(b). Note that in the 2nd level,

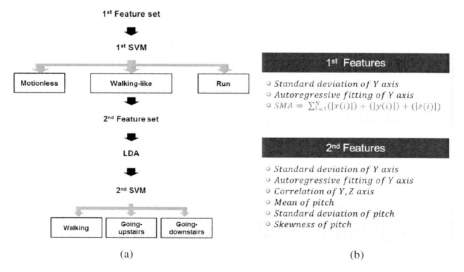

Fig. 3. (a) A hierarchically structured recognition system and (b) features used in each level of SVM

linear discriminant analysis (LDA) is utilized for dimensionality reduction before recognition [8].

2.3 Energy Expenditure Generation Module

Once each activity is recognized, energy expenditure (EE) for each activity gets estimated. To compute EE, the values of Metabolic Equivalents (METS) are utilized which are obtained from the equations and measured values [4] as given in table 1.

Table 1. METS values for five human activities

Activities	METS Values
Walking	0.072 x speed (m/min) +1.2
Running	0.093 x speed (m/min) – 4.7
Others	Motionless = 2.0; Upstairs = 8.0; Downstairs = 3.0

Then EE gets computed using the following equation,

$$EE\ (kcal) = 1.05 \times METS \times duration(hour) \times weight(kg) \qquad (1)$$

In the computation of speed (i.e., distance/duration) for walking and running, total distance is computed as a summation of stride length. Thus,

$$distance = \sum_{i=step} c(i) \qquad (2)$$

where the stride length c is estimated from a stride angle γ. The stride angle is computed from two extremum values of e and a leg length a which is obtained from the body ratio information as described in figure 4 along with equations (3) and (4).

$$c(i) = 2a^2(1 - cos\ (\gamma(i))) \qquad (3)$$

$$\gamma(i) = |e(i) - e(i-1)| \qquad (4)$$

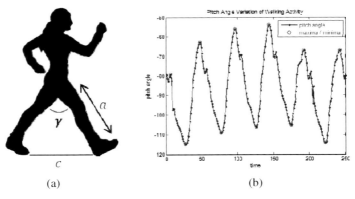

(a)　　　　　　　　　　　　(b)

Fig. 4. (a) A walking silhouette and (b) pitch angle information

3 Experiments and Results

In this study, activity recognition was performed utilizing motion sensors embedded in an Android smart phone. An Android application was developed to obtain the accelerometer and orientation data of the sensors. Figure 5(a) shows an application through which activity signals were collected. Motion sensor signals were gathered at the sampling frequency of 50 Hz from the following five activities: namely walking, going-upstairs, going-downstairs, running, and motionless. Experiments were performed while positioning the smart phone in user's trouser-pocket as shown in figure 5(b).

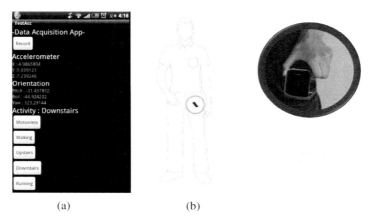

Fig. 5. (a) An Android application for data acquisition and (b) device position during experiments

The data sets were collected form four subjects between ages of 26 and 30 years old. All data of the five activities were measured in a building environment as shown in figure 6. The data sets include short duration of activities for training and long duration of randomly performed activities for continuous HAR.

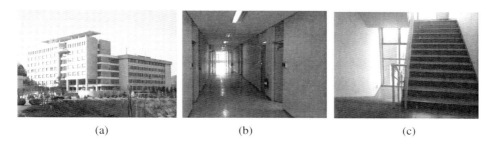

Fig. 6. A building environment: (a) exterior, (b) corridor, and (c) stairways

3.1 Subject-Independent Recognition

To evaluate the performance of the system, we conducted the subject-independent recognition. Four subjects were divided into two groups: each group consisted of two subjects. One group's data were used for training, another for testing and vice versa for cross validation. Table 2 shows the averaged confusion matrices for the 1st and 2nd recognition stages. An average recognition rate of 98.26% was achieved.

Table 2. Averaged confusion matrices from the cross-validation results of subject-independent recognition: (left) recognition results at the 1st stage and (right) the 2nd recognition stage.

Activity	Walking-like	Running	Motionless
Walking-like	100.0%	0.0	0.0
Running	0.0	100.0	0.0
Motionless	0.6	0.0	99.4

Activity	Walking	Upstairs	Downstairs
Walking	99.05%	0.95	0.0
Upstairs	1.8	98.2	0.0
Downstairs	5.35	0.0	94.65

Table 3. Confusion matrix of continuous activity recognition from a single subject

Activity	Walking	Upstairs	Downstairs	Running	Motionless
Walking	100%	0.0	0.0	0.0	0.0
Upstairs	0.0	100.0	0.0	0.0	0.0
Downstairs	15.4	0.0	84.6	0.0	0.0
Running	5.6	0.0	0.0	94.4	0.0
Motionless	0.0	5.8	0.0	0.0	94.2

Table 4. An example of estimated energy expenditures and exercise information

Activity	Duration (sec)	No. of Steps	Energy Expenditure (kcal)	Distance (m)	Average Speed (m/sec)
Motionless	245	-	11.15	-	-
Walking	179	228	13.66	243.14	1.43
Upstairs	140	159	25.48	-	-
Downstairs	110	158	7.51	-	-
Running	85	159	17.62	210.36	2.47
Total Energy Consumption: 75.42 kcal					
Total Elapsed Time: 750 sec					

3.2 Continuous Activity Recognition and Energy Expenditure

Table 3 shows the continuous activity recognition results from a single subject using the classifiers trained from the data of three other subjects. The overall accuracy of 94.6% was obtained. As shown a few downstairs activities were misclassified as walking, resulting in the reduced recognition rate of 84.5% for going-downstairs. Upon the recognition of the activities, EEs were estimated along with other exercise information as summarized in Table 4.

4   Conclusions

In this work, we have presented a human activity recognition and energy expenditure generation system utilizing motion sensors embedded in a smart phone. Our HAR system recognizes five general human activities, and generates energy expenditure and exercise information. Our proposed system should be useful in investigating building designs and environments in terms of resident's activities and energy consumptions.

Acknowledgements. This work was supported by the National Research Foundation of Korea (NRF) grant funded by the Korea government (MEST) (No. 2012-0000609).

References

1. Rassia, S.T., Hay, S., Beresford, A., Baker, N.V.: Movement dynamics of office environments. In: Proceedings of SASBE (2009)
2. Rassia, S.T., Alexandros, A., Baker, N.V.: Impacts of office design characteristics on human energy expenditure: developing a KINESIS model. Energy Systems 2, 33–44 (2010)
3. Lee, M.-W., Khan, A.M., Kim, T.-S.: A single tri-axial accelerometer-based real-time personal life log system capable of activity classification and exercise information generation, Pers. Ubiquit. Comput. 15, 887–898 (2011)
4. Ryu, N., Kawahawa, Y., Asami, T.: A calorie count application for a mobile phone based on METS value. Sensor, Mesh and Ad Hoc Communications and Networks, 583–584 (2008)
5. Khan, A.M., Truc, P.T.H., Lee, Y.-K., Kim, T.-S.: A tri-axial accelerometer-based physical-activity recognition via augmented-signal features and a hierarchical recognizer. IEEE Transactions on Information Technology in Biomedicine 14, 1166–1172 (2010)
6. Baek, J., Lee, G., Park, W., Yun, B.-J.: Accelerometer Signal Processing for User Activity Detection. In: Negoita, M.G., Howlett, R.J., Jain, L.C. (eds.) KES 2004. LNCS (LNAI), vol. 3215, pp. 610–617. Springer, Heidelberg (2004)
7. Chang, C.-C., Lin, C.-J.: LIBSVM: a library for support vector machines. ACM Transactions on Intelligent Systems and Technology 2, 1–27 (2011)
8. Belhumeur, P.N., Hespanha, J.P., Kriegman, D.J.: Eigenfaces vs. fisherfaces: Recognition using class specific linear projection. IEEE Transactions on Pattern Analysis and Machine Intelligence 19, 711–720 (1997)

Chapter 88
Cost and CO_2 Analysis
of Composite Precast Concrete Columns

Keun Ho Kim, Chaeyeon Lim, Youngju Na, Jeong Tai Kim, and Sunkuk Kim[*]

Department of Architectural Engineering, Kyung Hee University, Yongin 446-701, Korea
kimskuk@khu.ac.kr

Abstract. Green Frame is developed not only to reduce costs and construction duration and improve safety and constructability, but also to enhance environmental-friendliness resulting from reduction of CO_2 emission. Green Frame is a column-beam system composed of the Green Column and Green Beam, which are the composite precast concrete members. There are 5 types of Green Columns with different cost required and CO_2 emission. The importance of cost and CO_2 emission varies depending on the characteristics of a given project. Thus, this research is intended to perform cost and CO_2 analysis in order to help engineers select the most appropriate Green Column type. As a result, it is drawn out that a specific type is not superior in all aspects. If a wide range of performances including productivity, constructability, construction safety and construction period are analyzed in the future, it will be likely to suggest a meaningful guideline to engineers in selecting a suitable Green Column for Green Frame.

1   Introduction

Green Frame(GF) is a column-beam system that uses composite Precast Concrete(PC) members which is Green Column(GC) and Green Beam(GB). Previous studies have proven this system to be not only structurally safe, constructible, and economically feasible, but also environmentally-friendly.[1, 2]. Hong et al.[3, 4] proposed the concept of GF and 3 types of GC. Lee et al.[2] improved the columns suggested by Hong et al.[3, 4] and came up with a bolt connection type. With constant studies on its type, a coupler type with thread reinforced bars(rebars) was developed. As described herein, 5 GC types have varying cost required and CO_2 emission. The importance of cost and CO_2 emission varies depending on the characteristics of a given project. Thus, this research is intended to perform cost and CO_2 Analysis in order to help engineers select the most appropriate GC type.

The procedures of this research are as shown in Figure 1.

---

[*] Corresponding author.

A. Håkansson et al. (Eds.): *Sustainability in Energy and Buildings*, SIST 22, pp. 995–1002.
DOI: 10.1007/978-3-642-36645-1_88   © Springer-Verlag Berlin Heidelberg 2013

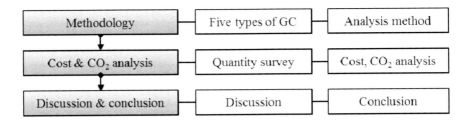

Fig. 1. Procedure of this study

First, the characteristics of each column type shall be identified and a method to calculate the cost and CO_2 emission shall be proposed.

Second, the cost and CO_2 emission of respective GC type, five types in total, shall be drawn out.

Third, conclusion is drawn out based on the calculated cost and CO_2 emission.

2 Methodology

2.1 Five types of Green Columns

As previously stated, GF is a column-beam system composed of GC and GB. As demonstrated in Figure 2, GC is embedded with steel frames for beam-joint of each floor. There are 5 types of GC depending on the steel frame type and column-column joint method as shown in the same figure.

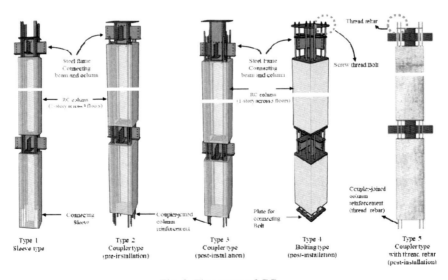

Fig. 2. Five types of GC

Type 1 is a sleeve type in which the column is post-installed. A post-installation type is to install a column after installing beams at the lower column and pouring slabs. Since the column is installed after the slab pouring, a grouting work is necessary [5].

Type 2 is a column pre-installation type in which an upper column is installed before slab pouring and after installing beams at the lower column, so-called a coupler type(pre-installation). Type 3 is similar to Type 2 in that a coupler is used to connect column rebars, yet the steel frame embedded in the column is extended for post-installation of columns[5].

Type 4 is applied with the screw thread-type rebar ends in order for bolt-joint of columns, and additional steel frames are embedded for joint of the lower section of upper columns[6].

Type 5 is applied with the standard steel frames and thread rebars for convenient, quick coupler joints, which is recently developed.

2.2 Cost and CO_2 Analysis

Figure 3 represents the method to calculate the cost and CO_2 emission based on the material quantity survey data.

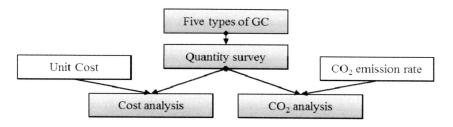

Fig. 3. Cost and CO_2 analysis process

Firstly, based on the installation order of five CG types, material quantity per column shall be calculated. Only the relevant material quantity is calculated since the difference of material quantity is shown in the column-column joint part. It targets columns that use 12 main reinforcements with 500mm x 500mm standard, and there is no difference in concrete quantity so it shall not be calculated. Any subsidiary materials that are applied to different types shall be considered to calculate the installation cost.

Secondly, the total cost is calculated by applying the unit price to the calculated material quantity.

The unit prices for cost calculation are based on the GF field-applied price and the market price defined by the Construction Association of Korea as of January, 2012.

Thirdly, the total CO_2 emission shall be calculated by applying CO_2 requirement per resource. The CO_2 requirement of construction materials analyzed with the LCA method by Kim et al.[7] shall be used for calculation of the total CO_2 emission[6].

3 Analysis on Cost and CO_2

3.1 Quantity Survey

The section describes the quantity calculated by analyzing the installation order per CG type.

1) Type 1; Sleeve Type

A sleeve joint part is composed of steel frames and sleeves to connect columns and beams. High-pressure shrinkage compensating grouting is performed for 12 holes that are 30mm diameter and 72.5cm long per column. Thus, the steel frame of sleeve-type installation is 80.6kg and the grouting is $0.02m^3$.

2) Type 2; Coupler Type(Pre-installation)

Type 2 requires 12 couplers instead of a grouting work. Just like the sleeve-type, the steel frame is 80.6kg and it needs 12 couplers.

3) Type 3; Coupler Type(Post-installation)

Type 3 that was developed to improve constructability of Type 2 is applied with a different steel frame, and the couplers and upper columns are installed after slab pouring. Thus, the steel frame increases to 89.6kg, and it requires grouting of $0.05m^3$.

4) Type 4; Bolting Type(Post-installation)

Type 4 uses the screw-thread type rebars instead of couplers to connect columns with bolts. To do so, steel frames are added to the lower section of upper columns, and it requires a grouting work. Moreover, it needs nuts to go with the bolts. Thus, it requires 95.6kg of steel frames, $0.02m^2$ of grouting, 12(EA) couplers, 24(EA) screw-thread rebars and 24(EA) nuts.

5) Type 5; Coupler Type with Thread Rebar(Post-installation)

Type 5 is applied with thread rebars to improve constructability of Type 3 and reduce the quantity of steel frames. Accordingly, the quantity of steel frame is 80.6kg which is equivalent to that of the sleeve-type and the quantities required for grouting and couplers are the same as those of Type 3.

The overall quantity of materials is as represented in Table 1.

3.2 Cost Analysis

The unit cost applied for cost analysis are as described in Table 2.

In reference to the GF field-applied unit prices, the thread-curved couplers and reinforced bars shall be 7,300 won and 5,000 won respectively per location. The cost calculated by applying the unit cost specified in Table 2 to Table 1 shall be as shown in Table 3.

Table 1. Quantity survey

Item	Unit	Type 1	Type 2	Type 3	Type 4	Type 5
Steel	kg	80.6	80.6	89.6	95.6	80.6
Grouting	m^3	0.0202	-	0.0463	0.0236	0.463
Sleeve	EA	12	-	-	-	-
Coupler	EA	-	12	12	-	12
Screwed rebar	EA	-	-	-	24	-
Nut	EA	-	-	-	24	-

Table 2. Unit cost of resources

Item	Unit	Unit price (won)
Steel	won/kg	1,160
Grouting	won/m^3	53,130
Sleeve	won/EA	7,000
Coupler	won/EA	7,300
Screwed rebar	won/EA	3,500
Nut	won/EA	288

Table 3. Cost analysis (won)

Item	Type 1	Type 2	Type 3	Type 4	Type 5
Steel	93,479	93,479	103,937	110,890	93,479
Grouting	1,074	-	2,457	1,253	2,457
Sleeve	84,000	-	-	-	-
Coupler	-	87,600	87,600	-	87,600
Screwed rebar	-	-	-	84,000	-
Nut	-	-	-	6,912	-
total	178,553	181,079	193,995	203,055	183,537

As a result of cost calculation, Type 1 was the most inexpensive. However, Type 1 and 2 are not the most favorable types for they lack constructability and construction safety when compared to the other types. Type 4 was the most costly one. This is due to the fact that the quantity of steel frames increased and it requires thread-curving of the rebar ends. The cost difference between the least expensive Type 1 and the most expensive Type 4 was 24,502 won, and it will show the construction cost difference of approximately 35,000,000 won when applied to a project implemented with eight 15-floor buildings.

3.3 CO_2 Analysis

The CO_2 emission rate for CO_2 analysis is as stated in Table 4.

The CO_2 emission rate of concrete is applied for grouting. In case of sleeves, couplers and nuts, there is no corresponding CO_2 emission rate, so the most similar steel product data is converted into rate per EA for application. For Screwed rebars, the CO_2 emission rate of rebar compression is applied. The CO_2 emission calculated by applying the CO_2 emission rate of Table 4 to the quantity specified in Table 1 is as shown in Table 5.

Table 4. CO2 emission rate

Item	Unit	CO_2 emission rate
Steel	kg-CO_2/kg	4.166
Grouting	kg-CO_2/m^3	140.43
Sleeve	kg-CO_2/EA	6.33
Coupler	kg-CO_2/EA	0.0135
Screwed rebar	kg-CO_2/EA	3.25
Nut	kg-CO_2/EA	0.0005

Table 5. CO2 analysis (kg-CO_2)

Item	Type 1	Type 2	Type 3	Type 4	Type 5
Steel	335.720	335.720	373.279	396.248	335.720
Grouting	2.838	-	6.495	3.312	6.495
Sleeve	75.960	-	-	-	-
Coupler	-	0.162	0.162	-	0.162
Screwed rebar	-	-	-	78.021	-
Nut	-	-	-	0.012	-
Total	414.517	335.882	379.936	479.594	342.377

As a result of analysis, Type 2 showed the least CO_2 emission. However, as previously mentioned, Type 1 and 2 are not favorable when it comes to constructability and construction safety, so they cannot be the most suitable type simply considering the CO_2 emission. Type 4 presented the most CO_2 emission. The difference in CO_2 emission between Type 2 and Type 4 is around 144kg. When these 2 different GC types are applied to a project on eight 15-floor buildings, it will show a CO_2 emission difference of approximately 207 tons.

4 Discussion

The overall result of cost and CO_2 emission is shown in Table 6.

Table 6. Cost and CO_2 analysis

Item	Unit	Type 1	Type 2	Type 3	Type 4	Type 5
Cost	won	178,553	181,079	193,995	203,055	183,537
CO_2	kg-CO_2	414.517	335.882	379.936	479.594	342.377

This research is conducted to analyze the cost and CO_2 emission of 5 GC types, and it is discovered that a single type is not superior in all aspects. Thus, constructors shall select the most suitable CG type taking into account of the project characteristics and circumstances. For instance, it is likely to use Type 1 on site where cost reduction is the most important factor to consider. On the other hand, Type 2 shall be appropriate when CO_2 reduction is the most significant factor. However, as described in section 3, Type 1 and 2 show relatively low constructability and construction safety when compared to the other types, so it is desirable to choose the type with better constructability aspect for projects that prioritize cost and CO_2 emission less than constructability or construction safety.

5 Conclusion

There are 5 GC types depending on the column-column joint types. Since each type has different cost requirement and CO_2 emission, a suitable GC shall be selected considering the project conditions. Therefore, this research analyzed the cost and CO_2 emission of 5 GC types and the conclusion is as follows:

First, 5 different types of GC were defined with description.

Second, the procedures for quantity survey and calculation of relevant cost and CO_2 emission were suggested. In addition, the cost and CO_2 emission were calculated for column-column joints where it shows differences for each GC type. In conclusion, it is discovered that a single type does not demonstrate the most superior performance, both in terms of cost requirement and CO_2 emission.

Third, Type 1 is the most cost-effective one, whereas Type 2 is the most suitable one in terms of CO_2 emission. Yet, Type 1 and 2 are not the most favorable ones. Since they show relatively low constructability and construction safety, other different types are developed.

This research is focused only on the cost and CO_2 emission, so it is insufficient to establish the criteria in selecting the most suitable column in accordance with the project characteristics. If a wide range of performances including productivity, constructability, construction safety and construction period are analyzed in the future, it is likely to suggest a meaningful guideline to engineers in selecting a suitable GC for GF construction.

Acknowledgements. This work was supported by the National Research Foundation of Korea (NRF) grant funded by the Korea government (MEST) (No. 2012-0000609).

References

1. Lim, C.Y., Joo, J.K., Lee, G.J., Kim, S.K.: In-situ Production Analysis of Composite Precast Concrete Members of Green Frame. Journal of the Korea Institute of Building Construction 11(5), 501–514 (2011),
   http://dx.doi.org/10.5345/JKIBC.2011.11.5.501

2. Lee, S.H., Joo, J.K., Kim, J.T., Kim, S.K.: An Analysis of the CO2 Reduction Effect of a Column-Beam Structure Using Composite Precast Concrete Members. Indoor and Built Environment 21(1), 150–162 (2012), doi:10.1177/1420326X11423162
3. Hong, W.-K., Kim, J.-M., Park, S.-C., Lee, S.-G., Kim, S.-I., Yoon, K.-J., Kim, H.-C., Kim, J.T.: A new apartment construction technology with effective CO2 emission reduction capabilities. Energy 35(6), 2639–2646 (2010), doi:10.1016/j.energy.2009.05.036
4. Hong, W.K., Park, S.C., Kim, M.M., Kim, S.I., Lee, S.G., Yun, D.Y., Yoon, T.H., Ryoo, B.Y.: Development of Structural Composite Hybrid Systems and their Application with regard to the Reduction of CO2 Emissions. Indoor and Built Environment 19(1), 151–162 (2010), doi:10.1177/1420326X09358142
5. Lee, S.K.: The Development & Feasibility Study of Apartment Building for Low-carbon emissions & Long-service life, Kyung Hee University Master's Thesis, 1–136 (2010)
6. Lee, S.H., Kim, S.K.: A Composite Frame Concept for the Long Life of Apartment Buildings, International Conference on Construction Engineering Project Management, sydney (2011)
7. Kim, S.W., Kim, J.Y., Lee, J.S., Park, K.H., Kim, T.W., Hwang, Y.W.: The environmental load unit composition and program development for LCA of building: The second annual report of the Construction Technology R&D Program. Korea Institute of Construction & Transportation Technology Evaluation and Planning (2004)

Chapter 89
A Field Survey of Thermal Comfort in Office Building with a Unitary Heat-Pump and Energy Recovery Ventilator

Seon Ho Jo, Jeong Tai Kim, and Geun Young Yun

Department of Architectural Engineering, Kyung Hee University, Yongin 446-701,
Republic of Korea
ssuno62@khu.ac.kr

Abstract. Air-conditioning plays an important role in creating comfortable indoor environment. This paper reports the field monitoring campaign and questionnaire survey of office building with a unitary heat-pump and energy recovery ventilator in Suwon, Korea, from 16 April to 20 April 2012. This study investigated the patterns of indoor thermal conditions, their relationship with the controls of the unitary heat-pump, and actual thermal satisfaction of building occupants. Although the setting temperature of each interior unit differed significantly from each other, the indoor temperature remained relatively constant and showed a pattern that the perimeter zone was 1°C lower than the interior zone on average. It is found that there was a big discrepancy between the actual thermal satisfaction of the occupants and predicted thermal satisfaction by PMV. The theoretical predictions by PMV indicated that only 42% of the occupants would feel thermally comfortable, although the proportion of the thermally comfortable occupants was observed 95%.

Keywords: Field survey, Unitary heat-pump, Thermal comfort, Adaptive comfort, Thermal satisfaction.

1 Introduction

Global warming is ongoing and the world is experiencing a grave world-wide climate changes including South Korea. According to the 'Extreme Weather Condition Report' in 2011 [1], South Korea is experiencing abnormally high and low temperature as well as cold wave. The meteorological observation has been making a new record each year with increased number of people with thermal disorder.

Along with global warming, the rising demand in improvement of indoor environment and increased amount of time spent indoor resulted in more time of both cooling and heating times. Consequently, the energy consumption has also been continuously increasing and many studies have been conducted to save these energy as well as to provide comfortable and healthy environment [2].

According to the architectural statistics, the office buildings take up to 17% of the total number of buildings in South Korea. In addition, the importance in energy

A. Håkansson et al. (Eds.): *Sustainability in Energy and Buildings*, SIST 22, pp. 1003–1010.
DOI: 10.1007/978-3-642-36645-1_89   © Springer-Verlag Berlin Heidelberg 2013

saving in office buildings has been increasing [3-5]. In case of office building, the use of personal air-conditioning systems increases to improve the comfortableness of the occupants. Therefore, through personal air-conditioning systems save the energy and unitary heat pump which assure indoor comfortableness is becoming popular. Thus in this study, by measuring the indoor thermal environment in an office building with the installed unitary heat pump and energy recovery ventilator, it reviews the characteristics of indoor thermal environment and the ability to implement an indoor environment of air-conditioner. And by conducting questionnaire surveys about the thermal environment during residence time, the actual satisfaction of the occupants were evaluated.

2 Methodology

2.1 Building Description

In order to review the indoor thermal environment of the office buildings and to analyze the thermal comfort, the S office was selected. S office building(Figure 1) is on the third floor of the selected building which is located in Maetan Dong, Youngtong Gu, Suwon city, Kyunggi Do with its height of 7.5m and designed with open plan type. It is facing the southeast direction and the windows are toward the northwest direction. Outline of the building is as shown in Table 1.

Table 1. Building Description

Location	Suwon, Republic of Korea
Type	Open plan office
Orientation	Southeast
Volume	$11913.83m^3$
Floor area	$1588.51m^2$

2.2 Air-Conditioner System

The applied air-conditioner system of the office is unitary heat pump with electric heat pump (EHP), Four Way Cassette and energy recovery ventilator (ERV). The unitary heat pump consists of 3 EHP, 28 Four Way Cassette and 5 ERV.

2.3 Data Acqusition

2.3.1 Physical Factors
Field monitoring was conducted from 16 April to 20 April 2012 for over 5 days in order to measure the physical elements of indoor thermal environment. HOBO U12-012 sensors were installed at the nearby partition of the occupant to measure the indoor temperature, indoor humidity, globe temperature and illuminance at 10 minutes intervals. Details on measurement and measuring devices are as shown in Table 2.

Table 2. Specification of measuring instrument

Measuring parameters	Measuring instrument	Measuring intervals	EA	Installation Image
Indoor temperature, Indoor Humidity, Globe temperature, Illuminance	HOBO U12-012	10minute	13	

Based on the location of the windows, the room was divided into three areas such as perimeter, semi-interior, interior and thirteen HOBO U12-012 sensors were positioned on each zone. Figure 1 is describing the installed location of the measuring device.

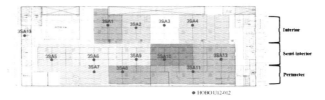

Fig. 1. Floor plan showing the monitoring points and zone division

2.3.2 Questionnaire Surveys

Questionnaire surveys were conducted to evaluate the occupants satisfaction. The survey was completed once a day from 16 April to 20 April 2012. The questionnaire surveys were to evaluate the indoor temperature, humidity and preferred temperature and humidity of the office where the occupant resides. 25 occupants in total have participated in the survey. Survey contents are as shown in Table 3.

Table 3. Specification of surveys and scale

Scale	Indoor temperature	Prefer indoor temperature	Indoor Humidity	Prefer indoor Humidity
-3	Cold	-	Too dry	-
-2	Cool	Much cooler	Dry	Much drier
-1	Slightly cool	A bit cooler	Slightly dry	A bit drier
0	Neutral	No change	Neutral	No change
1	Slightly warm	A bit warmer	Slight humid	A bit more humid
2	Warm	Much warmer	Humid	Much more humid
3	Hot	-	Very humid	-

2.3.3 Investigated Occupancy Schedules

To analyze the thermal environment of the subject office, daily occupancy patterns were calculated. Occupancy patterns using HOBO U12-012 sensors, the earliest time of lighting was the beginning time and the time when the light was turned off it was the closing time [6]. Table 4 shows the start and end of occupancy.

Table 4. Start and end of daily occupancy time during weekdays

Occupancy	4/16	4/17	4/18	4/19	4/20
Start	6:30	6:30	6:30	6:30	5:50
End	24:50	24:40	25:50	28:30	27:40

3 Field Study Results

3.1 Analysis of Indoor Thermal Environment

3.1.1 Indoor Temperature and Humidity

Average indoor temperature are distributed over 25.2~27.7℃ range. Recorded values of indoor average humidity are all below 30%, resulting in a low humidity distribution.

3.1.2 Ability to Implement an Indoor Environment of Air-Conditioner

The temperature during period of field monitoring was set different based on respective Four Way Cassette (internal unit) and time. Frequency of setting temperature were analyzed accordingly, and the most frequent setting temperature were represented as 20 ℃, 24℃, 30℃. Table 5 shows each setting temperature frequency rate.

Table 5. Setting temperature rate

Setting temperature (℃)	18℃	20℃	21℃	22℃	23℃	24℃	26℃	28℃	30℃
Rate(%)	5.9%	38.8%	1.8%	5.5%	1.1%	20.3%	2.2%	1.3%	23.1%

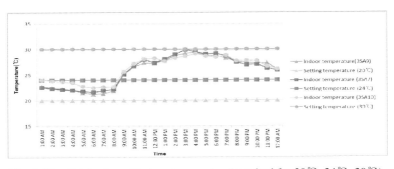

Fig. 2. The distribution of indoor temperature(standard for 20℃, 24℃, 30℃)

On 16 April 2012 was chosen as a standard day to examine the capacity of office's air conditioning system to realize indoor environment. First, HOBO U12-012 sensors (3SA7, 3SA9, 3SA10) that are vertically closest to Four Way Cassette (internal units), which were chosen among the 28 Four Way Cassette and set with highest frequency setting temperature(20℃, 24℃, 30℃), were contrasted 1:1 to examined the capacity of air conditioning system to realize the internal environment. The result is shown in Figure 2. It is such that the office's indoor temperature is not largely affected by the differently adjusted setting temperature.

3.2 Analysis of Thermal Comfort (PMV)

Occupant's thermal comfort was evaluated based on PMV (Predicted Mean Vote) [7], a evaluation index of thermal environment. Figure 3 is a histogram of PMV value's frequency distribution depending on perimeter, semi-interior, and interior.

Having evaluated thermal comfort of the target office using PMV, average PMV values are computed as 0.79 (SD 0.32) for perimeter, 0.85 (SD 0.48) for semi-interior, 0.76 (SD 0.32) for interior, and 0.80 (SD 0.45) for the entire area. As a result, the office as a whole shows 65.02% of thermal comfort satisfaction, and this does not satisfy the comfortable condition of thermal environment (satisfaction over 80%) as specified in the ASHRAE Standard 55-2004 [8].

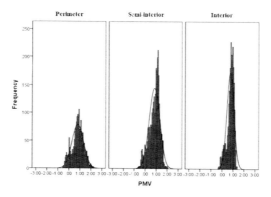

Fig. 3. PMV distribution under perimeter, semi-interior and interior

3.3 Questionnaire Surveys

The questionnaire surveys were executed to examine the office occupants' actual response to the thermal environment, and its result became a basis for an analysis of occupants' thermal environment satisfaction. Satisfaction standards of this subjective evaluation was based on ASHRAE, and if respondents were to choose −1(slightly cool), 0(Neutral), +1(slightly warm), they were thought to be satisfied. Table 6 shows result of questionnaire surveys.

3.4 Comparison of Satisfaction for the Subjective Evaluation and PMV

PMV from the field monitoring and questionnaire survey results for satisfaction was compared to see if there were any difference between the thermal sensations of occupants.

Table 6. Summary of questionnaire surveys

Division	N	Min	Max	Mean	SD
Indoor temperature	85	-1	2	0.32	0.66
Prefer indoor temperature	85	-2	1	-0.39	0.66
Indoor humidity	85	-3	1	-0.55	0.82
Prefer indoor humidity	85	-1	2	0.36	0.69

First, analysis of occupants' actual satisfaction to thermal environment through a questionnaire survey result of occupants' responses to their indoor thermal environment. Second, a sensor in charge of each occupant's position, and respective time and position of occupants at the moment of participation in the questionnaire survey, were matched to analyze thermal comfort satisfaction(PMV) based on data recorded in the sensor. In the meantime, if any written response to the survey lacked the time of participation, it was excluded from the analysis of results. Table 7 is a result of the satisfaction analysis to the indoor thermal environment, examined by both subjective evaluation and a survey.

Table 7. Comparision of thermal comfort for the subjective evaluation and PMV

Division	Subjective evaluation	PMV
Perimeter	100%	25%
Semi-interior	89.5%	31.6%
Interior	100%	100%
Total	94.7%	42.1%

The subjective evaluation by questionnaire survey showed 100% satisfaction for perimeter, 89.5% for semi-interior, 100% for interior, and 94.7% satisfaction for the entire area. On the other hand, the PMV through field monitoring showed 25% satisfaction for perimeter, 31.6% for semi-interior, 100% for interior, and 42.1% satisfaction for the entire area.

Satisfaction of PMV and subjective evaluation shows discrepancy. Such discrepancy is thought be mainly caused by the behavioral, psychological, and physiological adaptation that occupants developed to the target office's thermal environment, which in turn led to high satisfaction level unlike the PMV theory which estimates based on combination of physical components.

4 Conclusions

This research studied indoor thermal environment as well as unitary heat pump's ability to realize indoor environment through a field monitoring, which targets open plan office space where system air conditioner is installed. Then, evaluations of theoretical thermal comfort and subjective responses were executed to contrast and analysis predicted values(PMV) and satisfaction by actual occupants. Results of this study are as follows.

1) The target office showed difference in average temperatures according to perimeter, semi-interior, and interior. In particular perimeter showed a high temperature due to strong solar radiation during afternoon period. Average indoor humidity was below 30%, meaning somewhat dry indoor environment. This is judged to be affected by inflow of dry outdoor air from continued operation of energy recovery ventilatior (ERV).

2) Studying ability of the target office's air conditioning system to realize indoor environment, in case of the target office's open plan type, although the setting temperature might be set differently, the thermal sensations at the position of occupants did not show considerable difference. In order to create a comfortable indoor thermal environment, more investigations are required to study effects of setting temperature in relation to areas of occupants (perimeter, semi-interior and interior).

3) Comparing results from PMV and questionnaire survey, there was a discrepancy between satisfaction by occupants and predicted satisfaction from PMV. This can be explained by inaccuracy of PMV [9], and sufficient interactions between occupants and indoor environment that led to positive influence on thermal comfort satisfaction.

Later on, in order to create a comfortable indoor thermal environment, an algorithm to optimize control of indoor setting temperature will be developed, based on outcomes from analysis of indoor thermal environment at the target space with system air conditioner.

Acknowledgements. This work was supported by the National Research Foundation of Korea (NRF) grant funded by the Korea government (MEST) (No. 2012-0002212), (No. 2012-0000609)

References

1. Climate Change Information Center, Extreme Weather Condition Report (2011), http://www.climate.go.kr/index.html
2. Kong, H.J., Yun, G.Y., Kim, J.T.: A Field Survey of Thermal Comfort in Office Building with Thermal Environment Standard. Journal of the Korea Institute of Ecological Arichitecture and Environment 11(3), 37–42 (2011)
3. Yun, G.Y., Kong, H.J., Kim, J.T.: A field survey of visual comfort and lighting energy consumption in open plan offices. Energy and Buildings 46, 146–151 (2012)
4. Steemers, K., Yun, G.Y.: Household energy consumption: a study of the role of occupants. Building Research and Information 37(5&6), 625–637 (2009)
5. Yun, G.Y., Kong, H.J., Kim, J.T.: A Field Survey of Occupancy and Air-Conditioner Use Patterns in Open Plan Offices. Indoor and Built Environment 20(1), 137–147 (2011)
6. Steemers, K., Yun, G.Y.: Household energy consumption: a study of the role of occupants. Building Research and Information 37(5&6), 625–637 (2009)

7. Fanger, P.O.: Thermal Comfort Analysis and Application in Environmental Engineering, Copenhagen, Denmark, Danish Technical press (1970)
8. ASHRAE Standard 55-2004, Thermal environment conditions for human occupancy, Atlanta, American Society of Heating, Refrigerating and Air-conditioning Engineers, Inc. (2004)
9. Humpherys, M.A., Fergus Nicol, J.: The validity of ISO-PMV for predicting comfort votes in every-day thermal environments. Energy and Buildings 34, 667–684 (2002)

Chapter 90
An Analysis of Standby Power Consumption of Single-Member Households in Korea

JiSun Lee[1], Hyunsoo Lee[1,*], JiYea Jung[1], SungHee Lee[1], SungJun Park[1], YeunSook Lee[1], and JeongTai Kim[2]

[1] Department of Housing and Interior Design, YonSei University, Seoul 120-749, Korea
hyunsl@yonsei.ac.kr
[2] Department of Architectural Engineering, Kyung Hee University, Yongin 446-701, Korea

Abstract. This purpose of this study is to investigate the standby power consumption behaviors of live-alones. The research was performed by a survey that included general demographic characteristics like age, gender, and type of housing, and electrical energy consumption factors to analyze the standby power consumption of single-person households. Analyses were performed to find power consumption characteristics by households including the ratio of standby power consumption to total electrical energy consumption and respective waiting time and standby power consumption for each electrical appliance. In addition, the actual practice of plugging/unplugging, which in one of the consuming behaviors that directly affects standby power consumption, was investigated. Finally, the correlation of the amount of standby power and 8 consumption factors including: house size, time spent at home, number of appliances, plugging ration of electrical appliances, and 4 other factors categorized by the use of appliances. Here is the summary of the analysis results. Firstly, average standby power consumption of single-member households accounted for 6.1% of overall monthly electricity consumption. Secondly, the significant factors affecting standby power of single-person households proved to be 'number of appliances', 'plugging ratio', and 'size of the house'. Lastly, we divided respondents of questionnaire into two groups; one was those who consumed standby power more than average and the other was those consuming standby power less than average. Three variables, video, fax and dishwasher impact on standby power consumption in the first group and three variables, computer, monitor and Vacuum influence on standby power consumption in the second group. In terms of correlation test, the Video turned out to be one of the most important variables determining standby consumption in the first group. In the second group, computer and monitor were the most important variables determining standby consumption. This study is believed to have verified that standby power saving contributes significantly to total electrical energy savings of households.

---

* Corresponding author.

A. Håkansson et al. (Eds.): *Sustainability in Energy and Buildings*, SIST 22, pp. 1011–1024.
DOI: 10.1007/978-3-642-36645-1_90 © Springer-Verlag Berlin Heidelberg 2013

1 Introduction

Energy saving is one of the most important issues under the eco-friendly paradigm, and it needs to pay attention to the problem for the conservation of the sustainable earth. There are many ways of saving energy. Since houses are one of the building types with much energy consumptions, energy savings in residential sectors will demonstrate a huge effect in overall savings. Diverse ways for domestic energy savings are mentioned, it will be an important practice to analyze the energy consumption behavior and provide the proper solutions [1, 2]. There were many attempts in precedent researches that suggested technical methods to save energy, but studies focused on energy consumption behaviors are not many. In this effect, it is the purpose of this study to analyzed energy consumption patterns for energy savings. The research was performed on the hypothesis that reducing standby power consumption among energy consumption behaviors will reduce overall energy consumptions. Standby power is defined as the electricity consumed by end-use electrical equipment when it is switched off or not performing its main function. It is currently estimated to account for about 3 to 10 percent of home electricity use [3, 4, and 5]. Currently, there are adequate technical skills to limit standby power consumption in Korea and abroad but there are limitations since businesses have no choice but to consider the cost of production first. Above all things, the leading cause of hindrance to reduce standby power is lack of public awareness towards cutting of standby power [6, 7, 8, and 9]. According to Korea Electro Technology Research Institute, total power wasted by standby power consumption in residential sector is 209kWh annually, which accounts for 6.1% of total electricity consumption. If converted, each household is wasting about 20 dollars yearly, totaling to yearly electricity production by a 500MW-thermal power plant if assumed each household uses the same amount of energy [10]. Especially studies on energy consumption behaviors of one-member households among all types of households are scarce. Accordingly, this study is focused on the analysis energy consumption behavioral pattern of live-alones to derive important information in setting the direction to energy savings. The reason we focus on live-alones is that the number of one-member households is increasing rapidly in Korea. As for the capital city, Seoul, one-member household type became the number 1 type of household accounting for 24.4% [11]. In addition, according to The National Statistical Office, there is no significant difference in the electricity consumption between one-person and two-person households. One-member households used 23.9% of total electricity energy in 2010, while two-member households used 24.3% [12]. Therefore, this study was initiated in that a provision for saving energy based on the energy consumption pattern of single-person households will have a great social ripple effect. The final goal of this research is to analyze the standby power consumption patterns of live-alones. Sub-goals to accomplish the final goal are as follows. First, analyze the waiting time and standby power for electrical equipments used at home. Second, analyze the consumption pattern of each household. Third, analyze the correlation of standby power consumptions and consumption factors of homeowners.

The scope and Method of Research
The scope of this study was limited to 20-30 year old single-member households and electrical energy consumption. This variable was chosen because electrical energy usage is increasing while other types of energy usage continue on a downward trend, and the

increase of single-member households is the main reason [2]. Therefore, it is needed to see how single-member households consume standby power in order to effectively reduce electrical energy use in homes. A survey was used a main method of research of standby power consumption and consumption patterns of single-person households. The questionnaire was composed of following items. First, what are the electrical appliances used in each household, and where are they located? Second, what are the daily mean time of use of the electrical appliances, and are they plugged in or not when not used? Third, what are the monthly (as of April 2012) electricity consumption, electricity tax charged, and daily mean of stay at home? Finally, what are the socio-demographic factors including age, gender and housing factors such as housing type, floor area, number of rooms and type of ownership.

2 Energy Consumption Behaviors and Standby Power

One of the factors affecting energy consumption is the characteristics of physical environment. Insulation performance of the building, arrangement, and window area are some of the many physical elements affecting energy consumption. Therefore, there are precedent studies claiming positive correlation of physical properties if the building and the energy consumption [13]. However, there are some literatures suggesting that users' energy consumption pattern affects the actual energy consumption more than buildings' physical characteristics [7]. This study implies the energy consumption behavioral pattern affects the total energy consumption as much as physical characteristics of buildings. Therefore, the analysis of energy consumption behaviors is very important to suggest the direction for saving energy consumption. Some precedent researchers emphasizing the importance of energy consumption behaviors are Yigzaw Goshu Yohanis (2012), Robert J. Meyers et al. (2010) and Olivia Guerra Santin (2011). Yigzaw Goshu Yohanis discussed domestic energy use and energy behavior. It shows some improvement in domestic energy consumption and adoption of good energy practice [8]. Robert J. Meyers took a broad look at how information technology-enabled monitoring and control systems could assist in mitigating energy use in residences by more efficiently allocating the delivery of services by time and location [9]. Olivia Guerra Santin statistically determined behavioral patterns associated with the energy spent on heating and identified household and building characteristics that could contribute to the development of energy-User Profiles [1]. As may be seen in the above-mentioned studies, studies on energy consumption behaviors were mostly limited to common types of households and buildings. This study is focused on the electrical energy consumption behaviors of single–person households. There are many precedent studies on standby power including one's by Paolo Bertoldi et al. (2002), Alan Meier (2004) and Kristen Gram Hanssen (2009). Paolo Bertoldi presented the most recent figures on standby power consumption in OECD countries and China [3]. Alan K. Meier presented the first study on power use and the saving potential in China [4]. Kristen Gram Hanssen analyzed ten in-depth interviews with families participating in a project aimed at reducing standby power consumption [7]. As stated above, studies on standby power have been performed multilaterally, very little was about the investigation of individual standby power of main electrical appliances used at home. Therefore, respective standby power consumption will be calculated for each appliance and such electrical appliances will be identified as to be most related to entire standby power consumption.

3 Results

Standby Power Consumption for Electrical Appliances

There were 34 electric appliances commonly used in household [14]. Standby power each electric appliance consumes averagely was as follows.

Table 1. Standby power consumption of electrical appliances [10, 14]

Type	Electrical Appliances	Standby power(W)
Kitchen appliances	Rice cookers	3.5
	Refrigerator	Always on
	Kimchi Refrigerator	Always on
	Microwave	2.2
	Electric Frying Pan	1
	Toaster	0
	Coffee Machine	1
	Dishwasher	1
	Blender	0
	Water Purifier	5,8
	Oven	0.6
Office machine & Recreational goods	Computer	2.6
	Monitor	2.6
	Laptop	1
	TV	1.3
	Set top box	12.3
	Audio	4.4
	Video	4.9
	Telephone	0.2
	Printer	2.6
	Fax	3.4
	Scanner	2.9
Health Support &Personal hygiene	Iron	0
	Washing machine	1
	Humidifier	3.5
	Air Cleaner	0.7
	Hair Drier	0
	Vacuum	0
	Bidet	2.2
	Vibrator	8
Air conditioning and heating & ETC	Electric pad	0.59
	Electric fans	0.22
	Air conditioner	5.8
	Desk lamp	0.4

We categorized them into five types by their function: Kitchen appliances, Office machine & Recreational goods, Health support & Personal hygiene, Air conditioning & Heating and etc.

Table 1 presents standby power consumption for each appliance. We found in the homes in which it was impossible to unplug the refrigerators or Kimchi refrigerators. The appliances with the highest standby power consumption were set-top box. On the other hand, the appliances with the lowest standby power consumption were heating equipment such as blenders, electric fans, irons and hair driers. Generally the heating equipments are very low or zero in standby power consumption even when they are plugged.

Analysis of Mean Waiting Time & Standby Power by Appliances

Table 2 shows mean waiting time & standby power of each appliance. Appliances with greatest mean waiting time are in the order of washing machine, air conditioner, TV and microwave; with greatest standby power, air conditioner and set top box.

Table 2. Mean waiting time & mean standby power by appliances

Electrical Appliances	Weight of Standby power(%)	Average Standby power(kWh)	Average Waiting time(h)
Air conditioner	21.4	2.4	13.7
Set top box	18.7	2.1	5.4
Microwave	7.8	0.9	13.1
Washing machine	5.4	0.6	20.1
Monitor	5.2	0.6	7.4
Rice cookers	5.0	0.6	5.4
Computer	5.0	0.6	7.2
TV	4.7	0.5	13.5
Audio	3.6	0.4	3.0
Printer	3.3	0.4	4.7
Water Purifier	3.2	0.4	2.1
Video	2.8	0.3	2.1
Vibrator	2.5	0.3	1.2
Bidet	2.5	0.3	4.2
Laptop	2.4	0.3	8.9
Scanner	1.3	0.1	1.6
Desk lamp	0.9	0.1	8.5
Coffee Machine	0.9	0.1	3.3
Fax	0.6	0.1	0.7
Oven	0.4	0.1	2.8
Electric fans	0.4	0.0	7.6
Dishwasher	0.4	0.0	1.6
Electric Frying Pan	0.3	0.0	1.2
Telephone	0.3	0.0	5.7
Humidifier	0.3	0.0	0.3

Air cleaner	0.3	0.0	1.5
Electric pad	0.2	0.0	1.4
Toaster	0.0	0.0	2.8
Blender	0.0	0.0	1.2
Iron	0.0	0.0	0.7
Hair drier	0.0	0.0	8.0
Vacuum	0.0	0.0	5.0

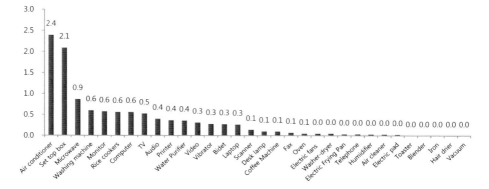

Fig. 1. Mean standby power by appliances

General Characteristics of the Respondents

This study involves an investigation of the general demographic characteristics of the single-member household who participated in it, including their age, sex, type of housing, and housing ownership, which are listed in Table 3.

Of the 101 respondents, 37(36.6%) were male and 64(63.4%) were female. With 68 respondents aged between 20 and 29, they represented bigger age group (67.3%). There were 33 respondents in their thirties (32.7%). 39.6% of respondents lived in studio flat, 24 lived in apartment and the other 21 lived in villa. 14 respondents lived in multiplex dwelling, and only 2 person lived in Detached house. A plurality of these single-member household (47, 46.5%) lived at a rental, or lease (38, 37.6%). 16 respondents were home owners. Lastly 48.5% of respondents lived in houses that were between 33 and 66 square meters in area. 21 respondents lived in houses that were below 33 square meters in area. And 19.8% of respondents lived in houses that were between 66 and 99 square meters.

Analysis of Standby Power Consumption Patterns

Each home's monthly standby power and electricity consumption based on utility bills for the month of April is displayed in Table 4. Standby power of each appliance that is normally left plugged in was calculated by the following equation using the mean use time collected by the survey.

Table 3. General characteristics of respondents

n = 101

Variable	Value	N	%
Age	Twenty	68	67.3
	Thirty	33	32.7
	Total	101	100
Sex	Male	37	36.6
	Female	64	63.4
	Total	101	100
Type of housing	Apartment	24	23.8
	Studio flat	40	39.6
	Villa	21	20.8
	Multiplex dwelling	14	13.9
	Detached house	2	2.0
	Total	101	100
Housing ownership	Own	16	15.8
	Lease	38	37.6
	Monthly rent	47	46.5
	Total	101	100
House size	Below 33sqm	21	20.8
	33-66sqm	49	48.5
	66-99sqm	20	19.8
	99sqm or greater	11	10.9
	Total	101	100
Occupation	Specialized job	28	27.7
	Office job	35	34.7
	Student	31	30.7
	Other	7	6.9
	Total	101	100

Table 4. Characteristics of power consumption by household

No	Standby Power (kWh)	Electricity Consumption (kWh)	Ratio of Standby power (%)	The number of applian- ces (N)	Ratio of Plugging (%)	No	Standby Power (kWh)	Electricity Consumption (kWh)	Ratio of Standby power (%)	The number of applian- ces (N)	Ratio of Plugging (%)
1	5.22	215	2.4	16	81.3	52	5.85	222	2.6	13	53.8
2	16.09	245	6.6	24	50.0	53	2.83	128	2.2	16	25.0
3	10.47	179	5.8	20	35.0	54	2.27	187	1.2	15	46.7
4	9.85	179	5.5	16	50.0	55	0.00	201	0.0	14	14.3
5	8.82	189	4.7	14	85.7	56	7.39	350	2.1	13	84.6
6	4.19	96	4.4	11	54.5	57	1.70	114	1.5	11	54.5
7	6.82	185	3.7	22	36.4	58	6.55	121	5.4	20	45.0

Table 4. (*continued*)

8	7.07	293	2.4	21	42.9	59	16.61	85	19.5	15	66.7
9	6.07	161	3.8	19	57.9	60	4.71	115	4.1	11	63.6
10	3.64	272	1.3	20	45.0	61	15.13	245	6.2	22	63.6
11	5.60	90	6.2	18	38.9	62	2.27	93	2.4	9	33.3
12	14.46	400	3.6	35	48.6	63	4.09	63	6.5	12	50.0
13	8.71	240	3.6	22	40.9	64	10.82	214	5.1	16	31.3
14	7.77	186	4.2	15	60.0	65	1.72	96	1.8	6	66.7
15	10.08	273	3.7	15	60.0	66	8.01	333	2.4	14	57.1
16	13.82	163	8.5	17	58.8	67	1.53	114	1.3	10	60.0
17	9.71	82	11.8	18	27.8	68	9.21	237	3.9	22	63.6
18	6.41	176	3.6	18	33.3	69	6.89	94	7.3	11	90.9
19	2.66	50	5.3	11	54.5	70	7.79	160	4.9	14	71.4
20	8.44	280	3.0	18	22.2	71	1.75	84	2.1	13	38.5
21	10.91	195	5.6	19	78.9	72	10.97	330	3.3	12	75.0
22	0.85	68	1.2	5	80.0	73	1.39	175	0.8	11	27.3
23	7.98	141	5.7	13	46.2	74	8.45	293	2.9	18	66.7
24	14.72	199	7.4	31	41.9	75	12.25	93	13.2	13	76.9
25	7.81	144	5.4	15	46.7	76	19.60	138	14.2	21	71.4
26	15.82	221	7.2	12	75.0	77	10.56	118	9.0	14	57.1
27	2.66	78	3.4	10	60.0	78	2.86	140	2.0	6	66.7
28	14.59	95	15.4	19	73.7	79	6.38	85	7.5	10	70.0
29	5.46	307	1.8	19	47.4	80	1.27	76	1.7	11	27.3
30	17.80	192	9.3	18	77.8	81	5.36	137	3.9	12	41.7
31	5.44	134	4.1	15	60.0	82	9.91	141	7.0	17	70.6
32	4.49	240	1.9	21	47.6	83	10.60	14	75.7	15	60.0
33	20.85	199	10.5	23	56.5	84	22.80	120	19.0	14	78.6
34	2.23	145	1.5	11	54.5	85	0.71	100	0.7	8	25.0
35	1.43	85	1.7	11	27.3	86	1.57	118	1.3	9	66.7
36	0.90	200	0.5	10	30.0	87	8.46	160	5.3	16	68.8
37	9.57	170	5.6	14	71.4	88	13.50	200	6.8	24	66.7
38	31.03	360	8.6	28	75.0	89	15.41	180	8.6	18	72.2
39	15.40	162	9.5	26	65.4	90	10.04	201	5.0	16	43.8
40	0.68	169	0.4	13	30.8	91	5.43	201	2.7	16	50.0
41	16.25	121	13.4	24	66.7	92	7.52	307	2.5	14	100.0
42	12.36	124	10.0	21	52.4	93	7.41	140	5.3	21	47.6
43	4.66	136	3.4	16	56.3	94	11.38	193	5.9	15	53.3
44	13.60	112	12.1	16	62.5	95	23.12	198	11.7	17	88.2
45	0.98	145	0.7	15	20.0	96	10.86	141	7.7	17	70.6
46	17.40	165	10.5	24	75.0	97	17.77	223	8.0	22	72.7
47	4.68	111	4.2	10	90.0	98	38.04	326	11.7	31	83.9
48	0.57	249	0.2	12	25.0	99	27.00	323	8.4	15	93.3
49	2.86	120	2.4	9	55.6	100	15.33	225	6.8	17	64.7
50	11.97	120	10.0	17	64.7	101	4.02	120	3.4	9	66.7
51	9.92	260	3.8	21	57.1					

Table 4. (*continued*)

Mean		
	Standby Power (kWh)	9.0
	Electricity Consumption (kWh)	174.2
	Ratio of standby power (%)	6.1
	The number of appliances (N)	16.1
	Ratio of plugging (%)	57.0

Standby power = [(standby power of each appliance that is left plugged) x (24hrs-daily time of use) x 30(days/month)]/1000

Percent weight was calculated for each appliance by dividing the calculated standby power by the total household electricity energy consumption. Standby power differs according to the plugging behaviors of the users, so the plugging ratio was calculated for each electrical appliance. We compared monthly electricity consumption and the ratio of plugging to measure standby power. We found that the average standby power in the 101 homes was 9.0kWh, and average monthly electricity use was 174.2kWh. Standby power accounted for 6.1% of total monthly electricity consumption. The households have an average of 16.1 appliances. An average of the ratio of plugging was 57.0%.

Correlation Analyses between Standby Power Consumption and Variables Relevant to Resident's Electrical Energy Consumption Behavior

In this section, correlations analyses are carried out to determine the relationship between the resident variables and standby power consumption. The statistics can be seen in Table 5.

Table 5. Correlation analyses between Standby power consumption and Factors of Electrical energy consumption behavior

Electrical energy consumption factors	Mean and SD		Correlation standby power consumption (+)
House Size	M=17.15	SD=9.24	r=0.416**
Time spent at home	M=10.00	SD=3.17	r=0.004
The number of appliances	M=16.07	SD=5.44	r=0.636**
The ratio of plugging	M=57.02	SD=18.48	r=0.496**
Duration of use, Kitchen appliances	M=546.25 SD=640.65		r=0.026
Duration of use, Office machine & Recreational	M=511.06 SD=365.68		r=0.399**
Duration of use, Health Support & Personal hygiene	M=157.58 SD=262.00		r=0.227*
Duration of use, Air conditioning & heating and Others	M=293.85 SD=274.64		r=0.133

* p < .05.   ** p < .01.   *** p < .001.

The 'number of appliances (.636)' in the dwelling turned out to be one of the most important variables determining standby power consumption. Also 'the ratio of plugging (.496)', 'house size (.416)' and 'duration of use, office machine & recreational (.399)' are correlated with standby power consumption. In addition to those factors, 'duration of use, health support & personal hygiene (.227)' showed correlations with standby power. As was discussed earlier, it would be the value of this study to quantify the relationship of 'number of appliances' and the standby power consumption. Habitual change is required to unplug the appliances that are not used for a while, and reducing the number of appliances is also necessary to diminish the standby power consumption.

Comparison Analysis of Standby Power Consumption by Consumption Groups

Fig. 2 shows standby power consumption of each household and their average. Averagely, the standby power consumption of respondents was 9.0kWh, and the highest was 38.0kWh, lowest was 0kWh. The majority of the households in this survey have a standby power below 10kWh.

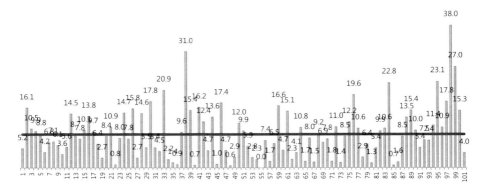

Fig. 2. Standby power per household and average

We divided respondents of questionnaire into two groups. First group was people who consume standby power more than average. Second group was people who consume standby power less than average. Therefore we compared both groups in their variables.

In this table 6 and 7, correlations analysis is carried out to determine the relationship between occupant variables and standby power consumption of each group.

In the first group, Video (.680) turned out to be one of the most important variables determining standby consumption. In addition, Fax (.593) and Dishwasher (.473) presented a slightly higher correlation. Also, computer (.325), monitor (.337), set top box (.329) and audio (.382) presented a slightly higher correlation in the first group. Second group, computer (.464), monitor (.450) and Vacuum (.420) turned out to be the most important variables determining standby consumption.

Table 6. The result of correlation between Electronic appliances of occupant and Standby power Consumption of First group

Electronic appliances	Mean and SD		Correlation standby power consumption (+)
Rice cookers	M=462.09	SD=589.68	r=0.01
Microwave	M=16.83	SD=31.88	r=0.171
Electric Frying Pan	M=0.93	SD=4.79	r=-0.062
Toaster	M=5.32	SD=11.43	r=0.083
Coffee Machine	M=5.23	SD=11.23	r=0.244
Dishwasher	M=3.11	SD=11.65	r=0.473**
Blender	M=4.34	SD=10.18	r=-0.044
Water Purifier	M=14.8	SD=47.51	r=0.026
Oven	M=5.72	SD=14.37	r=-0.135
Computer	M=87.16	SD=104.90	r=0.325*
Monitor	M=91.11	SD=1062.14	r=0.337*
laptop	M=145.34	SD=173.72	r=-0.018
TV	M=163.48	SD=119.74	r=0.079
Set top box	M=96.93	SD=107.33	r=0.329*
Audio	M=8.86	SD=18.57	r=0.382*
Video	M=2.44	SD=10.31	r=0.680*
Telephone	M=28.7	SD=65.32	r=0.041
Printer	M=6.32	SD=13.83	r=0.001
Fax	M=0.67	SD=3.18	r=0.593**
Scanner	M=2.97	SD=10.27	r=0.003
Iron	M=8.60	SD=14.40	r=0.116
Washing machine	M=53.67	SD=43.40	r=-0.048
Humidifier	M=34.55	SD=128.79	r=-0.035
Air Cleaner	M=96.74	SD=321.89	r=0.068
Hair Drier	M=12.67	SD=13.13	r=-0.143
Vacuum	M=13.74	SD=14.82	r=-0.002
Bidet	M=11.39	SD=38.57	r=-0.042
Vibrator	M=2.95	SD=7.60	r=0.056
Electric pad	M=60.00	SD=139.80	r=-0.212
Electric fans	M=184.88	SD=164.67	r=0.12
Air conditioner	M=61.06	SD=65.59	r=0.244
Desk lamp	M=15.06	SD=82.86	r=-0.157

* $p < .05$.   ** $p < .01$.   *** $p < .001$.

Table 7. The result of correlation between Electronic appliances of Occupant and standby power consumption of Second group

Electronic appliances	Mean and SD		Correlation standby power consumption (+)
Rice cookers	M=534.51	SD=658.15	r=0.191
Microwave	M=11.82	SD=19.69	r=0.005
Electric Frying Pan	M=4.29	SD=16.99	r=-0.11
Toaster	M=1.41	SD=4.46	r=0.191
Coffee Machine	M=2.74	SD=7.07	r=0.075
Dishwasher	M=1.12	SD=7.89	r=-0.061
Blender	M=3.31	SD=7.79	r=0.215
Water Purifier	M=2.82	SD=10.12	r=-0.014
Oven	M=4.82	SD=24.28	r=0.011
Computer	M=66.37	SD=134.29	r=0.464**
Monitor	M=66.72	SD=134.10	r=0.450**
laptop	M=150.68	SD=179.13	r=-0.384**
TV	M=107.41	SD=120.91	r=0.164
Set top box	M=9.31	SD=50.01	r=0.083
Audio	M=2.55	SD=11.28	r=0.06
Video	M=2.06	SD=15.76	r=0.049
Telephone	M=11.03	SD=44.94	r=-0.221
Printer	M=2.36	SD=8.96	r=0.286*
Fax	M=0.13	SD=0.76	r=-0.02
Scanner	M=1.20	SD=7.91	r=0.199
Iron	M=3.53	SD=7.24	r=0.176
Washing machine	M=57.93	SD=49.35	r=0.004
Humidifier	M=8.01	SD=42.18	r=-0.011
Air Cleaner	M=5.43	SD=39.38	r=0.12
Hair Drier	M=13.39	SD=14.51	r=0.12
Vacuum	M=8.98	SD=12.42	r=0.420**
Bidet	M=2.34	SD=9.18	r=0.248
Vibrator	M=1.03	SD=7.88	r=0.049
Electric pad	M=31.63	SD=78.45	r=0.099
Electric fans	M=132.75	SD=210.72	r=-0.141
Air conditioner	M=52.32	SD=73.66	r=-0.128
Desk lamp	M=25.55	SD=60.06	r=-0.012

* p < .05.   ** p < .01.   *** p < .001.

4 Conclusion

The correlation analysis between standby power and variables related to occupant's electrical energy consumption will provide important information setting directions to save power consumption. Since the single-member households show similar standby power consumption as other commonplace households (commonplace households: 6.1% of total power consumption; single-member households: 6.1%), the increase in the number of single-person households leads to the increase in the waste of electrical energy nationally, so it is expected that this study would imply to the society that a standby power cut-off system is imminent in the design of single-person housing. The results of this study are as follows.

Firstly, the analysis of mean waiting time of each appliance affecting standby power shows that the appliances with longest mean waiting time are washing machine Air conditioner, TV and Microwave; those with greatest standby power consumption are air conditioner and set top box. Especially, the set-top boxes, recently appeared to provide high-quality image, turned out to be the new main culprit of the waste of electrical energy.

Secondly, average standby power consumption of single-member households presents 9.0kWh. It accounted for 6.1% of total monthly electricity consumption. On average, households have 16.1 appliances with a plug in; ratio of 57.0%. Since the cost of usage 10cents per kilowatt hour, electricity which is wasted as a result of standby power per household monthly is 0.95dollars (11.4 dollars for a year). It seems insignificant, when the entire population in country is taken into account, the amount becomes huge.

Thirdly, correlations analyses are carried out to determine the relationship between the resident variables and standby power. The number of appliances in the dwelling turned out to be one of the most important variables determining standby power consumption. Also, 'house size', 'the ratio of plugging', 'duration of use, office machine & recreational' and 'duration of use, health support & personal hygiene' are correlated with the standby power consumption.

Lastly, we divided respondents of questionnaire into two groups. The first group was people who consumed standby power more than average, the second group was people who consumed standby power less than average. As the result of correlation test, the video and fax turned out to be one of the most important variables determining standby consumption in the first group. In the second group, computer and monitor turned out to be most important variables determining standby consumption.

According to the result of this study, the easiest way of reducing the standby power is to unplug the consents. According to the analysis, the ratio of standby power to electrical energy consumption in the households is 6.1%. If 3%, which is half of the total, is saved by the entire 18 million households of Korea, an yearly 210 billion won may be saved [2, 15]. It is not easy to change one's long accustomed behavioral patterns like plugging consents. Therefore, it is necessary to enhance occupants' perception first through various education and promotion. It would be another measure to grant diverse benefits to the occupants to encourage these behaviors. In addition, a new technology may be applied in house design stage to automatically cut-off standby power.

This study is more of a pilot study and does not have sufficient number of subjects of 101. For this reason, it has a limitation to be generalized. However, the saving effect can bring considerably positive rippling effect if the derived output is followed. Future studies are invited to be able to suggest generalized results with more number of sampled subjects to investigate domestic energy savings based on the electrical energy consumption patterns of occupants.

Acknowledgements. This work was supported by the National Research Foundation of Korea (NRF) grant funded by the Korea government (MEST) (No. 2012-0000609)

References

1. Santin, O.G.: Behavioural Patterns and User Profiles Related to Energy Consumption for Heating. Energy and Buildings 43 (2011)
2. Korea Energy Economics Institute, Estimate of End-Use Energy Consumption in Residential Sector (2010)
3. Bertoldi, P., et al.: Standby Power Use: How Big is the Problem? What Polices and Technical Solution Can Address It (2002)
4. Meier, A.K.: A Worldwide Review of Standby Power Use in Homes (2001)
5. Meier, A., et al.: Standby Power Use in Chinese Homes. Energy and Buildings 36 (2004)
6. Lee, E.-Y.: Analysis Result for the Technical Development Reducing Standby Power in Domestic Major Electric Appliances (2009)
7. Hanssen, K.G.: Standby Consumption in Households Analyzed With a Practice Theory Approach. Journal of Industrial Ecology 14(1) (2009)
8. Yohanis, Y.G.: Domestic Energy Use and Householders' Energy Behavior. Energy Policy 41 (2012)
9. Meyers, R.J., et al.: Scoping the Potential of Monitoring and Control Technologies to Reduce Energy Use in Homes. Energy and Buildings 42 (2010)
10. Korea Electro-Technology Research Institute, 2011 National Standby Power Actual Survey (2012)
11. Information System Planning Bureau, Family Structure in Seoul by Statistics (2012)
12. The Korea National Statistical Office, 2001~2010 Energy Consumption per Household in Residential Sector (2010)
13. Chan, Mingyin, et al.: The effects of external wall insulation thickness on annual cooling and heating energy uses under different climates. Applied Energy 97 (2012)
14. Korea Power Exchange, 2011 Report on Electric Appliance penetration rate & Electricity consumption behavior for home use (2011)
15. The National Statistical Office, Population and housing census (2010)

Chapter 91
A Classification of Real Sky Conditions for Yongin, Korea

Hyo Joo Kong and Jeong Tai Kim[*]

Department of Architectural Engineering, Kyung Hee University, Yongin 446-701,
Republic of Korea
{hjk0905,jtkim}@khu.ac.kr

Abstract. Information on sky condition classification is important to the energy-efficient and sustainable building designs for predicting the energy consumption and daylight performance. Sky conditions are commonly classified into overcast, partly cloudy, and clear sky. IESNA's Sky Ratio (SR), Perez's Clearness Index (CI), Li's Clearness Index (Kt) are representative sky conditions classification. However, reliable and accurate daylight climatic data are lacking in Korea. Thus, this paper presents a classification of real sky condition for Yongin, Korea from data recorded during a period of one year at the KHU station. For the research, illuminance and irradiance data were obtained from 1st February 2011 to 31st January 2012. Hourly frequency of sky conditions occurrence and hourly global horizontal illuminance were analysed according to sky conditions. It was found that sky conditions in Yongin, Korea are classified, respectively, as clear and overcast skies 34% and 17% of the time in Yongin, Korea. These findings indicate that sky condition classification and illuminance information gathered at KHU could be applied for various daylight design applications and energy-efficient building design in Yongin, Korea.

Keywords: Sky conditions, Clear sky, Partly cloudy sky, Overcast sky, Illuminance.

1 Introduction

An understanding of the prevailing sky condition is an important part of the building design process [1]. Also, the consideration of meteorological parameters is essential in the design and study of solar energy conservation devices [2]. Generally, sky conditions of the same category have similar features and are very useful for energy-efficient building designs by energy simulation tool [1]. For example, a computer simulation tool for predicting the performance of a solar photovoltaic (PV) system can be analyzed under clear sky condition [3]. In addition, the prevailing local sky conditions can affect both air-conditioning plant sizing and electric lighting

---

[*] Corresponding author.

A. Håkansson et al. (Eds.): *Sustainability in Energy and Buildings*, SIST 22, pp. 1025–1032.
DOI: 10.1007/978-3-642-36645-1_91   © Springer-Verlag Berlin Heidelberg 2013

consumption [1]. For evaluation of daylighting performance, the overcast and clear sky daylight factor method is dominantly used in contemporary building energy simulation programs to model artificial lighting control [4]. Therefore, classification of sky condition and illuminance values for Yongin, Korea is definitely needed.

Generally, there are three sky conditions, as divided by CIE (Commission Internationale de l'Eclairage): clear sky, overcast sky, and partly cloudy sky [5]. These conditions can be categorized by the following parameters: IESNA's sky ratio (SR) [6], Perez's clearness index (CI) [7], and Li's clearness index (Kt) [8]. These parameters have been used to establish sky conditions over Hong Kong [1], Italy [9], India [10], and the UK [7]. Previous research has revealed that the frequency of occurrence might be different in different seasons and in regions with different climates. However, their compatibility with illuminance and irradiance data by sky clearness indices in Korean climate has not yet been verified. Thus, the aim of this paper is to suggest the sky condition classifications and illuminance value for Yongin, Korea.

2 Daylight Measurement System at KHU

In 2008, a daylight measurement station was set up at Kyung Hee University in Korea to measure the global horizontal illuminance and irradiance, global vertical illuminance and irradiance, and sky luminance. The station can be classified as a research station in accordance with the International Daylight Measurement Program (IDMP) of the Commission Internationale de l'Eclairage (CIE) [11].

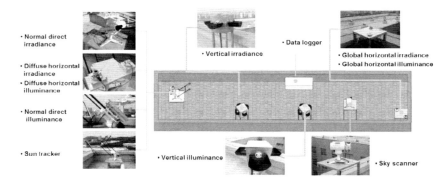

Fig. 1. Layout of daylight measurement systems

There were 16 measuring elements in total with the three largest sections of illuminance, irradiance, and luminance. The solar irradiance measuring instruments were supplied by MS-802 of EKO, Japan and used to measure the global, diffuse, and vertical irradiance. The illuminance measuring equipment was also supplied by ML-020s-o EKO Japan and used to measure the global, diffuse, and vertical illuminance. All equipment was connected to a CR1000 data logger (Campbell) to collect the data. A CHP 1 Pyheliometer produced by KIPP & ZONEN, Netherlands was used for the normal direct irradiance. The station is located on the roof of a thirteen-story building at KHU. There are no tall buildings or structures in the site causing obstruction. The site is at latitude N37.14° and longitude E127.4° with clock time of Korea, GMT +9. In this study, data were collected for 1 year (Feb. 2011-Jan. 2012). Also, the data with solar elevation below 20° and global horizontal irradiance of 4W/m^2 were eliminated from the analysis.

3   Sky Condition Classification

Sky clearness indexes based on these approaches are provided in Table 1, which summarizes the definitions for clear, partly cloudy, overcast sky using the SR, CI, Kt methods.

Table 1. Classification of sky condition

Sky Type	SR	CI	Kt
Clear	SR≤0.3	5<CR	0≤Kt≤0.3
Partly Cloudy	0.3<SR<0.8	2<CR≤5	0.3<Kt≤0.65
Overcast	SR≥0.8	CR≤2	0.65≤Kt≤1

3.1   IESNA's Sky Ratio

Sky ratio method has been used to classify sky conditions. The sky ratio is determined by dividing the diffuse horizontal irradiance by the global horizontal irradiance [5]. Sky ratio is defined as,

$$SR = \frac{Eed}{Eeg} \tag{1}$$

Where, Eed = Diffuse horizontal irradiance (W/m^2)
Eeg = Global horizontal irradiance (W/m^2)

3.2   Perez's Clearness Index

Perez's clearness index use normal direct and diffuse horizontal irradiance values to divide sky condition [6].
Perez's clearness index is defined as,

$$\varepsilon = \frac{[(Eed+Ees)]/(Eed+1.041)Z^3}{[1+1.041Z^3]} \tag{2}$$

Where, ε = Clearness index
Ees = Normal direct irradiance (W/m^2)
Eed = Diffuse horizontal irradiance (W/m^2)
Z = The zenith angle (rad)

3.3 Li's Clearness Index

Li's clearness index (Kt) is determined by dividing the global horizontal irradiance by the extraterrestrial irradiance [7].

Li's clearness index is defined as,

$$Kt = \frac{Eg}{Eeg} \qquad (3)$$

Where, Kt = Li's clearness index (Kt)
Eg = Extraterrestrial irradiance (W/m^2)
Eeg = Global horizontal irradiance (W/m^2)

4 Classification of Sky Conditions

The global and diffuse horizontal irradiance data measured for sky ratio at Kyung Hee University during one year from 2011 to 2012 were analysed. The frequency of occurrence is shown in Fig. 2. The yearly relative frequency of occurrence under clear sky based upon the sky ratio in Korea is assumed to be 36%. In the case of SR from 0.3 to 0.8, corresponding to a partly cloudy sky, made up the largest part with 57%. Overcast sky, from 0.8 to 1, accounted for 14%. The average global horizontal illuminance under clear sky is 71,265lux. The average global horizontal illuminance under partly cloudy sky is 50,987lux. The average global horizontal illuminance under overcast sky is 24,408lux.

Hourly Perez's clearness index data from the one year measurements taken in this study were analysed and the frequency of occurrence is shown in Fig 3. A peak of just

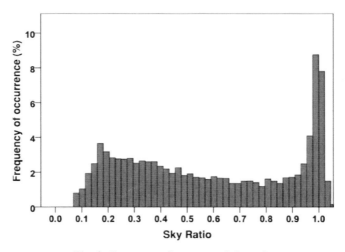

Fig. 2. Occurrence frequency of sky ratio

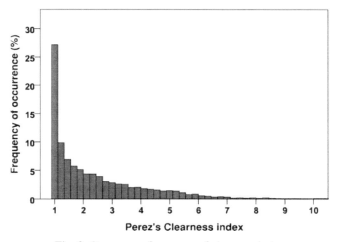

Fig. 3. Occurrence frequency of clearness index

over 52% is observed under partly cloudy sky. The second most frequent clearness index is overcast sky, accounting for just over 17%. At approximately 31%, clear sky is the least frequent clearness index. For the remaining time, partly cloudy sky conditions prevail. In the case of clear sky, the average illuminance is 73, 410lux. Under partly cloudy sky, the yearly average global horizontal illuminance is 53,433lux. The average global horizontal illuminance under overcast sky is 25,232lux.

The frequency of occurrence for the Kt obtained from the KHU in an interval of 15 minutes is presented in Fig 4. In analysing Kt annual data using equation 3, when Kt ranged from 0 to 0.3, the corresponding condition of clear sky was 34%. In the case of

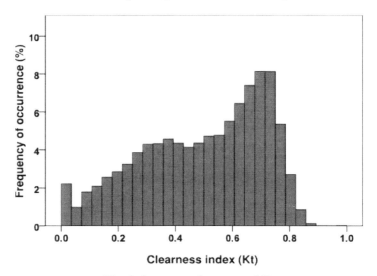

Fig. 4. Occurrence frequency of Kt

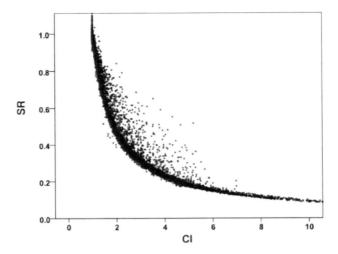

Fig. 5. Correlation between SR and CI

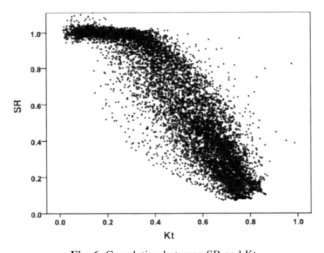

Fig. 6. Correlation between SR and Kt

Kt from 0.3 to 0.65, the corresponding partly cloudy sky condition accounted for the largest portion with 45%. Overcast sky from 0.65 to 1 was 20%. The global horizontal illuminance is 15,272lux, 40,304lux, and 75,917lux under overcast sky, partly cloudy sky, and clear sky.

Fig. 5 shows the correlation between IESNA's sky ratio (SR) and Perez's clearness index (CI). It can be seen that SR varies almost linearly with CI over 5 and up to 2. This means that the SR and CI values are very close to each other, indicating that under and clear and overcast sky. Figs. 6 and 7 show the correlation between Li's clearness index (Kt) and CI, SR. The correlation between Kt and CI, SR analysis, and

CI and SR becomes more scattered. This is due to the different variables of the three clearness indices. However, the average global horizontal illuminance under three sky conditions is very similar. This indicates that the relationship of three indices is useful for predicting the other illuminance data and index when one index is known.

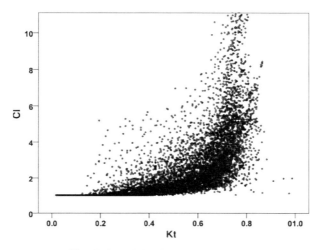

Fig. 7. Correlation between CI and Kt

5　Conclusions

The classification of three sky condition in Yongin, South Korea during from February to January has been presented. The results show that 34% of the total time in Korea is classified as clear sky. On the other hand, overcast sky and partly cloudy conditions occurred 17% and 49% of the time. The average global horizontal illuminance was 73,530lux, 48,241lux, and 21,637lux under clear, partly cloudy, and overcast sky. Thus, the annual average global illuminance at working hour is 38,644lux. The strong correlation between SR and CI showed that the sky condition can be categorised according to three CI ranges: $0 \leq CI < 0.2$ (overcast sky), $0.2 \leq CI < 0.5$ (partly cloudy sky), and $0.5 \leq CI$ (clear sky). As a result, the global horizontal illuminance and sky condition classification information provided by this research can be used for designers to estimate the local daylight and energy consumption. Further work will be necessary to evaluate the efficacy model with illuminance, irradiance and the sky radiance and luminance distribution models for all sky conditions. Long-term daylight weather measurement for Yongin, Korea will be conducted in the future.

Acknowledgments. This work was supported by the National Research Foundation of Korea (NRF) grant funded by the Korea government (MEST) (No. 2012-0000609).

References

1. Li, D.H.W., Lam, J.C.: An analysis of climatic parameters and sky condition classification. Building and Environment 36, 435–445 (2001), doi:10.1016/S0360-1323(00)00027-5
2. Hepbasli, A.: Prediction of solar radiation parameters through clearness index for Izmir. Turkey, Energy Sources 24, 773–785 (2002), doi:10.1080/00908310290086680
3. Sukamongkol, Y., Chungpaibulpatana, S., Ongsakul, W.: A simulation model for predicting the performance of a solar photovoltaic system with alternating current loads. Renewable Energy 27(2), 237–258 (2002), doi:10.1016/S0960-1481(02)00002-2
4. Janak, M.: Coupling Building Energy and Lighting Simulation. In: Proceedings of the fifth International IPBSA Conference (1997)
5. CIE, Spatial Distribution of Daylight – CIE Standard General Sky. CIE CIE, Austria (2003)
6. IESNA, Lighting Handbook 10th Edition. Illuminating Engineering Society, USA (2011)
7. Perez, R., Ineichen, P., Seals, R., Michalsky, J., Stewart, R.: Modeling Daylight Availability and Irradiance Components from Direct and Global Irradiance. Solar Energy 44(5), 271–289 (1990), doi:10.1016/0038-092X(90)90055-H
8. Lam, J.C., Li, D.H.W.: Study of Solar Radiation Data for Hong Kong. Energy Conversion and Management 37(3), 343–351 (1996), doi:10.1016/0196-8904(95)00179-4
9. Barbaro, S., Cannata, G., Coppolino, S., Leone, C., Sinagra, E.: Diffuse solar radiation statistics for Italy. Solar Energy 26(5), 429–435 (1981), doi:10.1016/0038-092X(81)90222-X
10. Choudhury, N.K.D.: Solar radiation at New Delhi. Solar Energy 7(2), 44–52 (1963), doi:10.1016/0038-092X(63)90004-5
11. CIE, Guide to recommended practice of daylight measurement. CIE, Austria (1994)

Chapter 92
Influence of Application of Sorptive Building Materials on Decrease in Indoor Toluene Concentration

Seonghyun Park[1], Jeong Tai Kim[2], and Janghoo Seo[3,*]

[1] Graduate student, College of Engineering, Chosun University, Gwangju 501-759, Korea
[2] Department of Architectural Engineering, Kyung Hee University, Yongin 446-701, Korea
[3] Department of Architectural Engineering, Chosun University, Gwangju 501-759, Korea
seo@chosun.ac.kr

Abstract. In order to improve indoor air quality, the interest in and the use of sorptive building materials that decrease the concentration of an indoor air pollutant have increased. The use of sorptive building materials is one way to decrease the concentration of an indoor pollutant that can adversely affect human health. In this study, we evaluated the effects of sorptive building materials applied to a wall on the decrease in the concentration of toluene emitted from the flooring. We also examined how the air exchange rate of the room, the loading factor of the sorptive materials, and the mass transfer coefficient influenced the sorptive performance; these effects were well reproduced experimentally with computational fluid dynamics (CFD) simulations. The results show that sorptive building materials have a fairly strong effect on the decrease in toluene concentrations in rooms and that this effect can be expected in real-world scenarios.

1 Introduction

Recently, the use of new chemical construction materials as well as heat insulation and air-tightness of buildings has led to fatal diseases such as the sick building syndrome. Further, it has been reported that economic loss can be caused by the presence of indoor air pollutants as they pose health risks and adversely affect the work efficiency of the occupants [1]. According to a previous study, there are many pollutants such as formaldehyde and toluene in the indoor environment [2]. These pollutants are emitted from construction materials, fuel combustion, cosmetics, and adhesives; they are also introduced from the outdoors to an indoor environment by the polluted outdoor air [3]-[5]. Hence, people have now started paying increased attention to the indoor air quality in order to improve their health. Therefore, the use of sorptive building materials with permanent ventilation and bake-out has increased as a method to decrease the concentration of pollutants in an indoor environment. From among these building materials, sorptive building materials have several advantages: their use not only helps to reduce the amount of energy consumed while

---

[*] Corresponding author.

A. Håkansson et al. (Eds.): *Sustainability in Energy and Buildings*, SIST 22, pp. 1033–1041.
DOI: 10.1007/978-3-642-36645-1_92   © Springer-Verlag Berlin Heidelberg 2013

operating ventilation equipment but also helps to decrease the concentration of indoor pollutants and this approach does not require the use of any special equipment [6].

In this study, through a computational fluid dynamics (CFD) analysis, we evaluate the extent to which sorptive building materials applied to building walls decrease the concentration of toluene emitted from the flooring. Further, the effect of the attachment force of sorptive building materials on the surface area of materials, position of materials, and the air diffuser is examined on the basis of three factors: the air exchange rate of the room, the loading factor of the sorptive building materials, and the mass transfer coefficient [7].

2 Evaluation of Sorption Flux for Sorptive Building Materials

In this study, the concentration of toluene emitted from the flooring was the same as the saturation concentration of toluene at indoor temperatures. Further, air containing a low pollutant concentration was flown into the indoor environment in which a sorptive building material was used. The performance of the sorptive building material was evaluated in terms of the sorption flux and sorption amount. The sorption flux was calculated from the difference in the mean toluene concentration depending on the sorptive building material applied to the building walls, the ventilation rate of the indoor environment, and the surface area of the sorptive building material [7]. The sorption flux was defined as follows:

$$F = \frac{(C_b - C_a) \times Q_c}{A} \tag{1}$$

Here, F is the sorption flux (g/m^2·h); C_b, the mean toluene concentration before the application of the sorptive material (g/m^3); C_a, the mean toluene concentration after the application of the material (g/m^3); Q_c, the air exchange rate (m^3/h); and A, the surface area of the building material (m^2).

3 CFD Analysis

The model used for the CFD analysis is illustrated in Figure 1 and has the following dimensions: $6 \times 4 \times 3$ m^3. The room has one air inlet and one outlet, both of which are in the ceiling and have the same size of 0.4×0.6 m^2. The fresh air is supplied only by the mechanical ventilation through the inlet; the air flows of the room through the outlet. A window is installed on the inner wall of the room and is closed. A fixed quantity of radiant heat is present in the room because of the window and the human model in the room. Toluene is emitted only from the floor in the room, and the human model is located in the center of the room. Table 1 shows the details of the cases considered in the CFD analysis. The different placements of the air diffuser and the openings with respect to the human model used in the CFD analysis are shown in Figure 2. Cases 1–3 correspond to the situations dependent on the air exchange rates. Cases 4–6 have the same air exchange rate strategy as that of Case 2; however, they differ in terms of the surface area of the sorptive material.

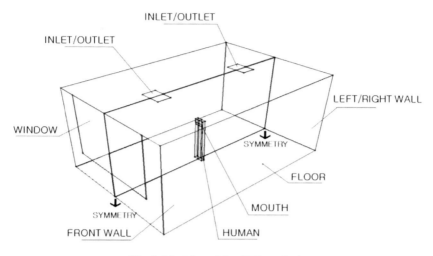

Fig. 1. Model used for CFD analysis

Table 1. Cases considered in CFD analysis

Case	Temp (K)	Air exchange rates (n/h)	Saturation concentration (g/m³)	Type of air diffuser	Location of sorptive material application	Area of sorptive material application (m²)
1	293.15	1.0	158	No. 3	Right	6
2	293.15	1.5	158	No. 3	Right	6
3	293.15	2.0	158	No. 3	Right	6
4	293.15	1.5	158	No. 1	-	-
5	293.15	1.5	158	No. 1	Left	6
6	293.15	1.5	158	No. 1	Left, Front	24
7	293.15	1.5	158	No. 2	Right	6
8	293.15	1.5	158	No. 4	Right	6

In Cases 2, 5, 7, and 8, the location of the air diffuser and the openings influence the performance of the sorptive material. Further, if the surface area of the sorptive material is 6 m², the material is applied to an area opposite the openings.

The boundary conditions for the CFD analysis are given in Table 2. We considered the radiation and the temperature difference between the outdoor and the indoor environments at the window. The velocity of the supplied air was 8.3×10^{-1}, 1.25×10^{-1} and 1.67×10^{-1} m/s, respectively, for each of the following air exchange rates: 1.0, 1.5, and 2.0 (n/h). The flow field was analyzed by using a three-dimensional analysis based on a low-Reynolds-number-type k-e model, the Abe-Nagano model. After analyzing the airflow field, we set the boundary condition of the evaporation that dominated the building materials on the side where the sorptive material was applied,

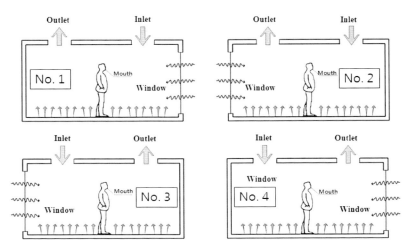

Fig. 2. Different placements of air diffuser and openings

Table 2. Conditions of numerical analysis

Turbulent flow model	Low-Reynolds-number-type k-ε model (Abe-Nagano model)
Number of meshes	Around 650,000
Scheme	Space difference: Second-order upwind
Inflow boundary	$U_{y,in}$ = 8.3 × 10^{-2} m/s, 1.25 × 10^{-1} m/s, 1.67 × 10^{-1} m/s, $U_{x,in}$ = 0, $U_{z,in}$ = 0, k_{in} = 3/2·$(U_{in} × 0.05)^2$, ε_{in} = C_μ·$k_{in}^{3/2}$/L_{in}, L_{in} = 1/7L_o, L_o = 0.3 m
Outflow boundary	U_{out} = outflow (mass flow conservation), k_{out}, ε_{out} = free slip
Breath boundary	14.4 L/min, human standard [8]
Flux boundary	H_{human} = 20 W/m², H_{window} = K_{window} · (θ_R - θ_O) θ_R = 303.15 K, θ_O = 293.15 K, K_{window} = 6.39 W/m²·K
Wall boundary	No slip
Analysis of diffusion field	Considering three-dimensional symmetry, we analyzed only one half of the space. After the airflow field analysis, the water (distilled water) saturation concentration on the surface of the building material was set.

as a model of the release of the chemical materials. The diffusion was analyzed assuming an isothermal state (293.15 K). It was assumed that water was a passive pollutant, and the water (vapor) diffusion in the air was expressed as the diffusion equation (3). The water (vapor) diffusion coefficient in the air was calculated using equations (4)–(6) [9][10].

$$\frac{\partial \overline{C_1}}{\partial t} + \frac{\partial \overline{U_j} \overline{C_1}}{\partial x_j} = \frac{\partial}{\partial x_j}\left(\left(D_a + \frac{v_t}{\sigma_t}\right)\frac{\partial \overline{C_1}}{\partial x_j}\right) \qquad (3)$$

$$\log_{10} P_w = \frac{A-B}{(C+T)} - 3 \qquad (4)$$

$$C_0 = \rho_a \frac{M_1}{M_2} \frac{P_w}{P - P_w} \tag{5}$$

$$D_a = \frac{6.7 \times 10^{-8} \times T^{1.83}}{P} \times \left[\left(\frac{T_{c1}}{P_{c1}} \right)^{\frac{1}{3}} + \left(\frac{T_{c2}}{P_{c2}} \right)^{\frac{1}{3}} \right]^{-3} \sqrt{\frac{1}{M_1} + \frac{1}{M_2}} \tag{6}$$

where C_1 is the pollutant concentration at a spatial point ($\mu g/m^3$); D_a, the molecular pollutant diffusion coefficient (m^2/s); U_j, the wind velocity (m/s); v_t, the eddy viscosity (m^2/s); σ_t, the turbulent Schmidt's number (-); and P_w, the water vapor pressure (Pa). Further, A, B, and C are the empirical constants 7.7423, 1554.16, and 219, respectively. T denotes the temperature (K); C_0, the saturation concentration (g/m^3); and ρ_a, the air density (g/m^3). M_1 and M_2 are the molecular weights, P is the atmospheric pressure in the chamber (Pa), T_{c1} and T_{c2} are the critical temperatures (K), and P_{c1} and P_{c2} are the critical pressure values (Pa).

The water (vapor) diffusivity D_a in the air was calculated to be $0.76 \times 10^{-5} m^2/s$ (293.15 K) using equations (4)–(6).

4   Results of CFD Analysis

The results of the CFD analysis are given in Table 3 and illustrated in Figures 3–6. Some of the toluene emitted as a pollutant from the indoor floor was either diluted by fresh air or adsorbed by the sorptive materials; the remainder was naturally flown into the outside environment through the outlet.

Figure 3(a) shows the mean toluene concentration in the indoor environment in Cases 1, 2, and 3. The effect of the use of a sorptive material on the reduction of the mean toluene concentration was increased by increasing the air exchange rate in the indoor environment, irrespective of the sorptive material used. Hence, we concluded that by changing only the air exchange rate, we could increase the effect of the use of a sorptive material on the reduction of the toluene concentration. Further, Figure 3(b) shows the change in the performance of the sorptive material in Cases 1, 2, and 3 with respect to the air exchange rate. The sorption amount in Cases 1, 2, and 3, where the air exchange rate was 1.0 n/h, 1.5 n/h, and 2.0 n/h, was 287 g/h, 371 g/h, and 445 g/h, respectively. The greater the increase in the air exchange rate, the larger was the sorption amount of the sorptive material. However, the rate of decrease in the mean toluene concentration was almost the same in each case. This implied that the increase in the air exchange rate led to an increase in the velocity distribution near the surface of the floor that emitted toluene as well as the surface of the sorptive material.

Figure 4 shows the toluene concentration distribution in the indoor environment with respect to the surface area and in cases with and without the application of a sorptive material. In Case 4, in which a sorptive material was not used, the mean toluene concentration in the indoor environment was 71 g/m^3. When the surface area of the sorptive material increased to 6 m^2 and 24 m^2, the toluene concentration in the indoor environment decreased by around 9.8% and 37.4%, respectively, as compared to the reference concentration. It was verified that an increase in the surface area of the sorptive material decreased the toluene concentration.

Meanwhile, the toluene concentration near the inlet was constant at below 8 g irrespective of the case in which the sorptive material was used because the toluene concentration near the inlet was diluted by fresh air. Further, a concentration boundary layer was observed near the surface of the sorptive material because of the sorption of toluene.

Table 3. Results of CFD analysis

Case	1	2	3	4	5	6	7	8
Air exchange rates (n/h)	1.0	1.5	2.0	1.5	1.5	1.5	1.5	1.5
Area of sorptive material application (m^2)	6	6	6	-	6	24	6	6
Type of air diffuser (No)	3	3	3	1	1	1	2	4
Toluene concentration before application of sorptive material (g/m^3)	78	71	66	71	71	71	71	71
Toluene concentration after application of sorptive material (g/m^3)	70	64	60	-	64	45	60	60
Sorption flux ($g/m^2 \cdot h$)	48	61.9	74	-	63	60	101	102
Sorption amount (g/h)	287	371	445	-	377	1431	608	613
Rate of decrease in concentration (%)	10.3	9.8	9.6	-	9.8	37.4	15.9	16.1

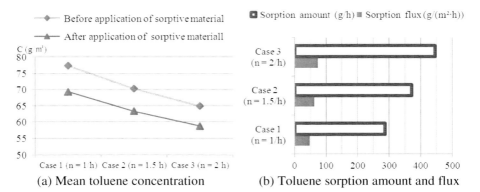

(a) Mean toluene concentration (b) Toluene sorption amount and flux

Fig. 3. Toluene concentration change of indoor environment in Cases 1, 2, and 3

92 Influence of Application of Sorptive Building Materials on Decrease 1039

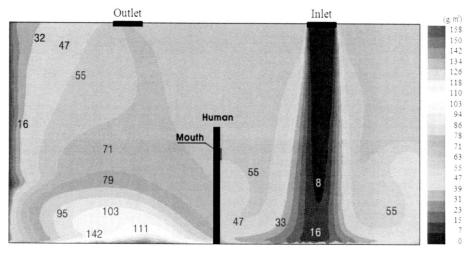

Fig. 4. Toluene concentration distribution in indoor environment in Case 5 (Surface area of sorptive material is 6 m^2)

Let us now consider cases 2, 5, 7, and 8 that have the same conditions except for the placement of the diffuser. The decrease in the mean toluene concentration in Cases 7 and 8, where the sorptive material was applied on the wall near the inlet, was around 16%; this decrease was higher than the decrease observed in Cases 2 and 5. This was because of the difference shown in Figure 5. Meanwhile, in Cases 2 and 5, in which the distance between the sorptive material applied on the wall and the inlet was large as compared to that in Cases 7 and 8, the velocity of the airflow at a spot 1 cm away from the sorptive material (Line A) was almost as low as less than 0.1 × 10^{-3} m/s, irrespective of the height. However, in Cases 7 and 8, a marked difference in the distribution of the airflow velocity due to the occurrence of a localized turbulence in the boundary layer where the wall met the floor, was observed.

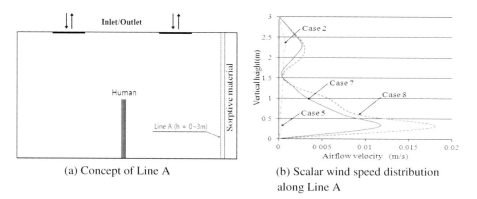

(a) Concept of Line A (b) Scalar wind speed distribution along Line A

Fig. 5. Scalar wind speed distribution in Cases 2, 5, 7, and 8

5 Conclusion

In this study, we examined the effect of the use of a sorptive material on the reduction of the mean toluene concentration in an indoor environment, by conducting a CFD analysis when toluene was emitted uniformly from the floor. The factors affecting the performance of the sorptive material were identified to be the air exchange rate, surface area of the sorptive material, and placement of the air diffuser.

A reduction in the concentration of toluene in an indoor environment could be brought about only by increasing the air exchange rate when a sorptive material was not used; however, a high air exchange rate can cause some discomfort to human beings. Hence, the consideration of the air exchange rate was necessary. Results also revealed that the decrease in the mean toluene concentration was around 10% despite the increase in the air exchange rate. Therefore, it was difficult to considerably decrease the mean toluene concentration only by increasing the air exchange rate. The decrease in the toluene concentration increased with an increased in the surface area of the applied sorptive material. However, it was necessary to consider the surface area of the applied sorptive material from the viewpoint of economy and efficiency because sorptive materials are expensive. When the air exchange rate was the same, sorption by the sorptive material was noticeably higher when the sorptive material was applied at a high air velocity distribution than at a low air velocity distribution. Hence, controlling the air velocity near the surface of the sorptive material and the place where toluene was emitted from the floor was expected to improve the reduction of the pollutant concentration in the case when a sorptive material was used. In this study, the results of the CFD analysis carried out to examine the reduction of the toluene concentration were confirmed by using a sorption model in which the surface concentration of the sorptive material was assumed to be zero (0).

In the future, we plan to experimentally examine the mass transfer coefficient on the surface of a commercially available sorptive material. Further, by using the CFD technique, we intend to analyze the effect of a sorptive material on the reduction of the pollutant concentration on the basis of the experimentally obtained values.

Acknowledgments. This work was supported by the National Research Foundation of Korea (NRF) grant funded by the Korea government (MEST) (No. 2012-0000609).

References

1. Seo, J.: Study on performance test and examination of sorptive building materials reducing indoor air pollutant using experiments and CFD analysis. Journal of the Architectural Institute of Korea 28(9), 287–294 (2008)
2. Ataka, Y., Kato, S., Murakami, S., Zhu, Q., Ito, K., Yokota, T.: Study of effect of adsorptive building material on formaldehyde concentrations: development of measuring methods and modeling of adsorption phenomena. Indoor Air 14(8), 51–64 (2004)
3. Reiser, R., Meile, A., Hofer, C., Knutti, R.: Indoor air pollution by volatile organic compounds (VOC) emitted from flooring material in a technical university in Switzerland. In: Proceedings of the Conference on Indoor Air, pp. 1004–1009 (2002)

4. Kim, H., Song, K., Lee, Y.: A study on the Mock up test for reduction of HCHO using the functional gypsum board. Korea Journal of Air-Conditioning and Refrigeration Engineering 20(12), 814–819 (2008)
5. Bornehag, C.G., et al.: The Association Between Asthma and Allergic Symptoms in Children and Phthalates in House Dust; A Nested Case Control Study. Environmental Healthy Perspectives 112(14), 1393–1397 (2004)
6. Seo, J., Kato, S., Ataka, Y., Nagao, S.: Measurement and CFD analysis on decteasing effect of toluene concentration with activated carbon and adsorptive building material. Proceedings of Healthy Building 5, 101–106 (2006)
7. Seo, J., Kato, S., Ataka, Y., Chino, S.: Performance test for evaluating the reduction of VOCs in rooms and evaluating the lifetime of sorptive building materials. Building Environment 44(1), 207–215 (2009)
8. Murakami, S., et al.: Combined simulation of airflow, radiation and moisture transport for heat release from a human body. Building and Environment 35, 489–500 (2000)
9. Seo, J., Kato, S., Ataka, Y., Yang, J.: Influence of Environmental Factors on Performance of Sorptive Building Materials. Indoor and Built Environment 19, 413–421 (2010)
10. Society of Chemical Engineers. Handbook of Chemistry, The Society of Chemical Engineers, Japan (1999)

Chapter 93
Perceived Experiences on Comfort and Health in Two Apartment Complexes with Different Service Life

Mi Jeong Kim[1], Myung Eun Cho[1], and Jeong Tai Kim[2,*]

[1] Department of Housing and Interior Design, Kyung Hee University, Seoul 130-701,
Republic of Korea
{mijeongkim,mecho}@khu.ac.kr
[2] Department of Architectural Engineering, Kyung Hee University, Yongin 446-701,
Republic of Korea
jtkim@khu.ac.kr

Abstract. Residents' responses and illnesses related to the new houses become currently a popular subject of 'sick building syndrome' research. Compared to the research dealing with varied symptoms experienced by residents in new built houses, there has been little attention to residents' health life for a rather long period in old houses with different service life. Thus, we investigate residents' perceived comfort and health in two apartments with different service life and see if there is any difference among them depending on the service life. Contrary to our expectation, the results show that the physical properties of the apartments do not much influence residents' perceived comfort further the energy efficiency does not affect their perceived comfort either. In terms of health, residents' perceived symptoms in the 33 years old apartment are a bit more severe compared to those in the 11 years old apartment.

1 Introduction

There have been numerous reports on 'sick building syndrome (SBS)' with a focus on newly built houses [1,2,3]. Residents' responses and illnesses related to the new houses become currently a popular subject of SBS research, thus the new terminology 'sick house syndrome (SHS)' recently emerged. The Ministry of Environment of Korea government legislated for maintaining optimal indoor air quality of newly built apartment houses in 2005 and Seoul Metropolitan City adopted a certified indoor air quality system in 2008 that require the measurement of air quality in newly built houses to meet the standards of them [4, 5]. Compared to the research dealing with varied symptoms experienced by residents in new built houses, there has been little attention to residents' health life for a rather longer period in existing houses with different service life [4].

To provide sustainable healthy houses, several sustainable building criteria such as Energy Performance of Building Directive (EPBD) in E.U., Leadership in Energy and

---

* Corresponding author.

A. Håkansson et al. (Eds.): *Sustainability in Energy and Buildings*, SIST 22, pp. 1043–1053.
DOI: 10.1007/978-3-642-36645-1_93   © Springer-Verlag Berlin Heidelberg 2013

Environmental Design (LEED) in USA and GBTool (international collaboration framework) have been introduced [6]. These criteria usually include three important related stakes: good energy performance (reducing the environmental impacts of the building), good indoor environment quality (IEQ) and health of the occupants. A building cannot be good if it fails in one of them [7,8]. In Health Optimization Protocol for Energy-efficient Building (HOPE) research project, Bluyssen et al. [9] argued that a healthy and energy-efficient building does not cause illnesses to the building occupants, rather assures a high level of comfort to them and minimizes the use of non-renewable energy.

It is generally believed that the energy efficiency of old apartments is decreased, thus residents' comfort and health in such old houses would be getting worse according to the service life of the apartments. Our research starts from a question on this anecdotal view. We investigate residents' perceived comfort and health in two apartments with different service life and see if there is any difference among them depending on the service life. This study aims to obtain an insight into residents' perceived experience in existing apartments, rather than new ones. Compared to previous research focusing on quantitative evaluation of physical comfort and harmful substances in indoor environment, this research has significance in analyzing residents' perceived comfort and health qualitatively.

2 Case Study: Methodology and Findings

We selected two apartments with 22 years-gap in service life and investigated residents' perceived experience on comfort and health. To reduce the impact of the characteristics of the households, 12 households with similar features in age, family size, income and unit size were firstly selected from the two apartments. Each household has four family members showing similar lifestyles and the subjects of the survey were housewives in the 40s who stay in a unit for a similar time. Three survey methods were used in this study: field survey inspecting each household' characteristics with the information on the apartments and their surrounding environment; in-depth interview investigating the pattern of energy use of each household and; questionnaire identifying how residents feel and perceive the indoor environment [7] and health [10].

2.1 Characteristics of Apartments and the Surrounding Environments

The old apartment complex in Banpo, Seoul has 15 flat-type buildings built in 1978, each one 12 stories facing south, where 1164 households live in different sizes of units with a corridor access. They adopt a district heating system using liquefied natural gas (LNG) in addition to electricity and city gas for cooking. In the surrounding of the complex, the express bus terminal and a transfer station meeting three subway lines are located, thus the floating population become large along the department stores and underground shopping malls around the transportation hub. Further, the Han river is next to the apartment complex, thus residents can access to the river park and swimming pools easily for leisure and recreation. Only six households in units of 107 m² were selected for the investigation.

93 Perceived Experiences on Comfort and Health in Two Apartment Complexes   1045

Table 1. Characteristics of Apartments and the Surrounding Environments

33 years old Apartment Complex	11 years old Apartment Complex
Appearance & Floor Plan	

Built year, Size, Orientation, Heating system, Award	
October, 1978	October, 2000
15 buildings in 12 stories, 1164 units	13 buildings in 14~25 stories, 1170 units
107 m² with three bedrooms	109 m² with three bedrooms
South facing	South-east, South-west facing
a district heating system using LNG	city gas, landscaping award
Surrounding environments	
Express bus terminal	Youth center
Department stores, Underground shopping malls	Sport center
Park, Soccer field	Baseball field
River leisure park, Swimming pool	Sport park, Eco-park

On the other hand, the apartment complex in Gwangjang, Seoul was built in 2000, where 1170 households live in 13 buildings ranging from 14 to 25 stories with different size of units. Buildings are stair access types facing south-east or south-west and adopt city gas for an individual heating system, cooking and hot water in addition to electricity. The apartment complex received a landscaping award by the city Seoul, thus walkways, benches, water spaces etc are well landscaped. It is surrounded by other apartment complexes with no public transportation facility and shopping centers in the vicinity of them, thus the residential area of the complex is rather quiet. There are schools, athletic facilities, parks etc., further the Acha mountain is near the complex, thus residents can access to the eco-park in the mountain. Six households in units of 109 m² were selected.

2.2   Control of Indoor Environment

Based on the assumption that that there some differences in the energy efficiency of the apartments according to the service life, their energy uses of households were investigated with an in-depth interview.

1) Thermal Environment
Heating
Households in the 33 years old complex cannot control the heating temperature individually due to the district heating system, where the maintenance office can

manipulate the temperature of the entire units as a single zone for continuous 24 hours or intermittent heating in a day. On the other hand, households in the 11 years old complex can control the heating temperature ranging from 21°C to 28°C because each boiler is installed in each unit. The average temperature for one year is 24.6°C. They feel colder in winter, thus they use auxiliary heating apparatus such as electric blankets and heaters. All households in two complexes expanded their balconies, where residents feel relatively cold due to no heating provided. In the 33 years old complex there are two units without insulation sash in the extension of the balcony, where residents feel so cold even in the living room because of no heated floor and cold air from the external walls.

Cooling
All households in the 33 years old complex have one air conditioner in their living room whereas four households of the 11 years old complex installed air conditioners in each room in addition to the living room. It seems that units in the 33years old complex are located facing the hallway and the capacity of the electricity is limited, thus it is not easy to install additional air conditioners in a room. Similarly, the number of electric fans in the 11 years old units is larger than that of the 33 years old units (the average number is 2). This result suggests that the consumption of the electricity in the 11 years old complex is larger than that in the 33 years old complex. For the specific period in summer, people have a difficulty in getting sleep deeply at night because the minimum temperature is over 25°C. All households in both complexes turned on air conditioners or electronic fans all the night for 1 week to 1 month when they sleep.

2) Indoor Air quality
For the indoor air quality of the households, we investigated the use of appliances and the way of the ventilation. Households in the 33 years old apartment feel uncomfortable with cold air through the external wall in winter, thus 2 households use sealing papers, 1 household use styrofoam and 2 households use thick curtains for blocking cold air from the outside. On the contrary, households in the 11 years old apartment have less natural ventilation and only 2 households which expanded their balconies adopt sealing papers in the extended wall. In winter, all households in the 33years old complex use humidifiers whereas only one household in the 11years old complex uses a humidifier. This implies that residents' perceive experience on dryness in the 33 years old apartment is larger that of those in the 11 years old apartment. Regarding the frequency of the ventilation, compared to residents in the 11 years old apartment (1.53 times in a day), those of the 33 years old apartment perform more ventilation (2.33 times in a day). However, the duration of the ventilation for both complexes is almost same (20 mins). For the way of the ventilation in cooking, 10 households in two complexes use both of the natural ventilation way and fan ventilator and 2 households use only fan ventilator.

3) Lighting Environment
All households in the 33 years old apartment face south whereas in the 11 years old apartment 2 households face south-east and 4 households face south-west. Since all

households of the 33 years old complex face south orientation, the amount of solar radiation seems to be a lot in winter and the natural lighting condition seems to be great throughout the year. In terms of the artificial lighting, households in the 33 years old apartment adopt only general lighting for rest and reading in their living room while those in the 11 years old apartment adopt localized lighting on the walls in addition to the general lighting in the center of the ceiling. Accordingly, the general lighting in the 33 years old apartment do not illuminate the living room uniformly, thus it provides uneven intensity of illumination. There is task lighting on desks or tables in children's rooms, thus the intensity of illumination is rather higher. In kitchens, there is fixed localized lighting over the dining tables in addition to general lighting, thus it is hard to change the location of the dining tables and the intensity of illumination is low in the night time. There was no household using halogen lamps in the 33 years old complex whereas the installation rate of the halogens is rather higher in the 11 years old complex. However, many households do not turn on halogen lamps often because they are afraid that the electricity fee would be increased. The mean of scales for households in the 33 years old complex is 4.33 whereas that of households in the 11 years old complex is 4.17.

2.3 Residents' Perceived Comfort

To investigate how residents perceive on temperature, air quality, lighting and noise, we developed a customized metric with five point Likert scale. There are some differences in the average of the responses to the categories. In order to further see if there is any significance statistically in their satisfactions for the categories, we conducted T-test with the responses. The result shows that there are significant differences in the perception of lighting and noise between two apartments. In terms of lighting, the mean of scales for households in the 33 years old complex is 4.33 whereas that of households in the 11 years old complex is 4.17. This suggests that the lighting condition in the 33 years old complex is rather better than that of the 11 years old complex. For the noise condition, the mean of scales for households in the 33 years old complex is 2.83 whereas that of households in the 11 years old complex is 3.67. This implies that the noise condition in the 11 years old complex is better than that of the 33 years old complex. However, there is no significant difference statistically in overall satisfaction on the indoor environment between two complexes. The mean of scales for households in the 33 years old complex (3.67) is a bit higher than that of households in the 11 years old complex (3.33).

Table 3 shows residents' perception of the indoor environment in detail. Through the T-test, it was identified that there is almost no significant difference in the responses to sub-categories between two apartments. There are slightly differences in the perception of the air and lighting conditions. Residents in the 33 years old apartment feel more humid in summer and dryness in winter whereas residents in the 11 years old apartment perceive more quiet regarding the exterior noise and the indoor noise between floors.

1048 M.J. Kim, M.E. Cho, and J.T. Kim

Table 2. Comparison of Overall Perception of Indoor Environment between Two Apartments

Perception of	33 years old apartment		11years old apartment		T- value
	Mean	Std.Devi ation	Mean	Std.Devi ation	
Temperature	2.33	0.30	2.89	0.52	-2.259
Air quality	3.00	0.00	3.00	0.63	-
Lighting	4.33	0.52	4.17	0.98	0.368*
Noise	2.83	0.41	3.67	1.37	1.431*
Overall Comfor t	3.67	0.52	3.33	0.52	1.118

*P<0.05, Scale from 1=never satisfied, 5=satisfied much.

Table 3. Comparison of Perception of Indoor Environment between Two Apartments through T-test

Perception of indoor environment quality		Satisfaction		33 years old		11 years old		T- value
		Point 1	Point 5	Mean	Std. Devia tion	Mean	Std. Devia tion	
Thermal comfort	Comfort temperature in summer	Never	Very comfortable	2.67	0.57	2.50	0.55	0.830
	Comfort temperature in winter	Never	Very comfortable	2.00	0.89	2.67	0.82	-0.869
	Indoor temperature in summer	Too hot	Too cold	1.67	0.52	2.12	0.98	-1.067
	Indoor temperature in winter	Too hot	Too cold	3.67	1.37	3.67	0.82	0.093
	Temperature change during the day	Too variable	Too stable	1.67	0.52	3.67	1.03	-3.747
	Draught	Never	Too draught	2.33	1.37	2.67	1.37	-0.543
Air	Indoor air quality in summer	Too dry	Too humid	4.33	0.52	2.50	1.38	3.706*
	Indoor air quality in winter	Too dry	Too humid	1.00	0.00	1.83	0.75	- 2.369**
	Freshness	Never	Very fresh	2.33	0.52	2.83	0.41	-1.583
	Unpleasant fragrance	Smelly	Odorless	4.00	0.89	3.50	0.55	1.304
	Smell of cigarette	Smelly	Odorless	5.00	0.00	3.50	1.05	1.365
	Dust	Too dust	Never	2.33	0.52	2.83	0.75	-1.137
Light	Natural light	Never	Very comfortable	4.67	0.52	3.83	0.75	1.807
	Glare from sun and sky	Never	Very comfortable	3.33	1.03	2.83	1.67	0.759
	Artificial light	Never	Very comfortable	3.33	1.37	3.83	0.75	-0.664
	Glare from artificial light	Never	Very comfortable	4.00	0.89	3.50	0.55	1.304

Table 3. (*continued*)

Noise	Outside noise	Too noisy	Too silent	2.67	1.03	3.33	0.52	-0.830
	Noise from housing system	Too noisy	Too silent	3.33	1.03	3.33	0.82	-0.113
	Noise between floors	Too noisy	Too silent	2.33	1.37	2.83	0.41	-0.726*
	Vibration	Too Vibration	Never Vibration	4.67	0.52	3.67	1.03	2.068

2.4 Residents' Perceived Symptom

Figure 1 shows residents' perceived symptoms in two apartments with different service life. Such symptoms in health are related to SHS and allergies. Residents in the 33 years old complex feel more uncomfortable in terms of fatigue, dry eyes, irritated or stuffy nose, runny nose and hoarse/dry throat compared to those in the 11 years old complex.

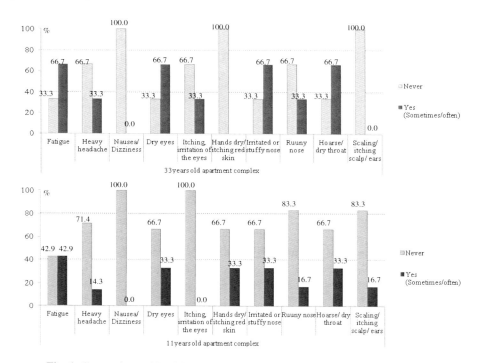

Fig. 1. Comparison of Residents' Perceived Symptoms between Two Apartments

In order to further see if there is any significance statistically in their perceived symptoms, we conducted T-test with the means of scales as shown in Table 4. Residents in the 11 years old apartment feel more uncomfortable with the index of hands dry and itching red skin whereas those in the 33 years old apartment perceive more irritated or stuffy noise.

Table 4. Comparison of Perceived Symptoms between Two Apartments through T-test

Symptom	33 years old apartment		11 years old apartment		T- value
	Mean	Std. Dev.	Mean	Std. Dev.	
Fatigue	2.33	1.03	2.33	1.63	-0.325
Heavy headache	1.67	1.03	1.33	0.82	0.452
Nausea/ Dizziness	1.00	0.00	1.00	0.00	-
Dry eyes	3.00	1.79	1.67	1.03	1.304
Itching, irritation of the eyes	2.33	2.07	1.00	0.00	1.430
Hands dry/ itching red skin	1.00	0.00	1.67	1.03	-1.809***
Irritated or stuffy nose	2.33	1.03	1.67	1.03	0.830***
Runny nose	1.67	1.03	1.33	0.82	0.452
Hoarse/ dry throat	2.33	1.03	1.67	1.03	0.830
Scaling/ itching scalp/ ears	1.00	0.00	1.33	0.82	-1.108

Scale from 5=never, 3=sometimes, 5=often.

Table 5. Respondents' Belief on the Causes of the Symptoms

Symptom	The symptoms might be caused by the physical properties of the apartments	
	Frequency (person)*	Percent (%)*
Dry eyes	5	41.7
Itching, irritation of the eyes	2	16.7
Hands dry/ itching red skin	1	8.3
Irritated or stuffy nose	3	25.0
Runny nose	2	16.7
Hoarse/ dry throat	5	41.7

* 'Yes' response rate of the total 12 respondents.

The respondents seem to think that their symptoms might be caused by the physical characteristic of the apartments. As shown in Table 5, they believe that those health problems in the 6 categories may be derived from the physical status of their units. In particular, the most frequencies of the responses are shown in the index of dry eyes and hoarse/dry throat to the questions.

3 Discussion

This research started from the assumption that there are differences in residents' perceived experiences on comfort and health depending on the degree of physical deterioration of the apartments, thus residents in two apartments with different service life were compared for the case study. The results are as follows; firstly, there is a difference in the energy use of households in two apartments with different service life. In particular, the energy efficiency of the 33 years old apartment is decreased

93 Perceived Experiences on Comfort and Health in Two Apartment Complexes 1051

with low insulation and density qualities of the building. Secondly, there are some differences in households' perceived comforts between two apartments with different service life. Households in the 33 years old apartment are satisfied with the lighting condition whereas those in the 11 years old apartment show high satisfaction on the noise condition. Speaking specifically, all households in the 33 years old apartment face south, thus they show high satisfaction in natural lighting. Households in the 11 years old complex feel more comfortable with the external noise and the internal noise between floors. Thirdly, there are some differences in residents' perceived health problems depending on the service life. Residents in the 33 years old complex feel uncomfortable regarding several sub-categories compared to those in the 11 years old complex.

However, we questioned if the differences in residents' perceived experiences on health might be from the physical properties of the apartments with different service life or the surrounding environments of the complexes. Thus we investigated further into the outdoor air qualities of the two apartment complexes through the statistical data given by the city Seoul. Figure 2 shows the degree of outdoor air pollutions in summer and winter for the two locations. The degree of outdoor air pollution in the surrounding of the 33 years old complex is a bit higher than that in the surrounding of the 11 years old complex. In particular, the index of SO2 and CO is rather higher in summer compared to those in winter.

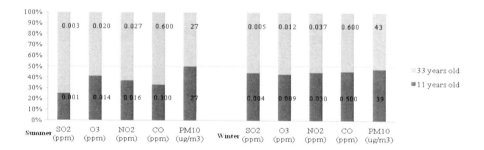

Fig. 2. Comparison of Air Pollution in Two Surrounding Environments

According to the statistical data on the outdoor air quality in the surrounding environments, it seems that the differences in residents' perceived symptoms might be from the surroundings of the complexes rather than the physical characteristics of the apartments with different service life. In the surrounding of the 33 years old complex, the express bus terminal and a transfer station meeting three subway lines are located, thus the floating population become large whereas the 11 years old complex is surrounded by other apartment complexes, thus the residential area of the complex is quiet. Accordingly, the frequent ventilation in the 33 years old apartments might introduce outdoor polluted air into the indoor environment, causing more allergic illnesses to residents.

4 Conclusion

This research investigated residents' perceived experiences on comfort and health in indoor environments from the residents' perspective. It seems that the physical properties of the apartments do not much influence residents' perceived comfort contrary to our expectation. Further the energy efficiency does not affect their perceived comfort either. Residents perform some energy conservation behaviors for reducing the energy loss in heating at a unit level however the more systematic approach to the reduction of the energy consumption would be needed at a complex level in order to produce more effects on the energy use. In terms of health, residents' perceived symptoms in the 33 years old apartment are a bit more severe. The reasons for them might be varied ones, one of which can be the introduction of the outdoor pollutant air into the indoor environment of the complex. Eventually, the causes of symptoms should be identified in detail through further studies in order to protect the residents from the health problems. The quality of outdoor air in apartment complexes should be investigated and noticed to the residents for the control of the natural ventilation because the introduction of outdoor air could affect the quality of the indoor air directly. In addition to the regulations for the newly built apartments, additional laws for apartments with longer service life need to be legislated for enabling residents to live in a healthy residential environment continuously. Further, those laws need to be adjustable or adaptable according to the changing service life of the apartments. For example, they would include desirable management ways, recommendable renovation methods, and some standard levels or criteria for supporting healthier housing life in apartments with different service life.

Acknowledgements. This work was supported by the National Research Foundation of Korea (NRF) grant funded by the Korea government (MEST) (No. 2012-0000609).

References

1. Kim, S.: Effect of furniture and interior finishing materials to sick house syndrome. In: SHB 2010 - 3rd International Symposium on Sustainable Healthy Buildings, Seoul, Korea (2010)
2. Saijo, Y., Reiko, K., Sata, F., Katakura, Y., Urashima, Y., Hatakeyama, A., Mukaihara, N., Kobayashi, S., Jin, K., Iikura, Y.: Symptoms of sick house syndrome and contributory factors; study of general dwellings in Hokkaido. Nihon Koshu Eisei Zasshi 49(11), 1169–1183 (2002)
3. Engvall, K., Wickman, P., Norbäck, D.: Sick building syndrome and perceived indoor environment in relation to energy saving by reduced ventilation flow during heating season: a 1 year intervention study in dwellings. Indoor Air 15(2), 120–169 (2005)
4. Yoo, B.H., Park, N.R.: Influential factors and characteristics on perception of indoor air quality based on apartment residents' responses. Journal of Architectural Institute of Korea 26(1), 349–356 (2010)
5. Chang, J.H.: Perceived sick building syndrome and responses of residents in newly built apartments. Master Thesis, Yonsei University, Seoul, Republic of Korea (2005)

6. Hodgson, M.: Indoor environmental exposures and symptoms. Environment Health Perspect 110(4), 663–667 (2002)
7. Roulet, C., Flourentzou, F., Foradini, F., Bluyssen, P., Aizlewood, C.C.: Perceived health and comfort in relation to energy use and building characteristics. Building Research & Information 34(5), 467–474 (2006)
8. Roulet, C., Flourentzou, F., Foradini, F., Bluyssen, P., Aizlewood, C.C.: Multicriteria analysis of health, comfort and energy efficiency in buildings. Building Research & Information 34(5), 475–482 (2006)
9. Bluyssen, P.M., Cox, C., Seppänen, O., Fernandes, E.D.O., Clausen, G., Müller, B., Roulet, C.-A.: Why, when and how do HVAC-systems pollute the indoor environment and what to do about it? The European AIRLESS project. Building and Environment 38, 209–225 (2003)
10. Molina, C., Pkckering, C.A.C., Valbjbrn, O., Bortoli, M.D.: European concerted action indoor air quality & Its impact on man: environment and quality of life. Commission of the European Communities. Joint Research Centre. Report No. 4 (1989)

Chapter 94
Impact of Different Placements of Shading Device on Building Thermal Performance

Hong Soo Lim[1], Jeong Tai Kim[2], and Gon Kim[2,*]

[1] Department of Architecture, Kangwon National University, Chuncheon, Gangwon-do, 200-701, South Korea
[2] Department of Architectural Engineering, Kyung Hee University, Yongin, Gyeonggi-do, 446-701 South Korea
youaresun@gmail.com

Abstract. In America, the method of receiving the certification for environmental architecture through renovation is activated but in South Korea, it is difficult to find except for the minor renovations occurred to change the previous construction material built in early 2000. It is more efficient to raise the value of an old building by putting over a double skin façade on the outside to make the building more energy efficient than to build a new building. Thus, in this study the energy efficiency of the shading device, the most frequently used composition in designing the double skin façade, is evaluated. Also, the energy efficiency of the typical horizontal blind and the shading device with V-shape slat were compared and used to evaluate the cooling and heating load according to the position changes

1 Introduction

Nowadays, in South-Korea, the restoration movement of 4 major rivers, Korean New town project and the relaxation of the real estate regulation project have been conducted for the real estate activation. However, the value of real estate has been decreasing until now, and as a result, household' loan on real property has been increased. Consequently, it increased 'house poor' and bankruptcy of construc-tion companies. Therefore, this study suggests a new type of construction such as combining shading devices and double skin façade. This method does not destroy the existing buildings nor built new one as new-town project or reconstruction. This research proposes having new outfit in existing buildings attaching new envelopes with shading device to increase the value of buildings and reduce energy consumption.

2 Energy and Daylight with Shading Devices

The most important objects in raising the energy efficiency of an old building are the double skin façade and the design of shading device, by putting on a double skin

---

* Corresponding author.

A. Håkansson et al. (Eds.): *Sustainability in Energy and Buildings*, SIST 22, pp. 1055–1059.
DOI: 10.1007/978-3-642-36645-1_94   © Springer-Verlag Berlin Heidelberg 2013

façade and creating a new exterior to improve not only the technical performance but also the aesthetic aspect.

Recently, many researchers have studied on the green building and the demands increased continuously. Although there are diverse existing standards, the daylight is the base of the green construction. The daylight which has an effect on the zoning that divides perimeter and core is one of the important elements in the environmental architecture. Since energy consumption, occupant's satisfaction, productivity and health depend on which shading device was used on the envelopes.

Also, Solar radiation and the outdoor temperature are essential factors influencing the optical and thermal conditions in a building via thermal gains/losses and incoming light. Commonly, thermal and illuminance are not in coordination and hard to control appropriately [1] Direct sunlight for room lighting is hard to be applied to all structures because its value of illuminance is too high and cause several problems such as overheating, uncomfortable visual environment and view performance. For these reasons, use of shading devices is coming to the fore again. External shading devices have been used extensively in residential and commercial buildings to control the amount of daylight coming into buildings. They are designed with the solar geometry in mind and their configurations are closely related to the sun path. Also, internal shading device is useful to control daylighting performance and shown good maintenance. Especially, venetian blinds reduce glare and provide the view to outdoor, by adjusting slat angles. Figure 1 shows the tendency of illuminance with different slat angles in summer and winter. The result shows that slat angles allow daylight to be transmitted with venetian blinds. By adjusting slat angles, daylighting

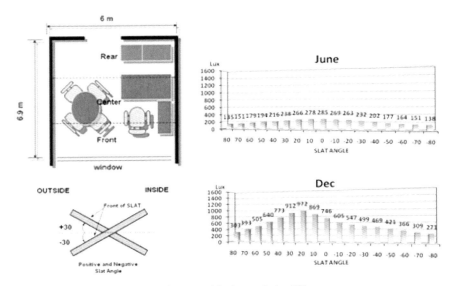

Fig. 1. Illuminance with slat angle in different seasons

performance is changed in the rear space. As a result, the blind slat of 20° might be the most appropriate angle of slats.

In this study, the objective is to use the shading device with V shape slat which blocks and transmits light to compare the result of cooling and heating load with that of the horizontal blind system. As for the variables, horizontal blind which is typically used in double skin façade was compared to the shading device with V-shape slat.

Since the energy performance could change depending on the position of shading device installation, the case was divided into three categories such as indoor, outdoor and intermediate space. In Figure 2, it shows the exterior and the installation location of the V-shape slat.

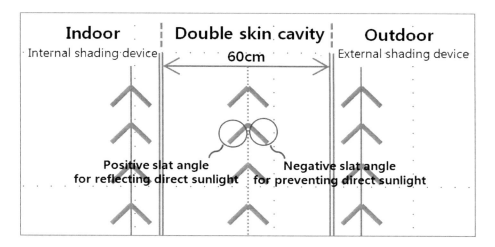

Fig. 2. V-shape slat and Installing position

3 Methodology and Analysis Tool

IES_VE (Virtual Environment), the building energy simulation program, integrated with various 3rd party applications carried out the thermal performance of the curtain wall configuration. Especially, Apache-sim is used to calculate the heating and cooling load in the process of the energy analysis.[2] In order to reduce the data noise and for more effective simulation, three layers have been equipped to mediate the negative impact of sol-air temperature with direct solar radiation. Also the final outcome of the second floor was suggested to eliminate the impact of geothermal heat. For performing simulation, we had to know the U-value of constructed material properties. The information is shown in Table 1. The dimension of experimental space is 8m X 8m with south-facing window.

Table 1. Boundary condition for Virtual Environment simulation

Construction	Description	U-value (W/m²K)
Exposed floor	Concrete(180mm)+bid-insulation(65mm)+cellular-concrete(40mm)	0.41
Ceiling	Concrete(180mm)+bid-insulation(65mm)+cellular-concrete(40mm)	0.41
Internal patition	Plaster(13mm)+brick(105mm)+plaster(11mm)	1.69
External wall	Concrete (200mm) + bid-insulation (75mm)	0.39
External glazing	Clear glass(6mm+6mm) double layers	1.46
Double-skin glazing	Clear glass(6mm+6mm) single layer	2.74

4 Thermal Performance with Shading Device

The result of the simulation is as shown in Figure 3. When shading device with the V shape was installed indoor, the estimated cooling load was 0.79 kWh/m² but when the double skin façade was installed outdoor the result was reduced down to 0.65 kWh/m². With the reduction of half the cooling load, the result indicates improved cooling load in case of outdoor installation. On the other hand, heating load showed different results. When indoor, the measurement was 1.8kWh/m², whereas, the result increased up to 10 times when installed outdoor. As a result, when the shading device is installed at double skin façade, it is possible to control the cooling and heating load appropriately. In comparison with the typical horizontal blind used in intermediate space, there was no particular difference in cooling and heating load but the amount of reflection light inflow increased due to the positive angle slat.

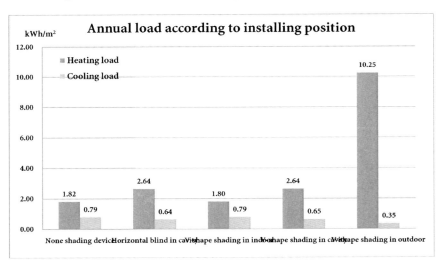

Fig. 3. Comparison of annual heating and cooling load

5   Conclusion

This research is aimed at exploring the usefulness of V-shape slat on the energy consumption of office building module in South Korea. The IES_V.E(Virtual environment) software was used to predict the energy consumption of different conditions. Consequently, it showed no particular difference in cooling and heating load both with the v shape blind and horizontal blind.

However, in case of V-shape slat, due to its slat angle which transmits the daylight, it is expected to have better result than with the horizontal blind. In analysis of V-shape blind according to its installation position, it had advantages in cooling load because it doesn't inflow long wavelength infrared light but it had disadvantages regarding the heating load of indoor shading device in winter.

Acknowledgements. This work was supported by the National Research Foundation of Korea (NRF) grant funded by the Korea government (MEST) (No. 2012-0000609).

References

[1] Kristl, Z., Košir, M., Lah, M.T., Krainer, A.: Fuzzy control system for thermal and visual comfort in building. Renewable Energy 33, 694–702 (2008)
[2] Kim, G., Lim, H.S., Lim, T.S., Schaefer, L., Kim, J.T.: Comparative advantage of an exterior shading device in thermal performance for residential buildings. Energy and Buildings 46, 105–111 (2012)

Chapter 95
Daylighting and Thermal Performance of Venetian Blinds in an Apartment Living Room

Ju Young Shin[*], Yoon Jeong Kim, and Jeong Tai Kim

Department of Architectural Engineering, Kyung Hee University, Yongin 446-701,
Republic of Korea
{jtkim,jyshin,kimyj22}@khu.ac.kr

Abstract. This paper investigated the daylighting and thermal performance of manually operated venetian blinds in an apartment living room. For the purpose the Mock-up model room with 7m length, 5.2m width and 2.4m ceiling height was used. Thirty-two subjects were asked to control the blinds angle and height to eliminate glare and maintain visual comfort from 9 a.m to 5p.m. The monitored window luminance and indoor horizontal illuminance were used to evaluate daylighting performance. Also indoor solar radiation was predicted with Ecotect software to analyze thermal performance. The results showed that adjusting venetian blind in winter times reduced the daylighting and thermal performance about 30% compared to without venetian blind conditions. However using venetian blind could increase the daylighting and thermal performance up to 40% at 15 and 16 hours.

1 Introduction

Venetian blinds are widely used as interior shading devices for controlling the amount of light and heat to maintain a comfortable environment [1,2]. However previous research has shown that venetian blinds are adjusted infrequently and users leave at the same position for days and weeks [3,4]. Inappropriate use of blinds increases the artificial lighting consumption as well as the heating and cooling loads, whereas a properly controlled blinds system not only increases the energy savings but also create comfortable daylit indoor environment [4,5].

Several studies have been proposed to predict the performance of blind control. Some studies developed an equation of blind height and angle control that could block the direct sunlight [4] and others developed the regression blind control model by using subjective responses and physical parameters [6]. Also, some studies developed new types of automatic blind that could enhance the performance of blinds [6,7]. However, the relationship between the set points of luminous environment (Global horizontal illuminance, window luminance and illuminance) and occupant behavior of blind control is still unclear. Thus, this paper investigated the subjective response of

---

[*] Corresponding author.

A. Håkansson et al. (Eds.): *Sustainability in Energy and Buildings*, SIST 22, pp. 1061–1069.
DOI: 10.1007/978-3-642-36645-1_95   © Springer-Verlag Berlin Heidelberg 2013

using venetian blind by controlling blind height and angle in hourly terms. According to the blind control strategies, daylighting and thermal performance was evaluated to define energy saving potential of optimally controlled venetian blind.

2 Methodology

2.1 Field Measurement Setting

An experimental room was located in Yongin, Korea (latitude 37.17N, longitude 127.01E). The test room was designed as typical apartment living room in Korea. The length of the room was 7.0m, width was 5.2m and the height was 2.4m. The window size and configuration was also designed as general type with 1.8m width and 2.0m height. The type of glass in widow was low-e glass.

Internal illuminance was measured with HOBO Data Logger(U12-012), per 1m up to 4m from the window. Total of 15 spots have been installed to measure for 1 min and average floor-plan illuminance was evaluated based on the actual depth. Window luminance was measured with High Dynamic Range(HDR) photograph technique via Photolux 2.1 version. Global horizontal illuminance was measured with EKO ML-0205-5 sensor. The monitoring period was winter season in Korea, from January 16, 2011 to February 21, 2011. From these periods, only clear sky condition with cloud ratio of 0 to 2% was analyzed.

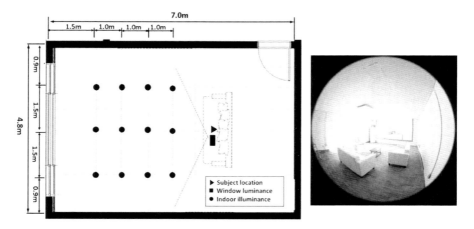

Fig. 1. Mock-up and measurement points

For this study, 32 subjects participated in the experiment and performed a blind control according to the luminous environments. The blind system was installed in three different parts (left, center, right) of windows. Also, the height of blind and its angles were able to adjust. The evaluation took place when the occupant directly faces the window (4m distance from window). The blind adjustment has been divided into 10 equal parts to give data from 0(fully open) to 10(fully closed), and was able to

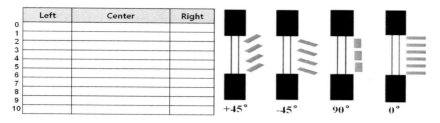

Fig. 2. Venetian blind control setting

adjust in 4 slat angles such as 90°(upward), 45°(upward and downward), and 0 (horizontal) according to subject's preference.

2.2 Simulation Setting

The computer simulation was conducted to evaluate the thermal performance of venetian blinds. Table 1 shows the material characteristics for the simulation.

Table 1. The material characteristics for the simulation

	Material	Thickness(mm)	Conductivity(W/m.K)	U-value(W/m2.K)
Walls	Brick Masonry Medium	110	0.71	
	Polystyrene Foam	50	0.03	0.49
	Brick Masonry Medium	110	0.71	
	Plaster	10	0.43	
Ceiling	Soil	1500	0.83	
	Concrete	100	0.40	0.24
	Concrete Screed	5	0.25	
	Linoleum	10	0.19	
Floor	Concrete	100	0.40	
	Concrete Screed	5	0.25	
	Polystyrene Foam	80	0.03	0.31
	Linoleum	10	0.19	

The simulation was focused on measuring indoor solar radiation according to the hourly blinds control. The simulation setting of space dimension and window size was identical with field experiment setting. The reflectance of the ceiling, walls and floor was assumed to be 0.65, 0.58 and 0.23 respectively. The transmittance of the window in the simulation setting was 0.95 and the reflectance of the venetian blind was 0.4. The time and date of the simulation were set on January and February from 9 to 17 hours which are identical with experimental period. The analysis was performed by Ecotect software (2011 version).

3 Results

3.1 Occupants Behavior

Figure 3 shows the average window luminance distribution against the horizontal global illuminance. Horizontal global illuminance and window luminance showed positive linear relationship with R square value of 0.494. The experiment period was winter season (January 16 2011 to February 21 2011), so that the horizontal global illuminance values were lower than the 80,000lx. At 9 hours, horizontal global illuminance was about 30,000lx and increased about 1.3times until 12 hours. From 12 hours, the horizontal global illuminance was decreased about 0.7 times and the illuminance value was about 7000lx at 17 hours. During these times, average window luminance was ranged in 5000-6000cd/m^2, and when the horizontal global illuminance increased by 10000lux, the window luminance follows by 1000cd/m^2.

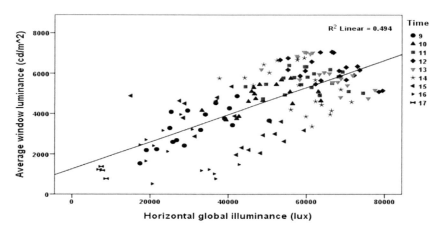

Fig. 3. Average window luminance against horizontal global illuminance during the experiment period

Figure 4 and 5 shows the hourly frequency of occupant's behavior in blind height control and blind angle control. Most of occupants did not fully roll down the blinds at all daylight hours. When the horizontal global illuminance and window luminance reached highest (13hours), most of the occupants rolled down the blinds to a half points of window height. When the average window luminance ranged in 5000-6000cd/m^2 (10, 11, 12, 14 hours), the majority rolled down the blind to 3 and 4. When the average window luminance ranged below 4000cd/ m^2 (9, 15, 17 hours), the majority controlled the blind height to 2. The controlled blind height was about the same for different section of window (left, center, right) at all times.

An analysis of blind slat angle control showed that controlling slat angle to upward 45 degree was relatively higher at all times except 16 hours. Especially, about 60% of subjects controlled to upward 45 degree at 14 hours. At 16 hours, highest controlled slat angle was downward 45 degree. In comparison with different window section, similar control patterns were shown at all times.

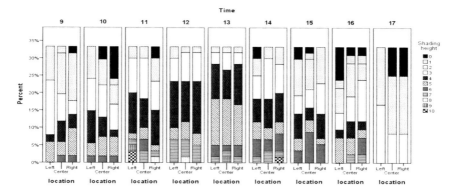

Fig. 4. Occupant's behavior of blind height control

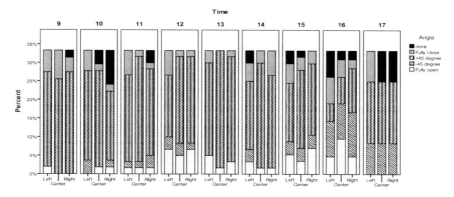

Fig. 5. Occupant's behavior of blind angle control

3.2 Daylighting Performance

Figure 6 shows the window luminance variation according to the blinds control. The average window luminance was decreased 1.8times from $4600cd/m^2$ to $2400cd/m^2$ when the blinds were used. Most of the luminance values are ranged over $4000cd/m^2$ when the blinds are not used whereas most of the luminance values are distributed in $2000\text{-}3000cd/m^2$ after the blinds is used. This indicates that subject's feel too bright when the luminance values are over $4000 cd/m^2$, and below the $2000 cd/m^2$ is too dark and tend to feel discomfort.

Figure 7 shows the variation of hourly indoor illuminance according to the blinds control. The maximum average illuminance value when the blinds were not used was about 19000lx at 12 hours and a minimum value was 1000lx at 17 hours.

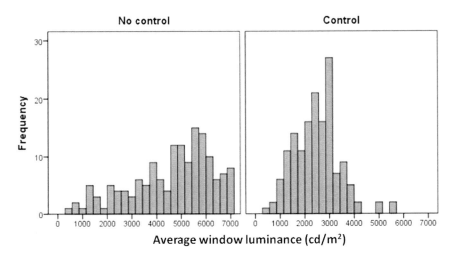

Fig. 6. Window luminance distribution according to venetian blind control

The overall average illuminance was about 9500lx. When the blinds are used, subjects controlled the blinds below the 13000lx and over the 800lx. The maximum illuminance value was about 13000lx at 12 and 13 hours, and a minimum value was 800lx at 17 hours. The overall average value was 7200lx which is 3.7 times lower than the no control conditions.

In all times (blinds used and not used), the indoor illuminance values were satisfied with the lighting recommendations of Illuminating Engineering Society(IESNA 2011)[1] and Korean industrial Standards (KS 2003)[2] in residential living room.

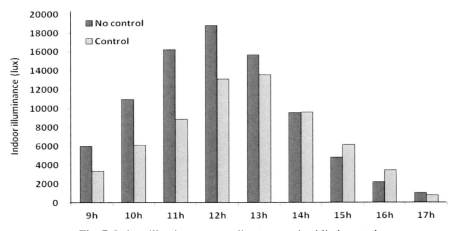

Fig. 7. Indoor illuminance according to venetian blind control

---

[1] IESNA lighting recommendation of residential living room (General: 30lx, Task: 50-300lx).
[2] KS lighting recommendation of residential living room (General: 30lx, Task: 200-400lx).

3.3 Thermal Performance

Figure 8 shows the hourly average indoor solar radiation. The average indoor solar radiation from 9 to 17 hours was 39Wh when the blinds are used. When the blinds are not used, it dropped about 1.3 times and the value was 31Wh. The maximum solar radiation value was shown at 10hours for both blind used and not used. For the no control condition, the solar radiation was reached to 73Wh and 57Wh for the blind used environment. The minimum solar radiation was shown at 16 hours for the no control condition with 1.22Wh, and 17 hours for the blind used condition with 0.3Wh.

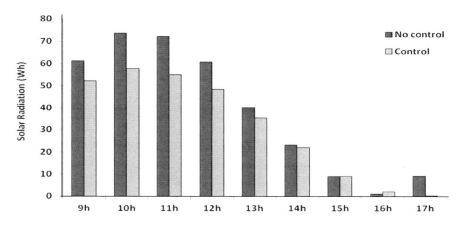

Fig. 8. Solar radiation according to venetian blind control

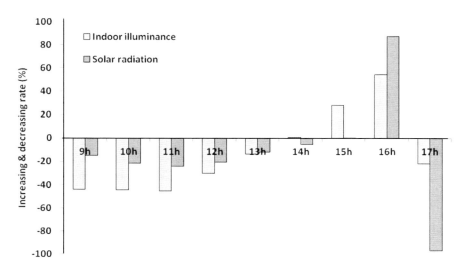

Fig. 9. Variation of Indoor illuminance and solar radiation according to the blind control

3.4 Correlation between Daylighting and Thermal Performance

Figure 9 shows the hourly variation rate of indoor illuminance and solar radiation according to the blinds usage. At most of the times, indoor illuminance and solar radiation was decreased when the blinds are used in comparison with no blind conditions. During the morning hours, the indoor illuminance was decreased about 40% whereas solar radiation was declined about 20%. At 12 to 14 hours, similar variations were shown with the illuminance and solar radiation.

However, at 15 and 16 hours blind used environment showed higher indoor illuminance and solar radiation. These values are about 40% greater than the no blind conditions. This indicates that after 15 hours, the redirecting of indoor direct solar radiation and daylight in indoor space is much greater with the venetian blind than the window glass. In comparison with increasing rate between the solar radiation and illuminance in blind used environment, solar radiation increased 4 times greater than the illuminance increasing percentage.

4 Conclusion

This study presented the daylighting and thermal performance of venetian blind based on occupant behavior of controlling blind height and angle in winter season. The results showed that people tends to control the blind to provide more light in the back of the room by adjusting blind angle to upward 45 degree at most of the times. The optimal window luminance range by controlling venetian blind was 2000-3000cd/m^2.

The results also indicated that controlled venetian blind environments reduced the daylighting and thermal performance at most of daylight hours compared to the reference condition. However, using venetian blind at certain hours could significantly increase the daylighting and thermal performance due to the reflection of direct solar from venetian blind to the internal space.

Acknowledgements. This work was supported by the National Research Foundation of Korea (NRF) grant funded by the Korea government (MEST) (No. 2012-0000609).

References

[1] Tzempelikos, A.: The impact of venetian blind geometry and tilt angle on view, direct light transmission and interior illuminance. Solar Energy 82, 1172–1191 (2008), doi:10.1016/j.solener.2008.05.014
[2] Tzempelikos, A., Athienitis, A.: The impact of shading design and control on building cooling and lighting demand. Solar Energy 81, 369–382 (2007), doi:10.1016/j.solener.2006.06.015
[3] Rea, M.S.: Window blind occlusion: a pilot study. Building and Environment 19, 133–137 (1984), doi:10.1016/0360-1323(84)90038-6

[4] Zhang, S., Birru, D.: An open-loop venetian blind control to avoid direct sunlight and enhance daylight utilization. Solar Energy 86, 860–866 (2012), doi:10.1016/j.solener.2011.12.015

[5] Inkarojrit, V.: Balancing Comfort: Occupants' control of window blinds in private offices. Dissertation, University of California (2005)

[6] Clear, R.D., Inkarojrit, V., Lee, E.S.: Subject responses to electrochromic windows. Energy and Buildings 38, 758–779 (2006), doi:10.1016/j.enbuild.2006.03.011

[7] Olbina, S., Jia, H.: Daylighting and thermal performance of automated split-controlled blinds 56, 127–138 (2012), doi:10.1016/j.buildenv.2012.03.002

Chapter 96
Environmentally-Friendly Apartment Buildings Using a Sustainable Hybrid Precast Composite System

Ji-Hun Kim, Won-Kee Hong[*], Seon-Chee Park, Hyo-Jin Ko, and Jeong Tai Kim

Department of Architectural Engineering, Kyung Hee University, Yongin 446-701, Korea
hongwk@khu.ac.kr

Abstract. Recently, as part of an effort to comply with the low-carbon green growth policy adopted in Korea, the building of new long-life apartment buildings has been encouraged to replace existing apartment buildings with bearing walls. The regulations imposed regarding floor area ratios, height, and available sunlight can all be alleviated when apartment buildings are built using the Rahmen structural frame instead of conventional bearing walls, which are difficult to remodel. However, a Rahmen structural frame with reinforced concrete increases the floor height due to the increased beam depth, resulting in economic issues. This paper introduces a hybrid precast composite structural system. Apartment buildings optimized using the hybrid precast composite Rahmen structural system was compared with concrete Rahmen structural frames. The results show that the lower material quantity used in the hybrid precast composite structural system reduces carbon emissions and as well as energy inputs related to construction. It is expected that the hybrid precast composite Rahmen structural system will play a significant role in building sustainable and healthy long-life apartment buildings.

1   Introduction

1.1   Background

Apartments with bearing walls account for 50% or more of existing Korean apartment buildings [1]. The bearing wall structure type is advantageous for its superior constructability and economical feasibility. However, it is difficult to respond to the rapidly changing housing preferences of residents due to the limitations in remodeling this type of wall.

Apartment buildings with bearing walls are reconstructed not because of structural degradation [2], but because the difficulties in remodeling result in the early reconstruction of buildings, resulting in significant energy loss and generation of construction waste [3].

The Korean government announced a plan to grant incentives on floor area ratio if apartment buildings are built using Rahmen structures rather than bearing walls. The

---

[*] Corresponding author.

A. Håkansson et al. (Eds.): *Sustainability in Energy and Buildings*, SIST 22, pp. 1071–1081.
DOI: 10.1007/978-3-642-36645-1_96   © Springer-Verlag Berlin Heidelberg 2013

Acceptability Standard of Apartment Housing by the Architecture Committee alleviated restrictions imposed on apartment buildings, including restrictions on floor area ratios, height, and sunlight availability when they are built using the Rahmen structure type.

1.2 Identifying the Problems

The incentive policy is accelerating the construction of apartment buildings which incorporate a reinforced concrete Rahmen structure. However, the requirement for beams deep enough to satisfy design loads makes it difficult to build to the same floor height as conventional apartment buildings with bearing walls, thus these buildings are forced to lose a number of stories.

1.3 Solution and Objectives

Studies of new structure types for use in apartment buildings have been conducted to address the problems of concrete Rahmen frames. The Modularized Hybrid System developed by Hong et al. [4] solved the floor height issue of apartment buildings with a reinforced concrete Rahmen frame and proved its structural and economic efficiency and constructability through tests and simulations [5]. Since a composite frame system is able to reduce the quantity of material needed, CO_2 emissions and energy consumption during construction can also be reduced using this method [6]. This paper compares a reinforced concrete Rahmen frame and a Hybrid Precast Composite Rahmen frame (hereinafter referred to as a Hybrid frame), and presents the environmentally friendly aspects of the Hybrid frame.

2 The Hybrid Frame Concept

A Hybrid frame for use in apartment buildings consists of a column-tree unit and a beam unit as shown in Fig. 1. Each unit has CT section shape steel, precast concrete and reinforcement. This frame has the advantages of using steel frames and concrete as supplements of disadvantages of each construction material. The column-tree units and beam units are manufactured in off-site plants and brought onto the construction site. After 2-3 story column-tree units are erected on site, three floors can be installed in one cycle, reducing construction periods [7].

The Hybrid frame was optimized based on previous studies for the location and type of horizontal steel used in the beam units which is connected to the column-tree units, thereby reducing the steel frame quantity and increasing material efficiency [8]. As demonstrated in Fig. 2, the lengths of the horizontal steel sections of the column-tree and beam units are determined by considering loadings imposed on the beam unit. The construction of connections would also become easier if they were manufactured in plants.

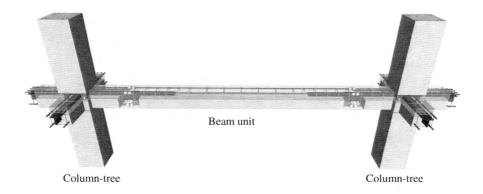

Fig. 1. Hybrid frame

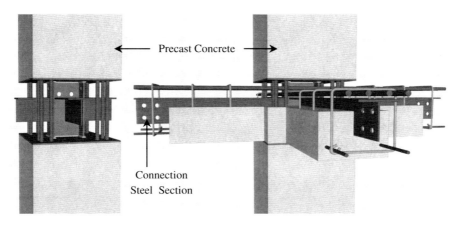

Fig. 2. Enhanced connections of a column-tree

3 Analysis

Apartment buildings using the reinforced concrete Rahmen structure and the Hybrid frame were compared in a 15-floor linear apartment whose total floor area was 6533.9 m². The number of resident units per floor was four and the floor height was designed to be 2.7 m. As shown in Fig. 3, we also compared two different lengths of horizontal steel sections, 25% and 12.5% of the span length, which are required as a connection between two units.

3.1 Analysis of the Floor Height

As shown in Fig. 4, the main beam section is 350 mm × 500 mm for the reinforced concrete Rahmen frame and 300 mm × 350 mm for the Hybrid frame. The floor

1074   J.-H. Kim et al.

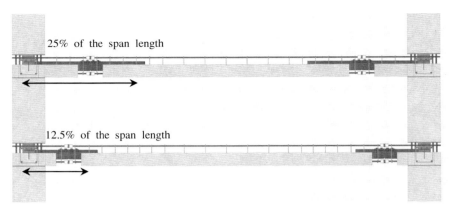

Fig. 3. The length of the steel frames

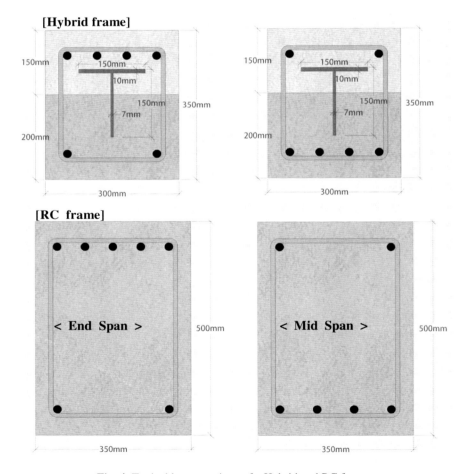

Fig. 4. Typical beam sections of a Hybrid and RC frame

height of apartment buildings using the reinforced concrete Rahmen structure increases by 150 mm per floor in order to maintain ceiling height, and the total increased height of 15-story apartment buildings is 2.25 m. Apartment buildings built using the reinforced concrete Rahmen structure would thus lose one floor. However, the Hybrid frame, which has better structural performance, is capable of providing the same floor height as that of buildings constructed using bearing walls and thus maintains the same number of floors.

Table 1. Material quantity comparisons according to frame systems

Material	Section	RC frame	Hybrid frame I (L/4)	Hybrid frame II (L/8)
Concrete	Beam	$650.91\ m^3$ ($0.100\ m^3/m^2$)	$443.52\ m^3$ ($0.068\ m^3/m^2$)	$457.48\ m^3$ ($0.070\ m^3/m^2$)
	Column	$364.64\ m^3$ ($0.056\ m^3/m^2$)	$288.49\ m^3$ ($0.044\ m^3/m^2$)	$287.22\ m^3$ ($0.044\ m^3/m^2$)
	Wall	$810.21\ m^3$ ($0.124\ m^3/m^2$)	$726.96\ m^3$ ($0.111\ m^3/m^2$)	$726.96\ m^3$ ($0.111\ m^3/m^2$)
	Slab	$757.24\ m^3$ ($0.116\ m^3/m^2$)	$757.24\ m^3$ ($0.116\ m^3/m^2$)	$757.24\ m^3$ ($0.116\ m^3/m^2$)
	Stair	$64.25\ m^3$ ($0.010\ m^3/m^2$)	$60.62\ m^3$ ($0.009\ m^3/m^2$)	$60.62\ m^3$ ($0.009\ m^3/m^2$)
	Sub total	$2647.26\ m^3$ ($0.405\ m^3/m^2$)	$2276.83\ m^3$ ($0.348\ m^3/m^2$)	$2289.52\ m^3$ ($0.350\ m^3/m^2$)
Steel Rein -forcement	Beam	$692.9kN$ ($0.11kN/m^2$)	$607.6kN$ ($0.09kN/m^2$)	$615.4kN$ ($0.09kN/m^2$)
	Column	$479.0kN$ ($0.07kN/m^2$)	$359.8kN$ ($0.06kN/m^2$)	$375.8kN$ ($0.06kN/m^2$)
	Wall	$597.4kN$ ($0.09kN/m^2$)	$536.0kN$ ($0.08kN/m^2$)	$536.0kN$ ($0.08kN/m^2$)
	Slab	$260.6kN$ ($0.04kN/m^2$)	$260.6kN$ ($0.04kN/m^2$)	$260.6kN$ ($0.04kN/m^2$)
	Stair	$108.7kN$ ($0.02kN/m^2$)	$102.6kN$ ($0.02kN/m^2$)	$102.6kN$ ($0.02kN/m^2$)
	Sub Total	$2138.7kN$ ($0.33kN/m^2$)	$1866.6kN$ ($0.29kN/m^2$)	$1890.4kN$ ($0.29kN/m^2$)
Steel Section		$0kN$ ($0kN/m^2$)	$447.1kN$ ($0.07kN/m)^2$	$233.4kN$ ($0.04kN/m^2$)
Form work		$18283\ m^2$ ($2.798\ m^2/m^2$)	$11353m^2$ ($1.738\ m^2/m^2$)	$11353\ m^2$ ($1.738\ m^2/m^2$)

3.2 Analysis of the Material Quantity

Table 1 compares the material quantities required for an apartment building constructed using the different Rahmen frame systems. The frame systems compared include the RC frame system, the Hybrid frame I, in which the horizontal steel frame is L/4 of the beam length at each end, and the Hybrid frame II, in which the horizontal steel frame is L/8 of the beam length at each end.

The results of the analysis of the RC frame show that the quantity of concrete is 2647.26 m^3 (about 0.405 m^3/m^2), steel reinforcement is 2138.7 kN (0.33 kN/m^2), and the formwork is 18283 m^2 (2.798 m^2/m^2). In the case of the Hybrid frame I, the quantity of concrete is 2276.83 m^3 (0.348 m^3/m^2), steel reinforcement is 1866.6 kN (0.29 kN/m^2), the steel frame is 447.1 kN (0.07 kN/m^2) and the formwork is 11353 m^2 (1.738 m^2/m^2). In the case of Hybrid frame II, the quantity of concrete is 2289.52 m^3 (approximately 0.350 m^3/m^2), steel reinforcement is 1890.4 kN (0.29 kN/m^2), the steel frame is 233.4 kN (0.04 kN/m^2), and the formwork is 11353 m^2 (1.738 m^2/m^2).

The amount of materials used in construction of apartment buildings with Hybrid frame I and II decreased to 86.2% for concrete, 87.8% for reinforcement steel, and 62.1% for formwork, and amounts of steel frame were increased when compared to the RC frame. When compared with Hybrid frame I, Hybrid frame II reduces the steel frame to around 52.2%, as its length is decreased. The quantities of concrete and reinforcement steel in Hybrid frame I increase about 0.55% and 1.27%, respectively, compared with Hybrid frame II.

3.3 Analysis of Carbon Emissions and Energy Consumption

Carbon emissions and energy consumption related to construction were analyzed based on the quantities of main construction materials calculated from the comparison of apartment buildings. Table 2 shows the carbon emission and energy consumption coefficients of the main construction materials [9].

Table 2. Embodied energy and carbon coefficients in construction materials [9]

Material	Embodied Carbon	Embodied Energy
Concrete	0.043 kgC/kg	1.11 MJ/kg
Reinforcing Steel	0.73 kgC/kg	36.4 MJ/kg
Steel Section	0.757 kgC/kg	36.8 MJ/kg
Plywood	0.221 kgC/kg	15 MJ/kg

Table 3 and Fig. 5 show the carbon emissions of each structural system. The carbon emitted from concrete is 261.81 ton-C/kg for the RC frame, 225.18 ton-C/kg for the Hybrid frame I and 226.43 ton-C/kg for the Hybrid frame II. The carbon generated from reinforcement steel is 156.12 ton-C/kg for the RC frame, 136.26 ton-C/kg for the Hybrid frame I and 138 ton-C/kg for the Hybrid frame II. The carbon emitted from the steel frame is 33.85ton-C/kg for the Hybrid Frame I and 17.67 ton-C/kg for the Hybrid frame II. The carbon generated from formwork is 4.04 ton-C/kg for the RC frame and 2.51 ton-C/kg for Hybrid frames I and II.

The RC frame emits 421.98 ton-C/kg in total. These results show that the carbon emissions of Hybrid frame I are reduced to 397.79 ton-C/kg (94.26% of the RC frame) and of Hybrid frame II to 384.61 ton-C/kg (91.14% of the RC frame). In the case of the Hybrid frame II, the quantities of concrete and reinforcement steel used slightly increase when compared to Hybrid frame I.

Table 3. Embodied carbon (dioxide) comparison

Material	Section	RC frame	Hybrid frame I (L/4)	Hybrid frame II (L/8)
Concrete	Beam	64.375 ton-C/kg	43.86 ton-C/kg	45.24 ton-C/kg
	Column	36.063 ton-C/kg	28.53 ton-C/kg	28.41 ton-C/kg
	Wall	80.130 ton-C/kg	71.90 ton-C/kg	71.90 ton-C/kg
	Slab	74.891 ton-C/kg	74.89 ton-C/kg	74.89 ton-C/kg
	Stair	6.355 ton-C/kg	6.00 ton-C/kg	6.00 ton-C/kg
	Sub Total	261.81 ton-C/kg	225.18 ton-C/kg	226.43 ton-C/kg
Steel Rein -forcement	Beam	50.582 ton-C/kg	44.35 ton-C/kg	44.92 ton-C/kg
	Column	34.968 ton-C/kg	26.27 ton-C/kg	27.43 ton-C/kg
	Wall	43.611 ton-C/kg	39.13 ton-C/kg	39.13 ton-C/kg
	Slab	19.024 ton-C/kg	19.02 ton-C/kg	19.02 ton-C/kg
	Stair	7.939 ton-C/kg	7.49 ton-C/kg	7.49 ton-C/kg
	Sub Total	156.12 ton-C/kg	136.26 ton-C/kg	138.00 ton-C/kg
Steel Section		0 ton-C/kg	33.85 ton-C/kg	17.67 ton-C/kg
Form work		4.04 ton-C/kg	2.51 ton-C/kg	2.51 ton-C/kg
Total		421.98 ton-C/kg	397.79 ton-C/kg	384.61 ton-C/kg
Percentage		100 %	94.26 %	91.14 %

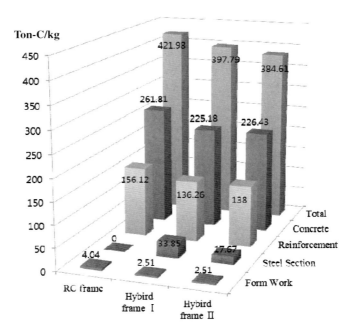

Fig. 5. Embodied carbon (dioxide) comparison

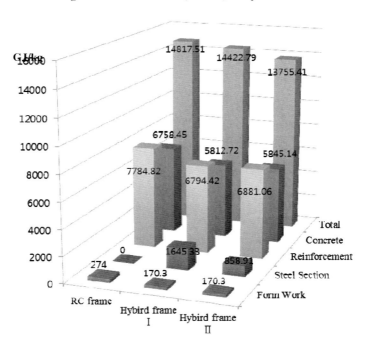

Fig. 6. Embodied energy comparison

Table 4. Embodied energy comparison

Material	Section	RC Frame	Hybrid frame I (L/4)	Hybrid frame II (L/8)
Concrete	Beam	1661.77 GJ/kg	1132.31 GJ/kg	1167.95 GJ/kg
	Column	930.93 GJ/kg	736.51 GJ/kg	733.27 GJ/kg
	Wall	2068.47 GJ/kg	1855.93 GJ/kg	1855.93 GJ/kg
	Slab	1933.23 GJ/kg	1933.23 GJ/kg	1933.23 GJ/kg
	Stair	164.05 GJ/kg	154.76 GJ/kg	154.76 GJ/kg
	Sub Total	6758.45 GJ/kg	5812.72 GJ/kg	5845.14 GJ/kg
Steel Rein -forcement	Beam	2522.16 GJ/kg	2211.66 GJ/kg	2240.06 GJ/kg
	Column	1743.61 GJ/kg	1309.67 GJ/kg	1367.91 GJ/kg
	Wall	2174.59 GJ/kg	1951.04 GJ/kg	1951.04 GJ/kg
	Slab	948.58 GJ/kg	948.58 GJ/kg	948.58 GJ/kg
	Stair	395.87 GJ/kg	373.46 GJ/kg	373.46 GJ/kg
	Sub Total	7784.817 GJ/kg	6794.42 GJ/kg	6881.06 GJ/kg
Steel Section		0 GJ/kg	1645.33 GJ/kg	858.91 GJ/kg
Form work		274.00 GJ/kg	170.30 GJ/kg	170.30 GJ/kg
Total		14817.51 GJ/kg	14422.79 GJ/kg	13755.41 GJ/kg
Percentage		100 %	97.33 %	92.83 %

Table 4 and Fig. 6 show the values calculated in the comparison of construction-related energy consumed for each structural system. The energy consumption of concrete is 6758.45 GJ/kg for the RC frame, 5812.72 GJ/kg for the Hybrid frame I and 5845.14 GJ/kg for the Hybrid frame II. The energy consumption of reinforcement steel is 7784.817 GJ/kg for the RC frame, 6794.42 GJ/kg for the Hybrid frame I and 6881.06 GJ/kg for the Hybrid frame II. The energy consumption of the steel frame is 1645.33 GJ/kg for the Hybrid frame I and 858.91 ton-C/kg for the Hybrid frame II.

The energy consumption of formwork is 274.00 GJ/kg for the RC frame and 170.30 GJ/kg for Hybrid frames I and II.

The total energy consumption of an RC frame is 14817.51 ton-C/kg. The energy consumption of the Hybrid frame I is reduced to 14422.79 GJ/kg (97.33% of the RC frame) Hybrid frame II is reduced to 13755.41 GJ/kg (92.83% of the RC frame). Energy consumption was also found to be highly dependent on the quantity of steel frame. This analysis shows that materials can be efficiently utilized as the quantity of horizontal steel frame applied to the Hybrid frame is reduced.

4 Conclusions

This study presented the following conclusions by applying the Hybrid frame to apartment buildings in an attempt to solve the problems of apartment buildings constructed with RC frames.

- The Hybrid frame is able to maintain the same floor height and the same number of floors as conventional apartment buildings with bearing walls.
- Construction of apartment buildings using a Hybrid frame reduces the required quantities of concrete (-13.8%), reinforcement (-12.2%) and formwork (-37.9%) when compared with use of the RC frame. A horizontal steel section of 12.5% of total beam length can be installed at each end of the beam.
- If the horizontal steel frame is 25% of the span, its carbon emissions are reduced to 24.19 ton-C/kg (-5.74%), and, its carbon emissions are reduced to 37.37 ton-C/kg (-8.86%) if it is 12.5% of the span when compared to that of an RC frame.
- The length of the horizontal steel frame can be efficiently designed, with the energy consumption of construction of Hybrid frame I decreasing to 394.72 GJ/kg (-2.67%) and that of Hybrid frame II to 1062.1 GJ/kg (-7.17%) compared to an RC frame.

This study showed that apartment buildings incorporating the Hybrid frame are superior in terms of their economic and ecological aspects when compared to the conventional RC frame, ultimately enabling the achievement of sustainable, healthy apartment buildings.

Acknowledgements. This work was supported by the National Research Foundation of Korea (NRF) grant funded by the Korea government (MEST) (No. 2012-0000609).

References

1. Lee, S.H., Kim, S.E., Kim, G.H., Joo, J.K., Kim, S.K.: Analysis of Structural Work Scheduling of Green Frame. Journal of the Korea Institute of Building Construction 11(3), 301–309 (2011)
2. Lee, B.R., Kim, S.A., Hwang, E.K.: The Study on the Systematic Improvement Plans for Facilitating Long-Life Housing. Architectural Institute of Korea 24(3), 3–10 (2008)

3. Hong, W.K., Park, S.C., Jeong, S.Y., Lim, G.T.: Evaluation of the Energy Efficiencies of Pre-cast Composite Columns. Indoor and Built Environment 21(1), 176–183 (2012)
4. Hong, W.K., Park, S.C., Kim, J.M., Lee, S.G., Kim, S.I., Yoon, K.J., Lee, H.C.: Composite Beam Composed of Steel and Pre-cast Concrete (Modularized Hybrid System, MHS) Part I: Experimental Investigation. Structural Design of Tall and Special Buildings 19(3), 275–289 (2010)
5. Hong, W.K., Park, S.C., Lee, H.C., Kim, J.M., Kim, S.I., Lee, S.G., Yoon, K.J.: Composite Beam Composed of Steel and Pre-cast Concrete (Modularized Hybrid System, MHS) Part III: Application for a 19 Story Building. Structural Design of Tall and Special Buildings 19(6), 679–706 (2010)
6. Hong, W.K., Kim, J.M., Park, S.C., Lee, S.G., Kim, S.I., Yoon, K.J., Kim, H.C., Kim, J.T.: A new apartment construction technology with effective CO_2 emission reduction capabilities. Energy 35(6), 2639–2646 (2010)
7. Hong, W.K., Kim, S.I., Park, S.C., Kim, J.M., Lee, S.G., Yoon, K.J., Kim, S.K.: Composite Beam Composed of Steel and Pre-cast Concrete (Modularized Hybrid System, MHS) Part IV: Application for Multi-Residential Housings. Structural Design of Tall and Special Buildings 19(7), 707–727 (2010)
8. Hong, W.K., Park, S.C., Kim, J.M., Kim, S.I., Lee, S.G., Yune, D.Y., Yoon, T.G., Ryoo, B.Y.: Development of Structural Composite Hybrid Systems and their Application with regard to the Reduction of CO2 Emissions. Indoor and Built Environment 19(1), 151–162 (2010)
9. Hammond, G., Jones, C.: Embodied energy and carbon in construction materials. The Institution of Civil Engineers 161(2), 87–98 (2008)

Chapter 97
A System for Energy Saving in Commercial and Organizational Buildings

Hamid Abdi[*], Michael Fielding, James Mullins, and Saeid Nahavandi

Center for Intelligent Systems Research, Deakin University,
Geelong Waurn Ponds Campus, Victoria 3217, Australia
hamid.abdi@deakin.edu.au

Abstract. Energy consumption in commercial and organizational buildings with shared electricity produces a considerable amount of greenhouse gas emissions worldwide. Sustainable reduction of greenhouse gas emissions in these building remains to be a challenge and further research is required to address this problem due to the complexity of human behavior. The present paper introduces distributed meters for these buildings in order to achieve a sustainable energy saving. The method provides a direct control to a humans' behavior that is essential for effectiveness of the energy saving. It is shown that by using distributed meters, the system can actively engage humans in the energy saving process. The hardware and software required to implement this concept are explored and the sustainability of the proposed method is discussed.

Keywords: Organizational buildings, commercial buildings, energy saving, human behavior, public building, energy efficiency.

1 Introduction

Climate change is an increasing life-threatening problem worldwide. There have been serious impacts and consequences of climate change over the past decades with noticeable increases in environmental pollution as well as extreme weather conditions around the world. In particular, a lot of research is focusing on energy efficiency and saving due to the wide influence energy efficiency has on greenhouse gas emission in various applications. Energy saving in commercial and organizational buildings with shared electricity is more complicated than private buildings because of the complex behavior and varying nature in which people consume energy. The energy that is used in these buildings produces a considerable amount of greenhouse gas emissions worldwide, not to mention the financial cost of inefficient buildings and systems [1, 2]. It is clear that in commercial and organizational buildings with shared electricity, energy use per person is commonly higher than the energy use in private houses [1]. This could stem from a number of reasons such as access to consumption information, or the fact that the homeowner as an individual will be financially liable for the

---

[*] Corresponding author.

A. Håkansson et al. (Eds.): *Sustainability in Energy and Buildings*, SIST 22, pp. 1083–1091.
DOI: 10.1007/978-3-642-36645-1_97   © Springer-Verlag Berlin Heidelberg 2013

energy consumed is simply justified by knowing that people in their homes have to pay the energy bills from their own pocket. In commercial and organizational buildings with shared electricity, the energy bills do not have a direct impact on the individuals, therefore performing energy saving is more complicated. Other research has tried to address human behavior on energy saving in public or organizational buildings [3] but the existing technologies for saving energy in these buildings [4] have been unable to adequately address the human behavior and therefore is not an efficient or sustainable solution.

Commercial, educational, organizational, governmental, medical, and residential buildings are responsible for a considerable percentage of greenhouse gas emissions worldwide. Therefore, energy saving in these buildings can noticeably reduce the greenhouse gas emission, which has environmental benefits and can decrease the energy expenses of the building owners. Energy saving will be more efficient if the people in these building are actively engaged in energy saving process.

Environmental management systems (EMS) deals with the impact of the organization on environment [5]. Presently, ISO 14000 is widely used for environmental compatibility [5] and defines a green building as being environmentally responsible and resource-efficient throughout the life cycle of the building [6, 7]. But one of the main challenges for EMSs is the engagement of the people in energy saving within these buildings. Most of the existing methods for energy saving in commercial and organizational buildings with shared electricity are unable to actively engage occupants in energy saving. Previous studies that have focused on energy efficiency/saving for a number of such buildings can be categorized into three groups that include statistical/psychological analysis of energy consumption in these buildings [1, 3, 8, 9], different technologies and process for energy saving in these buildings [2, 4, 10, 11], and different some case studies on energy saving in buildings [12-14]. The connection between the first two groups of research is challenging as it requires a technology and process that impacts on an occupants behavior. Such complexity has been observed by Doukas et al [10] where an intelligent system has been proposed and a high level of intelligence is required to address the human behavior. The proposed system is unable to effectively engage the individuals in energy saving process. Poortinga et al [15] performed a psychological analysis on humans behavior but there is a lack of a technology that could have a direct and measurable influence on humans energy consumption behaviors. Figure 1 shows the existing model of energy use. Each person consumes energy and the sum of the energy consumption will determine the total energy costs for the building.

The present paper discusses energy saving for commercial and organizational buildings with shared electricity and proposes a method that can actively engage individuals in energy saving. The paper proposes using Distributed Electricity Meter (DEM) technology to enable the measurement an individual's energy consumption. Similarly, the DEM technology also allows calculation of an individual's energy saving. Based on this configuration, we propose an energy rewarding method to ensure sustainability of individuals' energy saving behavior. The concept of using DEM is to allow feedback to the individuals' behavior by closing the loop back to the point of consumption, as shown in Figure 2.

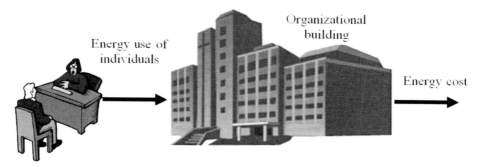

Fig. 1. Individual energy consumption in commercial and organizational buildings with shared electricity

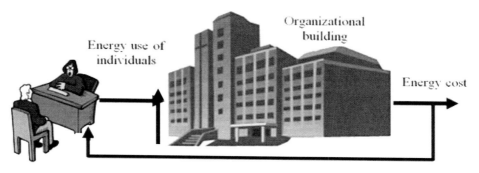

Fig. 2. Distributed Electricity Meters (DEM) is an enabling technology that could have a direct influence on an individual's energy consumption behavior

The present paper is organized into 6 sections. Section 1 contains the introduction. In section 2, medium and large size building energy consumption and their produced greenhouse gas emissions are discussed. In section 3, a concept for Distributed Electricity Meters (DEMs) for commercial and organizational buildings with shared electricity is proposed. An example of the system is presented in Section 4. In section 5, the sustainability of the proposed method is discussed and finally, in Section 6, the concluding remarks are presented.

2 Energy in Commercial and Organizational Buildings with Shared Electricity

Eco-sustainability of commercial and organizational buildings with shared electricity requires a further reduction of greenhouse gas emissions. Such reduction will be easier and cheaper to achieve if people are actively engaged in energy saving process. Currently, in commercial and organizational buildings with shared electricity, it is not always clear how much energy is used by any individuals or different sections because the energy meters output energy consumption data of the entire building (See Figure 4). Therefore, for buildings with many employees it is not possible to accurately determine energy consumption at an individual level. Using this method of metering, the dominant factor in an individuals energy consumption behavior is the personal attitudes toward efficient and responsible energy use. Also, there isn't any feedback mechanism to influence an individuals' energy consumption behaviors, especially for those who do not recognize value in reducing consumption.

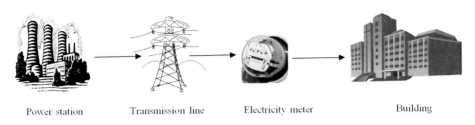

Fig. 3. Common electricity meters only meter the total electricity

Influencing human behavior has been discussed in [16] and a mobile application to provide a comparative energy information has been proposed. Further engagement of individuals in energy saving in organizational buildings needs a more effective action (rather than only a comparative). This problem can be tackled, if the individuals' energy consumption is metered or estimated. The concept of DEM has a great potential for energy saving in commercial and organizational buildings by providing local measurement and reporting.

3 Distributes Electricity Meters

Currently, a meter is used within buildings to measure the total electricity consumption as shown in Figure 3. For the DEM method, a small power meter is used to measure individual electricity consumption at the room or worktable level. An example of the DEM is shown for a part of an organizational building in Figure 4, where digital power meters are installed for different rooms and worktables. The meter for worktables is only for shared offices/rooms.

Using this method, it is possible for individuals to track their own usage, and would likely lead to an increased sense of responsibility. The information of the meters can be used to provide an efficient control action to maintain or encourage an energy saving behavior. Currently, we propose an energy saving reward or support for individuals if they contribute to the energy saving process. The reward could be a percentage proportional of energy saved, or is a fraction of saved energy expense. By this method, it is possible to actively engage individuals in the energy saving process.

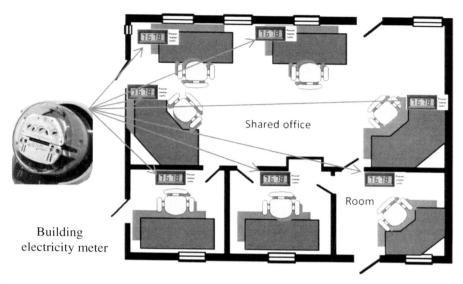

Fig. 4. Distributed electricity meters on each worktable or room to meter the individual's energy consumption

4 Enabling Technology

The enabling technology on which DEMs are based requires a 1) the addition of an in-line or hardwired power consumption monitoring unit, 2) a small display capable

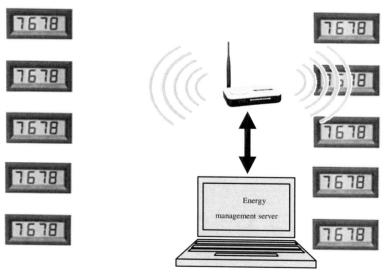

Fig. 5. Wireless distributed electricity meters communicating with the energy management server

of showing various consumptions data, 3) an electricity management server, and 4) a DEM software program. There is also a requirement for a communication medium for automatically transferring the meters reading to the server. The software records the energy consumption of each meter within a database and performs various analyses such as calculation of the total energy consumption, average consumed energy, saved energy, etc.

The DEMs can be designed in a way that they wirelessly communicate with the servers, however for the sample study the meter reading can be recorded manually daily or weekly and entered into the database. The information in the database can be used for analysis as well as for calculation of the energy reward/support of individuals. The reward can be a fraction (let say 50%) of the benefit of the individual's effort for energy saving for their organization.

This DEM idea has entrepreneurial potential and therefore, it is important to study the economic sustainability of this concept in order to ensure the applied research could have real application.

5 Economic Sustainability Analysis

5.1 Sustainability Chains

The energy consumption in commercial and organizational buildings with shared electricity consists of four chains as shown in Figure 6, including individuals, organization, environment, and government.

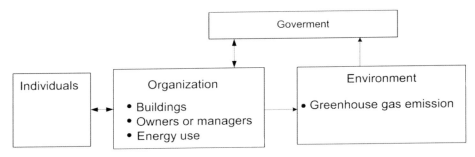

Fig. 6. Sustainability analysis chains of energy consumption in commercial and organizational buildings with shared electricity

The individual chain addresses the effect of the DEM on individuals and the organization chain represents the effect on the building, owners/managers, and total energy usage. The environment chain represents the impact of the DEM on the environment and can be assessed based on the greenhouse gas emissions of the buildings. The government chain interacts with organizations and their impact on the environment (only in countries that have carbon mitigation plan). To ensure the economic sustainability of the proposed DEM system, it is important to show that all the chains experience a benefit from the proposed energy saving method.

5.2 Sustainability Analysis for Individuals

In the proposed method, the individual will receive a direct benefit from their efforts to save energy in their organization. The benefit is somehow proportional to the actual saved energy of the individuals and therefore if they save more they would receive more benefits. Such system can actively engage individuals in the energy saving process. Furthermore, the there is an increasing awareness of climate change among public, and the awareness may result in people adopting other eco-friendly habits, both at work and at home. Having an accurate measure for how people are performing in terms of energy consumption and providing detailed information to the individual could encourage them for further improvement their energy consumption behavior.

5.3 Sustainability Analysis for Organizations

A lower consumption of energy translates directly into financial savings for the organization by means of reducing the power bill. Additionally, the method can help organizations to decrease their greenhouse gas emission more conveniently and with lower investment, especially in the case of existing infrastructure. The environmentally responsible practice presents further opportunity for green-marketing, and energy-conscious customers may come to favour businesses that themselves subscribe to these values. If these building belongs to businesses then, the normally customers would prefer them rather other business because of their eco-friendly. For example, customers when choosing which hotel to stay at, in 85% of cases preferred to hotels that were 'green' hotels [17]. This can help the businesses to increase market share.

5.4 Environment Sustainability

The proposed energy saving method contributes to environmentally sustainability by reducing the energy consumption in commercial and organizational buildings.

5.5 Sustainability Analysis for Government

There is an increasing awareness and action by governments around the world to address the climate change. For example, the Australian government encourages and provides supports to businesses that can demonstrate to decrease their carbon footprint. The DEM method aligns with these funding opportunity and the government support can be used to minimize the initial investment required for implementation of the hardware and software, as well as to further fund research within this field.

6 Conclusions

Energy consumption in commercial and organizational buildings with shared electricity produces a considerable amount of greenhouse gas emissions worldwide. A person's behavior toward energy saving in these buildings is widely varied and therefore encouraging energy saving practices is challenging. The present paper introduced distributed electricity metering systems for energy saving in commercial and organizational buildings with shared electricity. The method allows the metering of room-level or desk-level energy consumption, and energy saving behavior was used to provide direct feedback of the energy used by individuals. It was shown that using distributed electricity meters, an active engagement of people in energy saving processes could be obtained. The hardware and software requirements to implement the method were discussed and the sustainability of the method was evaluated both from a theoretical perspective and in context of Australian Federal incentives for the reduction of energy consumption. In future, the authors aim to 1) improve on and test the proposed distributed electricity meter hardware design, 2) further development of the energy management software, and 3) perform statistical analysis on the behavior of individuals before and after the implementation of the system.

References

[1] Lam, J.C.: Energy analysis of commercial buildings in subtropical climates. Building and environment 35, 19–26 (2000)

[2] Pantong, K., Chirarattananon, S., Chaiwiwatworakul, P.: Development of Energy Conservation Programs for Commercial Buildings based on Assessed Energy Saving Potentials. Energy Procedia 9, 70–83 (2011)

[3] Vlek, C., Steg, L.: Human Behavior and Environmental Sustainability: Problems, Driving Forces, and Research Topics. Journal of social issues 63, 1–19 (2007)

[4] Ruan, Y., Liu, Q., Zhou, W., Firestone, R., Gao, W., Watanabe, T.: Optimal option of distributed generation technologies for various commercial buildings. Applied Energy 86, 1641–1653 (2009)

[5] Tibor, T., Feldman, I.: SO 14000 a guide to the new environmental management standards. IRWIN professional publishing, BURR RIDGE, IL, USA (1995)

[6] U. S. G. B. Council, Green building facts,US GreenBuilding Council (2008)

[7] Kibert, C.J.: Sustainable construction: green building design and delivery. Wiley (2008)

[8] Lee, W.S.: Benchmarking the energy efficiency of government buildings with data envelopment analysis. Energy and Buildings 40, 891–895 (2008)

[9] Kazandjieva, M., Gnawali, O., Heller, B., Levis, P., Kozyrakis, C.: Identifying Energy Waste through Dense Power Sensing and Utilization Monitoring. Technical Report CSTR 2010-03, Stanford University (2010)

[10] Doukas, H., Nychtis, C., Psarras, J.: Assessing energy-saving measures in buildings through an intelligent decision support model. Building and environment 44, 290–298 (2009)

[11] Akbari, H., Konopacki, S., Pomerantz, M.: Cooling energy savings potential of reflective roofs for residential and commercial buildings in the United States. Energy 24, 391–407 (1999)

[12] Lee, W., Yik, F., Jones, P., Burnett, J.: Energy saving by realistic design data for commercial buildings in Hong Kong. Applied Energy 70, 59–75 (2001)

[13] Konopacki, S., Akbari, H., Gartland, L.: Cooling energy savings potential of light-colored roofs for residential and commercial buildings in 11 US metropolitan areas. Lawrence Berkeley Lab, CA (United States) (1997)

[14] Kaygusuz, K.: Energy situation, future developments, energy saving, and energy efficiency in Turkey. Energy Sources 21, 405–416 (1999)
[15] Poortinga, W., Steg, L., Vlek, C., Wiersma, G.: Household preferences for energy-saving measures: A conjoint analysis. Journal of Economic Psychology 24, 49–64 (2003)
[16] Petkov, P., Köbler, F., Foth, M., Medland, R., Krcmar, H.: Engaging energy saving through motivation-specific social comparison, pp. 1945–1950 (2011)
[17] Millar, M., Baloglu, S.: Hotel guests' preferences for green hotel attributes. Hospitality Management, pp. 1–11 (2008)

Chapter 98
A Comparative Analysis of Embodied and Operational CO_2 Emissions from the External Wall of a Reconstructed Bosphorus Mansion in Istanbul

Fatih Yazicioglu and Hülya Kus

Istanbul Technical University, Turkey

Abstract. In 2012 a Bosporus Mansion, which was demolished because of a fire in 1995, is reconstructed. Although it's facades were constructed the same as the original building, it's structural system component was made of reinforced concrete. But today doing so is forbiden by the new legislations in Istanbul, Turkey. It's structural system components must also be constructed with the original kind of materials and techniques. In the paper; firstly the embodied CO2 emissions of the reconstructed external wall and an eventual reconstruction of it according to the new legislation are calculated; secondly the U-Values of both alternatives are calculated; and thirdly operational CO2 emissions are calculated. And lastly embodied and operational CO2 emissions are compared and contrasted.

1 Introduction

Efficient usage of energy is one of the most crucial matters of this time because of the various direct and indirect costs of energy. One of the most important indirect costs of energy is related with environmental issues. Energy is a convertible concept, and it can be converted to CO2 emissions. Reducing the CO2 emissions is very important in achieving a sustainable and healthy environment. That is why, the embodied energies in the construction of buildings and manufacturing of building materials are popular research areas. The embodied energy is an evolving criterion for buildings and materials. There are some manufacturers who give the total embodied energies and CO2 emissions of their materials but many others are not specified yet. There are several researches which studies building materials in terms of energy and emissions. For example; Kus, H. et al. studied the embodied energy of masonry wall units regarding manufacturing process[1]. Another example is a detailed study of Schmidt, A.C., about rock wool thermal insulation [2].

The operational energy usage of buildings is also another popular research area. The study of Maile, T. et al. is a good example for operational energy usage in buildings [3]. They studied the energy used in a building and compare it with simulated energy usage which was obtained in the design processes. Protecting the energy used in buildings is also very important and it is mostly related with the U-value of the external envelope. Suleiman, B. M., studied the operational energy

---

A. Håkansson et al. (Eds.): *Sustainability in Energy and Buildings*, SIST 22, pp. 1093–1105.
DOI: 10.1007/978-3-642-36645-1_98   © Springer-Verlag Berlin Heidelberg 2013

usage in buildings using the U-values of the external walls [4]. The positive effect of insulation materials on the U-values of the external envelopes is significant. Friess, W. A., studied the effect of appropriate usage of thermal insulation on the building's energy consumption [5].

Neither embodied, nor operational energies are enough to assess the environmental performance of buildings. Buildings are complex structures with life cycles. There are also several studies assessing the environmental performance of buildings. The study of Hacker, J. N. et.al. is a good example for these kinds of studies [6]. They studied the embodied and operational CO2 emissions of housing. In this study, the embodied CO2 emissions of the external wall of a reconstructed building is analysed, energy losses from those walls are calculated and the data obtained is compared with the data calculated for the original building details.

During its long history, Istanbul served as the capital of the Roman Empire, the Eastern Roman (Byzantine) Empire, the Latin Empire, and the Ottoman Empire. Although several architectural monuments of all those civilisations can be seen in the entire city, Ottoman architecture is significant in the Bosphorus region. Bosphorus, dividing Istanbul into two parts between minor Asia and Europe, is about 30 km long and connects Black Sea with the Sea of Marmara. On the Bosphorus coastal zone, there are three main types of buildings; monumental palaces, "Yalı" buildings, and Mansions. Monumental palaces were the dwellings of the Ottoman dynastic family. The main difference between a "Yalı" and a mansion is their location. "Yalı" buildings were constructed on the coast of Bosphorus by the sea whereas the mansions on the inner parts. Mansions are also classified as "köşk" if they were built with wood in gardens in resort districts, and "konak" if they were built in the city and usually are masonry types [6]. Bosphorus has its unique architecture and after 1983 it has been protected by a special law through which new building constructions are forbidden. It is only possible to make restoration for old buildings and to construct a new building only if it can be proved that there used to be an original historical building in that place.

A study of inventory analysis on a case mansion building, located in the coastal district of Bosphorus and originally dated to 1900's, is presented in this paper. The restoration works of the mansion have recently completed. A comparative assessment of embodied and operational CO_2 emissions is made between the external walls of the reconstructed mansion according to the rules given in the old legislation and the same project if it was today renovated with its original details according to the new legislation valid at the present time.

2 The Case

The case building is located in Kanlica, one of the typical Bosphorus coastal quarters. In Figure 1, the aerial view of the Bosphorus is seen from south to east. And in Figure 2 a photograph of the front façade of the case building is seen.

Fig. 1. Aerial view of the Bosphorus

Fig. 2. The reconstructed building according to the old legislation

2.1 Legislation

Construction in the coastal zone of the Bosphorus is administrated by special legislations. Since 1983, construction of new buildings is generally forbidden in the coastal zone including the hills on both sides which can be seen from the Bosphorus. New buildings can only be built if either it's a public building, which is needed vitally, or a building for tourism purposes. Reconstruction of old buildings is also possible, but those buildings to be reconstructed should represent a special historical value. Destruction of those buildings is forbidden as well, so reconstruction is possible only if the building was destructed with a disaster, etc. Most of the building construction in the coastal zone of the Bosphorus is basic repairs and maintenance. There is a special organisation, "Board for the Protection of Cultural Heritage", in charge of assessing the old buildings and approving the architectural projects of them. Once a building or ruins of a building is determined as a Cultural Heritage and the necessary restoration projects are prepared, the approvals are then granted by another board, "Bosphorus Construction Legislation Affairs" which also grants permission to reconstruct or maintain the building and controls and approves the construction works.

The "Board for the Protection of Cultural Heritage" determines a numeric rank to determine the degree of heritage significance. The heritage is ranked as "first degree historical heritage" if it is culturally very important. Monumental buildings, like palaces and mosques, are some examples for buildings in this category. The heritage is ranked as "second degree historical heritage" if it is a historical heritage which was typically constructed at its original construction time. Most of the mansions in Bosphorus are examples for buildings in this category. The degree of a heritage determines the way a building can be used, maintained, and reconstructed.

Heritage buildings in the first category cannot be given a different function from its original function. They should be maintained considering the original construction techniques and original materials. If a reconstruction is going to be made, it should be reconstructed similar as the original in terms of architectural layout, building and construction techniques and materials. Whereas a differentiation in the function is possible if the heritage belongs to the second degree. For example a mansion can be transformed into a hotel. The facades and dimensions of the heritage should strictly be the same as the original building but the architectural layout can be changed. Until the year 2005, the structural system of this type of buildings could also be changed if it was going to be reconstructed. For example, reinforced concrete could be used in the structural system of the building if the facades and the dimensions of the building were constructed as the original details. However, after the year 2005, the legislation was changed and changing the original structural system was forbidden afterwards. Now, a heritage can only be reconstructed if it is done by the original structural materials and systems. The building, which was studied as the case in this paper, is a second degree historical heritage. All the necessary documents of the building were prepared and the necessary permissions were granted before the year 2005 which has resulted the building's structural system to be reconstructed with reinforced concrete skeleton. In order to assess the environmental impacts of the constructed building's facade and compare it with the scenario in which the building is constructed according to the new legislation, new details for the facade with wooden structure were designed. The constructed building's details are compared with the generated details considering the new legislation in terms of environmental performance.

2.2 The Original Architectural Features and the Structural System Characteristics

The case building represents all the original specialities of historical Bosphorus mansions. It's a three storey high wooden building. Because of the sloped topography, there is a semi basement which was used to be a storage house for firewood or coal. The semi basement and the footings were built as masonry. It has a common living space (sofa) at the centre of each floor and four rooms surrounding this sofa. The front façade of the building is directed to north-west from which two European side Bosphorus villages, "Emirgan" and "İstinye" can be seen. The building has a terraced garden at the backside. There is also a water well in the garden of the building which is another typical speciality of the Bosphorus mansions.

The mansion building is a traditional Turkish house built up with load-bearing platform walls having lightweight timber structural system, which can also be seen in particular places around Turkey and the Balkans. The main structural components of the wall system are; posts/studs, bottom and top plates, braces/diagonals, headers and sills for window and door openings. The roof form is gable with wooden structural system usually having a 33% slope with mission clay roof tiles.

The main façade characteristics of the Bosphorus mansions are the large ratio of window openings and the projections on the upper floors. The external wall openings, which are 32 % of the total façade area, are dominated by the window sizes of two-to-one, and are many in number. The windows are vertical slider type with counter balances which are operated with a pulley system. The openings are smaller on the ground floor, because of the privacy needs at that period. There was not any cantilevered floor at the front façade of the building as it was adjacent to the street. But, on the rear façade, all floors are projected. Almost the entire building was made of wood above ground. The opaque wall finishings were wood siding on the exterior side and lath and plaster in the interior.

2.3 The Reconstruction Project According to the Old Legislation

In 1996, the building was damaged entirely because of a fire. The necessary architectural drawings for restoration were then prepared, and by the year 2004, the necessary permissions were granted. In 2009, the reconstruction works was started and it is planned to be completely finished at the end of 2012. The external wall core was built of autoclaved aerated concrete. Gypsum plastering was applied internally and a wooden siding externally. The entire facade was constructed similar as the original façade visually, in terms of the type of the main material, which is wood. Double glazed glass, new lock systems, and new counter balance systems were preferred for windows, as the primary differences from the original details. The structural system of the roof was also changed into steel in order the attic to be used. Clay roof tiles were applied as the roof covering. But, the new roof system was detailed to have a thermal insulation and waterproofing membrane.

In Figure 3, the reconstructed building elevation (a) and the section detail (b) of the façade are demonstrated. The details of the wall and the RC structural system components can be seen in Figure 3-b. In the section, different parts of the façade, including the structural components and the masonry infill wall, together with the places coinciding with the wooden vertical laths for external siding, are marked with aI, aII, aIII, and aIV, in order to be separately taken into consideration in the calculations.

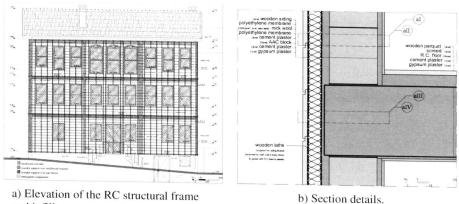

a) Elevation of the RC structural frame and infill masonry wall

b) Section details.

Fig. 3. Reconstructed building in accordance with the old legislation

2.4 The Reconstruction Project According to the New Legislation

The external walls are composed primarily of wood studs as the major elements of the lightweight structural system of the original building. The gaps between the studs are filled with mineral wool and the wall is covered with wood siding externally and plasterboard internally. In Figure 4, elevation (a) and the section detail (b) of the eventual reconstruction project, in accordance with the new legislation, are demonstrated. The details of the wall and floor components can be seen in Figure 4-b. In the section, different parts of the façade, including the structural components and the thermal insulation filling, are marked with aI, aII, and aIII, in order to be separately taken into consideration in the calculations.

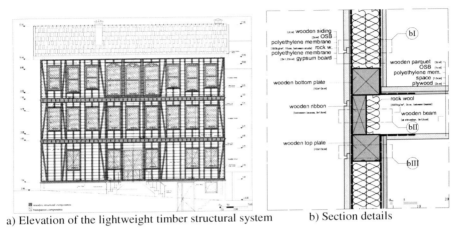

a) Elevation of the lightweight timber structural system   b) Section details

Fig. 4. Eventual reconstruction project according to the new legislation

3 Performance Assessment

The comparative performance assessment of alternative reconstruction projects, i.e. according to the old and new legislation, is made in terms of; (i) embodied CO_2 emissions, (ii) U-values of external walls and (iii) associated operational CO_2 emissions.

3.1 Review of the Embodied CO2 of the Details

The Embodied CO2 (ECO2) is the mass of embodied carbon dioxide per unit mass or volume of material, usually expressed as kilograms of CO_2 per tonne or m3 of material (kgCO2/t or kgCO2/m3) (Hacker J. N. et al.). ECO2 values of all materials used in the case were examined and the total CO_2 released from the external envelope was calculated both for the reconstructed building and the eventual reconstruction according to new legislation. The inventory list is presented in Table 1, together with the unit CO_2 equivalent emissions compiled from either directly from the manufacturers data sheets or from the most appropriate references. The main difference between the two alternative walls appears to be the different materials of the structural system.

Table 1. Inventory list

	Density kg/m3	CO2 Emission	Reconstructed building according to Old Legislation		Eventual reconstruction according to New Legislation	
			Total Usage	Total CO2	Total Usage	Total CO2
Autoclaved Aerated Concrete (AAC) [8]	500	191.6 kg/m³	12,56 m3	2406,5	-	
Polymer modified cementitious thin bed adhesive for AAC [8]	1500	0.248 kg/kg	156,02 kg	38,7	-	
Wood (sawn spruce) [9]	550	0.55 kg/kg	2,84 m3 1562kg	859,1	15,4 m3 8470kg	4658,5
PU foam [8, 9]	30	191.54 kg/m3	1,071 m3	205,1	-	
Gypsum plaster [8]	1300-1800	0.198 kg/kg	0,993 kg	0,2	-	
Cement mortar [8]	1900	0.2 kg/kg	4,376 m3 9189,6kg	1837,9	-	
Rock wool [8]	(70) 25-200	1.61 kg/kg	2,69 m3 188,3kg	303,1	4,6 m3 322 m2	518,4
PE membrane [9]	(360) 940	1.6 kg/kg	332 m2 61gr/m2 20,252kg	32,4	332 m2 61gr/m2 20,252kg	32,4
Reinforced Concrete RC35 [10]	2400	0.18 kg/kg	6,06 m3 14544kg	2617,9	-	
Plasterboard [10]	(664) 800	0.38 kg/kg			2,075 m3 1660kg	630,8
Water-based Paint [11]	1300	2.5 kg/liter	0,717 m3 932,1kg 166 m2	2330,2	0,717 m3	2330,2
OSB sheathing [10]	640	0.96 kg/kg			166 m2 1,66m3 1062,4kg	1019,9
TOTAL				10631,1		9190,2

3.2 Review of the U-Values of External Walls

The U-values of the external walls are calculated according to the details given in Figures 2-b and 3-b and the lambda values obtained from the standard TS825 [12]. The equations used to calculate the U-Values are obtained from TS825 and can be seen below [12].

$$R \qquad = d_1/\lambda_1 + d_2/\lambda_2 + \ldots + d_n/\lambda_n \qquad (1)$$

$$1/U \qquad = R_i + R + R_e \qquad (2)$$

R .... : Heat transmission resistance .... (m2K/W)
d_n .... : Width of the material .... (cm)
λ_n .... : Heat conductivity value .... (w/m2K)
R_i .... : Heat transmission res. of the internal surface .... (m2K/W)
R_e .... : Heat transmission res. of the external surface .... (m2K/W)

The details of the different parts in the external walls of the reconstructed external wall according to old legislation and their U values are listed below:

- a1. It is the typical wall detail which passes through the masonry wall structure with an area of 53,8 m2. This part does not comprise any reinforced concrete structural system components. It consists of external wooden siding, polyethylene membrane, rock wool, polyethylene membrane, cement plaster, AAC block, cement undercoat plastering and gypsum top coat plastering, respectively. The resulting U value is 0,394 W/m2K.
- a2. This part is structurally very similar to the part a1, but passes through vertical wooden laths mounted to the masonry surface, and its area is 9 m2. The only difference is there is no rock wool in this part because rock wool is placed in between the vertical wooden laths which support the wooden sidings. The resulting U value is 0,609 W/m2K.
- a3. This part passes through the reinforced concrete structural system components covered with rock wool externally, its area is 18,1 m2. The resulting U value is 0,568 W/m2K.
- a4. This part is structurally very similar to the a3, but passes through vertical wooden laths its area is 2,1 m2. The only difference is that there is no rock wool in this part since the rock wool is placed in between the vertical wooden laths which support the wooden sidings. The resulting U value is 1,158 W/m2K.

The details of the different parts in the external walls of the eventual reconstruction of external wall according to new legislation and their U values are listed below:

- b1. It is the typical wall detail which passes through the wooden wall structure with an area of 46 m2. This part does not comprise any wooden structural system components. It consists of external wooden siding, OSB, polyethylene membrane, rock wool, polyethylene membrane and gypsum board, respectively. The resulting U value of it is 0,320W/m2K.
- b2. This part passes through the wooden structural system components of the floor level and its area is 2,92 m^2. It consists of external wooden siding, OSB, polyethylene membrane, wooden ribbon between the wooden beams and rock wool behind the ribbons, respectively. The resulting U value of it is 0,472 W/m2K.
- b3. This part passes through the vertical and horizontal structural system components with an area of 33,08m2. External wooden siding, OSB, polyethylene membrane, wooden studs or wooden top-bottom plates, respectively. The resulting U value of it is 0,820 W/m2K.

In Figure 5, the U-values (W/m2K) at different parts of the reconstructed external wall according to old legislation and the eventual reconstruction of external wall according to new legislation presented. In the graphics, the fractional U-values are shown in dark colours and the total U-values and the effect of the fractional U-values on the total U-value are shown in light colour.

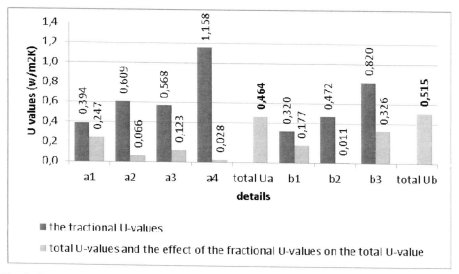

Fig. 5. U values at different parts (Figure 3-b, 4-b) of the reconstructed details and eventual reconstruction details.

3.3 Review of the Operational CO2 of the External Wall

In this part of the paper operational CO2 emissions due to the heat loses from the opaque parts of the external wall alternatives are going to be examined. The heat loses from the reconstructions according to the old and new legislations are going to be calculated and compared. The calculation is based on the equations given in the standard EN 832 [13].

$$Q_{year} = \Sigma Q_{month} \quad (3)$$

$$Q_{month} = [H \ (T_{i,month} - T_{d,month}) - \eta_{month} \ (\varphi_{i,month} + \varphi_{g,month})] \cdot t \quad (4)$$

Q_{year} ... : Total heat energy need in a year .... (Joule)
Q_{month} . : Total heat energy need in a month .... (Joule)
H .... : Specific heat loss of the building .... (W/K)
$T_{i,month}$: Average internal temperature .... (°C)
$T_{d,month}$: Average external temperature .... (°C)
H_{month} . : Monthly average usage factor for heat gains .... (unitless)
$\varphi_{i,month}$: Average internal heat gains .... (W)
$\varphi_{g,month}$: Average solar heat gains .... (W)
t .... : time, (a month in seconds(= 86400 x 30) .... (s)

Since the heat gain from windows is the same for both alternatives, it is neglected and therefore equation (5) is used in the calculations.

$$Q_{month} = H \cdot (T_{i,month} - T_{d,month}) \cdot t \quad (5)$$

Average internal temperature is accepted as 19^0C. The heat energy loses are calculated in joule units and then it is converted to kwh (1kwh = 3600000 joule). The heating system of the building is supplied with natural gas. 1 m3 natural gas provides 10,64 kwh energy, and in turn, 0,582 kgCO2e/kwh is released to the atmosphere [12, 13].

$$
\begin{aligned}
Q_{a\ year} &= 0,394 \text{ x } (12,4+12,4+10,6+6,3+1,6+0+0+0+2,6+7,1+10,5) \text{ x } (86400 \text{ x } 30) + \\
&+ 0,609 \text{ x } (12,4+12,4+10,6+6,3+1,6+0+0+0+2,6+7,1+10,5) \text{ x } (86400 \text{ x } 30) + \\
&+ 0,568 \text{ x } (12,4+12,4+10,6+6,3+1,6+0+0+0+2,6+7,1+10,5) \text{ x } (86400 \text{ x } 30) + \\
&+ 1,158 \text{ x } (12,4+12,4+10,6+6,3+1,6+0+0+0+2,6+7,1+10,5) \text{ x } (86400 \text{ x } 30) + \\
&= 8002087757 \text{ joule} = 2223 \text{ kwh} \\
&= 1293 \text{ CO2e}
\end{aligned}
$$

$$
\begin{aligned}
Q_{b\ year} &= 0,320 \text{ x } (12,4+12,4+10,6+6,3+1,6+0+0+0+2,6+7,1+10,5) \text{ x } (86400 \text{ x } 30) + \\
&+ 0,472 \text{ x } (12,4+12,4+10,6+6,3+1,6+0+0+0+2,6+7,1+10,5) \text{ x } (86400 \text{ x } 30) + \\
&+ 0,820 \text{ x } (12,4+12,4+10,6+6,3+1,6+0+0+0+2,6+7,1+10,5) \text{ x } (86400 \text{ x } 30) \\
&= 8884646346 \text{ joule} = 2468 \text{ kwh} \\
&= 1436 \text{ CO2e}
\end{aligned}
$$

4   Results

The results are examined in four topics; architectural design related, U-value related, operational CO2 emission related, and embodied CO2 related results.

4.1   Architectural Design / Legislation Related Results

The new legislation about the historical building reconstructions in Istanbul compels the reconstruction to be done with its original structural material. It is possible to change the plan schemas of a historical building in both old and new legislations. At the end the facades reconstructed according to both legislations should be the same as the original façade. That's why only compelling the usage of original structural material is not enough for preserving historical buildings successfully and meaningfully. Preserving the original plan schemas is as important as the usage of original materials. On the other hand, reinforced concrete historical reconstructions make the width of the external wall increase which effect the plan schemas indirectly.

4.2   U-Value Related Results

The reconstruction alternatives, designed according to the old and new legislations, have different U-values. The possible reasons of that and proposals for improving both designs are listed below.

- The reconstruction made according to the new legislation makes the mean U-value increase 0,51 W/m^2K from the reconstruction made according to the old legislation.
- Although the spaces between the wooden structural studs are filled with excessive amount of thermal insulation material, the continuity of the insulation was cut with the studs which results many thermal bridges.

- As the wooden reconstruction's external wall does not have any kind of material having heat storage capacity, it is not suitable for the dwelling function of the building.
- The R.C. structured reconstruction also has thermal bridges at the intersection points with the wooden studs carrying the wooden siding. But as the number and amount of the structural components are small in R.C., it doesn't effect the mean U-value much.
- The main problem decreasing the mean U-value in R.C. reconstruction is the wooden studs, which carries the wooden siding, screwed to the external wall core. They cut the continuity of the thermal insulation. But as the wall core, which is AAC, has appropriate heat insulation performance that did not effect the overall values very much.
- In R.C. reconstruction one of the main critical points decreasing the mean U-value is the usage of same amount of heat insulation both in front of AAC and R.C. parts of the external wall. The amount of heat insulation material in front of the R.C. parts should be increased.
- In R.C. reconstruction, a different material, which has a small surface area or a better heat transmission value, may be used for studs carrying the wooden siding. For example, the usage of U shaped light section steel will both decrease the area of the thermal bridge and increase the continuity of the thermal insulation.

4.3   Operational CO_2 Related Results

External wall is one of the major components of the building envelope. Most of the energy loses from the buildings are resulted from the thermal loss from the external walls. Thermal loss from buildings not only effects the operational costs, but also effects the environmental costs of the buildings negatively. The environmental operational costs, in terms of CO2e, of two alternative reconstruction methods according to two different legislations of different times are studied and the data derived are listed below:

- 245 kwh more energy is lost by the wooden structured reconstruction each year from a single façade of the building.
- Natural gas is used as the heating source in the building. 245 kwh equivalent natural gas makes about 143 kg more CO2 emission to the atmosphere each year. (about 270 kg more CO2 if coal was used and about 119 kg more CO2 if electricity was used – *Turkey conditions)

4.4   Embodied CO_2 Related Results

The inventory list together with the unit and total CO2 equivalent emissions was presented previously in Table 1. The results related with the embodied CO2 are summarized in Figure 6 and data derived are listed below:

- The total embodied CO2 emission for the reconstruction according to the old legislation is 10631,1 kgCO2e and according to the new legislation is 9190,2 kgCO2e.
- The reconstruction according to the old legislation gives 1440,9 kg more CO2e to the atmosphere.
- Structural system components, reinforced concrete and wood are the most CO2e intensive components in both alternatives.
- The unit CO2 emission of OSB is greater than wood's and usage of OSB, increased the CO2 emission of the reconstruction alternative significantly.
- Paint, which is used in both alternatives with the same amounts, is the fourth most CO2 embodied material.
- The effect of thermal insulation on the embodied CO2 emissions is relatively smaller than the other major components of the alternatives.

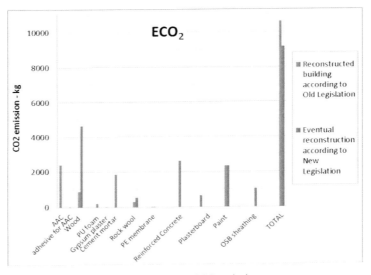

Fig. 6. Embodied CO2 emissions

5 Conclusion

The embodied and operational CO2 emissions of the external wall of a reconstructed building in Bosphorus are calculated in this study and they are compared with the calculations done for an eventual reconstruction considering the new building legislation valid in Bosphorus. It is derived from the comparison that, the embodied CO2 emissions is improved with the external wall details generated according to the new legislation (1440,9 kg less CO2 is embodied in with new legislation). But, since the heat loses from the external walls are greater in the details generated according to the new legislation, heating system of the building gives 143 kg more CO2 to the atmosphere each year. In the long run, considering 30 years of Life Cycle of the building, the external wall details generated according to the old legislation is more environmentally friendly.

References

1. Kus, H., et al.: Comparative environmental assessment of masonry wall units regarding manufacturing process. In: The Proceedings of the Sustainable Building 2008 World Conference, pp. 278–289 (2008)
2. Schmidt, A.C.: Comparing Life-Cycle Greenhouse Gas Emissions from Natural Gas and Coal. Life Cycle Assesment Journal 9(1A), 53–66 (2004)
3. Maile, T., et al.: A method to compare simulated and measured data to assess building energy. Building and Environment 56, 241–251 (2012)
4. Suleiman, B.M.: Estimation of U-value of traditional North African houses. Applied Thermal Engineering 31, 1923–1928 (2011)
5. Friess, W.A.: Wall insulation measures for residential villas in Dubai: A case study in energy efficiency. Energy and Buildings 44, 26–32 (2012)
6. Hacker, J.N., et al.: Embodied and operational carbon dioxide emissions from housing: A case study on the effects of thermal mass and climate change. Energy and Buildings 40, 375–384 (2008)
7. http://www.tuncaykaraca.com/2011/01/02/selim-ileriyi-okuyu-okuyuverin/ (last accesed on June 25, 2012)
8. Berge, B.: The Ecology of Building Materials, 2nd edn. Architectural Press, Amsterdam (2009); Translated by Butters, C. and Henley, F., Architectural Press, Amsterdam (2000)
9. Institut Bauen und Umwelt, http://bau-umwelt.de/hp2/Home.htm (last accessed May 14, 2012)
10. Hammond, G., Jones, C.: Inventory of Carbon and Energy (ICE), Version 1.6a. University of Bath (2008), http://www.greenspec.co.uk/embodied-energy.php (last accessed May 14, 2012)
11. http://www.jotun.com/www/uk/20020181.nsf/viewunid/C55888384CC2778080257497003 7E39A/$file/Optimal.pdf (last accesed on June 25, 2012)
12. TS825 Standart for Turkish Regulation of Heat Insulation in Buildings, Istanbul (2008)
13. EN 832 Standart for Thermal performance of buildings - Calculation of energy use for heating - Residential buildings

Author Index

Abdi, Hamid   435, 669, 1083
Abela, Alan   491
Aboubou, Abdennacer   875
Acquaye, Adolf   209
Albano, Leônidas   337, 347
Ali, Mona Aal   837
Alonso, Fernando   229
Alonso, Luis   229
André, Elisabeth   147
Andrii, Rassamakin   119
Arshad, Naveed   425
Aumônier, Simon   555
Aziz, Naamane   787
Azzem, Soufiane Mebarek   875

Baborska-Narozny, Magdalena   581, 591
Bac, Anna   581, 591
Bailey, David   95
Baniotopoulos, Charalambos C.   25
Barrett, Mark   381, 413
Becherif, Mohamed   875
Bedoya, C.   229
Belusko, Martin   887
Berry, Stephen Robert   1
Binarti, Floriberta   157
Boris, Rassamakin   119
Bosche, Jerome   179
Boßmann, Tobias   71
Boukhris, Yosr   827
Boxem, Gert   445, 511, 705
Brown, Nils   613

Cánovas, Alba   659
Capizzi, Alfonso   865

Carbonnier, Kevin   289
Carmona-Andreu, Isabel   545
Carr, Stephen J.W.   261
Çelik, Serdar   357
Chinyio, Ezekiel   679, 757
Cho, Jin-Ho   987
Cho, Myung Eun   1043
Chollacoop, Nuwong   251
Ciravoğlu, Ayşen   299
Coch, Helena   337
Cole, Wesley   601
Collins, Michael   555
Costantino, Domenico   975
Counsell, John   601
Creighton, Doug   669

Dantsiou, Dimitra   567
Davarifar, Mehrdad   179
Dessouky, Yasser G.   797
de Trocóniz y Revuelta, Alberto Jose
    Fernández   239
de Trocóniz y Rueda, Alberto Xabier
    Fernández   239
Diez, Raquel Mucientes   531
Durany, Xavier Gabarrell   313
Dyer, M.   271
Dyer, Mark   373

Ebrahimpour, Abdolsalam   167
Efthekhari, Mahroo   479
Ekren, Orhan   357
El Azzi, Charles   743
Elshimy, Hisham   323
Elsland, Rainer   71

1108   Author Index

Emmanouloudis, Dimitrios   25
Emziane, Mahieddine   837, 847, 859, 875
Evans, Nick   623

Farah, Sleiman   887
Fares, Zaoui   875
Favre, Bérenger   137
Fazelpour, Farivar   199
Femenías, Paula   45
Fielding, Michael   435, 1083
Frostell, Björn   107

Gagliano, Antonio   865
Galesi, Aldo   865
Gassó, Santiago   659
Gauthier, Stephanie   401
Geens, Andrew   601
Genoese, Massimo   71
Gharbi, Leila   827
Ghrab-Morcos, Nadia   827
Gillott, Mark   95, 391, 453
Gnann, Till   71
Goodhew, Steve   491
Greenough, Rick   209
Grey, Tom   373
Guerra, José   337
Gupta, Rajat   567
Gusig, Lars-O.   261
Guwy, Alan J.   261

Hajjaji, Ahmed EL   179
Hancock, Mary   545
Hartel, Rupert   71
Hatherley, Simon   601
Havtun, Hans   717, 731, 743
Hesaraki, Arefeh   189
Hisarligil, Hakan   59
Holmberg, Sture   189
Hong, Won-Kee   1071
Hoque, Mohammad Rashedul   313
Hoxley, Mike   491
Hsieh, Yao-Tsung   847
Huerta, Miguel Ángel Gálvez   239
Hunpinyo, Piyapong   251

Ibn-Mohammed, Taofeeq   209
Ippolito, Mariano Giuseppe   975
Isalgué, Antoni   337
Istiadji, Agustinus D.   157

Iswanto, Priyo T.   157
Izadi, Roozbeh   743

Jang, Marianne   85
Javed, Fahad   425
Jo, Seon Ho   1003
Johnsson, Helena   35
Jung, JiYea   1011

Kamalisarvestani, Masoud   813
Kaziolas, Dimitrios N.   25
Khorasany, Mohsen   281
Kim, Gon   1055
Kim, Jeong Tai   987, 995, 1003, 1011, 1025, 1033, 1043, 1055, 1061, 1071
Kim, Ji-Hun   1071
Kim, Keun Ho   995
Kim, Mi Jeong   1043
Kim, Sunkuk   995
Kim, Tae-Seong   987
Kim, Yoon Jeong   1061
Kinnane, Oliver   271, 373
Ko, Hyo-Jin   1071
Kong, Hyo Joo   1025
Kramers, Anna   127
Kugler, Michael   147
Kus, Hülya   1093

Lauret, Benito   229
Lee, Hyunsoo   1011
Lee, JiSun   1011
Lee, SungHee   1011
Lee, YeunSook   1011
Lewis, John   903
Liaqat, Muhammad Dawood   425
Lidelöw, Sofia   35
Lim, Chaeyeon   995
Lim, Hong Soo   1055
Littlewood, John R.   601, 623, 637
Liu, Linn   463
Lorimer, Stephen   85
Lu, Hai   955

Maaijen, Rik   511
Maassen, Wim   649
Madadnia, Jafar   965
Maddy, J.   261
Malevit, Eva   913
Mambo, D. Abdulhameed   479

Martinac, Ivo   931, 947, 955
Martínez, Francisco Javier Rey   531
Masoodian, Masood   147
McGill, Gráinne   367
McGrath, Paddy   491
Mekhilef, Saad   813
Méndez, Gara Villalba   313
Mounir, Aksas   787
Moussa, Mona F.   797
Msirdi, Nacer   779
Mullins, James   435, 1083

Na, Youngju   995
Naamane, Aziz   779
Nagijew, Eldar   391
Nahavandi, Saeid   435, 669, 1083
Nararatuksa, Phavanee   251
Neto, Alberto Hernandez   347
Nocera, Francesco   865
Nooraei, Masoudeh   623
Nordström, Gustav   35

Olga, Alforova   119
Osaji, Emeka E.   679, 757
Oyedele, Lukumon   367
Ozawa-Meida, Leticia   209

Palme, Massimo   337
Pana-Suppamassadu, Karn   251
Park, Seon-Chee   1071
Park, Seonghyun   1033
Park, SungJun   1011
Patania, Francesco   865
Pelzer, Ruben   649
Penny, Tom   555
Peuportier, Bruno   137
Phdungsilp, Aumnad   947, 955
Pierre, Xavier   179
Premier, Giuliano C.   261

Qin, Menghao   367

Rabhi, Abdelhamid   179
Rahbari, Omid   199
Ramchurn, Kay   555
Reinhart, Florian   147
Rima, Zouagri   787
Riva_Sanseverino, Eleonora   975
Riva_Sanseverino, Raffaella   975

Rodrigues, Lucelia   453
Rogers, Bill   147
Romero, Marta   347
Rostyslav, Musiy   119

Sabouri, Vahid   45
Saidur, Rahman   813
Sajjadian, Seyed Masoud   903
Sala, Cristina Sendra   313
Saman, Wasim   887
Samet, Haidar   281
Satwiko, Prasasto   157
Schlieper, Kevin   147
Seo, Janghoo   1033
Sergii, Khairnasov   119
Sharples, Stephen   903
Shea, Andy   501
Shin, Ju Young   1061
Shipworth, David   401
Sinnett, Nigel   601
Spataru, Catalina   381, 413
Stafford, Anne   521
Stavroulakis, Georgios E.   25
Steffen, Thomas   479
Stevenson, Fionn   545
Sukkathanyawat, Hussanai   251
Suresh, Subashini   679, 757

Taygun, Gökçe Tuna   299
Taylor, Simon   209
Thanapalan, Kary K.T.   261
Thiele, Terry   555
Thielen, Korinna   85
Thoresson, Josefin   463
Tina, Giuseppe M.   15
Toledo, Linda   637
Törnqvist, Caroline   717, 731
Treacy, C.   271
Treado, Stephen   289
Tsitsiriggos, Christos   913
Tungkamani, Sabaithip   251

Vaccaro, Valentina   975
Vafaeipour, Majid   199
Vahed, Yousef Karimi   167
Vakiloroaya, Vahid   965
Valizadeh, Mohammad H.   199
van der Velden, Joep   445
van Houten, Rinus   445
Ventura, Cristina   15

Villegas, Ricardo Ramírez 107
Vissers, Derek 705

Wall, Katharine 501
Wang, Qian 931
Wietschel, Martin 71
Wilson, Robin 95, 391

Yazicioglu, Fatih 1093
Yousif, Charles 531
Yun, Geun Young 1003

Zah, Rainer 659
Zeiler, Wim 445, 511, 649, 705
Zhang, Fan 261
Zygomalas, Iordanis 25

Printed by Books on Demand, Germany